SURVEY METHODOLOGY

WILEY SERIES IN SURVEY METHODOLOGY
Established in Part by WALTER A. SHEWHART AND SAMUEL S. WILKS

Editors: *Robert M. Groves, Graham Kalton, J. N. K. Rao, Norbert Schwarz, Christopher Skinner*

A complete list of the titles in this series appears at the end of this volume.

SURVEY METHODOLOGY

Second Edition

Robert M. Groves
Floyd J. Fowler, Jr.
Mick P. Couper
James M. Lepkowski
Eleanor Singer
Roger Tourangeau

A JOHN WILEY & SONS, INC., PUBLICATION

Copyright © 2009 by John Wiley & Sons, Inc. All rights reserved.

Published by John Wiley & Sons, Inc., Hoboken, New Jersey.
Published simultaneously in Canada.

No part of this publication may be reproduced, stored in a retrieval system or transmitted in any form or by any means, electronic, mechanical, photocopying, recording, scanning, or otherwise, except as permitted under Section 107 or 108 of the 1976 United States Copyright Act, without either the prior written permission of the Publisher, or authorization through payment of the appropriate per-copy fee to the Copyright Clearance Center, Inc., 222 Rosewood Drive, Danvers, MA 01923, (978) 750-8400, fax (978) 646-8600, or on the web at www.copyright.com. Requests to the Publisher for permission should be addressed to the Permissions Department, John Wiley & Sons, Inc., 111 River Street, Hoboken, NJ 07030, (201) 748-6011, fax (201) 748-6008 or online at http://www.wiley.com/go/permission.

Limit of Liability/Disclaimer of Warranty: While the publisher and author have used their best efforts in preparing this book, they make no representations or warranties with respect to the accuracy or completeness of the contents of this book and specifically disclaim any implied warranties of merchantability or fitness for a particular purpose. No warranty may be created or extended by sales representatives or written sales materials. The advice and strategies contained herein may not be suitable for your situation. You should consult with a professional where appropriate. Neither the publisher nor author shall be liable for any loss of profit or any other commercial damages, including but not limited to special, incidental, consequential, or other damages.

For general information on our other products and services or for technical support, please contact our Customer Care Department within the U.S. at (800) 762-2974, outside the U.S. at (317) 572-3993 or fax (317) 572-4002.

Wiley also publishes its books in a variety of electronic formats. Some content that appears in print may not be available in electronic format. For information about Wiley products, visit our web site at www.wiley.com.

Library of Congress Cataloging-in-Publication Data:

Survey methodology / Robert Groves ... [et al.]. — 2nd ed.
   p. cm.
  Includes bibliographical references and index.
  ISBN 978-0-470-46546-2 (paper)
1. Surveys—Methodology. 2. Social surveys—Methodology. 3. Social sciences—Research—Statistical methods. I. Groves, Robert M.
  HA31.2.S873 2009
    001.4'33--dc22   2009004196

CONTENTS

PREFACE TO THE FIRST EDITION   xv

PREFACE TO THE SECOND EDITION   xix

ACKNOWLEDGMENTS   xxi

CHAPTER 1. AN INTRODUCTION TO SURVEY METHODOLOGY

1.1	Introduction	2
1.2	A Brief History of Survey Research	3
	1.2.1 The Purposes of Surveys	3
	1.2.2 The Development of Standardized Questioning	5
	1.2.3 The Development of Sampling Methods	6
	1.2.4 The Development of Data Collection Methods	7
1.3	Some Examples of Ongoing Surveys	7
	1.3.1 The National Crime Victimization Survey	8
	1.3.2 The National Survey on Drug Use and Health	14
	1.3.3 The Surveys of Consumers	17
	1.3.4 The National Assessment of Educational Progress	20
	1.3.5 The Behavioral Risk Factor Surveillance System	24
	1.3.6 The Current Employment Statistics Program	27
	1.3.7 What Can We Learn From the Six Example Surveys?	29
1.4	What is Survey Methodology?	30
1.5	The Challenge of Survey Methodology	32
1.6	About this Book	34
Keywords		35
For More In-Depth Reading		35
Exercises		36

CHAPTER 2. INFERENCE AND ERROR IN SURVEYS

2.1	Introduction	39
2.2	The Lifecycle of a Survey from a Design Perspective	41
	2.2.1 Constructs	41
	2.2.2 Measurement	43
	2.2.3 Response	43
	2.2.4 Edited Response	44
	2.2.5 The Target Population	44
	2.2.6 The Frame Population	45
	2.2.7 The Sample	45
	2.2.8 The Respondents	46
	2.2.9 Postsurvey Adjustments	47
	2.2.10 How Design Becomes Process	48
2.3	The Lifecycle of a Survey from a Quality Perspective	49
	2.3.1 The Observational Gap between Constructs and Measures	50
	2.3.2 Measurement Error: the Observational Gap between the Ideal Measurement and the Response Obtained	52
	2.3.3 Processing Error: the Observational Gap between the Variable Used in Estimation and that Provided by the Respondent	53
	2.3.4 Coverage Error: the Nonobservational Gap between the Target Population and the Sampling Frame	54
	2.3.5 Sampling Error: The Nonobservational Gap between the Sampling Frame and the Sample	56
	2.3.6 Nonresponse Error: The Nonobservational Gap between the Sample and the Respondent Pool	59
	2.3.7 Adjustment Error	59
2.4	Putting It All Together	60
2.5	Error Notions in Different Kinds of Statistics	61
2.6	Nonstatistical Notions of Survey Quality	62
2.7	Summary	63
	Keywords	64
	For More In-Depth Reading	64
	Exercises	65

CHAPTER 3. TARGET POPULATIONS, SAMPLING FRAMES, AND COVERAGE ERROR

3.1	Introduction	69
3.2	Populations and Frames	69

3.3	Coverage Properties of Sampling Frames	72
	3.3.1 Undercoverage	72
	3.3.2 Ineligible Units	76
	3.3.3 Clustering of Target Population Elements Within Frame Elements	77
	3.3.4 Duplication of Target Population Elements in Sampling Frames	79
	3.3.5 Complicated Mappings between Frame and Target Population Elements	80
3.4	Alternative Frames for the Target Population of Households or Persons	81
	3.4.1 Area Frames	81
	3.4.2 Telephone Number Frames for Households or Persons	81
	3.4.3 Frames for Web Surveys of General Populations	83
3.5	Frame Issues for Other Common Target Populations	84
	3.5.1 Customers, Employees, or Members of an Organization	84
	3.5.2 Organizations	85
	3.5.3 Events	86
	3.6.4 Rare Populations	87
3.6	Coverage Error	87
3.7	Reducing Undercoverage	88
	3.7.1 The Half-Open Interval	88
	3.7.2 Multiplicity Sampling	90
	3.7.3 Multiple Frame Designs	91
	3.7.4 Increasing Coverage While Including More Ineligible Elements	93
3.8	Summary	94
	Keywords	95
	For More In-Depth Reading	95
	Exercises	95

CHAPTER 4. SAMPLE DESIGN AND SAMPLING ERROR

4.1	Introduction	97
4.2	Samples and Estimates	99
4.3	Simple Random Sampling	103
4.4	Cluster Sampling	106
	4.4.1 The Design Effect and Within-Cluster Homogeneity	110
	4.4.2 Subsampling within Selected Clusters	113

4.5	Stratification and Stratified Sampling	113
	4.5.1 Proportionate Allocation to Strata	116
	4.5.2 Disproportionate Allocation to Strata	122
4.6	Systematic Selection	123
4.7	Complications in Practice	125
	4.7.1 Two-Stage Cluster Designs with Probabilities Proportionate to Size (PPS)	127
	4.7.2 Multistage and Other Complex Designs	129
	4.7.3 How Complex Sample Designs are Described: The Sample Design for the NCVS	130
4.8	Sampling US Telephone Households	133
4.9	Selecting Persons Within Households	136
4.10	Summary	138
	Keywords	139
	For More In-Depth Reading	139
	Exercises	139

CHAPTER 5. METHODS OF DATA COLLECTION

5.1	Alternative Methods of Data Collection	150
	5.1.1 Degree of Interviewer Involvement	153
	5.1.2 Degree of Interaction with the Respondent	154
	5.1.3 Degree of Privacy	155
	5.1.4 Channels of Communication	156
	5.1.5 Technology Use	157
	5.1.6 Implications of these Dimensions	158
5.2	Choosing the Appropriate Method	159
5.3	Effects of Different Data Collection Methods on Survey Errors	160
	5.3.1 Measuring the Marginal Effect of Mode	160
	5.3.2 Sampling Frame and Sample Design Implications of Mode Selection	162
	5.3.3 Coverage Implications of Mode Selection	163
	5.3.4 Nonresponse Implications of Mode Selection	166
	5.3.5 Measurement Quality Implications of Mode Selection	168
	5.3.6 Cost Implications	173
	5.3.7 Summary on the Choice of Method	174
5.4	Using Multiple Modes of Data Collection	175
5.5	Summary	177
	Keywords	178
	For More In-Depth Reading	179
	Exercises	179

CHAPTER 6. NONRESPONSE IN SAMPLE SURVEYS

6.1	Introduction	183
6.2	Response Rates	183
	6.2.1 Computing Response Rates	184
	6.2.2 Trends in Response Rates Over Time	186
6.3	Impact of Nonresponse on the Quality of Survey Estimates	189
6.4	Thinking Causally About Survey Nonresponse Error	191
6.5	Dissecting the Nonresponse Phenomenon	192
	6.5.1 Unit Nonresponse Due to Failure to Deliver the Survey Request	193
	6.5.2 Unit Nonresponse Due to Refusals	197
	6.5.3 Unit Nonresponse Due to the Inability to Provide the Requested Data	201
6.6	Design Features to Reduce Unit Nonresponse	201
6.7	Item Nonresponse	208
6.8	Are Nonresponse Propensities Related to Other Error Sources?	210
6.9	Summary	210
Keywords		211
For More In-Depth Reading		211
Exercises		211

CHAPTER 7. QUESTIONS AND ANSWERS IN SURVEYS

7.1	Alternatives Methods of Survey Measurement	217
7.2	Cognitive Processes in Answering Questions	218
	7.2.1 Comprehension	220
	7.2.2 Retrieval	220
	7.2.3 Estimation and Judgment	222
	7.2.4 Reporting	223
	7.2.5 Other Models of the Response Process	223
7.3	Problems in Answering Survey Questions	225
	7.3.1 Encoding Problems	225
	7.3.2 Misinterpreting the Questions	226
	7.3.3 Forgetting and Other Memory Problems	229
	7.3.4 Estimation Processes for Behavioral Questions	234
	7.3.5 Judgment Processes for Attitude Questions	236
	7.3.6 Formatting the Answer	237
	7.3.7 Motivated Misreporting	240
	7.3.8 Navigational Errors	241
7.4	Guidelines for Writing Good Questions	242

7.4.1 Nonsensitive Questions About Behavior — 243
7.4.2 Sensitive Questions About Behavior — 246
7.4.3 Attitude Questions — 248
7.4.4 Self-Administered Questions — 251
7.5 Summary — 252
Keywords — 254
For More In-Depth Reading — 254
Exercises — 255

CHAPTER 8. EVALUATING SURVEY QUESTIONS

8.1 Introduction — 259
8.2 Expert Reviews — 260
8.3 Focus Groups — 261
8.4 Cognitive Interviews — 263
8.5 Field Pretests and Behavior Coding — 265
8.6 Randomized or Split-Ballot Experiments — 267
8.7 Applying Question Standards — 268
8.8 Summary of Question Evaluation Tools — 269
8.9 Linking Concepts of Measurement Quality to Statistical Estimates — 274
    8.9.1 Validity — 274
    8.9.2 Response Bias — 279
    8.9.3 Reliability and Simple Response Variance — 281
8.10 Summary — 286
Keywords — 287
For More In-Depth Reading — 287
Exercises — 288

CHAPTER 9. SURVEY INTERVIEWING

9.1 The Role of the Interviewer — 291
9.2 Interviewer Bias — 292
    9.2.1 Systematic Interviewer Effects on Reporting of Socially Undesirable Attributes — 292
    9.2.2 Systematic Interviewer Effects on Topics Related to Observable Interviewer Traits — 292
    9.2.3 Systematic Interviewer Effects Associated with Interviewer Experience — 294
9.3 Interviewer Variance — 295
    9.3.1 Randomization Requirements for Estimating

	Interviewer Variance	296
	9.3.2 Estimation of Interviewer Variance	297
9.4	Strategies for Reducing Interviewer Bias	300
	9.4.1 The Role of the Interviewer in Motivating Respondent Behavior	300
	9.4.2 Changing Interviewer Behavior	301
9.5	Strategies for Reducing Interviewer-Related Variance	302
	9.5.1 Minimizing Questions that Require Nonstandard Interviewer Behavior	303
	9.5.2 Professional, Task-Oriented Interviewer Behavior	304
	9.5.3 Interviewers Reading Questions as They Are Worded	305
	9.5.4 Interviewers Explaining the Survey Process to the Respondent	306
	9.5.5 Interviewers Probing Nondirectively	308
	9.5.6 Interviewers Recording Answers Exactly as Given	311
	9.5.7 Summary on Strategies to Reduce Interviewer Variance	312
9.6	The Controversy About Standardized Interviewing	312
9.7	Interviewer Management	315
	9.7.1 Interviewer Selection	315
	9.7.2 Interviewer Training	316
	9.7.3 Interviewer Supervision and Monitoring	317
	9.7.4 The Size of Interviewer Workloads	318
	9.7.5 Interviewers and Computer Use	318
9.8	Validating the Work of Interviewers	319
9.9	The Use of Recorded Voices (and Faces) in Data Collection	322
9.10	Summary	323
Keywords		324
For More In-Depth Reading		324
Exercises		325

CHAPTER 10. POSTCOLLECTION PROCESSING OF SURVEY DATA

10.1	Introduction	329
10.2	Coding	331
	10.2.1 Practical Issues of Coding	332
	10.2.2 Theoretical Issues in Coding Activities	334
	10.2.3 "Field Coding" – An Intermediate Design	334
	10.2.4 Standard Classification Systems	337

10.2.5 Other Common Coding Systems 341
10.2.6 Quality Indicators in Coding 342
10.2.7 Summary of Coding 344
10.3 Entering Numeric Data into Files 344
10.4 Editing 345
10.5 Weighting 347
10.5.1 Weighting with a First-Stage Ratio Adjustment 348
10.5.2 Weighting for Differential Selection Probabilities 349
10.5.3 Weighting to Adjust for Unit Nonresponse 350
10.5.4 Poststratification Weighting 352
10.5.5 Putting All the Weights Together 352
10.6 Imputation for Item-Missing data 354
10.7 Sampling Variance Estimation for Complex Samples 359
10.8 Survey Data Documentation and Metadata 363
10.9 Summary 365
Keywords 366
For More In-Depth Reading 367
Exercises 367

CHAPTER 11. PRINCIPLES AND PRACTICES RELATED TO ETHICAL RESEARCH

11.1 Introduction 371
11.2 Standards for the Conduct of Research 371
11.3 Standards for Dealing with Clients 374
11.4 Standards for Dealing with the Public 375
11.5 Standards for Dealing with Respondents 376
11.5.1 Legal Obligations to Survey Respondents 376
11.5.2 Ethical Obligations to Respondents 377
11.5.3 Informed Consent: Respect for Persons 379
11.5.4 Beneficence: Protecting Respondents from Harm 381
11.5.5 Efforts at Persuasion 383
11.6 Emerging Ethical Issues 384
11.7 Research About Ethical Issues in Surveys 384
11.7.1 Research on Informed Consent Protocols 385
11.7.2 Research on Confidentiality Assurances and Survey Participation 390
11.8 Administrative and Technical Procedures for Safe-Guarding Confidentiality 392
11.8.1 Administrative Procedures 392
11.8.2 Technical Procedures 393
11.9 Summary and Conclusions 398

Keywords	400
For More In-Depth Reading	400
Exercises	400

CHAPTER 12. FAQs ABOUT SURVEY METHODOLOGY

12.1 Introduction	405
12.2 The Questions and Their Answers	405

REFERENCES	421
INDEX	451

PREFACE TO THE FIRST EDITION

We wrote this book with a specific purpose in mind. We are all survey methodologists—students of the theories and practices of the various data collection and analysis activities that are called "survey research." Surveys (in a form that would be recognizable today) are approximately 60–80 years old. Over the past two decades, a set of theories and principles has evolved that offer a unified perspective on the design, conduct, and evaluation of surveys. This perspective is most commonly labeled the "total survey error" paradigm. The framework guides modern research about survey quality and shapes how practicing survey professionals approach their work. The field arising out of this research domain can appropriately be called "survey methodology."

We increasingly noticed a mismatch, however, between the texts related to surveys and how the science of surveys was evolving. Many survey research texts focused on the application of tools and deemphasized the theories and science underlying those tools. Many texts told students to do things that were no longer or never supported by the methodological research in the area. In short, there were books that emphasized "how to do" surveys but neglected the science underlying the practices that were espoused.

Most harmful we thought was the impression conveyed to those who read the texts that surveys were merely a recipe-like task; if step-by-step instructions were followed, high quality would be guaranteed. In contrast, we saw surveys as requiring the implementation of principles in unique ways to fit a particular substantive purpose for a particular target population.

These issues became particularly important to us when the demand for a one semester graduate level (and senior undergraduate level) course became obvious at the Joint Program in Survey Methodology (JPSM), a consortium graduate program funded by the U.S. Federal statistical agencies in which the authors teach. The students would often have advanced education in another field (e.g., economics, statistics, or psychology) but no formal exposure to the field of survey methodology. We planned a 14-week lecture course with exercises and examinations that began in the Fall of 1998, and we immediately suffered from the absence of a text that could accompany the lectures and motivate the exercises.

We began to envision a text describing the basic principles of survey design discovered in methodological research over the past years and the guidance they offered for decisions that are made in the execution of good quality surveys. We wanted to include exercises that would help integrate an understanding of the field. We wanted to convey that the field is based on experimental and other research findings and that practical survey design was not a mere matter of judgment and opinion but rather the result of a body of research findings.

We drafted this book over several years. After we wrote the first couple of chapters, we hit a dry spell, which was ended when our colleague Nancy

Mathiowetz kicked us back in gear. We appreciated her energy in getting us going again.

The manuscript profited greatly from the critique of our student colleagues. The text had a dry run in the Summer of 2003 in a class at the University of Michigan Survey Research Center Summer Institute in Survey Research Techniques, entitled "Introduction to Survey Research Techniques," taught by Maria Krysan and Sue Ellen Hansen. We thank these instructors for helping improve the manuscript. We learned much from the criticisms and ideas of both Krysan and Hansen and the students in the class: Nike Adebiyi, Jennifer Bowers, Scott Compton, Sanjay Kumar, Dumile Mkhwanazi, Hanne Muller, Vuyelwa Nkambule, Laurel Park, Aaron Russell, Daniel Spiess, Kathleen Stack, Kimiko Tanaka, Dang Viet Phuong, and Christopher Webb.

It is fair to say that this book strongly reflects the lessons taught by many of our own mentors. One deserves special mention. All of the authors were friends and students of Charlie Cannell (some formally; all informally). Charles F. Cannell began his survey career with Rensis Likert at the U.S. Department of Agriculture, Division of Program Surveys. Cannell later joined Likert and others in founding the University of Michigan Survey Research Center in 1946. He was the first director of field operations at the Center and had a long and distinguished career in survey methodology. In memory of Charlie and his work, the Institute for Social Research (the larger institute of which the SRC is part) established the Charles F. Cannell Fund in Survey Methodology. All royalties that result from the sales of this text will be contributed to this fund. The endowment from the fund is designated for support of young scholars developing their research careers in survey methodology. We can think of no better use.

We designed the text to be used in a class where the participants had taken one or more courses in statistics. The key relevant skill is the reading of statistical notation, including summation signs, notation for expected values, and simple algebraic manipulation of summed quantities. Some chapters present quantitative analyses using regression and logistic regression models, and students unfamiliar with linear modeling need some help in understanding these results.

This book has 12 chapters, in the order in which they are presented in the semester-length course on survey methodology called "Fundamentals of Survey Methodology" at the JPSM. We envision that instructors will want to assign additional readings, often from one or more review articles referenced in the chapters.

The first two chapters ("Introduction to Survey Methodology" and "Inference And Error In Surveys") are conceptual in nature. Chapter 1 presents six example surveys that are used throughout the book to illustrate various principles and practices. The instructor can supplement the text by displaying the Web pages of these surveys in the class and leading class discussions about the key design features and products of the surveys.

The second chapter presents the key components of the total survey error paradigm. Again, at this early stage in the class, we have found that providing students with examples of key error components by referencing the example surveys aids in student understanding. A defining characteristic of surveys as we see them is that they are designed to produce statistical descriptions of populations. Although there are computer programs that will calculate statistics, we think it is critical that a survey methodologist understand the calculations that underlie those statistics. Hence, the book routinely presents statistical notation along with a conceptual discussion of what is being calculated.

The treatment of Chapter 2 would be a good time to devote a class to statistical notation, which, once it is learned, will help the students be more comfortable throughout the rest of the book

Starting with Chapter 3 ("Target Populations, Sampling Frames, and Coverage Error"), each chapter deals with a different component of total survey error and the methodological research discoveries that guide best practices. The focus of these chapters is deliberately the research on which best practices in survey research are based. We have often found that students beginning the study of survey methodology have the perspective that their opinions on a specific design feature are diagnostic of the best practices. The material that is presented in Chapters 3–11 attempts to show that there are scientific studies of survey methods that inform best practice; opinions are of little value unless they are research-based. Some of these studies do not have intuitively obvious findings. Hence, a student of the field must review the past methodological literature and at times do novel research to determine good design. There are two devices in the text that can help convey this perspective of the field. One is the set of embedded references to research in the discussions. The other is the presentation of illustrative boxes that give short descriptions of classic research in the domain covered in the chapter. These are summaries that describe the design, findings, limitations, and impact of the research. The full articles on this research can be used as supplementary readings, which could be discussed in class. There are also suggested supplementary readings at the end of each chapter.

Chapter 4 ("Sample Design and Sampling Error") uses more statistical notation than most of the other chapters. When many participants in the course need remedial instruction in reading and understanding statistical notation, we have referred them to the small monograph by Kalton, *An Introduction to Survey Sampling* (Sage, 1983). In some editions of the course, we have spent three weeks on the coverage and sampling chapters.

Each of Chapters 5–10 is normally covered in one week of the course. We have found it useful to emphasize the parallels between equations expressing coverage error and nonresponse error. We have also emphasized how the basic principles of intraclass correlations apply both to sample clustering effects and interviewer variance.

Chapter 11 ("Principles and Practices Related to Scientific Integrity") is included not just as sensitivity training but because it includes both conceptual frameworks underlying ethical treatment of human subjects and also recent theory and practice regarding disclosure analysis of survey data. Again, we describe how research, as well as judgement, can affect decisions related to ethical issues.

We wrote Chapter 12 ("FAQs About Survey Methodology") in a very different style. It is a tradition in the course, in a review session prior to the final examination, to have an open question section. At this time, we found students asking the kind of questions that come from attempts to integrate their learning of specific lessons with their larger worldview. Hence, we constructed a "frequently asked questions" format including those global questions and offering our answers to them.

The manuscript was greatly improved by the editorial wisdom of Sarah Dipko and Sonja Ziniel. Adam Kelley assisted in computer-based processing of figures and tables. Lisa Van Horn at Wiley is a production editor with a wonderful sense of when intervention is needed and when it isn't. We thank them all.

It was great fun writing this book, assembling our views on key research areas, and debating how to convey the excitement of survey methodology as an area of knowledge. We hope you have as much fun as we did.

Ann Arbor, Michigan	ROBERT M. GROVES
Boston, Massachusetts	FLOYD J. FOWLER, JR.
Ann Arbor, Michigan	MICK P. COUPER
Ann Arbor, Michigan	JAMES M. LEPKOWSKI
Ann Arbor, Michigan	ELEANOR SINGER
College Park, Maryland	ROGER TOURANGEAU
March 2004	

PREFACE TO THE SECOND EDITION

We have been pleased by the acceptance of the first edition of *Survey Methodology*. It has now been used by instructors around the world and has been translated into several other languages. Some of these instructors and their students have graciously pointed out weaknesses and errors in some sections of the text. Some of them gave us great ideas to improve the text.

In addition, as survey methodologists actively conducting research in the field, we became increasingly aware that some of the lessons in the text were becoming out of date. This was most true of sections of the book that concern the role of survey nonresponse in the quality of survey estimates and the rapidly evolving new modes of data collection.

For those reasons, we assembled the group of coauthors and agreed to update parts of chapters that could most profit from changes. As the reader will see, there is increased discussion of sampling frame issues for mobile telephone and web surveys in Chapter 3. There is an integration of some of the example surveys into the presentation of sample designs in Chapter 4, along with a new section on selection of persons within households. The changes in Chapter 5 update the findings on mobile phone and web surveys. Chapter 6, on survey nonresponse, is radically changed, reflecting new insights into how nonresponse rates and nonresponse errors relate to one another. Chapter 8, on evaluating survey questions, highlights new research findings on effective questionnaire development techniques. Chapter 11, on ethical issues in survey research, is reorganized to emphasize the growing research results on privacy, informed consent, and confidentiality issues. The remaining chapters provide the reader with more recent methodological research findings, especially when they expand our understanding of survey errors. The chapters have about 50% more exercises, following feedback from instructors that such additions would benefit their use of the text.

Two assistants labored over this edition's manuscript: Michael Guterbock and Kelly Smid. Some Ph.D. students at Michigan read draft chapters (Ashley Bowers, Matthew Jans, Courtney Kennedy, Joe Sakshaug, and Brady West). When we signed the contract with Wiley, we demanded that Lisa Van Horn continue as our production editor. All of the above went beyond the call to make this edition a success. We thank them.

As with the last edition, we want to use the royalties of the text to help persons newly entering the field of survey methodology. They will be given to the Rensis Likert Fund for Research on Survey Methodology, which directly benefits graduate students in survey methodology.

Ann Arbor, Michigan	ROBERT M. GROVES
Boston, Massachusetts	FLOYD J. FOWLER, JR.
Ann Arbor, Michigan	MICK P. COUPER
Ann Arbor, Michigan	JAMES M. LEPKOWSKI
Ann Arbor, Michigan	ELEANOR SINGER
College Park, Maryland	ROGER TOURANGEAU
March, 2009	

ACKNOWLEDGMENTS

Reprinted tables and figures in *Survey Methodology* and their copyright holders are listed below. The authors appreciate permission to adapt or reprint them.

Figure 1.5c, from Mokdad, Ford, Bowman, Dietz, Vinicor, Bales, and Marks (2003) with permission of the American Medical Association. Copyright © 2003.

Table 5.1 from Groves (1989) with permission of John Wiley and Sons. Copyright © 1989.

Figure 6.5 from Groves (2006) and text surrounding from Groves and Peytcheva (2008) with permission of the American Association for Public Opinion Research. Copyright © 2006, 2008.

Table in box on page 185, from Merkle and Edelman in Groves, Dillman, Eltinge, and Little (2002) with permission of John Wiley and Sons. Copyright © 2002.

Figure 7.2, from Tourangeau, Rips, and Rasinski (2000) reprinted with permission of Cambridge University Press. Copyright © 2000.

Box on page 236, from Schwarz, Hippler, Deutsch, and Strack (1985) reprinted with permission of the American Association for Public Opinion Research. Copyright © 1985.

Figure 7.3, from Jenkins and Dillman in Lyberg, Biemer, Collins, de Leeuw, Dippo, Schwarz, and Trewin (1997) reprinted with permission of John Wiley and Sons. Copyright © 1997.

Box on p. 268, from Oksenberg, Cannell, and Kalton (1991) reprinted with permission of Statistics Sweden. Copyright © 1991.

Box on p. 293, from Schuman and Converse (1971) reprinted with permission of the American Association for Public Opinion Research. Copyright © 1971.

Box on p. 299, from Kish (1962) reprinted with permission of the American Statistical Association. Copyright © 1962.

Table 9.2, from Fowler and Mangione (1990) reprinted with permission of Sage Publications. Copyright © 1990.

Table 10.5, from Campanelli, Thomson, Moon, and Staples in Lyberg, Biemer, Collins, de Leeuw, Dippo, Schwarz, and Trewin (1997) reprinted with permission of John Wiley and Sons. Copyright © 1997.

CHAPTER ONE

AN INTRODUCTION TO SURVEY METHODOLOGY

> **A Note to the Reader**
>
> You are about to be exposed to a system of principles called "survey methodology" for collecting information about the social and economic world. We have written this book in an attempt to describe the excitement of designing, conducting, analyzing, and evaluating sample surveys. To appreciate this fully, use the devices we have placed in each chapter to enrich your memory of the material. Throughout the book, you will see boxes with illustrations and examples of key principles, terminology notes, and highlights of classic research studies in the field. In the outside margin of each page you will find key terms, at the point where they are defined. At the end of each chapter is a set of exercises that you can use to test your understanding of that chapter's material. The best strategy is to read the text through, then, at the end of each chapter, go back, read the boxes, and review the key terms.

At 8:30 AM on the day before the first Friday of each month, a group of economists and statisticians enter a soundproof and windowless room in a building at 2 Massachusetts Avenue, NE, in Washington, DC, USA. Once those authorized are present, the room is sealed.

Those in the room are professional staff of the U.S. Bureau of Labor Statistics (BLS), and their task is to review and approve a statistical analysis of key economic data. Indeed, they have spent the week poring over sets of numbers, comparing them, examining indicators of their qualities, looking for anomalies, and writing drafts of a press release describing the numbers. They write the press release in simple language, understandable by those who have no technical knowledge about how the numbers were produced.

At 8:00 AM the next day, a group of journalists assemble in a monitored room in the nearby main Department of Labor building, removed from any contact with the outside world. The BLS staff enter the room and then reveal the results to the journalists. The journalists immediately prepare news stories based on the briefing. At exactly 8:30 AM, they simultaneously electronically transmit their stories to their news organizations and sometimes telephone editors and producers.

The statistics revealed are the unemployment rate of the prior month and the number of jobs created in the prior month. The elaborate protections and security used prior to their release stem from the enormous impact the numbers can have on society. Indeed, in months when the numbers signal important changes in the health of the U.S. economy, thousands of stock market investors around the world make immediate buy and sell decisions. Within 45 minutes of the announcement, trillions of dollars can move in and out of markets around the world based on the two numbers revealed at 8:30 AM.

Both the unemployment rate and the jobs count result from statistical surveys. A household survey produces the unemployment rate; an employer survey, the jobs count. The households and employers surveyed have been carefully selected so that their answers, when summarized, reflect the answers that would be obtained if the entire population were questioned. In the surveys, thousands of individual people answer carefully phrased questions about their own or their company's attributes. In the household survey, professional interviewers ask the questions and enter the answers onto laptop computers. In the employer survey, the respondents complete a standardized questionnaire either on paper or electronically. Complex data processing steps follow the collection of the data, to assure internal integrity of the numbers.

These two numbers have such an impact because they address an important component of the health of the nation's economy, and they are credible. Macroeconomic theory and decades of empirical results demonstrate their importance. However, only when decision makers believe the numbers do they gain value. This is a book about the process of generating such numbers through statistical surveys and how survey design can affect the quality of survey statistics. In a real sense, it addresses the question of when numbers from surveys are credible and when they are not.

1.1 INTRODUCTION

This chapter is an introduction to survey methodology as a field of knowledge, as a profession, and as a science. The initial sections of the chapter define the field so that the reader can place it among others. At the end of the chapter, readers will have a sense of what survey methodology is and what survey methodologists do.

survey   A "survey" is a systematic method for gathering information from (a sample of) entities for the purposes of constructing quantitative descriptors of the attributes of the larger population of which the entities are members. The word "systematic" is deliberate and meaningfully distinguishes surveys from other ways of gathering information. The phrase "(a sample of)" appears in the definition because sometimes surveys attempt to measure everyone in a population and sometimes just a sample.

statistic   The quantitative descriptors are called "statistics." Statistics are quantitative summaries of observations on a set of elements. Some are "descriptive statistics," describing the size and distributions of various attributes in a population (e.g., the mean years of education of persons, the total number of persons in the hospital, the percentage of persons supporting the president). Others are "analytic statistics," measuring how two or more variables are related (e.g., a regression coefficient describing how much increases in income are associated with increases in years of education; a correlation between education and number of books read in the last year). That goal sets surveys apart from other efforts to describe people or events. The statistics attempt to describe basic characteristics or experiences of large and small populations in our world.

descriptive statistic

analytic statistic

A BRIEF HISTORY OF SURVEY RESEARCH

Almost every country in the world uses surveys to estimate their rate of unemployment, basic prevalence of immunization against disease, opinions about the central government, intentions to vote in an upcoming election, and people's satisfaction with services and products that they buy. Surveys are a key tool in tracking global economic trends, the rate of inflation in prices, and investments in new economic enterprises. Surveys are one of the most commonly used methods in the social sciences to understand the way societies work and to test theories of behavior. In a very real way, surveys are a crucial building block in a modern information-based society.

Although a variety of activities are called surveys, this book focuses on surveys that have the following characteristics:

1) Information is gathered primarily by asking people questions.
2) Information is collected either by having interviewers ask questions and record answers or by having people read or hear questions and record their own answers.
3) Information is collected from only a subset of the population to be described—a sample—rather than from all members.

Since "ology" is Greek for "the study of," survey methodology is the study of survey methods. It is the study of sources of error in surveys and how to make the numbers produced by surveys as accurate as possible. Here the word "error" refers to deviations or departures from the desired outcome. In the case of surveys, "error" is used to describe deviations from the true values applicable to the population studied. Sometimes, the phrase "statistical error" is used to differentiate this meaning from a reference to simple mistakes.

survey methodology

error

statistical error

The way each of the above steps is carried out—which questions are asked, how answers are collected, and which people answer the questions—can affect the quality (or error properties) of survey results. This book will describe how to conduct surveys in the real world and how to evaluate the quality of survey results. It will describe what is known, and not known, about how to minimize error in survey statistics. Most of all, this book will attempt to distill the results of 100 years of scientific studies that have defined the theories and principles, as well as practices, of high-quality survey research.

1.2 A BRIEF HISTORY OF SURVEY RESEARCH

Converse (1987) has produced an important account of the history of survey research in the United States, and we recount some of the highlights here. There are four perspectives on surveys that are worth describing: the purposes to which surveys were put, the development of question design, the development of sampling methods, and the development of data collection methods.

1.2.1 The Purposes of Surveys

Perhaps the earliest type of survey is the census, generally conducted by governments. Censuses are systematic efforts to count an entire population, often for

census

> **Schuman (1997) on "Poll" Versus "Survey"**
>
> What is the difference between a poll and a survey? The word "poll" is most often used for private-sector opinion studies, which use many of the same design features as studies that would be called "surveys." "Poll" is rarely used to describe studies conducted in government or scientific domains. There are, however, no clear distinctions between the meanings of the two terms. Schuman notes that the two terms have different roots: "'Poll' is a four letter word, generally thought to be from an ancient Germanic term referring to 'head,' as in counting heads. The two-syllable word 'survey,' on the other hand, comes from the French *survee*, which in turn derives from Latin *super* (over) and *videre* (to look). The first is therefore an expression with appeal to a wider public, the intended consumers of results from Gallup, Harris, and other polls. The second fits the needs of academicians in university institutes who wish to emphasize the scientific or scholarly character of their work." (page 7)

purposes of taxation or political representation. In the United States, the Constitution stipulates that a census must be conducted every ten years, to reapportion the House of Representatives reflecting current population residence patterns. This gives the statistics from a census great political import. Because of this, they are often politically contentious (Anderson, 1990).

A prominent early reason for surveys was to gain understanding of a social problem. Some people trace the origins of modern survey research to Charles Booth, who produced a landmark study titled *Life and Labour of the People of London* (1889–1903) (*http://booth.lse.ac.uk/*). As Converse recounts it, Booth spent his own money to collect voluminous data on the poor in London and the reasons why they were poor. He wrote at least 17 volumes based on the data he collected. He did not use methods like the ones we use today—no well-defined sampling techniques, no standardized questions. Indeed, interviewer observation and inference produced much of the information. However, the Booth study used quantitative summaries from systematic measurements to understand a fundamental societal problem.

In contrast to studies of social problems, journalism and market research grew to use surveys to gain a systematic view of "the man on the street." A particular interest was reactions to political leaders and preferences in upcoming elections. That interest led to the development of modern public opinion polling.

In a related way, market research sought knowledge about reactions of "real" people to existing and planned products or services. As early as the 1930s, there was serious research on what programs and messages delivered via the radio would be most popular. The researchers began to use surveys of broader samples to produce information more useful to commercial decision makers.

Over the early 20th century, public opinion polling and market research, sometimes done by the same companies, evolved to use mail surveys and telephone surveys. They often sampled from available lists, such as telephone, driver's license, registered voter, or magazine subscriber listings. They collected their data primarily by asking a fixed set of questions; observations by interviewers and proxy reporting of other people's situations were not part of what they needed. These features were directly tied to the most important difference between what they were doing and what those who had gone before had done; rather than collecting data about facts and objective characteristics of people, the polling and market research surveyors were interested in what people knew, felt, and thought.

The measurement of attitudes and opinions is a key foundation of the modern management philosophies that place much weight on customer satisfaction.

Customer satisfaction surveys measure expectations of purchasers about the quality of a product or service and how well their expectations were met in specific transactions. Such surveys are ubiquitous tools of management to improve the performance of their organizations.

Politicians and political strategists now believe that opinion polls are critical to good decisions on campaign strategy and messages to the public about important issues. Indeed, a common criticism of modern politicians is that they rely too heavily on polling data to shape their personal opinions, choosing to reflect the public's views rather than provide leadership to the public about an issue.

1.2.2 The Development of Standardized Questioning

The interest in measuring subjective states (i.e., characteristics that cannot be observed, internalized within a person) also had the effect of focusing attention on question wording and data collection methods. When collecting factual information, researchers had not thought it important to carefully word questions. Often, interviewers were sent out with lists of objectives, such as age, occupation, and education, and the interviewers would decide on how the questions would be worded. Experienced researchers often did the interviewing, with great confidence that they knew how to phrase questions to obtain good answers.

However, the market research and polling organizations were doing large numbers of interviews, using newly hired people with no special background in the social sciences. Of necessity, researchers needed to specify more carefully the information sought by the survey. Further, researchers found that small changes in wording of an attitude question sometimes had unusually large effects on the answers.

Thus, early in the development of opinion surveys, attention began to be paid to giving interviewers carefully worded questions that they were to ask exactly the same way for each interview. Also, as interviewers were used more to ask questions, it was found that how they asked questions and recorded answers could affect the results. This led eventually to researchers training and supervising interviewers more formally than earlier.

Question wording also was influenced as the academics started to pay some attention to what the commercial researchers were doing. Psychometricians, psychologists who quantify psychological states, had been interested in how to put meaningful numbers on subjective states. Measuring intelligence was the first effort in this direction. However, people such as Thurstone also worked on how to assign numbers to attitudes, feelings, and ratings (e.g., Thurstone and Chave, 1929).

For the most part, their approaches were extremely cumbersome and were used primarily when they could get captive college student volunteers to fill out lengthy, highly redundant questionnaires. Such instruments were not going to be useful for most survey interviews with representative samples; they took too long to measure one or a few attitudes. Rensis Likert in his PhD dissertation (Likert, 1932), however, demonstrated that a single, streamlined question, with a scaled set of answers, could accomplish much the same thing as a lengthy series of paired comparisons. Likert applied the work to surveys (and later founded the University of Michigan Survey Research Center in 1946).

1.2.3 The Development of Sampling Methods

Early researchers, such as Booth, essentially tried to collect data on every element of a defined population. Such censuses avoided problems of errors arising from measuring just a subset of the population, but were clearly impractical for large populations. Indeed, the difficulty of analyzing complete census data led to early efforts to summarize a census by taking a sample of returns. Early efforts to sample would study a "typical" town, or they would purposively try to collect individuals to make the samples look like the population—for example, by interviewing about half men and half women, and trying to have them distributed geographically in a way that is similar to the population.

Although the theory of probability was established in the 18th century, its application to practical sample survey work was largely delayed until the 20th century. The first applications were the taking of a "1 in N" systematic selection from census returns. These were "probability samples"; that is, every record had a known nonzero chance of selection into the sample.

probability sample

A big breakthrough in sampling came from people who did research on agriculture. In order to predict crop yields, statisticians had worked out a strategy they called "area probability sampling." This is just what it sounds like: they would sample areas or plots of land and find out what farmers were doing with those plots in the spring (for example, if they were planting something on them and, if so, what) in order to project what the fall crops would look like. The same technique was developed to sample households. By drawing samples of geographic blocks in cities or tracts of land in rural areas, listing the housing units on the blocks or rural tracts, then sampling the housing units that were listed, samplers found a way to give all households and, by extension, the people living in them, a chance to be sampled. The attraction of this technique included the elimination of the need for a list of all persons or all households in the population prior to drawing the sample.

area probability sample

The Depression and World War II were major stimuli for survey research. One of the earliest modern probability samples was drawn for the Monthly Survey of Unemployment, starting in December, 1939, led by a 29-year-old statistician, Morris Hansen, who later became a major figure in the field (Hansen, Hurwitz, and Madow, 1953). During the war, the federal government became interested in conducting surveys to measure people's attitudes and opinions, such as interest in buying war bonds, as well as factual information. Considerable resources were devoted to surveys during the war, and researchers who were recruited to work with the government during the war later came to play critical roles in the development of survey methods. When the war was over, methodologists understood that in order to produce good population-based statistics it was necessary to attend to three aspects of survey methodology: how questions were designed; how the data were collected, including the training of interviewers; and how samples were drawn.

Probability samples are the standard by which other samples are judged. They are routinely used by almost all government statistical agencies when data are used to provide important information for policy makers. They are used for surveys used in litigation. They are used for measurement of media audience sizes, which in turn determine advertising rates. In short, whenever large stakes ride on the value of a sample, probability sampling is generally used.

1.2.4 The Development of Data Collection Methods

The gathering of information in early surveys was only one step more organized than talking to as many people as possible about some topic. The qualitative interviews produced a set of verbal notes, and the task of summarizing them with statistics was huge. Surveys grew to be popular tools because of the evolution of methods to collect systematic data cheaply and quickly.

Mailed paper questionnaires offered very low costs for measuring literate populations. Indeed, by 1960 a formal test of census procedures based on mailed questionnaires succeeded to an extent that the 1970 census was largely a mailed questionnaire survey. Further, mailed questionnaires proved to be much cheaper than sending interviewers to visit sample cases. On the other hand, mailed surveys were subject to the vagaries of the postal service, which, even when it worked perfectly, produced survey periods that lasted months, not weeks.

With the spread of telephone service throughout the country, market researchers first saw two advantages of using the medium as a data collection tool. It was much faster than mail questionnaire surveys, and it was still cheaper than face-to-face surveys. For decades, however, the mode suffered from the clear lack of coverage of telephones among the poor and more transient members of the society. By the 1990s, however, almost all market research had moved away from the face-to-face survey and much scientific research was close behind in the abandonment of that mode. It was largely the federal government that continued to rely on face-to-face household surveys.

Like many fields of human activity, huge leaps in the efficiencies of surveys came from the invention of the computer. One of the first computers made in the United States was used in processing decennial census data. Survey researchers quickly recognized how computers could reduce the large amount of human resources needed to conduct surveys. Survey researchers first used computers to perform the analysis steps of a survey, then they began to use them to assist in checking the raw data for clerical errors, then to assist them in coding text answers, then in the data collection step itself. Now, computers (from handheld devices to networked systems) are used in almost every step of survey design, data collection, and analysis. The fastest growing application is the development of Web surveys.

As these various developments evolved, the field also developed a set of performance guidelines. Empirical studies demonstrated the value of various sample designs on the quality of statistics. Interviewer training guidelines improved the standardization of interviewers. Standards about computing and reporting response rates offered the field measures useful in comparing surveys.

In the 60 years following the advent of surveys, a great deal has been learned about how to design data collection systems to improve the quality of survey statistics. However, as can be seen from this short history, the basic elements of good survey methodology were defined in the first half of the 20th century.

1.3 SOME EXAMPLES OF ONGOING SURVEYS

One way to understand the range of survey methods and the potential of surveys to provide information is to give some examples. The following is a brief description of six surveys. We have chosen them to use as examples throughout the book for several reasons. First, they are all ongoing surveys. They are conducted year after year. By definition, that means that the sponsors think that there is a continuous need for the kind of information that they provide. That also means that someone thinks they are important. These are not particularly typical surveys. They do not include public opinion, political, or market research studies. They do not include any one-time surveys, which are highly prevalent. All of these surveys are funded by government sources.

However, they do differ from one another in numerous ways. One reason we chose this set is they do give a sense of the range of topics that are addressed by surveys and the variety of survey designs that are used. They also were chosen because they provide examples of excellence in survey research. Hence, they will provide opportunities for us to discuss how different methodological problems are addressed and solved.

In the brief summaries and charts provided, we describe some of the basic characteristics of each survey:

1) Their purposes
2) The populations they try to describe
3) The sources from which they draw samples
4) The design of the way they sample people
5) The use of interviewers
6) The mode of data collection
7) The use of computers in the collection of answers

Readers should think about these surveys in two different ways. First, think of them as information sources—what we can learn from them, what questions they answer, what policies and decisions they inform; in short, why they are conducted. Second, compare the design features above in order to see how different survey design features permit the surveys to achieve their different purposes.

1.3.1 The National Crime Victimization Survey

How much crime is there in the United States? Are crimes increasing in frequency or going down in frequency? Who gets victimized by crimes? Every society seeks answers to these questions. In the United States, the answers were sought through quantifying crime in a period of great public concern about organized crime. In the 1930s, the International Association of Chiefs of Police began a collection of administrative record counts. The method rested on the reporting of crimes to police in jurisdictions around the country, based on the administrative records kept by individual sheriffs, transit police, city police, and state police offices. Police chiefs had designed the record systems to have legal documentation of the circumstances of the crime, the victim of the crime, the offender, and any relevant evidence related to the crime. Individual staff members completed the paperwork that produced the administrative records. However, many crimes only come to the attention of the police if a citizen decides to report them. Often, the decision to produce a record or to label an incident was left to a relatively low-level police officer. For years, these records were the key information source on U.S. crime.

SOME EXAMPLES OF ONGOING SURVEYS

Over the years, several weaknesses in the statistics from police records became obvious. Sometimes a new mayor, fulfilling a pledge to reduce crime, created an environment in which more police officers chose not to label an incident as a crime, thus not producing an administrative record. Further, the statistics were tainted by different jurisdictions using different definitions for crime categories. When police believed that the crime would never be solved, they encouraged the resident not to file a formal report. There was growing evidence in some jurisdictions that relations between the public and the police were poor. Fear of the police among the public led to avoidance of reporting criminal incidents. The police officers themselves carried with them attitudes toward subpopulations that led to classifying an incident as a crime for one group but not for another group. It was becoming clear that, whereas major crimes like homicide were well represented in the record systems, the records tended to miss more minor, often unreported, crimes. Some jurisdictions kept very detailed, complete records, whereas others had very shoddy systems.

Thus, over the decades many began to distrust the value of the statistics to address the simplest question: "How much crime is there in the United States?" Further, the simple counts of crimes were not giving policy makers clear information about the characteristics of crimes and their victims, information helpful in considering alternative policies to reducing crime. The President's Commission on Law Enforcement and the Administration of Justice, established by President Johnson, noted in a task force report that, "If we knew more about the character of both offenders and victims, the nature of their relationships and the circumstances that create a high probability of crime conduct, it seems likely that crime prevention programs could be made much more effective" (President's Commission, 1967, as cited in Rand and Rennison, 2002).

In the late 1960s, criminologists began exploring the possibilities of using surveys to ask people directly whether they were a victim of a crime. This forced a conceptually different perspective on crime. Instead of a focus on the incident, it focused on one actor in the incident—the victim. This shift of perspective produced clear contrasts with the police reported crime record systems. Most obviously, homicide victims cannot report! Victimizations of young children might not be well reported by their parents (who may not know about incidents at school), and the children may not be good respondents. Crimes against companies present problems of who can report them well. For example, if someone starts a fire that burns down an apartment building, who are the victims—the owner of the property, the renters of the apartments, or the people visiting during the fire? From one perspective, all are victims, but asking all to report the arson as a victimization may complicate the counts of crimes. Victims can sometimes report that an unpleasant event occurred (e.g., someone with no right of entry entered their home). However, they cannot, as do police, gather the information that asserts the intent of the offender (e.g., he was attempting a theft of a television).

On the other hand, using individual reporters, a survey can cover victimizations reported to the police and those not reported to the police. This should provide a more complete picture of crime, if there are crimes not reported to the police. Indeed, self-reported victimizations might be a wonderful addition to statistics on police-reported crimes, as a way to compare the perceived victimization in a society with the officially reported victimization status. Moreover, the survey has the advantage of utilizing standard protocols for the measurement of victimizations across the country.

However, the notion of using a national survey to measure victimization faced other problems. Although all police agencies in the United States could be asked to report statistics from their record systems, it was financially impossible to ask all persons in the United States to report their individual victimizations. If a "representative" sample could be identified, this would make the victimization survey possible. Thus, in contrast to the record data, the survey would be subject to "sampling error" (i.e., errors in statistics because of the omission of some persons in the population).

sampling error

But could and would people really report their victimizations accurately? Survey methodologists studied the reporting problem in the late 1960s and early 1970s. In methodological studies that sampled police records and then went back to victims of those crimes, they found that, in large part, they provided reports that mirrored the data in the records. However, Gottfredson and Hindelang (1977) among others noted that one problem was a tendency for persons to misdate the time of an incident. Most incidents that were important to them were reported as occurring more recently than they actually did. By and large, however, the pattern of reporting appeared to justify the expense of mounting a completely separate system of tracking crime in the country.

When all the design features were specified for the survey approach, it was clear that there were inevitable differences between the police reports and the victim reports (Rand and Rennison, 2002). The abiding strength of the survey was that it could measure crimes that are not reported to the police or for which police do not make formal documentation. However, the survey would be subject to sampling error. On the other hand, the police-reported crimes include some victims who are not residents of the country, and these would be missed in a survey of the US household population. Further, the police report statistics include homicides and arson, but do not include simple assault. Police reports exclude rapes of males; the survey could include rapes of both sexes. Some crimes have multiple victims, who could report the same incident as occurring to them (e.g., household crimes, group thefts); the survey would count them as multiple crimes, and the police reports as one incident, generally. The police report statistics depend on voluntary partnerships between the federal government and thousands of jurisdictions, but jurisdictions vary in their cooperation; the survey (using traditional methods) would suffer from some omissions of persons who are homeless or transient and from those who choose not to participate. The methodological work suggested that the survey would tend to underreport crimes where the offender is well known to the victim. There was also evidence that the survey would underreport less important crimes, apparently because they were difficult to remember. Finally, both systems have trouble with repeated incidents of the same character. In the survey method, repeated victimizations of the same type were to be counted once as "series incidents" (e.g., repeated beatings of a wife by her husband); the police-reported series would have as many reports as were provided to the police. The National Crime Victimization Survey (NCVS) (Table 1.1) has its roots within the U.S. Department of Justice in 1972. The current Bureau of Justice Statistics has the mission of collecting and disseminating statistical information on crimes, criminal offenders, victims of crimes, and the operations of justice systems at all levels of government. The Bureau of Justice Statistics contracts with the U.S. Census Bureau to collect the data in the NCVS.

The NCVS asks people to report the crimes they have experienced in the 6 months preceding the interview. If they asked people to report for 12 months, the

SOME EXAMPLES OF ONGOING SURVEYS

Table 1.1. Example Survey: National Crime Victimization Survey (NCVS)

Sponsor	U.S. Bureau of Justice Statistics
Collector	U.S. Census Bureau
Purpose	Main objectives are to: • Develop detailed information about the victims and consequences of crime • Estimate the number and types of crimes not reported to the police • Provide uniform measures of selected types of crimes • Permit comparisons over time and by types of areas
Year Started	1973 (previously called the National Crime Survey, 1973–1992)
Target Population	Adults and children 12 or older, civilian and noninstitutionalized
Sampling Frame	U.S. households, enumerated through counties, blocks, listed addresses, lists of members of the household
Sample Design	Multistage, stratified, clustered area probability sample, with sample units rotating in and out of the sample over three years
Sample Size	About 41,800 households (78,600 persons)
Use of Interviewer	Interviewer administered
Mode of Administration	Face-to-face and telephone interviews
Computer Assistance	Paper questionnaire for 70% of the interviews, both face-to-face and telephone interviews; computer assistance for 30% of the interviews
Reporting Unit	Each person age 12 or older in household reports for self
Time Dimension	Ongoing rotating panel survey of addresses
Frequency	Monthly data collection
Interviews per Round of Survey	Sampled housing units are interviewed every six months over the course of three years
Levels of Observation	Victimization incident, person, household
Web Link	*http://www.ojp.usdoj.gov/bjs/cvict.htm*

researchers could learn about more events per interview; it would be a more efficient use of interview time. However, early studies showed that there is a marked drop-off in the accuracy of reporting when people are asked to remember events that happened more than 6 months in the past. In fact, there is underreporting of crimes even when the questions ask about the past 6 months. The accuracy of reporting would be higher if the questions asked about only one or two months, or, better yet, only a week or two. However, as the reporting period gets shorter, fewer and fewer people have anything to report, so more and more interviews provide minimal information about victimization. The designers of the survey chose 6 months as a reasonable point at which to balance reporting accuracy and the productivity of interviews.

The sample for the NCVS is drawn in successive stages, with the goal of giving every person 12 years old and older a known chance of selection and, thereby, producing a way to represent all age eligible persons in the United States. (The jargon for this is a "multistage, stratified clustered area probability sample," which will be described in Chapter 4.) The sample is restricted to persons who are household members, excluding the homeless, those in institutions, and in group quarters. (The survey methodologists judged that the cost of covering these subpopulations would be prohibitively expensive, detracting from the larger goals for the survey.) The sample is clustered into hundreds of different sample areas (usually counties or groups of counties) and the sample design is repeated samples of households from those same areas over the years of the study. The clustering is introduced to save money by permitting the hiring of a relatively small group of interviewers to train and supervise, who travel out to each sample household to visit the members and conduct interviews. Further, to save money all persons 12 years and over in the household are interviewed; thus, one sample household might produce one interview or many.

A further way to reduce costs of the survey is to repeatedly measure the same address. When the design randomly identifies a household to fall into the NCVS sample, the interviewer requests that, in addition to the first interview, the household be willing to be visited again six months later, and then again and again, for a total of seven interviews over a three-year period. In addition to saving money, this produces higher-quality estimates of change in victimization rates over years. This design is called a "rotating panel design" because each month, different people are being interviewed for the first time, the second time, the third time, the fourth time, the fifth time, the sixth time, and the last (seventh) time. Thus, the sample is changing each month but overlaps with samples taken six months previously.

rotating panel design

Each year, the NCVS collects interviews from about 42,000 households containing more than 76,000 persons. About 92% of the sample households provide one or more interviews; overall, about 87% of the persons eligible within the sample households provide an interview. In 2006, the most recent year with published estimates, 91% of households and 86% of persons in interviewed households were interviewed

The interviews contain questions about the frequency, characteristics, and consequences of criminal victimizations the households may have experienced in the previous six months. The interview covers incidents of household victimization and personal victimization: rape, sexual assault, robbery, assault, theft, household burglary, and motor vehicle theft. An interviewer visits those households and asks those who live there about crimes they have experienced over the past six months. One person in the household acts as the informant for all property crimes (like burglary, vandalism, etc.); each person then reports for him- or herself about personal crimes (e.g., assault, theft of personal items). The interview is conducted in person in the first wave; subsequent waves attempt whenever possible to use telephone interviewers calling from two different centralized call centers. Thus, the NCVS statistics are based on a mix of telephone (60%) and face-to-face interviews (40%).

The questionnaire asks the respondent to remember back over the past six months to report any crimes that might have occurred. For example, the questions in the box on page 13 ask about thefts.

SOME EXAMPLES OF ONGOING SURVEYS

> I'm going to read some examples that will give you an idea of the kinds of crimes this study covers.
>
> As I go through them, tell me if any of these happened to you in the last 6 months, that is since _ _(MONTH) _ _(DAY), 20_ _.
>
> Was something belonging to YOU stolen, such as:
> a) Things that you carry, like luggage, a wallet, purse, briefcase, or book
> b) Clothing, jewelry, or calculator
> c) Bicycle or sports equipment
> d) Things in your home, like a TV, stereo, or tools
> e) Things from a vehicle, such as a package, groceries, camera, or cassette tapes

If the respondent answers "yes, someone stole my bicycle," then the interviewer records that and later in the interview asks questions about details of the incident. Figure 1.1 shows the kind of statistics that can be computed from the NCVS. The percentages of households reporting one or more crimes of three different types (property crimes, vandalism, and violent crime) are displayed for the years 1994–2000. Note that the percentages are declining, showing a reduced frequency of crime over the late 1990s. Policy makers watch these numbers closely as an indirect way to assess the impact of their crime-fighting programs. However, other research notes that when the nation's economy is strong, with low unemployment, crime rates tend to decline.

"Crime at Lowest Point in 25 Years, Fed Says" reads the headline on CNN.com on December 27, 1998. "Fewer people in the United States were the

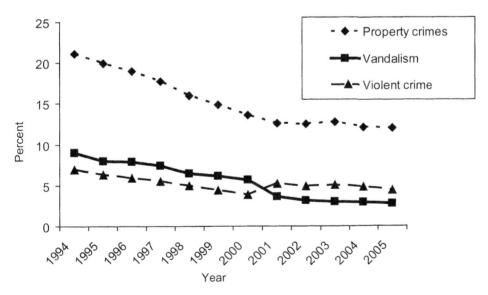

Figure 1.1 Percentage of U.S. households experiencing a crime by type, 1994–2005 National Crime Victimization Survey.
(Source: www.ojp.usdoj.gov/bjs/.)

victims of crimes last year than at any time since 1973, the Justice Department reported Sunday," reads the first line. Later in the story, "President Bill Clinton applauded the new crime figures Sunday. They 'again show that our strategy of more police, stricter gun laws, and better crime prevention is working,' he said in a statement." These findings resulted from the NCVS. The statement attributing drops in crime to policies promoted by the current administration is common, but generally without strong empirical support. (It is very difficult, given available information, to link changes in crime rates to policies implemented.)

NCVS data have informed a wide audience concerned with crime and crime prevention. Researchers at academic, government, private, and nonprofit research institutions use NCVS data to prepare reports, policy recommendations, scholarly publications, testimony before Congress, and documentation for use in courts (U.S. Bureau of Justice Statistics, 1994). Community groups and government agencies use the data to develop neighborhood watch and victim assistance and compensation programs. Law enforcement agencies use NCVS findings for training. The data appear in public service announcements on crime prevention and crime documentaries. Finally, print and broadcast media regularly cite NCVS findings when reporting on a host of crime-related topics.

1.3.2 The National Survey on Drug Use and Health

What percentage of people in the United States use illicit drugs? Are rates of usage higher among the poor and less educated than among others? How does drug use change as people get older? Are rates of usage changing over time? Do different groups tend to use different drugs? Is use of alcoholic beverages related to drug use? Are the rates of usage changing over time? Do different states have different drug usage patterns? Each year the National Survey on Drug Use and Health (NSDUH) draws samples of households from each state of the union. Interviewers visit each sampled home and ask questions of each sample person about their background and other relatively nonsensitive information. To collect data on drug usage, the interviewers provide a laptop computer to the respondent, who dons headphones to listen to a voice recording of the questions and keys in the answers on the laptop keyboard. Every year, the NSDUH produces estimates of rates of usage of several different kinds of drugs. The data are used to inform U.S. federal government drug policy, aimed at reducing demand and supply of illegal drugs.

The first national household survey attempt to measure drug use was conducted in 1971. The survey oversampled persons 12 to 34 years old in order to have enough cases to make separate estimates of that age group because it was suspected to have the highest prevalence of drug use. There was early concern about the U.S. population's willingness to be interviewed about drug use. The designers were concerned about response rates. In the early rounds, personal visit interviewers contacted sample households and asked questions of sample persons. When the interviewer came to sensitive questions about drug use, the interviewer noted that the data would be held confidential, and switched the questioning from oral answers to a self-administered mode, with respondents writing down their answers on an answer sheet, placing the completed form in a sealed envelope, and accompanying the interviewer to mail it.

Table 1.2. Example Survey: National Survey of Drug Use and Health (NSDUH)

Sponsor	Substance Abuse and Mental Health Services Administration (SAMHSA)
Collector	RTI International
Purpose	Main objectives are to: • Provide estimates of rates of use, number of users, and other measures related to illicit drug, alcohol, and tobacco use at the state and national level • Improve the nation's understanding of substance abuse • Measure the nation's progress in reducing substance abuse
Year Started	1971 (formerly named National Household Survey on Drug Abuse)
Target Population	Noninstitutionalized population of the United States aged 12 years old or older
Sampling Frame	U.S. households, enumerated through U.S. counties, blocks, and list of members of the households
Sample Design	Multistage, stratified clustered area probability sample within each state
Sample Size	141,487 housing units; 67,870 persons (2007 NSDUH)
Use of Interviewer	Interviewer-administered, with some self-administered questionnaire sections for sensitive questions
Mode of Administration	Face-to-face interview in respondent's home, with portions completed by respondent alone
Computer Assistance	Computer-assisted personal interview (CAPI), with audio computer-assisted self-interview (ACASI) component
Reporting Unit	Each person age 12 or older in household reports for self. Respondents may allow more knowledgeable family member to complete Health Insurance and Income sections of survey for them.
Time Dimension	Repeated cross-sectional survey
Frequency	Conducted annually
Interviews per Round of Survey	One
Levels of Observation	Person, household
Web Link	http://www.samsha.gov/

This first survey defined many of the ongoing features of the design. Like the NCVS, the NSDUH targets the household population and noninstitutional group quarters (e.g., shelters, rooming houses, dormitories). It covers persons who are civilian residents of U.S. housing units and 12 years of age or older. Over the years, the sample has grown and has recently been redesigned to provide independent samples of each of the 50 states, with about 70,000 persons measured each year. This permits the survey to provide separate estimates of drug use for eight of the largest states each year and for the other states by combining data over several years. The design also oversamples youths and young adults, so that each state's sample is distributed equally among three age groups (12 to 17 years, 18 to 25 years, and 26 years or older). In contrast to the NCVS, each sample person is interviewed only once. A new sample is drawn each year, and estimates for the year are based on that new sample. (This is called a "repeated cross-section design.") About 91% of households provide the screening interview (that measures household composition), and 79% of the selected individuals provide the full interview.

repeated cross-section design

Although the survey is sponsored by a federal government agency, the Substance Abuse and Mental Health Services Administration, it is conducted under contract by RTI International. This structure is common for large household surveys in the United States but is rarer in other countries. The United States developed over its history an ongoing partnership between survey research organizations in the private sector (both commercial, nonprofit, and academic) and the federal and state governments, so that much of the governmental survey information is collected, processed, and analyzed by nongovernmental staff.

The survey covers a wide set of drugs (e.g., alcohol, prescription drugs). The designers periodically update its measurement procedures, with special focus on how it could get more accurate self-reports of drug use. The interviewers now use laptop computers (the process is called "computer-assisted personal interviewing" or CAPI) to display questions and store answers. It also uses audio computer-assisted self-interviewing (ACASI), which has the respondent listen to questions via earphones attached to a laptop computer, see the questions displayed, and enter his/her responses using the keyboard. Increases in reporting of drug use appear to be produced by this technique versus having an interviewer administer the survey [e.g., 2.3 times the number of persons reporting cocaine use (Turner, Lessler, and Gfroerer, 1992, p. 299)]. The survey has increasingly tried to help guide policy on drug treatment and prevention.

CAPI
ACASI

By creating statistics from self-reports of the same drugs each year, NSDUH can provide the United States with net changes over time in drug use. Much of the methodological research of the NSDUH has been focused on the fact that drug use tends to be underreported. By comparing statistics over time, using the same method, the hope is that any underreporting problems remain relatively consistent over time and, thus, the change from year to year is an accurate gauge of differences over time. For example, Figure 1.2 shows that there are small increases between 1999 and 2001 in the use of marijuana, psychotherapeutic drugs, and hallucinogens. Such increases could be cause for reassessment of programs attempting to reduce the supply of drugs and to treat those abusing the drugs.

"Clinton Proposes Holding Tobacco Industry Accountable for Teen Smoking," is the headline in the CNN.com story of February 4, 2000. The first line reads, "To crack down on smoking by minors, President Bill Clinton wants to make the tobacco industry pay $3000 for every smoker under the age of 18, and also proposes raising the tax on cigarettes by another 25 cents per pack." Later, "The administration says there are currently 4.1 million teen smokers, according to the latest National Household Survey of Drug Abuse conducted by the Department of Health and Human Services." Policy makers watch the NSDUH as evidence of success or failure of current antidrug policies. Activist governments tend to propose new programs when the data suggest failure.

SOME EXAMPLES OF ONGOING SURVEYS

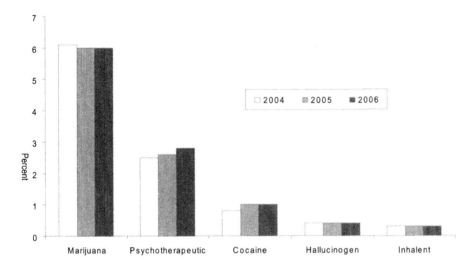

Figure 1.2 Percentage of persons reporting illicit drug use in past month, by drug type, 2004–2006 (Source: NSDUH.)

1.3.3 The Surveys of Consumers

Are people optimistic or pessimistic about their financial future? Are they planning to make major purchases (e.g., automobiles, refrigerators) in the near future? Do they think they are better off now than they were a few years ago? Are the rich more optimistic than the poor? Are levels of optimism changing over time?

In 1946, the economist George Katona discovered that asking people about their views of their personal and the nation's economic outlook could provide useful information about the future of the entire nation's economy. This stemmed from the basic finding that people's attitudes about the future affect their behavior on consumer purchases and savings. Further, the individual decisions on purchases and savings are an important component of the health of the economy. Since that time, the University of Michigan Survey Research Center has conducted an ongoing survey of consumer attitudes. The survey is not financed solely by the federal government (despite the fact that its statistics form part of the U.S. Leading Economic Indicators), but by a consortium of commercial entities and the Federal Reserve Board, the organization that makes monetary policy for the country.

Each month, the Surveys of Consumers (SOC) telephones a sample of phone numbers, locates the household numbers among them, and selects one adult from those in the households. The target is the full household population, but to reduce costs, only persons in households with telephones are eligible. [Those without telephones tend to be poorer, to live in rural areas, and to be more transient (see Chapter 3).] The sample is drawn using a "random digit dialing" design, which samples telephone numbers in working area codes and prefixes. (Not all of the resulting sample numbers are household numbers; some are nonworking, commercial, fax, or modem numbers.) Each month, about 500 interviews are conducted. Currently about 60% of selected adults in the households provide an interview. When a sample household is contacted, interviewers attempt to take an

random digit dialing

Table 1.3. Example Survey: Surveys of Consumers (SOC)

Survey Name	Survey of Consumers
Sponsor	University of Michigan
Collector	Survey Research Center, University of Michigan
Purpose	Main objectives are to: • Measure changes in consumer attitudes and expectations • Understand why such changes occur • Evaluate how they relate to consumer decisions to save, borrow, or make discretionary changes
Year Started	1946
Target Population	Noninstitutionalized adults in the coterminous United States (omits Hawaii and Alaska)
Sampling Frame	Coterminous U.S. telephone households, through lists of working area codes and exchanges
Sample Design	List-assisted random-digit dial sample, randomly selected adult
Sample Size	500 adults
Use of Interviewer	Interviewer administered
Mode of Administration	Telephone interview
Computer Assistance	Computer-assisted telephone interviewing (CATI)
Reporting Unit	Randomly selected adult
Time Dimension	Two-wave panel of persons
Frequency	Conducted monthly
Interviews per Round of Survey	Two; reinterview conducted six months after initial interview on subset of wave 1 respondents
Levels of Observation	Person
Web Link	*http://sca.isr.umich.edu/*

interview and then call again 6 months later for a follow-up interview. Thus, the SOC, like the NCVS, is a rotating panel design. After the data are collected, there are statistical adjustments made to the estimates in an attempt to repair the omission of nontelephone households and nonresponse. Each month, statistics about consumer confidence are computed, using a mix of cases from the first-wave interview and from the second-wave interview.

The interview contains about 50 questions about personal finance, business conditions, and buying conditions. The telephone interviewers working in a centralized telephone interviewing facility use a computer-assisted interviewing system, which uses desktop computers to display the questions, accept responses, and route the interviewer to the next appropriate question.

SOME EXAMPLES OF ONGOING SURVEYS

Each month, the Consumer Sentiment Index is published, which has been found to be predictive of future national economic growth. This indicator, combined with a set of others, forms the Index of Leading Economic Indicators, which is used by economic forecasters in advising on macroeconomic policy and investment strategies. It is interesting and awe-inspiring to many that simple questions on a survey can predict important shifts in something as complex as the U.S. economy. For example, Figure 1.3 shows a plot of two series of statistics. One is the answer to the question, "How about people out of work during the coming 12 months—do you think that there will be more unemployment than now, about the same, or less?" The figure compares an index based on the answers to this question (see left y axis) to the actual annual change in the unemployment rate (right y axis). The consumer expectation index is the dark line and the actual percentage point change in the unemployment rate is the light line. Note that consumer expectations have consistently anticipated changes in the unemployment rate months in advance. It has been shown that consumer expectations contain predictive information about future changes in unemployment that are not captured by other economic information (Surveys of Consumers, 2003).

Financial investment firms and stock markets watch the consumer confidence statistics. They appear to believe that they are credible predictors of later behaviors. On June 15, 2002, *The New York Times,* ran an article entitled, "Falling Consumer Confidence Sends the Dow Lower," describing sentiments of those investing in the U.S. stock exchanges, which led to a short-term decline in stock prices because of increased selling. To illustrate the complexity of the information impact on the markets, the same article lists a car bomb in Pakistan as another possible cause of the drop. There are many sources of information that affect financial decision makers' behaviors. It is clear that one of them is the ongoing measure of consumer confidence.

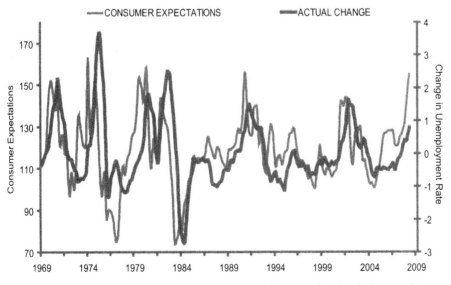

Figure 1.3 Consumer unemployment expectations and actual change in the U.S. unemployment rate, 1969–2009 (Source: Surveys of Consumers, 2008.)

1.3.4 The National Assessment of Educational Progress

What level of skills do elementary school children have in basic mathematics, reading, and writing? Do some schools foster more learning than others? Do children from lower-income families and racial/ethnic subgroups perform more or less well? Are there areas of the country or states where the children's achievement on mathematics and verbal tasks differ? How does the United States compare to other countries? Are achievement levels changing over time?

For many decades, there have been examinations that different school districts have given to their students from time to time. State boards of education often link state funding to the testing, so that whole states might administer the same test. Indeed, over the past few decades, politicians, with the support of parents, have urged more accountability by public schools, and this has led to widespread testing.

Unfortunately, despite the prevalence of testing, there was too little uniformity of the assessments. Each state, and sometimes each school district, makes independent decisions about which tests to use. Hence, the comparability of assessments over areas is low. Furthermore, not all schools use the standardized assessments. Often, districts with low levels of funding omit the assessments. Hence, the answers to the questions above are not coming systematically from administrative procedures used in the schools around the country. (Note how this resembles the situation with the NCVS and police reports of crime.)

The spur to uniform assessments of educational performance coincided with an increasing role of the federal government in education during the 1960s (Vinovskis, 1998). But creating a statistical survey to provide uniform statistics was politically controversial. This is a great example of how survey statistics can become so important that they themselves become a political issue.

First, some states did not want to be compared to others on assessment scores; hence, there was more support for a national survey limited to national estimates than one providing individual estimates for each state. Second, there was opposition to making the assessment a tool to provide individual student evaluations, for fear that the statistical purposes of the survey could be undermined by special efforts of schools, parents, and students to perform so well on the assessments that they would not reflect the national performance. There was also political opposition arguing that the money spent on assessments might be better spent on improving education itself. Finally, there were political ideologies that clashed. Some believed strongly in local control over schools and saw a national assessment of education as a threat that the federal government sought control over education.

Despite all these issues, the U.S. Department of Education launched the National Assessment of Educational Progress (NAEP) in 1969 to produce national assessment summaries, and state-based samples (permitting accurate estimates for each state) were added in 1990. The NAEP is really a collection of surveys with three separate assessments: the "main national" (which provides the annual national figures), "main state" (which provides the state estimates), and "trend" (which is used for assessing change over time). Each of these assessments consists of four components: Elementary and Secondary School Students Survey, School Characteristics and Policies Survey, Teacher Survey, and Students with Disabilities or Limited English Proficiency (SD/LEP) Survey (for the main NAEP), or Excluded Student Survey (for the trend NAEP). In years in which the main state is conducted, there is no longer a separate national sample; the national

Table 1.4. Example Survey: National Assessment of Educational Progress (NAEP)

Sponsor	National Center for Education Statistics, U.S. Department of Education
Collector	Westat
Purpose	Main objectives are to: • Assess the academic performance of fourth, eighth, and twelfth graders in a range of subjects • Reflect current educational and assessment practices • Measure change over time
Year Started	1969
Target Population	National NAEP – schoolchildren in grades 4, 8, and 12 State NAEP – schoolchildren in grades 4 and 8
Sampling Frame	U.S. elementary and secondary school children, through U.S. counties or groups of counties, listed schools and students within schools
Sample Design	Multistage, stratified clustered area probability sample of primary sampling units (PSUs); sample of schools within PSU drawn, classrooms of students
Sample Size	2,000 schools and 100,000 students (National NAEP) 100 schools and 2,500 students per subject grade (State NAEP sample size per state) [typical sample sizes]
Use of Interviewer	None; self-administered background questionnaires completed by students, teachers, and principals; cognitive assessments completed by students; proctored by survey administrators
Mode of Administration	Paper-and-pencil self-administered questionnaires and cognitive assessment instruments
Computer Assistance	None
Reporting Unit	Students, teachers, school principals
Time Dimension	Repeated cross-sectional survey
Frequency	Conducted annually
Interviews per Round of Survey	One
Levels of Observation	Student, class, school
Web Link	http://nces.ed.gov/nationsreportcard/

sample is the aggregate of the state samples. For the long-term trend NAEP, we are now administering the Students with Disabilities/English Language Learners (SD/ELL) Survey. In 1985, the Young Adult Literacy Study was also conducted nationally as part of NAEP, under a grant to the Educational Testing Service and Response Analysis Corporation; this study assessed the literacy skills of persons 21 to 25 years old. In addition, NAEP conducts a high school transcript study. In this book, we will concentrate our remarks on the main national NAEP.

The National Center for Education Statistics, a federal government statistical agency, sponsors NAEP, but the data are collected through a contract, like NSDUH. The Educational Testing Service, a company that conducts the Scholastic Aptitude Test (SAT) and other standardized tests, develops the assessments. Westat, a survey firm, designs and selects the sample, and conducts the assessments in sample schools. NCS Pearson, another testing and educational assessment firm, scores the assessments.

The national NAEP selects students enrolled in schools in the 50 states and the District of Columbia in grades 4, 8, and 12. The sample design is clustered into primary sampling units (counties or groups of counties), just like NCVS and NSDUH (this is a cost-saving effort for obtaining cooperation from school administrators and for the administration of the assessments). After the stage of sampling areas, schools are selected. Within each sampled school, students from the appropriate grades are selected directly, so that they are spread over different classes in the school. NAEP measures different subjects over the years. Further, the design assigns different assessments to different students in each selected school. The content of the tests vary randomly over students, while measuring the same subject.

Each assessment is built around a well-specified and evaluated conceptual framework. These frameworks define the set of knowledge that underlies various levels of sophistication of the student's understanding. A complicated and lengthy consensus process involving teachers, curriculum experts, parents, school administrators, and the general public produces the framework. Then experts in the field write and pretest individual questions representing the components of the framework. In contrast to the surveys reviewed above, there are multiple questions measuring the different components within the framework. For example, in the mathematics framework, there would be different types of mathematical knowledge within the framework (e.g., for 4th grade, addition, subtraction, multiplication, and division).

The NAEP is used as an important indicator of the performance of the nation's schools. Funding advocates for educational initiatives use the information to justify levels and targets of funds. NAEP provides results regarding subject-matter achievement, instructional experiences, and school environment for populations of students (e.g., fourth graders) and subgroups of those populations (e.g., female students, Hispanic students). The results are highly watched statistics within the political and policy domains. Figure 1.4 presents the average scores (scaled so that 500 is a perfect score) on the mathematics assessment for high school seniors by type of school. This shows a very common finding: that students in private schools tend to attain higher scores, that Catholic schools are in the middle, and that public schools have the lowest average scores. Much of this is related to what types of students attend these schools. There in no uniformly increasing trend in the scores of students; for example, the year 2000 scores appear to decline except among the Catholic school students. However, all

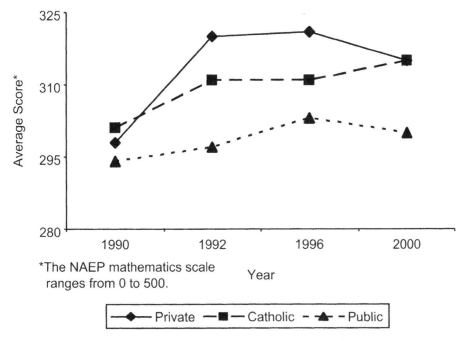

Figure 1.4 Average scale scores on grade 12 mathematics assessment, by year by type of school. [Source: U.S. Department of Education, Institute of Education Sciences, National Center for Education Statistics, National Assessment of Education Progress (NAEP), 1990, 1992, 1996, and 2000 Mathematics Assessments.]

were higher than their corresponding scores in 1990. When declines in the scores occur, there is usually some public discussion and policy debate about the alternative causes of the decline and the desirability of change in educational policies. For Figure 1.4, the participation rate of 12th graders in private schools was below the 70% minimum rate after 2000, so it was not reported.

A CNN.com story of June 20, 2003 led with "Are Little Kids Smarter than High Schoolers?" The story was:

> Fourth graders are showing they are better readers, while the skills of 12th graders are declining, the government said in a report Thursday.... Overall, less than one-third of fourth graders and eighth graders showed they could understand and analyze challenging material. That skill level, defined as proficient, is the focal point of the test. Among high school seniors, 36 percent hit the mark. Four years ago, 29 percent of the fourth grader were proficient. That increased to 31 percent in 2002. It is those younger students who are at the center of a national push to improve basic education. Among the seniors, the percentage of those who reached the highest skill level dropped from 40 percent.... "There are no scientific answers as to why our high school seniors have performed so poorly on this reading assessment, but we're still searching for solutions to these

daunting challenges," said Education Secretary Rod Paige. "At the same time, we know what works to teach youngsters to read, and we know that all children can learn."

These are complex results and require more analysis to disentangle whether there are group differences between the 12th graders of 2002 and the 12th graders of 1998, based on experiences they had in grammar school, or whether there are contemporaneous education policies producing the results. The absence of firm data on the causes produces many alternative speculations from policy makers about the alternative solutions. Despite the complexity of disentangling causes of the assessment levels, understanding these phenomena requires the kind of uniform measurement that NAEP offers.

1.3.5 The Behavioral Risk Factor Surveillance System

How many people exercise, smoke cigarettes, or wear seatbelts? Do the states within the United States vary on these phenomena? How do such health-related behaviors vary across the states? Do people exhibit more or fewer healthy behaviors as they get older? Is there change over time in health behaviors that are associated with public health education programs in the states?

Since 1965, the National Center for Health Statistics had been providing annual U.S. survey estimates on a wide variety of health behaviors and conditions (e.g., self-reported health conditions, frequency of visits to doctors, self-reported exercise, and other risk-related behaviors). The survey estimates, coupled with biomedical research, clearly indicated that individual personal behaviors affected premature morbidity and mortality. Much public health policy and oversight rests at a state level, however, and there were no comparable set of statistics at the state level. State health agencies have a lead role in determining resource allocation to reduce health risks linked to behaviors.

Since the early 1980s, the Behavioral Risk Factor Surveillance System (BRFSS) coordinated by the U.S. Centers for Disease Control and Prevention has provided state-level survey estimates of key health factors. In contrast to all the surveys described above, BRFSS involves a partnership of individual states with assistance from the Federal Centers for Disease Control. The states determine the questions and conduct the survey. The Federal Centers for Disease Control and Prevention coordinate the definition of a core set of questions, develop standards for the survey data collection, perform the postdata collection processing of the data, and distribute a national combined dataset. The core questions ask about current health-related perceptions, conditions, and behaviors (e.g., health status, health insurance, diabetes, tobacco use, selected cancer screening procedures, and HIV/AIDS risks) and questions on demographic characteristics.

Like the SOC, BRFSS uses random-digit-dialed samples of the telephone household population. Unlike SOC, each state draws an independent sample, and arranges for the data collection via computer-assisted telephone interviewing. Most states uses commercial or academic survey organizations under contract. Some collect the data themselves. The sample sizes vary across the states. In each sample household, an adult (18 years old or older) is selected to be interviewed.

The survey publishes annual estimates of rates of smoking and other risky health behaviors for each of the states. The estimates are used as social indicators

Table 1.5. Example Survey: Behavioral Risk Factor Surveillance System (BRFSS)

Sponsor	U.S. Centers for Disease Control and Prevention
Collector	Varies by state; for the 2007 BRFSS, 12 state or jurisdictional health departments collected their own data, whereas 42 used outside contractors
Purpose	The main objectives of the BRFSS are to: 1) Collect uniform, state-specific data on preventive health practices and risk behaviors that are linked to chronic diseases, injuries, and preventable infectious diseases in the adult population 2) Enable comparisons between states and derive national-level conclusions 3) Identify trends over time 4) Allow states to address questions of local interest 5) Permit states to readily address urgent and emerging health issues through addition of topical question modules
Year Started	1984
Target Population	U.S. adult household population
Sampling Frame	U.S. telephone households, through lists of working area codes and exchanges, then lists of household members
Sample Design	For the 2007 BRFSS all states use probability designs. All states use a disproportionate stratified random sample (DSS), but Guam, Puerto Rico, and U.S. Virgin Islands use a simple random sample
Sample Size	Average state sample size of 8,309 (2007 BRFSS)
Use of Interviewer	Interviewer-administered
Mode of Administration	Telephone interview
Computer Assistance	Computer-assisted telephone interview (CATI) in 54 areas; two areas use paper-and-pencil interviews (PAPI)
Reporting Unit	Randomly selected adult
Time Dimension	Repeated cross-sectional survey
Frequency	Conducted annually
Interviews per Round of Survey	One
Levels of Observation	Household person
Web Link	http://www.cdc.gov/brfss/

Figure 1.5a Percentage of state adults who are obese (body mass index ≥ 30) by state, 1994, BRFSS. (Source: BRFSS, CDC.)

Figure 1.5b Percentage of state adults who are obese (body mass index ≥ 30) by state, 2001, BRFSS. (Source: BRFSS, CDC.)

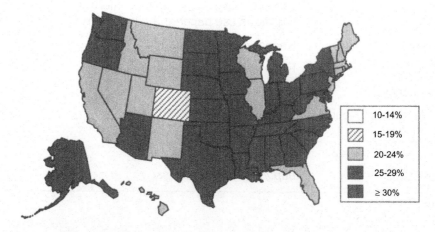

Figure 1.5c Percentage of state adults who are obese (body mass index ≥ 30) by state, 2007, BRFSS. (Source: BRFSS, CDC.)

SOME EXAMPLES OF ONGOING SURVEYS

of the health status of the population and guide government health policies that might affect health-related behavior. For example, the three Figure 1.5 maps show the dramatic increase in obesity in the United States in the 1990s. The statistic is presented separately for each state, using a different shading to denote the outcome. In 1994, there were no states with more than 20% of the adult population who were obese; that is, with a body mass index equal to 30 or more. By 2001, over half of the states had that prevalence of obesity. By tracking trends like this at a state level, state public health officials can both compare their state to others and compare trends over time in their state. States use BRFSS data to establish and track state health objectives, plan health programs, or implement a broad array of disease prevention activities.

The BRFSS data quickly enter the national debate. Although there were other indicators of obesity problems in the U.S. population, the BRFSS helped present a local perspective to the problem. The March 6, 2003 the *Washington Post* ran the story, "When Spotlight Shines, San Antonio Citizens Seem to Swell to Fill It—City in Texas Called Obesity Capital of US" detailing the data. Policy officials acted. On May 14, 2003, the same paper ran a story entitled, "Health Costs of Obesity Near Those of Smoking. HHS Secretary Presses Fast-Food Industry." Then on July 2, 2003, the *Washington Post* ran an article titled, "Slimming Down Oreos: Kraft Plans to Make Its Food Products Healthier," describing plans to reduce portion size, alter content, and change marketing strategies within U.S. school cafeterias.

Survey data like those of BRFSS can make a difference when they measure important issues and have the credibility to gain the attention of decision makers.

1.3.6 The Current Employment Statistics Program

How many jobs were created in the U.S. economy in the last month? Are some industries changing their employment counts more rapidly than others? Which industries are growing; which are declining? Are large employers producing most of the employment change or does the dynamic nature of the job market lie among new, small firms? Are there regional differences in growth or decline in jobs?

The Current Employment Statistics (CES) program of the U.S. Bureau of Labor Statistics is a focus of the story that opens this chapter. CES is one of two parallel surveys that attempt to provide monthly estimates of the employment situation. The CES is a survey of employers that asks for six different pieces of information: the total number of employees on the payroll, number of women employees, number of production workers, the total payroll of the production workers, the total hours worked by production workers, and the total number of overtime hours by production workers. The Current Population Survey (CPS) is a household survey that asks residents whether they are employed or looking for work. That survey produces a monthly rate statistic, the unemployment rate.

The sample of the CES is large—over 150,000 employers per month. The sample is drawn from a list of active employers who have registered their company with the state unemployment insurance agency. The sample design gives large employers very high probabilities of selection (some are permanently in the sample) and small employers small probabilities. Once an employer falls in the sample, it is repeatedly measured over months and years. Smaller employers rotate out of the sample and are "replaced" by other smaller employers; larger employers tend to stay in the sample for a longer time. CES is conducted as a part-

Table 1.6. Example Survey: Current Employment Statistics (CES)

Sponsor	U.S. Bureau of Labor Statistics, US Department of Labor
Collector	U.S. Bureau of Labor Statistics, State Employment Security Agencies
Purpose	The main objective of the CES is to produce monthly estimates of employment, hours, and earnings for the nation, states, and major metropolitan areas.
Year Started	1939
Target Population	U.S. employers
Sampling Frame	U.S. employers filing unemployment insurance tax records with State Employment Security Agencies
Sample Design	Originally quota sample; then a probability sample was fully implemented in 2002
Sample Size	About 150,000 business establishments
Use of Interviewer	Most self-administered; about 25% of respondents interviewed by telephone
Mode of Administration	Many sample units (about 10%) use touchtone data entry (TDE) to complete the interview by phone keypad, with prerecorded questions read to the respondent and answers entered by pressing touchtone phone buttons. Other modes used include mail, fax, Web entry, electronic data interchange, and computer-assisted telephone interviewing (CATI).
Computer Assistance	Touchtone data entry (TDE), electronic data interchange (EDI), Web entry, computer-assisted telephone interviewing (CATI)
Reporting Unit	Contact person at establishment
Time Dimension	Longitudinal panel survey of employers
Frequency	Conducted monthly
Interviews per Round of Survey	One
Levels of Observation	Establishment, employer
Web Link	http://www.bls.gov/ceshome.htm

nership between the federal Bureau of Labor Statistics and the state employment security agencies.

The CES differs from the other surveys described above in that it uses many different methods of collecting the data simultaneously. Indeed, an employer is given great freedom to choose the method of delivering the answers to the six questions each month. Paper forms can be completed and mailed or faxed back to the Bureau of Labor Statistics. The respondent can use touchtone data entry, whereby the employer uses a telephone keypad to enter the digits of the answers in response to a recorded voice. The respondent can enter the data on a special secure Web page. The respondent can send electronic records directly to BLS for processing. The respondent can orally deliver the data in response to a telephone interviewer. These different methods, each requiring some cost to develop, appear to fit different employers; the overall cooperation rate with the survey can be increased by allowing the sample employers to use a method easiest for them.

To illustrate how CES tracks the U.S. economy, Figure 1.6 shows the total number of jobs in nonfarm employers between 1941 and 2001. Looking at the trend over so many decades, it is easy to see periods of no or slow growth in jobs and periods of rapid growth. For example, the recessions of the early 1980 and early 1990s are followed by periods of high growth in number of jobs. The leveling off of the growth during the early 2000 recession is also obvious.

Statistics like these are important because they are used by the Federal Reserve, the White House, and by Congress to consider changes in economic policy. It is common that periods of recession contain many statements by politicians that the economy must be stimulated in order to create jobs for the country. Fairly or unfairly, political regimes rise and fall because of the values of such statistics.

1.3.7 What Can We Learn From the Six Example Surveys?

All of these are large-scale, national, repeated surveys. Five have lengthy histories—the Current Employment Statistics Program dates back to 1939, the Surveys of Consumers originated in 1946, NAEP began in 1969, and the NCVS and NSDUH were started in the early 1970s. All of these are used by government agencies and social scientists as indicators of the status of society or the economy.

The design features of the surveys are tailored to fit their purposes. The target populations for these surveys are varied—for example, the NCVS aims to describe adults and children age 12 or older, whereas the CES survey has a target population of employers. The NCVS, NSDUH, and NAEP surveys all draw samples of households in two steps, first sampling geographic areas, then sampling from created lists of households within those areas. The Surveys of Consumers

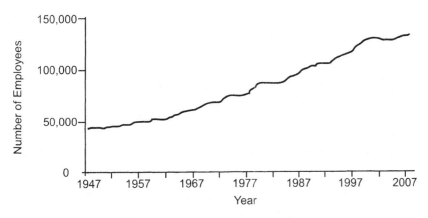

Figure 1.6 Number of employees of all nonfarm employers in thousands, annual estimates 1947–2007, Current Employment Statistics. (Source: www.data.bls.gov.)

and BRFSS draw samples of telephone numbers from all possible numbers within working area codes and exchanges. Then the numbers are screened to find those associated with households.

The example surveys use varying data collection methods. The NAEP survey does not use interviewers; respondents complete self-administered forms and assessments. The NCVS, BRFSS, and SOC all rely on interviewers to collect answers. The NSDUH uses both techniques, with some questions self-administered and others asked by interviewers. Another way in which the surveys vary is the use of computer assistance. The NAEP uses paper forms, followed by electronic processing of the forms. There is a pilot study in 2009 to do interactive computer tasks for a subsample of examinees. For 2011, NAEP is planning to administer writing online to 8th and 12th graders. The CES uses many different types of computer assistance, including touchtone data entry, electronic data interchange, and computer-assisted telephone interviewing.

All of the surveys are ongoing in nature. This is because they all are designed to measure change in the phenomena they study. All of the designs supply estimates of change at the national level, measuring how population means and totals have changed over time. Those that collect data from the same people more than once, such as the CES, SOC, and the NCVS, can measure changes in individuals or employers, as well as overall changes in the population.

Many of the qualities and costs of a survey are determined by these kinds of design features. The first questions to ask about a new survey are:

1) What is the target population (whom is it studying)?
2) What is the sampling frame (how do they identify the people who have a chance to be included in the survey)?
3) What is the sample design (how do they select the respondents)?
4) What is the mode of data collection (how do they collect data)?
5) Is it an ongoing survey or a one-time survey?

Reread the tables that accompany Sections 1.3.1–1.3.6 to get an overview of the purpose and design of each example of survey. Throughout this book, we will refer to these surveys to illustrate specific methodological issues.

1.4 What is Survey Methodology?

Survey methodology seeks to identify principles about the design, collection, processing, and analysis of surveys that are linked to the cost and quality of survey estimates. This means that the field focuses on improving quality within cost constraints, or, alternatively, reducing costs for some fixed level of quality. "Quality" is defined within a framework labeled the total survey error paradigm (which will be discussed further in Chapter 2). Survey methodology is both a scientific field and a profession.

Within the scientific side of surveys, the achievement of high-quality survey results requires applying principles from several traditional academic disciplines. Mathematics, especially the principles of probabilities or chance events, is key to knowing the relative frequency of various outcomes. A whole subfield of statistics is devoted to the principles of sampling and inference from sample results to population results. Thus, the elements of sampling and analysis are historically grounded in the mathematical sciences.

However, because human beings often are involved either as interviewers or respondents, a variety of principles from the social science disciplines also apply to surveys. When we turn to the study of how data collection protocols affect survey estimates, the psychologists have been the leading sources of knowledge. Furthermore, social psychology provides the framework for understanding how interviewer behaviors may influence the activities of respondents, both when they are recruited as respondents and during the survey interview. When it comes to understanding question design, cognitive psychology offers principles regarding how memories are formed, how they are structured, and what devices are helpful to the recall of memories relevant to the answers to survey questions. Sociology and anthropology offer principles of social stratification and cultural diversity, which inform the nature of reactions of subpopulations to requests for survey-based measurement or to particular questions. Computer science provides principles of database design, file processing, data security, and human–computer interaction.

Because survey methodology has this inherently multidisciplinary nature, it has only recently developed as a unified field. The scientific side of survey methodology was practiced largely outside traditional academic disciplines. For example, the important developments in probability sampling theory applied to surveys largely took place in large government survey organizations in the 1930s and 1940s. The major early texts in sampling were written by scientists with feet firmly planted in applied survey environments such as the U.S. Bureau of the Census (e.g., Hansen, Hurwitz, and Madow, *Sample Survey Methods and Theory*, 1953; Deming, *Some Theory of Sampling*, 1950). Many of the early scientific contributions regarding survey data collection came from staff in government agencies conducting surveys of the military and citizenry during World War II. Major early texts in interviewing similarly were written by scientists in academic survey organizations, grappling with practical survey problems (Kahn and Cannell, *The Dynamics of Interviewing*, 1958; Hyman, *Interviewing in Social Research*, 1954).

It is consistent that the research literature or survey methodology is widely scattered. There is a concentration of survey methods in statistical journals: *Journal of the American Statistical Association* (Applications section), *Journal of Official Statistics*, and *Survey Methodology*. An important outlet for reports of survey methods studies is the *Proceedings of the Survey Research Methods Section of the American Statistical Association*. However, an interdisciplinary journal, *Public Opinion Quarterly*, is another major outlet for survey methods papers, and there are many papers on survey methods in journals of the various academic disciplines and applied fields such as health, criminology, education, and market research.

The field also has active professional associations, which act as gathering places for both scientists and professionals. There are four professional associations for which survey methodology is a central focus. The American Statistical Association has a large membership in the Survey Research Methods Section. The American Association for Public Opinion Research (and its international counterpart, the World Association for Public Opinion Research) is an association of commercial, academic, and government survey researchers. The International Association for Survey Statisticians is a component of the International Statistical

Institute. The commercial survey organizations have created a trade association, the Council of American Survey Research Organizations, which promotes activities of benefit to the members. All of these organizations, including the Survey Research Methods Section of ASA, are interdisciplinary, including members with a wide array of formal training backgrounds in mathematical, social, or various applied sciences. Recently, a new organization, the European Survey Research Association, was founded.

Survey methodology is also a profession, a set of occupational classes devoted to the design, collection, processing, and analysis of survey data. Throughout the world, there are specialists in survey methods found in academia, governments, and commerce. In the United States, the academic sector contains survey researchers who use the method to investigate discipline-based questions in sociology, political science, public health, communications studies, psychology, criminology, economics, transportation studies, gerontology, and many other fields. On campuses in the United States, there are over 100 survey research centers that have permanent staffs, providing survey capabilities for faculty and staff on their campus. The federal government plays an even larger role than academia in collecting and commissioning surveys. The Bureau of the Census, the Department of Agriculture, and the Bureau of Labor Statistics all collect survey data themselves, and over 60 other agencies commission numerous surveys. Moreover, in the United States, the survey efforts of the private commercial sector, including polls, political surveys, and market research, are several times larger than the combined effort of the agencies of the federal government.

Because the field of survey methodology did not develop within a single academic discipline, education in the field historically was rather haphazard. Professionals in the field are not certified, although from time to time discussions regarding the desirability of certification are held. Formal training in survey and marketing research techniques has existed for some years. Training in survey methods can be found in undergraduate and graduate departments in all the social sciences and in professional schools of all kinds. For those who specialize in survey methodology, almost invariably an apprenticeship in an organization that conducts surveys is part of the training process. Practical experience in solving the real challenges that designing and executing surveys entail is a complement to formal classroom training. One way to think about the purpose of this book is that it is designed to provide the intellectual foundation needed to understand and address the problems that survey research projects pose.

1.5 THE CHALLENGE OF SURVEY METHODOLOGY

Surveys are not the only way to collect information about large populations. There is no "right" way. Administrative record systems of governments and businesses sometimes offer all the necessary data for good decision making for those covered by the systems. Qualitative investigations, involving trained ethnographers or sociologists, offer the possibility of rich encounters with deep understanding of the perspectives of the subjects. Observations of behaviors of persons can yield quantitative information about the frequency of events in public spaces. Randomized experiments in controlled settings can answer important questions about whether various stimuli cause behaviors.

However, in administrative record systems, the researchers have little control over the measurement, and their results may be tainted by data of poor quality. Ethnographic investigations often use small groups of informants, limiting the studies' ability to describe large populations. Observations of persons are limited to a tiny fraction of all human behaviors. Randomized experiments face challenges of applicability in the real world.

Conversely, surveys often limit their measures to those that can be standardized and repeated over large numbers of persons. Surveys are conducted in the uncontrolled settings of the real world and can be affected by those settings. Surveys gain their inferential power from the ability to measure groups of persons that form a microcosm of large populations, but rarely achieve perfection on this dimension. Part of the task of a survey methodologist is making a large set of decisions about thousands of individual features of a survey in order to improve it. Some of the most important decisions include:

1) How will the potential sample members be identified and selected?
2) What approach will be taken to contact those sampled, and how much effort will be devoted to trying to collect data from those who are hard to reach or reluctant to respond?
3) How much effort will be devoted to evaluating and testing questions that are asked?
4) What mode will be used to pose questions and collect answers from respondents?
5) If interviewers are involved, how much effort will be devoted to training and supervising interviewers?
6) How much effort will be devoted to checking the data files for accuracy and internal consistency?
7) What approaches will be used to adjust the survey estimates to correct for errors that can be identified?

Each of these decisions has the potential to affect the quality of estimates that emerge from a survey. Often, though not always, there are cost implications of how these decisions are made; making the decision that involves more effort or that has a better chance of minimizing error in the survey often costs more money.

There is a methodological research literature that provides the intellectual basis for understanding the implications of these various decisions on data quality. One of the most important goals of this book is to document what currently is known, and in some cases not known, about how these decisions affect data quality and the credibility of the data.

The second goal of this book is to convey an understanding of the concept of total survey error and how to use it. Essentially, all surveys involve some kinds of compromises with the ideal protocol. Some of those compromises are based on cost. Researchers have to decide how much to invest in each of the components of a survey listed above. When there is a finite budget, researchers often have to weigh the merits of spending more on one aspect of the survey, while reducing costs on another. For example, a researcher might decide to increase the size of the survey sample and offset those added costs by spending less on efforts to maximize the rate of response.

One challenge for the survey methodologist is to figure out how best to use the available resources—how to balance the investments in each of the components of a survey to maximize the value of the data that will result. Recognizing that each aspect of a survey has the potential to affect the results, the survey methodologist takes a total survey error approach. Rather than focusing on just one or a few of the elements of a survey, all the elements are considered as a whole. A survey is no better than the worst aspect of its design and execution. The total survey error approach means taking that broad perspective and ensuring that no feature of the survey is so poorly designed and executed that it undermines the ability of the survey to accomplish its goals.

In a similar way, in some cases there are only imperfect solutions to survey design problems. All the approaches available to solve a particular problem may have pluses and minuses. The survey methodologist has to decide which of a set of imperfect options is best. Again, the total survey error approach is to consider the various ways that the options will affect the quality of the resulting data and choose the one that, on balance, will produce the most valuable data.

Survey methodology is about having the knowledge to make these trade-off decisions appropriately, with as much understanding as possible of the implications of the decisions. When a well-trained methodologist makes these decisions, it is with a total survey error perspective, considering all the implications of the decisions that are at stake and how they will affect the final results.

1.6 ABOUT THIS BOOK

This book describes how the details of surveys affect the quality of their results. The second chapter is a discussion of what we mean by "error" in surveys. It turns out that this is a potentially confusing topic, because there is more than one kind of error that affects survey estimates. Different disciplines also use different words to describe the same or similar concepts. The focus of all the subsequent chapters is on how to minimize error in surveys. Understanding what is meant by error is an essential first step to understanding the chapters that follow.

The subsequent chapters take on different aspects of the design and execution of surveys. They lay out the decisions that face researchers, why they are potentially important, and what our science tells us about the implications of the available choices. Which choice is best often depends on the survey's goals and other features of the design. In a few cases, there is a clear best practice, an approach to carrying out surveys that consistently has been found to be critical to creating credible data. The chapters attempt to give readers an understanding of how these choices should be made. Throughout the book, we mix practical concrete examples (often drawn from the six example surveys described previously) with the general discussion of knowledge and principles, so readers can understand how the generalizations apply to real decisions.

The last two chapters cut across the others. In Chapter 11, we discuss issues related to the ethical conduct of survey research. In the final chapter, we answer a set of common questions that survey methodologists are asked. When readers have completed the book, we expect that they will have a solid foundation on which to build through reading, additional course work, and practical experience in order to become survey methodologists.

KEYWORDS

ACASI
analytic statistic
area probability sample
CAPI
census
descriptive statistic
error
probability sample

random-digit dialing
repeated cross-section design
rotating panel design
sampling error
statistic
statistical error
survey
survey methodology

FOR MORE IN-DEPTH READING

National Crime Victimization Survey

Pastore, Ann L. and Maguire, Kathleen (eds.) (2008), *Sourcebook of Criminal Justice Statistics* [Online]. Available: http://www.albany.edu/sourcebook/.

Taylor, Bruce M. and Rand, Michael R., "The National Crime Victimization Survey Redesign: New Understandings of Victimization Dynamics and Measurement," Paper prepared for presentation at the 1995 American Statistical Association Annual Meeting, August 13–17, 1995 in Orlando, Florida (http://www.ojp.usdoj.gov/bjs/ncvsrd96.txt).

National Survey of Drug Use and Health

Gfroerer, J., Eyerman, J., and Chromy, J. (eds.) (2002), *Redesigning an Ongoing National Household Survey: Methodological Issues*, DHHS Publication No. SMA 03-3768. Rockville, MD: Substance Abuse and Mental Health Services Administration, Office of Applied Statistics.

Turner, C. F., Lessler, J. T., and Gfroerer, J. C. (1992), *Survey Measurement of Drug Use*, Washington, DC: National Institute on Drug Abuse.

Surveys of Consumers

Curtin, Richard T. (2003), *Surveys of Consumers: Sample Design*, http://www.sca.isr.umich.edu/.

Curtin, Richard T. (2003), *Surveys of Consumers: Survey Description*, http://www.sca.isr.umich.edu/.

National Assessment of Educational Progress

U.S. Department of Education. National Center for Education Statistics. *NCES Handbook of Survey Methods,* NCES 2003-603, by Lori Thurgood, Elizabeth Walter, George Carter, Susan Henn, Gary Huang, Daniel Nooter, Wray Smith, R. William Cash, and Sameena Salvucci. Project Officers, Marilyn Seastrom, Tai Phan, and Michael Cohen. Washington, DC: 2003.

Vinovskis, Maris A. (1998), *Overseeing the Nation's Report Card: The Creation and Evolution of National Assessment Governing Board*, Washington, DC: U.S. Department of Education.

Behavioral Risk Factor Surveillance System

Centers for Disease Control (2005), *BRFSS User's Guide*, http://www.cdc.gov/brfss/pdf/userguide.pdf.

Centers for Disease Control (2008), *BRFSS Questionnaires*, http://www.cdc.gov/brfss/questionnaires/pdf-ques/2008brfss.pdf.

Current Employment Statistics

U.S. Bureau of Labor Statistics (2003), *BLS Handbook of Methods*, http://www.bls.gov/opub/hom/home.htm.

U.S. Bureau of Labor Statistics, *Monthly Labor Review*, http://www.bls.gov/opub/mlr/mlrhome.htm. (Website contains frequent articles of relevance to CES.)

EXERCISES

1) Find the Web page for the National Crime Victimization Survey (first go to http://www.ojp.usdoj.gov/ and look for the Bureau of Justice Statistics). Find the questionnaire of the NCVS and identify what types of victimizations are collected from the household respondent and what types of crimes are collected on each self-respondent.

2) Go to the BRFSS website (http://www.cdc.gov/brfss/) and complete the training regimen that interviewers are offered on the website.

3. Read the most recent annual report from the NSDUH (http://www.oas.samsha.gov/nhsda.htm). For what drugs did the prevalence of use increase over the recent years?

4) Go to the technical notes for the Current Employment Statistics program (http://www.bls.gov/web/cestn1.htm). What is the current distribution of respondents by the various methods of returning the data to BLS?

5) Read the latest press release on the consumer confidence index from the Survey of Consumers website (http://www.sca.isr.umich.edu/), then go to a news service website to search for "consumer confidence." Compare the news treatment of the statistics to the report from the survey. What is the same; what is different?

6) Go the website of the National Assessment of Educational Progress (http://nces.ed.gov/nationsreportcard/). Find the latest assessment report;

EXERCISES

read it to find comparisons with the past report. Try to find explanations for the changes that were observed. Do the authors cite reasons from the survey data themselves or do they cite possible external causes, not measured in the survey?

7) Consider a survey that you knew prior to reading this chapter. Examine the questions on page 33 and try to answer each of these questions for that survey. If you cannot find the answer to a specific question, write down where you searched for the answer.

8) Both repeated cross-section designs and rotating panel designs involve measurements at different time periods. What is the difference between the two designs? What types of estimates of change over time are permitted with each design?

9) How do the target populations and the sampling frames differ between the NCVS and the SOC surveys? For each survey, what speculations would you have about the gaps in the sampling frames relative to the target populations? How might those gaps affect the key statistics of the surveys?

CHAPTER TWO

INFERENCE AND ERROR IN SURVEYS

2.1 INTRODUCTION

Survey methodology seeks to understand why error arises in survey statistics. Chapters 3 through 11 describe in detail strategies for measuring and minimizing error. In order to appreciate those chapters, and to understand survey methodology, it is first necessary to understand thoroughly what we mean by "error."

As the starting point, let us think about how surveys work to produce statistical descriptions of populations. Figure 2.1 provides the simplest diagram of how they work. At the bottom left is the raw material of surveys—answers to questions by an individual. These have value to the extent they are good descrip-

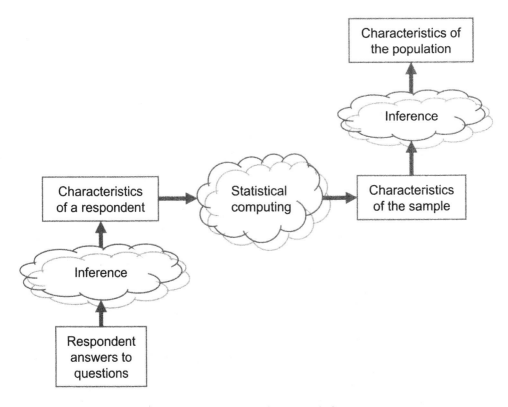

Figure 2.1 Two types of survey inference.

tors of the characteristics of interest (the next-higher box on the left). Surveys, however, are never interested in the characteristics of individual respondents per se. They are interested in statistics that combine those answers to summarize the characteristics of groups of persons. *Sample* surveys combine the answers of individual respondents in statistical computing steps (the middle cloud in Figure 2.1) to construct statistics describing all persons in the sample. At this point, a survey is one step away from its goal—the description of characteristics of a larger population from which the sample was drawn.

statistic

The vertical arrows in Figure 2.1 are "inferential steps." That is, they use information obtained imperfectly to describe a more abstract, larger entity. "Inference" in surveys is the formal logic that permits description of unobserved phenomena based on observed phenomena. For example, inference about unobserved mental states, like opinions, is made based on answers to specific questions related to those opinions. Inference about population elements not measured is made based on observations of a sample of others from the same population. In the jargon of survey methodology, we use an answer to a question from an individual respondent to draw inferences about the characteristic of interest to the survey for that person. We use statistics computed on the respondents to draw inferences about the characteristics of the larger population.

inference

These two inferential steps are central to the two needed characteristics of a survey:

1) Answers people give must accurately describe characteristics of the respondents.
2) The subset of persons participating in the survey must have characteristics similar to those of a larger population.

error

measurement error

error of observation

error of nonobservation

When either of these two conditions is not met, the survey statistics are subject to "error." The use of the term "error" does not imply mistakes in the colloquial sense. Instead, it refers to deviations of what is desired in the survey process from what is attained. "Measurement errors" or "errors of observation" will pertain to deviations from answers given to a survey question and the underlying attribute being measured. "Errors of nonobservation" will pertain to the deviations of a statistic estimated on a sample from that on the full population.

Let us give an example to make this real. The Current Employment Statistics (CES) program is interested in measuring the total number of jobs in existence in the United States during a specific month. It asks individual sample employers to report how many persons were on their payroll in the week of the 12th of that month. (An error can arise because the survey does not attempt to measure job counts in other weeks of the month.) Some employer's records are incomplete or out of date. (An error can arise from poor records used to respond.) These are problems of inference from the answers obtained to the desired characteristic to be measured (the leftmost vertical arrows in Figure 2.1).

The sample of the employers chosen is based on lists of units of state unemployment compensation rolls months before the month in question. Newly created employers are omitted. (An error can arise from using out-of-date lists of employers.) The specific set of employers chosen for the sample might not be a good reflection of the characteristics of the total population of employers. (An error can arise from sampling only a subset of employers into the survey.) Further, not all selected employers respond. (An error can arise from the absence of answers from some of the selected employers.) These are problems of inference from statistics on the respondents to statistics on the full population.

One's first reaction to this litany of errors may be that it seems impossible for surveys ever to be useful tools to describe large populations. Do not despair! Despite all these potential sources of error, carefully designed, conducted, and analyzed surveys have been found to be uniquely informative tools to describe the world. Survey methodology is the study of what makes survey statistics more or less informative.

Survey methodology has classified these various errors illustrated with the CES example above into separate categories. There are separate research literatures for each error because each seems to be subject to different influences and have different kinds of effects on survey statistics.

One way of learning about surveys is to examine each type of error in turn, or studying surveys from a "quality" perspective. This is a perspective peculiar to survey methodology. Another way of learning about surveys is to study all the survey design decisions that are required to construct a survey; identification of the appropriate population to study, choosing a way of listing the population, selecting a sampling scheme, choosing modes of data collection, and so on. This is an approach common to texts on survey research (e.g., Babbie, 1990; Fowler, 2001).

2.2 THE LIFE CYCLE OF A SURVEY FROM A DESIGN PERSPECTIVE

In this and the next section, we will describe the two dominant perspectives about surveys: the design perspective and the quality perspective. From the design perspective, discussed in this section, survey designs move from abstract ideas to concrete actions. From the quality perspective, survey designs are distinguished by the major sources of error that affect survey statistics. First, we tackle the design perspective.

A survey moves from design to execution. Without a good design, good survey statistics rarely result. As the focus moves from design to execution, the nature of work moves from the abstract to the concrete. Survey results, therefore, depend on inference back to the abstract from the concrete. Figure 2.2 shows that there are two parallel aspects of surveys: the measurement of constructs and descriptions of population attributes. This figure elaborates the two dimensions of inference shown in Figure 2.1. The measurement dimension describes what data are to be collected about the observational units in the sample: what is the survey about? The representational dimension concerns what populations are described by the survey: who is the survey about? Both dimensions require forethought, planning, and careful execution. observational unit

Because Figure 2.2 contains important components of survey methods, we will spend some time discussing it. We will do so by defining and giving examples of each box in the figure.

2.2.1 Constructs

"Constructs" are the elements of information that are sought by the researcher. The Current Employment Statistics survey attempts to measure how many new construct

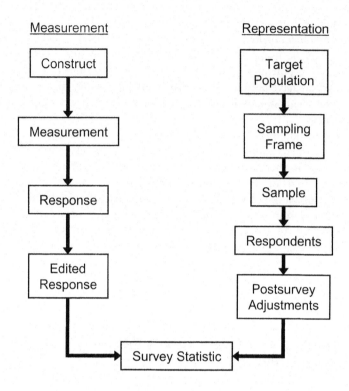

Figure 2.2 Survey lifecycle from a design perspective.

jobs were created in the past month in the United States, the National Assessment of Education Progress measures knowledge in mathematics of school children, and the National Crime Victimization Survey (NCVS) measures how many incidents of crimes with victims there were in the last year. The last sentence can be understood by many; the words are simple. However, the wording is not precise; it is relatively abstract. The words do not describe exactly what is meant, nor exactly what is done to measure the constructs. In some sense, constructs are ideas. They are most often verbally presented.

For example, one ambiguity is the identity of the victim of the crime. When acts of vandalism occur for a household (say, a mailbox being knocked down), who is the victim? (In these cases, NCVS distinguishes crimes against a household from crimes against a person.) When graffiti is spray painted over a public space, who is the victim? Should "victimization" include only those crimes viewed as eligible for prosecution? When does an unpleasant event rise to the level of a crime? All of these are questions that arise when one begins to move from a short verbal description to a measurement operation. Some constructs more easily lend themselves to measurements than others.

Some constructs are more abstract than others. The Survey of Consumers (SOC) measures short-term optimism about one's financial status. This is an attitudinal state of the person, which cannot be directly observed by another person. It is internal to the person, perhaps having aspects that are highly variable within and across persons (e.g., those who carefully track their current financial status may have well-developed answers; those who have never thought about it may have to construct an answer de novo). In contrast, the National Survey on Drug Use and Health (NSDUH) measures consumption of beer in the last month. This is a construct much closer to observable behaviors. There are a limited number of ways this could be measured. The main issues are simply to decide what kinds of drinks count as beer (e.g., does nonalcoholic beer count?) and what units to count (12-ounce cans or bottles is an obvious choice). Thus, the consumer optimism construct is more abstract than the construct concerning beer consumption.

2.2.2 Measurement

Measurements are more concrete than constructs. "Measurements" in surveys are ways to gather information about constructs. Survey measurements are quite diverse: soil samples from the yards of sample households in surveys about toxic contamination, blood pressure measurements in health surveys, interviewer observations about housing structure conditions, electronic measurements of traffic flow in traffic surveys. However, survey measurements are often questions posed to a respondent, using words (e.g., "During the last 6 months, did you call the police to report something that happened to YOU that you thought was a crime?"). The critical task for measurement is to design questions that produce answers reflecting perfectly the constructs we are trying to measure. These questions can be communicated orally (in telephone or face-to-face modes) or visually (in paper and computer-assisted self-administered surveys). Sometimes, however, they are observations made by the interviewer (e.g., asking the interviewer to observe the type of structure of the sample housing unit or to observe certain attributes of the neighborhood). Sometimes they are electronic or physical measurements (e.g., electronic recording of prices of goods in a sample retail store, taking a blood or hair sample in a health-related survey, taking a sample of earth in a survey of toxic waste, taking paint samples). Sometimes questions posed to respondents follow their observation of visual material (e.g., streaming video presentation of commercials on a laptop, presentation of magazine covers).

measurement

2.2.3 Response

The data produced in surveys come from information provided through the survey measurements. The nature of the responses is determined often by the nature of the measurements. When questions are used as the measurement device, respondents can use a variety of means to produce a response. They can search their memories and use their judgment to produce an answer [e.g., answering the question, "Now looking ahead, do you think that a year from now you (and your family living there) will be better off financially, or worse off, or just about the same as now?" from the SOC]. They can access records to provide an answer (e.g., looking at the employer's personnel records to report how many nonsupervisory employees were on staff on the week of the 12th, as in the CES). They can seek another person to help answer the question (e.g., asking a spouse to recall when the respondent last visited the doctor).

Sometimes, the responses are provided as part of the question, and the task of the respondent is to choose from the proffered categories. Other times, only

response

the question is presented, and the respondents must generate an answer in their own words. Sometimes, a respondent fails to provide a response to a measurement attempt. This complicates the computation of statistics involving that measure.

2.2.4 Edited Response

In some modes of data collection, the initial measurement provided undergoes a review prior to moving on to the next. In computer-assisted measurement, quantitative answers are subjected to range checks, to flag answers that are outside acceptable limits. For example, if the question asks about year of birth, numbers less than 1890 might lead to a follow-up question verifying the stated year. There may also be consistency checks, which are logical relationships between two different measurements. For example, if the respondent states that she is 14 years old and has given birth to 5 children, there may be a follow-up question that clarifies the apparent discrepancy and permits a correction of any errors of data. With interviewer-administered paper questionnaires, the interviewer often is instructed to review a completed instrument, look for illegible answers, and cross out questions that were skipped in the interview.

After all of the respondents have provided their answers, further editing of data sometimes occurs. This editing may examine the full distribution of answers and look for atypical patterns of responses. This attempt at "outlier detection" often leads to more careful examination of a particular completed questionnaire.

outlier detection

To review, edited responses try to improve on the original responses obtained from measurements of underlying constructs. The edited responses are the data from which inference is made about the values of the construct for an individual respondent.

2.2.5 The Target Population

target population

We are now ready to move to the right side of Figure 2.2, moving from the abstract to the concrete with regard to the representational properties of a survey. The first box describes the concept of a "target population." This is the set of units to be studied. As denoted in Figure 2.2, this is the most abstract of the population definitions. For many U.S. household surveys, the target population may be "the U.S. adult population." This description fails to mention the time extents of the group (e.g., the population living in 2004). It fails to note whether to include those living outside traditional households, fails to specify whether to include those who recently became adults, and fails to note how residency in the United States would be determined. The lack of specificity is not damaging to some discussions, but is to others. The target population is a set of persons of finite size, which will be studied. The National Crime Victimization Survey targets those aged 12 and over who are not in active military service and reside in noninstitutionalized settings (i.e., housing units, not hospitals, prisons, or dormitories). The time extents of the population are fixed for the month in which the residence of the sample person is selected.

LIFE CYCLE OF A SURVEY—DESIGN PERSPECTIVE

2.2.6 The Frame Population

The frame population is the set of target population members that has a chance to be selected into the survey sample. In a simple case, the "sampling frame" is a listing of all units (e.g., people and employers) in the target population. Sometimes, however, the sampling frame is a set of units imperfectly linked to population members. For example, the SOC has as its target population the U.S. adult household population. It uses as its sampling frame a list of telephone numbers. It associates each person to the telephone number of his/her household. (Note that there are complications in that some persons have no telephone in their household and others have several different telephone numbers.) The National Survey on Drug Use and Health uses a sampling frame of county maps in the United States. Through this, it associates each housing unit with a unique county. It then associates each person in the target population of adults and children age 12 or older with the housing unit in which they live. (Note that there are complications for persons without fixed residence and those who have multiple residences.)

> **Illustration—Populations of Inference and Target Populations**
>
> Often, survey statistics are constructed to describe a population that cannot easily be measured. For example, the Surveys of Consumers attempts to estimate consumer sentiment among U.S. adults in a specific month. Each minute, households are being formed through family or rent-sharing arrangements; being dissolved through death, divorce, and residential mobility; being merged together, and so on. The household population of a month is different at the beginning of the month than at the end of the month. Sometimes, the phrase "population of inference" is used for the set of persons who at any time in the month might be eligible. The "target population" describes the population that could be covered, given that the frame is set at the beginning of the month and contact with sample households occurs throughout the month.

2.2.7 The Sample

A sample is selected from a sampling frame. This sample is the group from which measurements will be sought. In many cases, the sample will be only a very small fraction of the sampling frame (and, therefore, of the target population).

sampling frame

2.2.8 The Respondents

In almost all surveys, the attempt to measure the selected sample cases does not achieve full success. Those successfully measured are commonly called "respondents" ("nonrespondents" or "unit nonresponse" is the complement). There is usually some difficulty in determining whether some cases should be termed "respondents" or "nonrespondents," because they provide only part of the information that is sought. Decisions must be made when building a data file about when to include a data record with less than complete information and when to exclude a respondent altogether from the analytic file. "Item missing data" is the term used to describe the absence of information on individual data items for a sample case successfully measured on other items. Figure 2.3 is a visual portrayal of the type of survey and frame data and the nature of unit and item nonresponse.

The figure portrays a data file; each line is a data record of a different sample person. The left columns contain data from the sampling frame, on all sample

respondents

non-respondents

unit non-response

item missing data

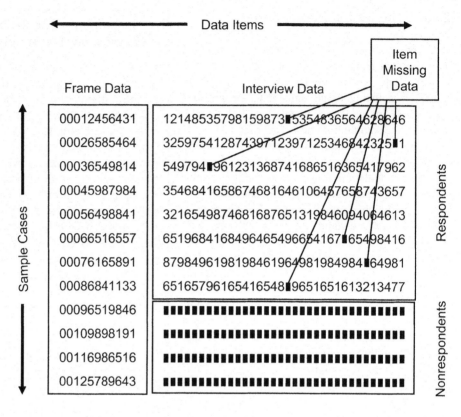

Figure 2.3 Unit and item nonresponse in a survey data file.

cases. Respondents have longer data records containing their answers to questions. The nonrespondents (at the end of the file) have data only from the sampling frame. Here and there throughout the respondent records are some individual missing data, symbolized by a "∎." One example of item missing data from the CES is missing payroll totals for sample employers who have not finalized their payroll records by the time the questionnaire must be returned.

2.2.9 Postsurvey Adjustments

After all respondents provide data and a set of data records for them is assembled, there is often another step taken to improve the quality of the estimates made from the survey. Because of nonresponse and because of some coverage problems (mismatches of the sampling frame and the target population), statistics based on the respondents may depart from those of the full population the statistics are attempting to estimate. At this point, examination of unit nonresponse patterns over different subgroups (e.g., the finding that urban response rates are lower than rural response rates) may suggest an underrepresentation of some groups relative to the sampling frame. Similarly, knowledge about the type of units not included in the sampling frame (e.g., new households in the SOC or new employers in the CES) may suggest an underrepresentation of certain types of target population

LIFE CYCLE OF A SURVEY—DESIGN PERSPECTIVE

members. We will learn later that "weighting" up the underrepresented in our calculations may improve the survey estimates. Alternatively, data that are missing are replaced with estimated responses through a process called "imputation." There are many different weighting and imputation procedures, all labeled as "postsurvey adjustments."

weighting

imputation

postsurvey adjustments

2.2.10 How Design Becomes Process

The design steps described above typically have a very predictable sequence. It is most common to array the steps of a survey along the temporal continuum in which they occur, and this is the order of most texts on "how to do a survey."

Figure 2.4 shows how the objectives of a survey help make two decisions, one regarding the sample and another regarding the measurement process. The decision on what mode of data collection to use is an important determinant of how the measurement instrument is shaped (e.g., "questionnaire" in Figure 2.4). The questionnaire needs a pretest before it is used to collect survey data. On the right-hand track of activities, the choice of a sampling frame, when married to a sample design, produces the realized sample for the survey. The measurement instrument and the sample come together during a data collection phase, during which attention is paid to obtaining complete measurement of the sample (i.e.,

mode of data collection

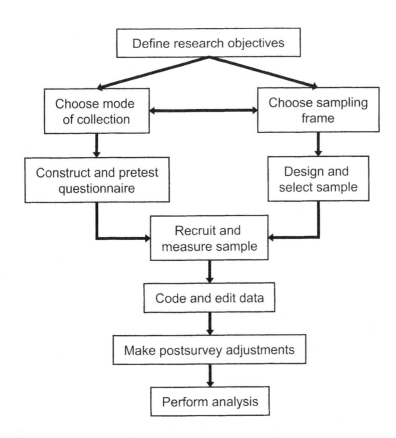

Figure 2.4 A survey from a process perspective.

avoiding nonresponse). After data collection, the data are edited and coded (i.e., placed into a form suitable for analysis). The data file often undergoes some post-survey adjustments, mainly for coverage and nonresponse errors. These adjustments define the data used in the final estimation or analysis step, which forms the statistical basis of the inference back to the full target population. This book takes the perspective that good survey estimates require simultaneous and coordinated attention to the different steps in the survey process.

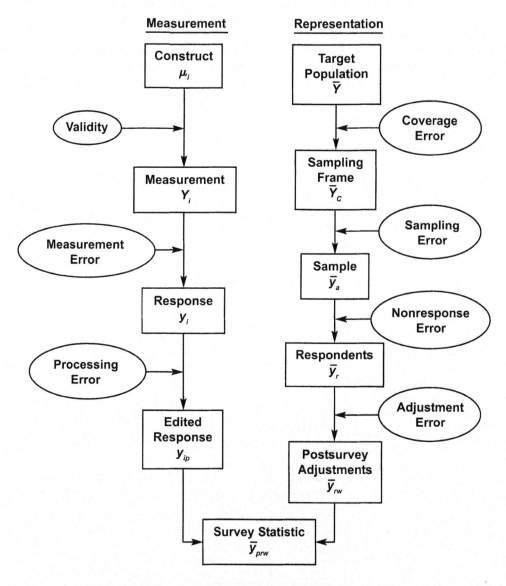

Figure 2.5 Survey life cycle from a quality perspective.

2.3 THE LIFE CYCLE OF A SURVEY FROM A QUALITY PERSPECTIVE

We used Figure 2.2 to describe key terminology in surveys. The same figure is useful to describe how survey methodologists think about quality. Figure 2.5 has added in ovals a set of quality concepts that are common in survey methodology. Each of them is placed in between successive steps in the survey process, to indicate that the quality concepts reflect mismatches between successive steps. Most of the ovals contain the word "error" because that is the terminology most commonly used. The job of a survey designer is to minimize error in survey statistics by making design and estimation choices that minimize the gap between two successive stages of the survey process. This framework is sometimes labeled the "total survey error" framework or "total survey error" paradigm.

total survey error

There are two important things to note about Figure 2.5:

1) Each of the quality components (ovals in Figure 2.5) has verbal descriptions and statistical formulations.
2) The quality components are properties of individual survey statistics (i.e., each statistic from a single survey may differ in its qualities), not of whole surveys.

The next sections introduce the reader both to the concepts of different quality components and to the simple statistical notation describing them. Since the quality is an attribute not of a survey but of individual statistics, we could present the statistics for a variety of commonly used statistics (e.g., the sample mean, a regression coefficient between two variables, estimates of population totals). To keep the discussion as simple as possible, we will describe the error components for a very simple statistic, the sample mean, as an indicator of the average in the population of some underlying construct. The quality properties of the sample mean will be discussed as a function of its relationship to the population mean.

We will use symbols to present a compact form of description of the error concepts, because that is the traditional mode of presentation. The Greek letter μ (mu) will be used to denote the unobservable construct that is the target of the measurement. The capital letter Y will be used to denote the measurement meant to reflect μ (but subject to inevitable measurement problems). When the measurement is actually applied, we obtain a response called y (lower case).

The statistical notation will be

μ_i = the value of a construct (e.g., reported number of doctor visits) for the ith person in the population, $i = 1, 2, \ldots, N$

Y_i = the value of a measurement (e.g., number of doctor visits) for the ith sample person, $i = 1, 2, \ldots, n$

y_i = the value of the response to application of the measurement (e.g., an answer to a survey question)

y_{ip} = the value of the response after editing and other processing steps.

In short, the underlying target attribute we are attempting to measure is μ_i, but instead we use an imperfect indicator, Y_i, which departs from the target because of imperfections in the measurement. When we apply the measurement, there are problems of administration. Instead of obtaining the answer Y_i, we obtain instead y_i, the response to the measurement. We attempt to repair the weakness in the measurement through an editing step, and obtain as a result y_{ip}, which we call the edited response (the subscript p stands for "postdata collection").

2.3.1 The Observational Gap between Constructs and Measures

The only oval in Figure 2.5 that does not contain the word "error" corresponds to mismatches between a construct and its associated measurement. Measurement theory in psychology (called "psychometrics") offers the richest notions relevant to this issue. Construct "validity" is the extent to which the measure is related to the underlying construct ("invalidity" is the term sometimes used to describe the extent to which validity is not attained). For example, in the National Assessment of Educational Progress, when measuring the construct of mathematical abilities of 4th graders, the measures are sets of arithmetic problems. Each of these problems is viewed to test some component of mathematical ability. The notion of validity is itself conceptual; if we knew each student's true mathematical ability, how related would it be to that measured by the set of arithmetic problems? Validity is the extent to which the measures reflect the underlying construct.

In statistical terms, the notion of validity lies at the level of an individual respondent. It notes that the construct (even though it may not be easily observed or observed at all) has some value associated with the ith person in the population, traditionally labeled as μ_i, implying the "true value" of the construct for the ith person. When a specific measure of Y is administered (e.g., an arithmetic problem given to measure mathematical ability), simple psychometric measurement theory notes that the result is not μ_i but something else:

$$Y_i = \mu_i + \varepsilon_i$$

That is, the measurement equals the true value plus some error term, ε_i, the Greek letter epsilon, denoting a deviation from the true value. This deviation is the basis of the notion of validity. For example, in the NAEP we might conceptualize mathematics ability as a scale from 0 to 100, with the average ability at 50. The model above says that on a particular measure of math ability, a student (i) who has a true math ability of, say, 57, may achieve a different score, say, 52. The error of that measure

---

The Notion of Trials

What does it mean when someone says that a specific response to a survey question is just "one trial of the measurement process?" How can one really ask the same question of the same respondent multiple times and learn anything valuable? The answer is that "trials" are a concept, a model of the response process. The model posits that the response given by one person to a specific question is inherently variable. If one could erase all memories of the first trial measurement and repeat the question, somewhat different answers would be given.

of math ability for the ith student is $(52 - 57) = -5$, because $Y_i = 52 = \mu_i + \varepsilon_i = 57 + (-5)$.

One added feature of the measurement is necessary to understand notions of validity: a single application of the measurement to the ith person is viewed as one of an infinite number of such measurements that could be made. For example, the answer to a survey question about how many times one has been victimized in the last six months is viewed as just one incident of the application of that question to a specific respondent. In the language of psychometric measurement theory, each survey is one trial of an infinite number of trials.

Thus, with the notion of trials, the response process becomes

$$Y_{it} = \mu_i + \varepsilon_{it}$$

Now we need two subscripts on the terms: one to denote the element of the population (i) and one to denote the trial of the measurement (t). Any one application of the measurement (t) is but one trial from a conceptually infinite number of possible measurements. The response obtained for the one survey conducted (Y_{it} for the tth trial) deviates from the true value by an error that is specific to the one trial (ε_{it}). That is, each survey is one specific trial, t, of a measurement process, and the deviations from the true value for the ith person may vary over trials (requiring the subscript t, as in ε_{it}). For example, on the math ability construct, using a particular measure, sometimes the ith student may achieve a 52 as above, but on repeated administrations might achieve a 59 or a 49 or a 57, with the corresponding error being +2 or –8 or 0, respectively. We do not really administer the test many times; instead, we envision that the one test might have achieved different outcomes from the same person over conceptually independent trials.

Now we are very close to defining validity for this simple case of response deviations from the true value. Validity is the correlation of the measurement, Y_i, and the true value, μ_i, measured over all possible trials and persons:

validity

$$E_{it}\left[(Y_{it} - \bar{Y})(\mu_i - \mu)\right] / \left[\sqrt{E_{it}(Y_{it} - \bar{Y})^2} \sqrt{E_{it}(\mu_i - \mu)^2}\right]$$

where μ is merely the mean of the μ_{it} over all trials and all persons and $\bar{Y}$ is the average of the Y_{it}. The E at the beginning of the expression denotes an expected or average value over all persons and all trials of measurement. When y and μ covary, moving up and down in tandem, the measurement has high construct validity. A valid measure of an underlying construct is one that is perfectly correlated to the construct.

expected value

Later, we will become more sophisticated about this notion of validity, noting that two variables can be perfectly correlated but produce different values of some of their univariate statistics. Two variables can be perfectly correlated but yield different mean values. For example, if all respondents underreport their weight by 5 pounds, then true weight and reported weight will be perfectly correlated, but the mean reported weight will be 5 pounds less than the mean of the true weights. This is a point of divergence of psychometric measurement theory and survey statistical error properties.

2.3.2 Measurement Error: The Observational Gap between the Ideal Measurement and the Response Obtained

measurement error

The next important quality component in Figure 2.5 is measurement error. By measurement error we mean a departure from the true value of the measurement as applied to a sample unit and the value provided. For example, imagine that the question from the National Survey on Drug Use and Health (NSDUH) is "Have you ever, even once, used any form of cocaine?" A common finding (see Sections 5.3.5 and 7.3.7) is that behaviors that are perceived by the respondent as undesirable tend to be underreported. Thus, for example, the true value for the response to this question for one respondent may be "yes," but the respondent will answer "no" in order to avoid the potential embarrassment of someone learning of his/her drug use.

To the extent that such response behavior is common and systematic across administrations of the question, there arises a discrepancy between the respondent mean response and the true sample mean. In the example above, the percentage of persons reporting any lifetime use of cocaine will be underestimated. In statistical notation, we need to introduce a new term that denotes the response to the question as distinct from the true value on the measure, Y_i, for the ith person. We call the response to the question y_i, so we can denote the systematic deviation from true values as $(y_i - Y_i)$. Returning to the CES example, the count for nonsupervisory employees might be 12 for some employer but the response to the question is 15 employees. In the terminology of survey measurement theory, a response deviation occurs to the extent that $y_i \neq Y_i$, in this case, $y_i - Y_i = 15 - 12 = 3$.

response bias

bias

In our perspective on measurement, we again acknowledge that each act of application of a measure is but one of a conceptually infinite number of applications. Thus, we again use the notion of a "trial" to denote the single application of a measurement. If response deviations described above are systematic, that is, if there is a consistent direction of the response deviations over trials, then "response bias" might result. "Bias" is the difference between the expected value (over all conceptual trials) and the true value being estimated. Bias is a systematic distortion of a response process. There are two examples of response biases from our example surveys. In the NSDUH, independent estimates of the rates of use of many substances asked about, including cigarettes and illegal drugs, suggest that the reporting is somewhat biased; that is, that people on average tend to underreport how much they use various substances. Part of the explanation is that some people are concerned about how use of these substances would reflect on how they are viewed. It also has been found that the rates at which people are victims of crimes are somewhat underestimated from survey reports. One likely explanation is that some individual victimizations, particularly those crimes that have little lasting impact on victims, are forgotten in a fairly short period of time. Whatever the origins, research has shown that survey estimates of the use of some substances and of victimization tend to underestimate the actual rates. Answers to the survey questions are systematically lower than the true scores; in short, they are biased. In statistical notation, we note the average or expected value of the response over trials as $E_t(y_{it})$, where, as before, t denotes a particular trial (or application) of the measurement. Response bias occurs when

$$E_t(y_{it}) \neq Y_i$$

In addition to systematic underreporting or overreporting that can produce biased reports, there can be an instability in the response behavior of a person, producing another kind of response error. Consider the case of a survey question in the SOC: "Would you say that at the present time, business conditions are better or worse than they were a year ago?" A common perspective on how respondents approach such a question is that, in addition to the words of the individual question and the context of prior questions, the respondent uses all other stimuli in the measurement environment. But gathering such stimuli (some of which might be memories generated by prior questions) is a haphazard process, unpredictable over independent trials. The result of this is variability in responses over conceptual trials of the measurement, often called "variability in response deviations." For this kind of response error, lay terminology fits rather well; this is an example of low "reliability" or "unreliable" responses. (Survey statisticians term this "response variance" to distinguish it from the error labeled above, "response bias.") The difference between response variance and response bias is that the latter is systematic, leading to consistent overestimation or underestimation of the construct in the question, but response variance leads to instability in the value of estimates over trials.

> **The Notion of Variance or Variable Errors**
>
> Whenever the notion of errors that are variable arises, there must be an assumption of replication (or trials) of the survey process. When the estimates of statistics vary over those replications, they are subject to variable error. Variability at the response step can affect individual answers. Variability in frame development, likelihood of cooperation with the survey request, or characteristics of samples, can affect survey statistics.
>
> Usually, variance is not directly observed because the replications are not actually performed.

reliability

response variance

2.3.3 Processing Error: The Observational Gap between the Variable Used in Estimation and that Provided by the Respondent

What errors can be introduced after the data are collected and prior to estimation? For example, an apparent outlier in a distribution may have correctly reported a value. A respondent in the National Crime Victimization Survey may report being assaulted multiple times each day, an implausible report that, under some editing rules, may cause a setting of the value to missing data. However, when the added information that the respondent is a security guard in a bar is provided, the report becomes more plausible. Depending on what construct should be measured by the question, this should or should not be altered in an editing step. The decision can affect processing errors.

Another processing error can arise for questions allowing the respondent to phrase his or her own answer. For example, in the Surveys of Consumers, if a respondent answers "yes" to the question, "During the last few months, have you heard of any favorable or unfavorable changes in business conditions?" the interviewer then asks, "What did you hear?" The answer to that question is entered into a text field, using the exact words spoken by the respondent. For example, the respondent may say, "There are rumors of layoffs planned for my plant. I'm worried about whether I'll lose my job." Answers like this capture the rich diversity of situations of different respondents, but they do not lend themselves to

processing error

quantitative summary, which is the main product of surveys. Hence, in a step often called "coding," these text answers are categorized into numbered classes. For example, this answer might be coded as a member of a class labeled "Possible layoffs at own work site/company." The sample univariate summary of this measure is a proportion falling into each class (e.g., "8% of the sample reported possible layoffs at own work site/company").

coding

What errors can be made at this step? Different persons coding these text answers can make different judgments about how to classify the text answers. This generates a variability in results that is purely a function of the coding system (e.g., coding variance). Poor training can prompt all coders to misinterpret a verbal description consistently. This would produce a coding bias.

In statistical notation, if we were considering a variable like income, subject to some editing step, we could denote processing effects as the difference between the response as provided and the response as edited. Thus, y_i = response to the survey question, as before, but y_{ip} = the edited version of the response. The processing or editing deviation is simply $(y_{ip} - y_i)$.

2.3.4 Coverage Error: The Nonobservational Gap between the Target Population and the Sampling Frame

The big change of perspective when moving from the left side (the measurement side) of Figure 2.2 to the right side (the representation side) is that the focus becomes statistics, not individual responses. Notice that the terms in Figure 2.5 are expressions of sample means, simple statistics summarizing individual values of elements of the population. Although there are many possible survey statistics, we use the mean as our illustrative example.

finite population

Sometimes, the target population (the finite population we want to study) does not have a convenient sampling frame that matches it perfectly. For example, in the United States there is no updated list of residents that can be used as a sampling frame of persons. In contrast, in Sweden there is a population register, an updated list of names and addresses of almost all residents. Sample surveys of the target population of all U.S. residents often use sampling frames of telephone numbers. The error that arises is connected to what proportion of U.S. residents can be reached by telephone *and* how different they are from others *on the statistics in question*. Persons with lower incomes and in remote rural areas are less likely to have telephones in their homes. If the survey statistic of interest were the percentage of persons receiving unemployment compensation from the government, it is likely that a telephone survey would underestimate that percentage. That is, there would be a coverage bias in that statistic.

undercoverage

Figure 2.6 is a graphical image of two coverage problems with a sampling frame. The target population differs from the sampling frame. The lower and left portions of elements in the target population are missing from the frame (e.g., nontelephone households, using a telephone frame to cover the full household population). This is labeled "undercoverage" of the sampling frame with respect to the target population. At the top and right portions of the sampling frame are a set of elements that are not members of the target population but are members of the frame population (e.g., business telephone numbers in a telephone frame trying to cover the household population). These are "ineligible units," sometimes labeled "overcoverage" and sometimes the existence of "foreign elements."

ineligible units

overcoverage

LIFE CYCLE OF A SURVEY—QUALITY PERSPECTIVE

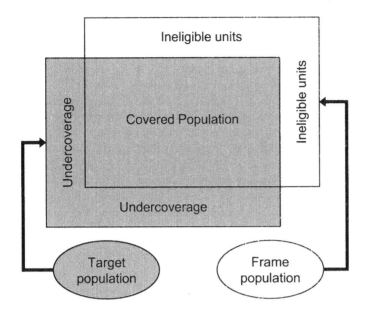

Figure 2.6 Coverage of a target population by a frame.

In statistical terms for a sample mean, coverage bias can be described as a function of two terms: the proportion of the target population not covered by the sampling frame, and the difference between the covered and noncovered population. First, we note that coverage error is a property of a frame and a target population on a specific statistic. It exists *before* the sample is drawn and thus is not a problem arising because we do a *sample* survey. It would also exist if we attempted to do a census of the target population using the same sampling frame. Thus, it is simplest to express coverage error prior to the sampling step. Let us express the effect of the mean of the sampling frame:

coverage bias

coverage error

$\bar{Y}$ = Mean of the entire target population
$\bar{Y}_C$ = Mean of the population on the sampling frame
$\bar{Y}_U$ = Mean of the target population not on the sampling frame
N = Total number of members of the target population
C = Total number of eligible members of the sampling frame ("covered" elements)
U = Total number of eligible members not on the sampling frame ("not covered" elements)

The bias of coverage is then expressed as

$$\bar{Y}_C - \bar{Y} = \frac{U}{N}\left(\bar{Y}_C - \bar{Y}_U\right)$$

That is, the error in the mean due to undercoverage is the product of the noncoverage rate (U/N) and the difference between the mean of the covered and noncovered cases in the target population. The left side of the equation merely shows that the coverage error for the mean is the difference between the mean of the covered population and the mean of the full target population. The right side is the result of a little algebraic manipulation. It shows that the coverage bias is a function of the proportion of the population missing from the frame and the difference between those on and off the frame. For example, for many statistics on the U.S. household population, telephone frames describe the population well, chiefly because the proportion of nontelephone households is very small, about 5% of the total population. Imagine that we used the Surveys of Consumers, a telephone survey, to measure the mean years of education, and the telephone households had a mean of 14.3 years. Among nontelephone households, which were missed due to this being a telephone survey, the mean education level is 11.2 years. Although the nontelephone households have a much lower mean, the bias in the covered mean is

$$\bar{Y}_C - \bar{Y} = 0.05(14.3 \text{ years} - 11.2 \text{ year}) = 0.16 \text{ years}$$

or, in other words, we would expect the sampling frame to have a mean years of education of 14.3 years versus the target population mean of 14.1 years.

Coverage error on sampling frames results in sample survey means estimating the $\bar{Y}_C$ and not the $\bar{Y}$ and, thus, coverage error properties of sampling frames generate coverage error properties of sample-based statistics.

2.3.5 Sampling Error: The Nonobservational Gap between the Sampling Frame and the Sample

One error is deliberately introduced into sample survey statistics. Because of cost or logistical infeasibility, not all persons in the sampling frame are measured. Instead, a sample of persons is selected; they become the sole target of the measurement. All others are ignored. In almost all cases, this deliberate "nonobservation" introduces deviation from the achieved sample statistics and the same statistic on the full sampling frame.

For example, the National Crime Victimization Survey sample starts with the entire set of 3067 counties within the United States. It separates the counties by population size, region, and correlates of criminal activity, forming separate groups or strata. In each stratum, giving each county a chance of selection, it selects sample counties or groups of counties, totaling 237. All the sample persons in the survey will come from those geographic areas. Each month of the sample selects about 8300 households in the selected areas and attempts interviews with their members.

As with all the other survey errors, there are two types of sampling error: sampling bias and sampling variance. Sampling bias arises when some members of the sampling frame are given no chance (or reduced chance) of selection. In such a design, every possible set of selections excludes them systematically. To the extent that they have distinctive values on the survey statistics, the statistics will depart from the corresponding ones on the frame population. Sampling variance arises because, given the design for the sample, by chance many

sampling error

sampling variance

sampling bias

LIFE CYCLE OF A SURVEY—QUALITY PERSPECTIVE

different sets of frame elements could be drawn (e.g., different counties and households in the NCVS). Each set will have different values on the survey statistic.

Just like the notion of trials of measurement (see Section 2.3.1), sampling variance rests on the notion of conceptual replications of the sample selection. Figure 2.7 shows the basic concept. On the left appear illustrations of the different possible sets of sample elements that are possible over different samples. The figure portrays S different sample "realizations," or different sets of frame elements, with frequency distributions for each (the x axis is the value of the variable and the y axis is the number of sample elements with that value). Let us use our example of the sample mean as the survey statistic of interest. Each of the S samples produces a different sample mean. One way to portray the sampling variance of the mean appears on the right of the figure. This is the sampling distribution of the mean, a plotting of the frequency of specific different values of the sample mean (the x axis is the value of a sample mean and the y axis is the number of samples with that value among the S different samples). The dispersion of this distribution is the measure of sampling variance normally employed. If the average sample mean over all S samples is equal to the mean of the sampling frame, then there is no sampling bias for the mean. If the dispersion of the distribution on the right is small, the sampling variance is low. (Sampling variance is zero only in populations with constant values on the variable of interest.)

realization

sampling distribution

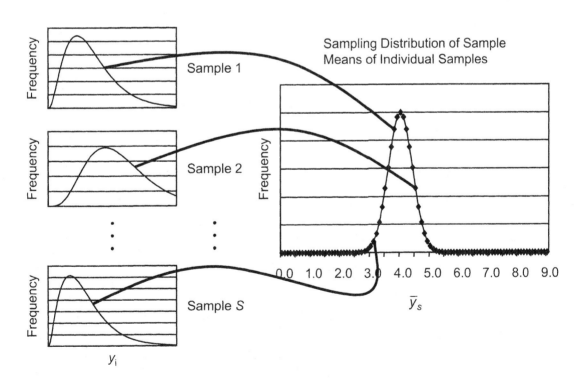

Figure 2.7 Samples and the sampling distribution of the mean.

The extent of the error due to sampling is a function of four basic principles of the design:

probability sampling
stratification

1) Whether all sampling frame elements have known, nonzero chances of selection into the sample (called "probability sampling")
2) Whether the sample is designed to control the representation of key subpopulations in the sample (called "stratification")

element sample
cluster sample

3) Whether individual elements are drawn directly and independently or in groups (called "element" or "cluster" samples)
4) How large a sample of elements is selected

Using this terminology, the NCVS is thus a stratified, clustered probability sample of approximately 8500 households per month.

Sampling bias is mainly affected by how probabilities of selection are assigned to different frame elements. Sampling bias can be easily removed by giving all elements an equal chance of selection. Sampling variance is reduced with big samples, with samples that are stratified, and samples that are not clustered.

In statistical terms,

$\bar{y}_s$ = Mean of the specific sample draw, sample s; $s = 1, 2, ..., S$
$\bar{Y}_c$ = Mean of the total set of C elements in the sampling frame

These means (in simple sample designs) have the form:

$$\bar{y}_s = \frac{\sum_{i=1}^{n_s} y_{si}}{n_s}, \text{ and } \bar{Y}_C = \frac{\sum_{i=1}^{C} Y_i}{C}$$

sampling variance

"Sampling variance" measures how variable the $\bar{y}_s$ are over all sample realizations. The common measurement tool for this is to use squared deviations of the sample means about the mean of the sampling frame, so that the "sampling variance" of the mean is

$$\frac{\sum_{s=1}^{S}(\bar{y}_s - \bar{Y}_C)^2}{S}$$

When sampling variance is high, the sample means are very unstable. In that situation, sampling error is very high. That means that for any given survey with that kind of design, there is a larger chance that the mean from the survey will be comparatively far from the true mean of the population from which the sample was drawn (the sample frame).

2.3.6 Nonresponse Error: The Nonobservational Gap between the Sample and the Respondent Pool

Despite efforts to the contrary, not all sample members are successfully measured in surveys involving human respondents. Sometimes, 100% response rates are

LIFE CYCLE OF A SURVEY—QUALITY PERSPECTIVE

obtained in surveys requiring sampling of inanimate objects (e.g., medical records of persons, housing units). Almost never does it occur in sample surveys of humans. For example, in the SOC, about 30–35% of the sample each month either eludes contact or refuses to be interviewed. In the main (nationwide) National Assessment of Educational Progress (NAEP), about 17% of the schools refuse participation, and within cooperating schools about 11% of the students are not measured, either because of absences or refusal by their parents. In even years, schools are required by law to participate in reading and mathematics components, grades 4 and 8. As a result, in 2007, there was 100% school participation.

Nonresponse error arises when the values of statistics computed based only on respondent data differ from those based on the entire sample data. For example, if the students who are absent on the day of the NAEP measurement have lower knowledge in the mathematical or verbal constructs being measured, then NAEP scores suffer nonresponse bias, they systematically overestimate the knowledge of the entire sampling frame. If the nonresponse rate is very high, then the amount of the overestimation could be severe. *(nonresponse error)*

Most of the concern of practicing survey researchers is about nonresponse bias, and its statistical expression resembles that of coverage bias described in Section 2.3.1:

$\bar{y}_s$ = Mean of the entire specific sample as selected
$\bar{y}_r$ = Mean of the respondents within the sth sample
$\bar{y}_m$ = Mean of the nonrespondents within the sth sample
n_s = Total number of sample members in the sth sample
r_s = Total number of respondents in the sth sample
m_s = Total number of nonrespondents in the sth sample

The nonresponse bias is then expressed as an average over all samples of

$$\bar{y}_r - \bar{y}_s = \frac{m_s}{n_s}(\bar{y}_r - \bar{y}_m)$$

Thus, nonresponse bias for the sample mean is the product of the nonresponse rate (the proportion of eligible sample elements for which data are not collected) and the difference between the respondent and nonrespondent mean. This indicates that response rates alone are not quality indicators. High response rate surveys can also have high nonresponse bias (if the nonrespondents are very distinctive on the survey variable). The best way to think about this is that high response rates reduce the *risk* of nonresponse bias. *(nonresponse bias)*

2.3.7 Adjustment Error

The last step in Figure 2.5 on the side of errors of nonobservation is postsurvey adjustments. These are efforts to improve the sample estimate in the face of coverage, sampling, and nonresponse errors. (In a way, they serve the same function as the edit step on individual responses, discussed in Section 2.3.3.)

The adjustments use some information about the target or frame population, or response rate information on the sample. The adjustments give greater weight to sample cases that are underrepresented in the final dataset. For example, some adjustments pertain to nonresponse. Imagine that you are interested in the rate of personal crimes in the United States and that the response rate for urban areas in the National Crime Victimization Survey is 85% (i.e., 85% of the eligible sample persons provide data on a specific victimization), but the response rate in the rest of the country is 96%. This implies that urban persons are underrepresented in the respondent dataset. One adjustment weighting scheme counteracts this by creating two weights, $w_i = 1/0.85$ for urban respondents and $w_i = 1/0.96$ for other respondents. An adjusted sample mean is computed by

$$\bar{y}_{rw} = \frac{\sum_{i=1}^{r} w_i y_{si}}{\sum_{i=1}^{r} w_i}$$

which has the effect of giving greater weight to the urban respondents in the computation of the mean. The error associated with the adjusted mean relative to the target population mean is

$$(\bar{y}_{rw} - \bar{Y})$$

which would vary over different samples and applications of the survey. That is, adjustments generally affect the bias of the estimates and the variance (over samples and implementations of the survey). Finally, although postsurvey adjustments are introduced to reduce coverage, sampling, and nonresponse errors, they can also increase them, as we will learn later in Chapter 10.

2.4 PUTTING IT ALL TOGETHER

The chapter started by presenting three perspectives on the survey. The first, portrayed in Figure 2.2, showed the stages of survey design, moving from abstract concepts to concrete steps of activities. Next, in Figure 2.4, we presented the steps of a survey project, from beginning to end. Finally, in Figure 2.5, we presented the quality characteristics of surveys that the field associates with each of the steps and the notation used for different quantities. The quality components focus on two properties of survey estimates: errors of observation and errors of nonobservation. Errors of observation concern successive gaps between constructs, measures, responses, and edited responses. Errors of nonobservation concern successive gaps between statistics on the target population, the sampling frame, the sample, and the respondents from the sample.

For some of these errors we described a systematic source, one that produced consistent effects over replications or trials (e.g., nonresponse). We labeled these "biases." For others, we described a variable or random source of error (e.g., validity). We labeled these as "variance." In fact, as the later chapters in the book will show, all of the error sources are both systematic and variable, and contain both biases and variances.

The reader now knows that the quantitative products of surveys have their quality properties described in quantitative measures. Look again at Figure 2.5, showing how the notation for a simple sample mean varies over the development of the survey. This notation will be used in other parts of the text. Capital letters will stand for properties of population elements. Capital letters will be used for the discussions about measurement, when the sampling of specific target population members is not at issue. In discussions about inference to the target population through the use of a sample, capital letters will denote population elements and lower case letters will denote sample quantities. The subscripts of the variables will indicate membership in subsets of the population (e.g., i for the ith person, or the existence of adjustment, such as w for weighting).

2.5 ERROR NOTIONS IN DIFFERENT KINDS OF STATISTICS

The presentation above focused on just one possible survey statistic—the sample mean—to illustrate error principles. There are many other statistics computed from surveys (e.g., correlation coefficients and regression coefficients).

Two uses of surveys, linked to different statistics, deserve mention:

1) Descriptive uses (i.e., how prevalent an attribute is in the population, how big a group exists in the population, or the average value on some quantitative measure)
2) Analytic uses (i.e., what causes some phenomenon to occur or how two attributes are associated with one another)

Many surveys are done to collect information about the distribution of characteristics, ideas, experiences, or opinions in a population. Often, results are reported as means or averages. For example, the NCVS might report that 5% of the people have had their car stolen in the past year.

In contrast, statements such as "Women are more likely to go to a doctor than men," "Republicans are more likely to vote than Democrats," or "Young adults are more likely to be victims of crime than those over 65" are all statements about relationships. For some purposes, describing the degree of relationship is important. So a researcher might say that the correlation between a person's family income and the likelihood of voting is 0.23. Alternatively, the income rise associated with an investment in education might be described by an equation, often called a "model" of the income generation process:

$$\ln(y_i) = \beta_0 + \beta_1 x_i + \beta_2 x_i^2$$

where y_i is the value of the ith person's income, and x_i is the value of the ith person's educational attainment in years.

The hypothesis tested in the model specification is that the payoff in income of educational investments is large for the early years and then diminishes with additional years. If the coefficient β_1 is positive and β_2 is negative, there is some support for the hypothesis. This is an example of using survey data for causal analysis. In this case, the example concerns whether educational attainment causes income attainment.

Are statistics like a correlation coefficient and a regression coefficient subject to the same types of survey errors as described above? Yes. When survey data are used to estimate the statistics, they can be subject to coverage, sampling, nonresponse, and measurement errors, just as can simple sample means. The mathematical expressions for the errors are different, however. For most survey statistics measuring relationships, the statistical errors are properties of crossproducts between the two variables involved (e.g., covariance and variance properties).

The literatures in analytic statistics and econometrics are valuable for understanding errors in statistics related to causal hypotheses. The language of error used in these fields is somewhat different from that in survey statistics, but many of the concepts are similar.

Thus, the kind of analysis one is planning, and the kinds of questions one wants to answer can be related to how concerned one is about the various kinds of error that we have discussed. Bias, either in the sample of respondents answering questions or in the answers that are given, is often a primary concern for those focused on the goal of describing distributions. If the main purpose of a survey is to estimate the percentage of people who are victims of particular kinds of crime, and those crimes are systematically underreported, then that bias in reporting will have a direct effect on the ability of the survey to achieve those goals. In contrast, if the principal concern is whether old people or young people are more likely to be victims, it could be that the ability of the survey to accomplish that goal would be unaffected if there were consistent underreporting of minor crimes. If the bias was really consistent for all age groups, then the researcher could reach a valid conclusion about the relationship between age and victimization, even if all the estimates of victimization were too low.

2.6 NONSTATISTICAL NOTIONS OF SURVEY QUALITY

In addition to the components of the total survey error perspective reviewed above, there are three additional notions that are relevant to assessing the quality of a survey estimate. These three notions arise from the overall desire to maximize the "fitness for use" of an estimate. "Fitness for use" acknowledges that different users of the same estimate may have different purposes for the information. For one, highly accurate estimates are necessary for a good decision based on the information; for another, rough orders of magnitude of the population value are sufficient. This viewpoint implies that a "good" estimate for the second user may not be a "good" estimate for the first. High fitness for use means the indicator provides the information needed for the specific use.

fitness for use

credibility

The first notion is "credibility," the extent to which the producer of the information is judged by the user to be free of any particular point of view, a perspective on the phenomena being measured that may influence an outcome of the survey in a known direction. Central government statistical agencies strive to achieve the image of neutrality and objectivity. Scientists using the survey method (a) document each step in their design and implementation, to facilitate replication of their results, then they (b) explicitly note the weaknesses in the estimates that may affect their conclusions. Both of these steps are intended to enhance the credibility of the estimates.

relevance

The second notion is "relevance." A survey estimate is relevant to a user if it measures a construct quite similar in meaning to the user's main concern.

Sometimes, there is gap between the construct measured by the survey and that needed by the user. For example, a user may wish to have a prevalence indicator of economic suffering, the extent to which emotional and physical discomfort arises from economic difficulties among persons. They may use the unemployment rate as an indicator of such suffering. The relevance of this indicator might be criticized by noting that some unemployment does not produce such suffering, due to government support mechanisms. The reader may note a similarity between notions of "construct validity" (see p. 49) and relevance. Relevance focuses on differences among constructs; construct validity relates to differences between a construct and a given measurement.

The third notion is "timeliness." One key determinant of whether the survey estimate is fit for a user's purpose is whether the estimate is available at a time needed for the decision based on the information. For example, a survey estimate of the Surveys of Consumers describing the confidence of the U.S. public in March, 2009 is of little value to macroeconomists a year later in March, 2010. Timeliness of an estimate is completely determined by its use.

timeliness

Indeed, all three of these notions—credibility, relevance, and timeliness—are ones that are well defined only when specific to a particular use of a survey estimate. The notions lie outside the paradigm of total survey error (and we will not discuss them further), and they have not influenced the field of survey methodology in the same way as others. Nonetheless, they are important notions when considering how the same estimates might be used in different ways by different users.

2.7 SUMMARY

Sample surveys rely on two types of inference – from the questions to constructs, and from the sample statistics to the population statistics. The inference involves two coordinated sets of steps: obtaining answers to questions constructed to mirror the constructs, and identifying and measuring sample units that form a microcosm of the target population.

Despite all efforts, each of the steps is subject to imperfections, producing statistical errors in survey statistics. The errors involving the gap between the measures and the construct are issues of validity. The errors arising during application of the measures are called "measurement errors." Editing and processing errors can arise during efforts to prepare the data for statistical analysis. Coverage errors arise when enumerating the target population using a sampling frame. Sampling errors stem from surveys measuring only a subset of the frame population. The failure to measure all sample persons on all measures creates nonresponse error. "Adjustment" errors arise in the construction of statistical estimators to describe the full target population. All of these error sources can have varying effects on different statistics from the same survey.

This chapter introduced the reader to these elementary building blocks of the field of survey methodology. Throughout the text, we will elaborate and add to the concepts in this chapter. We will describe a large and dynamic research literature that is discovering new principles of human measurement and estimation of large population characteristics. We will gain insight into how the theories being developed lead to practices that affect the day-to-day tasks of constructing and implementing surveys. The practices will generally be aimed at improving the quality of the survey statistics (or reducing the cost of the survey). Often, the practices will provide new measurements of how good the estimates are from the survey.

Keywords

bias
cluster sample
coding
construct
construct validity
coverage bias
coverage error
credibility
element sample
error
errors of nonobservation
errors of observation
expected value
finite population
fitness for use
imputation
ineligible units
inference
item missing data
measurement
measurement error
mode of data collection
nonrespondents
nonresponse bias
nonresponse error
observation unit

outlier detection
overcoverage
postsurvey adjustment
probability sampling
processing erro
realization
relevance
reliability
respondents
response
response bias
response variance
sampling error
sampling bias
sampling frame
sampling variance
statistic
stratification
target population
timeliness
total survey error
true values
undercoverage
unit response
validity
weighting

For More In-Depth Reading

Biemer, P. and Lyberg, L. (2003), *Introduction to Survey Quality*, New York: Wiley.

Groves, R. M. (1989), *Survey Errors and Survey Costs*, New York: Wiley.

Lessler, J. and Kalsbeek, W. (1992), *Nonsampling Error in Surveys*, New York: Wiley.

Weisberg, H. (2005), *The Total Survey Error Approach: A Guide to the New Science of Survey Research*, Chicago: University of Chicago Press.

EXERCISES

1) A recent newspaper article reported that "sales of handheld digital devices (e.g., Blackberries and PDAs) are up by nearly 10% in the last quarter, while sales of laptops and desktop PCs have remained stagnant." This report was based on the results of an on-line survey in which 9.8% of the more than 126,000 respondents said that they had "purchased a handheld digital device between January 1 and April 30 of this year."

   E-mails soliciting participation in this survey were sent to individuals using an e-mail address frame from the five largest commercial Internet service providers (ISPs) in the United States. Data collection took place over a 6-week period beginning May 1, 2002. The overall response rate achieved in this survey was 13%.

   Assume that the authors of this study wanted to infer something about the expected purchases of U.S. adults (18 years old and older).

   a) What is the target population? What is the population in the sample frame?
   b) Based on this chapter and your readings, briefly discuss how the design of this survey might affect the following sources of error:

      - Coverage error
      - Nonresponse error
      - Measurement error

   c) Without changing the duration or the mode of this survey (i.e., computer assisted or self-administered), what could be done to reduce the errors you outlined in 1b? For each source of error, suggest one change that could be made to reduce this error component, making sure to justify your answer based on readings and lecture material.
   d) To lower the cost of this survey in the future, researchers are considering cutting the sample in half, using an e-mail address frame from only the two largest ISPs. What effect (if any) will these changes have on sampling error and coverage error?

2) Describe the difference between coverage error and sampling error in survey estimates.

3) Given what you have read about coverage, nonresponse, and measurement errors, invent an example of a survey design in which attempting to reduce one error might lead to another error increasing. After you have constructed the example, invent a methodological study design to investigate whether the reduction of the one error actually does increase the other.

4) This chapter described errors of observation and errors of nonobservation.

   a) Name three sources of error that affect inference from the sample from which data were collected to the target population.
   b) Name three sources of error that affect inference from the respondents' answers to the underlying construct.

c) For each source of error you mentioned, state whether it potentially affects the variance of estimates, biases the estimates, or both.

5) For each of the following design decisions, identify which error sources described in your readings might be affected. Each design decision can affect at least two different error sources. Write short (2–4 sentences) answers to each point.

   a) The decision to include or exclude institutionalized persons (e.g., residing in hospitals, prisons, or military group quarters) from the sampling frame in a survey of the prevalence of physical disabilities in the United States.
   b) The decision to use self-administration of a mailed questionnaire for a survey of elderly Social Security beneficiaries regarding their housing situation.
   c) The decision to use repeated calls persuading reluctant respondents in a survey of customer satisfaction for a household product manufacturer.
   d) The decision to reduce costs of interviewing by using existing office personnel to interview a sample of patients of a health maintenance organization (HMO) and thereby increase the sample size of the survey. The topic of the survey is satisfaction with the medical care they receive.
   e) The decision to increase the number of questions about assets and income in a survey of income dynamics, resulting in a lengthening of the interview.
   f) The decision to extend interviewing on a survey of use of child care facilities by parents of young children from the originally scheduled period of January 1–May 1, to the new schedule of January 1–August 1.
   g) The decision to include prisons and hospitals in the sampling frame for a study of consumer expenditures.
   h) The decision to use an existing trained staff of female interviewers (instead of hiring and training some male interviewers) in a survey measuring attitudes toward an amendment to the constitution to provide equal rights under the law to females and males.
   i) The decision to change from a face-to-face interview design to a mailed questionnaire mode in a household survey of illegal drug usage.

6) For each of the following questions, state briefly what the construct is that you think the question is most likely designed to measure. In some cases, there may be more than one plausible measurement goal.

   a) How old are you?
   b) Are you married?
   c) Do you own a car?
   d) What is your income?
   e) Did you vote in the last election for U.S. President?
   f) Do you consider yourself to be a Democrat, a Republican, or an Independent?
   g) In the next 12 months, do you think that the economy will get better, get worse, or will it stay about the same as it is now?

EXERCISES

    h) Do you consider yourself to be a happy person?
    i) Has a doctor ever told you that you have high blood pressure?
    j) How would you rate your doctor in ability to diagnose and propose treatments for medical problems: excellent, good, fair or poor?
    k) In the past week, have you prepared any meals for yourself?

7) From an inference perspective, what is the central concern one should have about those who are sampled but do not respond to surveys?

CHAPTER THREE

TARGET POPULATIONS, SAMPLING FRAMES, AND COVERAGE ERROR

3.1 INTRODUCTION

Sample surveys describe or make inferences to well-defined populations. This chapter presents the conceptual and practical issues in defining and identifying these populations. The fundamental units of populations are referred to as "elements." The totality of the elements forms the full population. In most household populations, elements are persons who live in households; in NAEP and other school samples, the elements are often students within the population of schools; in business surveys like CES, the element is an establishment. Elements can be many different kinds of units, even in the same survey. For example, in a household survey, in addition to persons, the inference might also be made to the housing units where the persons live, or to the blocks on which they live, or to the churches that they attend. In short, statistics describing different populations can be collected in a single survey when the populations are linked to units from which measurements are taken.

elements

Among common research tools, surveys are unique in their concern about a well-specified population. For example, when conducting randomized biomedical experiments, the researcher often pays much more attention to the experimental stimulus and the conditions of the measurement than to the identification of the population under study. The implicit assumption in such research is that the chief purpose is identifying the conditions under which the stimulus produces the hypothesized effect. The demonstration that it does so for a variety of types of subjects is secondary. Because surveys evolved as tools to describe fixed, finite populations, survey researchers are specific and explicit about definitions of populations under study.

3.2 POPULATIONS AND FRAMES

The "target population" is the group of elements for which the survey investigator wants to make inferences by using the sample statistics. Target populations are finite in size (i.e., at least theoretically, they can be counted). They have some time restrictions (i.e., they exist within a specified time frame). They are observable (i.e., they can be accessed). These aspects of target populations are desirable

target population

for achieving a clear understanding of the meaning of the survey statistics and for permitting replication of the survey.

The target population definition has to specify the kind of units that are elements in the population and the time extents of the group. For example, the target population of many U.S. household surveys is persons 18 years of age or older, "adults" who reside in housing units within the United States. A "household" includes all the persons who occupy a housing unit. A "housing unit" is a house, apartment, mobile home, group of rooms, or single room that is occupied (or, if vacant, is intended for occupancy) as separate living quarters. Separate living quarters are those in which the occupants live and eat separately from any other persons in the building and which have direct access from the outside of the building or through a common hall. The occupants may be a single family, one person living alone, two or more families living together, or any other group of related or unrelated persons who share living arrangements. Not all persons in the United States at any moment are adults and not all adults reside in housing units (some live in prisons, long-term care medical facilities, or military barracks).

Not all U.S. national household surveys choose this target population. Some limit the target population to those living in the 48 coterminous states and the District of Columbia (excluding Alaska and Hawaii). Others add to the target population members of the military living on military bases. Still others limit the target population to citizens or English speakers.

Since the population changes over time, the time of the survey also defines the target population. Since many household surveys are conducted over a period of several days, weeks, or even months, and since the population is changing daily as persons move in and out of the U.S. households, the target population of many household surveys is the set of persons in the household population during the survey period. In practice, the members of households are "fixed" at the time of first contact in many surveys.

There are often restrictions placed on a survey data collection operation that limit the target population further. For example, in some countries it may not be possible to collect data in a district or region due to civil disturbances. These districts or regions may be small in size, and dropped from the population before sample selection begins. The restricted population, sometimes called a "survey population" is not the intended target population, and yet it is realistically the actual population from which the survey data are collected. For example, the CES target population consists of all work organizations with employees in a specific month. Its survey population, however, consists of employers who have been in business for several months (long enough to get on the frame). A survey organization may note the restriction in technical documentation, but users of available public-use data may not make a clear distinction between the target population (e.g., persons living in the country) and the survey population (e.g., persons living in the country, except for districts or regions with civil disturbances).

A set of materials, or "sampling frame," is used to identify the elements of the target population. Sampling frames are lists or procedures intended to identify all elements of a target population. The frames may be maps of areas in which elements can be found, time periods during which target events would occur, or records in filing cabinets, among others. Sampling frames, at their simplest, consist of a simple list of population elements. There are populations for which lists are readily available, such as members of a professional organization, business establishments located in a particular city or county, or hospitals, schools, and other kinds of institutions. There are registries of addresses or of persons in a number of countries that also serve as sampling frames of persons.

There are many populations, though, for which lists of individual elements are not readily available. For example, in the United States lists are seldom available in one place for all students attending school in a province or state, inmates in prisons, or even adults living in a specific county. There may be lists of members in a single institution or cluster of elements, but the lists are seldom collected across institutions or combined into a single master list. In other cases, lists may have to be created during survey data collection. For example, lists of housing units are often unavailable for household surveys. In area sampling, well-defined geographic areas, such as city blocks, are chosen in one or more stages of selection, and staff are sent to selected blocks to list all housing units. The cost of creating a list of housing is thus limited to a sample of geographic areas, rather than a large area.

When available sampling frames miss the target population partially or entirely, the survey researcher faces two options:

1) Redefine the target population to fit the frame better.
2) Admit the possibility of coverage error in statistics describing the original target population.

A common example of redefining the target population is found in telephone household surveys, where the sample is based on a frame of telephone numbers. Although the desired target population might be all adults living in U.S. households, the attraction of using the telephone frame may persuade the researcher to alter the target population to adults living in telephone households. Alternatively, the researcher can keep the full household target population and document that approximately 2% of U.S. adults are missed because they have no telephones. Using a new target population is subject to the criticism that the population is not the one of interest to the user of the survey statistics. Maintaining the full household target population means that the survey is subject to the criticism that there is coverage error in its statistics. Clearly, these are mostly labeling differences for survey weaknesses that are equivalent; the telephone survey will still be an imperfect tool to study the full adult household population.

A more dramatic example of the options above could affect the NAEP, the survey of students in U.S. schools. Imagine that the target population was all school children in the United States, but the sampling frame consisted only of children in public schools. Because children in private schools on average come from wealthier families, their mathematical and verbal assessment scores often exceed those of public school children. The choice of redefining the target population to fit the frame (and reporting the survey as describing public school children only) would be subject to the criticism that the survey fails to measure the full student population; in essence, the survey is not fully relevant to the population of interest to U.S. policy makers. Using the target population of all students (and reporting the survey as describing all students), but noting that there may be coverage error for private school students, leads to coverage errors. In short, the first option focuses on issues of relevance to different users of the survey; the second option focuses on statistical weaknesses of the survey operations.

3.3 COVERAGE PROPERTIES OF SAMPLING FRAMES

Although target and survey populations can be distinguished, a central statistical concern for the survey researcher is how well the sampling frame (the available materials for sample selection) actually covers the target population. In Figure 2.6 in Chapter 2, the match of sampling frame to target population created three potential outcomes: coverage, undercoverage, and ineligible units.

coverage

undercoverage

ineligible units

When a target population element is in the sampling frame, it is labeled as "covered" by the frame. There can be elements in the target population that do not, or cannot, appear in the sampling frame. This is called "undercoverage," and such eligible members of the population cannot appear in any sample drawn for the survey. A third alternative, "ineligible units," occurs when there are units in the frame that are not in the target population (e.g., business numbers in a frame of telephone numbers when studying the telephone household target population).

A sampling frame is perfect when there is a one-to-one mapping of frame elements to target population elements. In practice, perfect frames do not exist; there are always problems that disrupt the desired one-to-one mapping.

It is common to examine a frame to measure the extent to which each of four problems arises. Two of these have already been discussed briefly above: undercoverage and ineligible or foreign units. The other two concern cases in which a unit is present in the frame and it maps to an element in the target population, but the mapping is not unique, not one to one. "Duplication" is the term used when several frame units are mapped onto the single element in the target population. In sample surveys using the frame, the duplicated elements may be overrepresented. "Clustering" is the term used when multiple elements of the target population are linked to the same single frame element. In sample surveys using the frame, the sample size (in counts of elements) may be smaller or larger depending on the clusters selected. There are also cases in which multiple frame units map to multiple target population elements—many-to-many mappings. We consider this more complicated problem only briefly in this section, viewing the problem as a generalization of a combination of duplication and clustering.

duplication

clustering

3.3.1 Undercoverage

Undercoverage is the weakness of sampling frames prompting the greatest fears of coverage error. It threatens to produce errors of nonobservation in survey statistics from failure to include parts of the target population in any survey using the frame. For example, in telephone household surveys in which the target population is defined as persons in all households, undercoverage occurs because no telephone sampling frame includes persons in households without telephones. This is true for the BRFSS and SOC. There may be an additional coverage issue to discuss in the telephone sampling frame as the change from traditional landline telephones to cellular-telephone-only households continues. In many countries of the world, because telephone subscription requires ongoing costs, poor persons are disproportionately not covered. In countries in which mobile telephones are replacing fixed-line service, younger persons are likely to be uncovered by frames limited to line telephone numbers because they are adopting the new technology more quickly. As we will discuss in Section 3.6, the impact on survey statistics

COVERAGE PROPERTIES OF SAMPLING FRAMES

(whether based on censuses or surveys) of noncoverage depends on how those on the frame differ from those not on the frame.

The causes of coverage problems depend on the processes used to construct the sampling frame. Those processes may be under the control of the survey design, or they may be external to the survey (when a frame is obtained from an outside source). For example, in some household surveys, the survey sample is initially based on a list of areas, such as counties, blocks, enumeration areas, or other geographic units; then on lists of housing units within selected blocks or enumeration areas, and finally, on lists of persons within the households. These samples are called "area frame samples" or "area probability samples." Coverage problems can arise at all three levels.

In area probability designs, each selected area incurs a second frame development, in which survey staffs develop sampling frames of housing units, usually using addresses to identify them. Staffs are sent to sample areas, such as a block or group of blocks, and instructed to list all housing units in them. The task is considerably more difficult than it may appear. Boundaries such as streets, roads, railroad tracks, rivers, or other bodies of water are relatively fixed and readily identified. Whether a particular housing unit is in or out of the area, and should or should not be listed, is relatively easily determined. Boundaries based on "imaginary lines" based on lines of sight between natural features such as the top of a mountain or ridge, are open to interpretation, and more difficult to identify under field conditions. Whether a particular housing unit is in the area or not also becomes a matter of interpretation. Housing units that are widely separated from others may be left out of the listing because of boundary interpretation errors. These will be part of the noncovered population.

Housing unit identification is not a simple task in all cases. A housing unit is typically defined to be a physical structure intended as a dwelling that has its own entrance separate from other units in the structure and an area where meals may be prepared and served. Single family or detached housing units may be readily identified. However, additional units at a given location may not be easily seen when walls or other barriers are present. Gated communities or locked buildings may prevent inspection altogether. It is also possible to miss a unit in rural areas or under crowded urban conditions because it is not visible from public areas. Housing units located along alleyways or narrow lanes with an entrance not clearly visible from a public street can be easily missed during listing. Each missed unit may contribute to undercoverage of the target population.

Housing units in multiunit structures are also difficult to identify. External inspection may not reveal multiple units in a particular structure. The presence of mailboxes, utility meters (water, gas, or electricity), and multiple entrances are used as observational clues about the presence of multiple units. Hidden entrances, particularly those that cannot be seen from a public street, may be missed.

There are also living arrangements that require special rules to determine whether a unit is indeed a housing unit. For example, communal living arrangements are not uncommon in some cultures. A structure may have a single entrance, a single large communal cooking area, and separate sleeping rooms for families related by birth, adoption, or marriage. Procedures must be established for such group quarters, including whether the unit is to be considered a household, and whether the single structure or each sleeping room is to be listed.

Institutions must also be identified, and listing rules established. Some institutions are easily identified, such as prisons or hospitals. Caretaker housing units

area frame

area probability sample

on the institutional property must be identified, however, even if the institution itself is to be excluded. Other institutions may not be as easily identified. For example, prison systems may have transition housing in which prisoners still under the custody of the system live in a housing unit with other inmates. Procedures must be established for whether such units are to be listed as housing units, or group quarters, or excluded because of institutional affiliation. Similarly, hospitals or other health care systems may use detached housing units for care of the disabled or those requiring extended nursing care. To the extent that housing units are left off a list because staff is uncertain about whether to include them, coverage error in survey statistics might arise.

Another common concern about undercoverage in household surveys stems from the fact that sampling frames for households generally provide identification of the housing unit (through an address or telephone number) but not identifiers for persons within the household. (Countries with population registers often use the registry as a sampling frame, skipping the household frame step.) In a census or a survey using a frame of addresses, but with a target population of persons, a small sampling frame of persons in each household must be developed. Interviewers list persons living in the household, but if the listings are not accurate reflections of who lives in the household, coverage problems arise.

The frames of persons in household surveys generally list "residents" of the household. Residency rules must be established so that an interviewer can determine, based on informant reports, whether to include persons in the household listing. Two basic residency rules are used in practice. In the *de facto* rule used in census and some survey operations, persons who slept in the housing unit the previous night are included. This rule is typically reserved for shorter-term data collection activities to avoid overcoverage of individuals who may have frequent residence changes, could appear in more than one housing unit across a short time period, and be overrepresented in the sample. It is easy to apply, because the definition is relatively clear. Undercoverage may arise for individuals traveling and staying in institutions (such as a hotel) the previous evening, even though the person usually sleeps in the household in the evening.

de facto residence rule

A more common residency rule in surveys is the *de jure* rule, based on "usual residence," who usually lives in the housing unit. This rule can be straightforward to apply for many individuals, but there are also many circumstances in which the application of the rule is difficult. Usual residency for individuals whose employment requires travel, such as sales representatives, truck drivers, or airline pilots, may be unclear. If the informant says that the housing unit is their usual residence when not traveling, the rule uses the residence for the majority of some time period (such as the previous year or month). If the individual intends to use the housing unit as their residence (for those who have just moved into the unit) the de jure rule will include them as usual residents.

de jure residence rule

usual residence

Most U.S. household censuses and surveys use such procedures, and their coverage properties are well documented. Younger males (18–29 years old), especially those in minority racial groups, appear to have looser ties with households. They may live with their parents some days of the week and with friends on other days. Similarly, young children in poorer households, especially those without two parents, may live with their mother sometimes, their grandparents sometimes, and their father or other relative other times. In such housing units, when the interviewer asks the question, "Who lives here?" it appears that such persons are disproportionately omitted (see Robinson, Ahmed, das Gupta, and Woodrow, 1993) and are a source of undercoverage.

Sometimes, the set of persons residing in a housing unit is not approved by legal authority. For example, a rental agreement for an apartment may specify that only one family of at most five persons can occupy the unit. However, poor persons may share rental expenses among several families in violation of the agreement. They may be reluctant to report the additional family as residents of the unit. If social welfare rules limit eligibility to married couples, an unmarried woman may fail to report a male resident in the unit. This leads to systematic omissions of certain types of persons (de la Puente, 1993).

In some cultures, certain individuals are not considered to be part of the household, even though they fit the usual resident requirements of the de jure rule. Infants, for example, may not be considered residents, and left off the list. Indeed, the fit between the traditionally defined "household" and the population is a ripe area for research in survey methodology. So central has been the use of the household as a convenient sampling unit that most survey research employs it for person-level surveys. When people are only ambiguously related to housing units, however, the practice needs scrutiny.

In an establishment survey, the creation, merger, and death of establishments are important factors in undercoverage. The definition of an establishment, particularly with very large and very small firms, is difficult to apply in practice. Firms with many locations, such as franchised establishments, may have to be separated into multiple establishments based on geographic location. Firms with several offices or factories, warehouses, or shipping locations may also have to be listed separately. The distinction between a survey-defined establishment and a business unit in a firm may be difficult to determine.

Establishments may be in existence for very short periods of time, or may be so small that they are not included in available frames. For example, the CES misses newly established employers for a period of months. Establishment frames may be in part based on administrative registries, but these can be out of date or incomplete, particularly for newly created establishments. Mergers or subdivision of

---

Mulry (2007) on U.S. Decennial Census Coverage

Mulry summarizes a large set of evaluations of the coverage properties of the 2000 U.S. census.

Study design: A large sample survey with enhanced quality controls is matched to decennial census records to measure whether persons and households were included in the census count. In addition, "demographic analysis" uses administrative data on population change: births, deaths, immigration, and emigration. Finally, computerized matching of census records searches for duplicates.

Findings: In contrast to the common finding of undercounts in prior U.S. censuses, the study summarizes repeated evidence of net overcount, of 0.5% of the population. Many of the duplicates arose from the persons being reported on a mail-returned census questionnaire and one completed by an enumerator, due to residential moves, vacation homes, college student mobility, and shared custody of children. Those undercounted tended to be renters, non-Hispanic African-American, and male (aged 18–49).

Limitations of the study: The estimates of the net overcount focus on persons living in housing units, not those in group quarters. Demographic analysis methods showed a small undercount, in contrast to the survey approach, but differences were not large. There appeared to be reporting errors to the usual residence of the person, complicating the analysis.

Impact of the study: This study demonstrated how new technology to measure duplicate counting improved the understanding of how diverse household situations affect the ability to count persons and correctly place them in a single residential location.

firms complicate administrative record keeping, and may lead to overcoverage as well as undercoverage. Keeping establishment frames up to date is a sizable and ongoing task.

Undercoverage is a difficult problem to identify and to solve. If population elements do not appear in the frame, additional frames might be used to try to identify them (see multiple frame surveys in Section 3.7.3). In telephone household surveys, nontelephone households may be covered through the use of an area sampling frame that in principle covers all households, regardless of telephone subscription. There are techniques for expanding the coverage of the frame through respondent reports about other population elements (see multiplicity techniques in Section 3.7.2), but in U.S. household surveys, proxy reports about other households or persons are increasingly restricted due to human subjects' concerns about privacy.

3.3.2 Ineligible Units

Sometimes, sampling frames contain elements that are not part of the target population. For example, in telephone number frames, many of the numbers are nonworking or nonresidential numbers, complicating the use of the frame for the target population of households. In area probability surveys, sometimes the map materials contain units outside the target geographical area. When the survey staff visits sample areas to list housing units, they sometimes include unoccupied or business structures that appear to be housing units.

When interviewers develop frames of household members within a unit, they often use residence definitions that do not match the meaning of "household" held by the informant. Parents of students living away from home often think of them as members of the household, yet many survey protocols would place them at college. The informants might tend to exclude persons unrelated to them who rent a room in the housing unit. Studies show that children in shared custody between their father and mother are likely to be duplicated by appearing in each parent's household listing.

ineligible unit
foreign unit

Although undercoverage is a difficult problem, "ineligible" or "foreign" units in the frame can be a less difficult problem to deal with if the problem is not extensive. When foreign units are identified on the frame before selection begins, they can be purged at little cost. More often, foreign or ineligible units cannot be identified until data collection begins. If few in number, after sampling they can be identified in a screening step and dropped from the sample, with a reduction in sample size. If the prevalence of foreign units is known, even approximately, in advance, additional units can be selected, anticipating that some will be screened out. For example, it is known that approximately 15% of entries in residential portions of national telephone directories are numbers that are no longer in service. To achieve a sample of 100 telephone households, one could select a sample of $100/(1 - 0.15) = 118$ entries from the directory, expecting that 18 are going to be out-of-service numbers.

random digit dialing (RDD)

When the proportion of foreign entries is very large, the sampling frame may not be cost-effective to use. In telephone household surveys in the United States, for example, one frame contains all the known area code–prefix combinations (the first six digits of a U.S. 10-digit phone number). Surveys based on the frame are often called "random digit dialed surveys." Of all the possible 10-digit phone

COVERAGE PROPERTIES OF SAMPLING FRAMES

numbers in the frame, more than 85% of the numbers are not in service (foreign units). It is time-consuming to screen numbers with that many foreign units. Other sampling frames and sampling techniques have been developed that are more cost-effective for selecting telephone households (some of these are described in Section 4.8).

3.3.3 Clustering of Target Population Elements Within Frame Elements

As mentioned previously, multiple mappings of frame to population (clustering) or population to frame (duplication) are problems in sample selection. A sample of adults living in telephone households (the target population) using a telephone directory (the sampling frame) illustrates each of these problems.   **multiple mappings**

A telephone directory lists telephone households in order by surname, given name, and address. When sampling adults from this frame, an immediately obvious problem is the clustering of eligible persons that occurs. "Clustering" means that multiple elements of the target population are represented by the same frame element. A telephone listing in the directory may have a single or two or more adults living there.   **clustering**

Figure 3.1 illustrates clustering. The left side of the figure shows seven different target population elements, persons who live in telephone households. The Smith family (Ronald, Alicia, Thomas, and Joyce) lives in the same household, which has the telephone number 734-555-1000, the sampling frame element. All the Smith's are associated with only one frame element, even though together they may form four elements of the target population.

One way to react to clustering of target population elements is by simply selecting all eligible units in the selected telephone households (or all eligible units in a cluster). With this design, the probability of selection of the cluster applies to all elements in the cluster.

Clustering poses important issues that often lead to subsampling the cluster. First, in some instances it may be difficult to collect information successfully

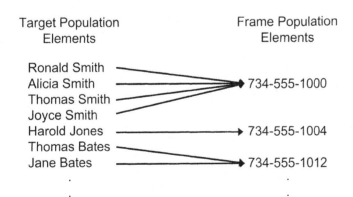

Figure 3.1 Cluster of target population elements associated with one sampling frame element.

from all elements in the cluster. In telephone surveys, nonresponse sometimes increases when more than one interview is attempted in a household by telephone. Second, when interviews must be conducted at more than one point in time, initial respondents may discuss the survey with later respondents, affecting their answers. In opinion surveys, an answer to a question may be different if the respondent is hearing the question for the first time, or has already heard the question from another person who has already completed the interview. Even answers to factual questions may be changed if respondents have talked among themselves. Third, if the clusters vary in size (as in the case of clusters of adults at the same telephone number), control of sample size may become difficult. The sample size of elements is the sum of the cluster sizes, and that is not under the direct control of the survey operation unless cluster size is known in advance.

To avoid or reduce these problems, a sample of elements may be selected from each frame unit sampled (the cluster of target population members). In the case of telephone household surveys of adults, one adult may be chosen at random from each sampled household. All efforts to obtain an interview are concentrated on one eligible person, contamination is eliminated within the household, and the sample size in persons is equal to the number of households sampled.

In the case of telephone and other household surveys in which a single eligible element is chosen, the development of the within-household frame of eligible persons, the selection of one of them, and the interview request is done in real time, at the time of data collection. Since the primary responsibility of the interviewer is data collection, and since interviewers are seldom trained in sampling statistics, simple procedures have been designed to allow selection during data collection of a single element rapidly, objectively, and with little or no departures from randomization. These are discussed in Section 4.9.

After sample selection, there is one other issue that needs to be addressed in this form of cluster sampling: unequal probabilities of selection. If all frame elements are given equal chances, but one eligible selection is made from each, then elements in large clusters have lower overall probabilities of selection than elements in small clusters. For example, an eligible person chosen in a telephone household containing two eligibles has a chance of one-half of being selected, given that the household was sampled, whereas those in a household with four eligibles have a one-in-four chance.

The consequence of this kind of sampling is that the sample ends up overrepresenting persons from households with fewer eligibles, at least relative to the target population. In other words, more eligibles are in the sample from smaller households than one would find in the target population. If, for some variables collected in the survey instrument, there is a relationship between cluster size and the variable, the sample results will not be unbiased estimates of corresponding target population results. For example, persons living in households with more persons tend to be victims of crime more often than persons living in smaller households.

Some compensation must be made during analysis of the survey data in order to eliminate this potential source of bias. Selection weights equal to the number of eligibles in the cluster can be used in survey estimation. Weighting and weighted estimates are described in detail in Chapter 10.

COVERAGE PROPERTIES OF SAMPLING FRAMES   79

3.3.4 Duplication of Target Population Elements in Sampling Frames

The other kind of multiple mapping between frame and target populations that arises is duplication. "Duplication" means that a single target population element is associated with multiple frame elements. In the telephone survey example, this may arise when a single telephone household has more than one listing in a telephone directory. In Figure 3.2, Tom Clarke, a target population member, has two frame elements associated with him: the telephone numbers 314-555-9123 and 314-555-9124. Multiple listings of the target population of households may occur because the household has more than one telephone number assigned to it, or because individuals within the household request and pay for additional listings in the directory. For example, in towns with universities and colleges, unrelated students often rent housing together, and acquire a single telephone number for the household. One listing in the directory is provided with the telephone subscription, and one person is given as the listed resident. Other residents may add listings for the same phone number under different names. This gives multiple frame listings for one household.

    The problem that arises with this kind of frame problem is similar to that encountered with clustering. Target population elements with multiple frame units have higher chances of selection and will be overrepresented in the sample, relative to the population. If there is a correlation between duplication and variables of interest, survey estimates will be biased. In survey estimation, the problem is that both the presence of duplication and the correlation between duplication and survey variables are often unknown.

    The potential for bias from duplication can be addressed in several ways. The sampling frame can be purged of duplicates prior to sample selection. For example, an electronic version of a telephone directory can be sorted by telephone number, and duplicate entries for the same number eliminated. When the frame cannot be easily manipulated, though, purging of duplicates may not be cost-effective.

duplication

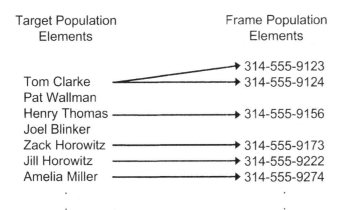

Figure 3.2 Duplication of target population elements by more than one sampling frame element.

Duplicate frame units may also be detected at the time of selection, or during data collection. A simple rule may suffice to eliminate the problem, designating only one of the frame entries for sample selection. Any other duplicate entries would be treated as foreign units and ignored in selection. For example, in one selection technique, only the first entry in the directory is eligible. At the time of contact with the telephone household, the household informant can be asked if there is more than one entry in the directory for the household. If so, the entry with the surname that would appear first is identified. If the selection is for another entry, the interview would be terminated because the selection was by definition a foreign unit.

Another solution, as in the case of clustering, is weighting. If the number of duplicate entries for a given population element is determined, the compensatory weight is equal to the inverse of the number of frame elements associated with the sampled target element. For example, if a telephone household has two phone lines and three total entries in the directory (identified during data collection by informant report), the household receives a weight of one-third in a sample using the directory frame and a weight of one-half in a sample using an RDD frame.

3.3.5 Complicated Mappings between Frame and Target Population Elements

It is also possible to have multiple frame units mapped to multiple population elements. For example, in telephone household surveys of adults, one may encounter a household with several adults who have multiple entries in the directory. This many-to-many mapping problem is a combination of clustering and duplication. For example, in Figure 3.3 the three member Schmidt household (Leonard, Alice, and Virginia) has two telephone number frame elements (403-555-5912 and 403-555-5919). They might represent three target population elements mapped onto two sampling frame elements. A common solution to this problem is to weight survey results to handle both problems simultaneously. The compensatory weight for person-level statistics is the number of adults (or eligibles) divided by the

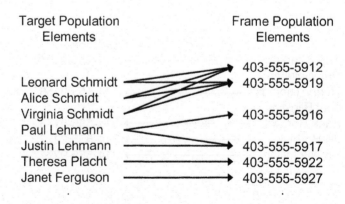

Figure 3.3 Clustering and duplication of target population elements relative to sampling frame elements.

FRAMES FOR HOUSEHOLDS OR PERSONS

number of frame entries for the household. In the example, the weight for the Schmidt household member selected would be three-halves. More complicated weighting methods may be required for more complicated many-to-many mappings occurring in other types of surveys.

3.4 ALTERNATIVE FRAMES FOR SURVEYS OF THE TARGET POPULATION OF HOUSEHOLDS OR PERSONS

Given the description of different frame problems above, we are now ready to describe some common target populations and frame issues they present. We begin with the target population of households and persons because it is such a common focus of the social sciences.

3.4.1 Area Frames

In the United States, the common sampling frames for households are area frames (lists of area units like census tracts or counties). The area frame, because it is based on geographical units, requires an association of persons to areas, accomplished through a residency linking rule (de facto or de jure). Such a frame requires multiple stages when used to sample persons. First, a subset of area units is selected; then listings of addresses are made. If good maps or aerial photographs are available, the frame offers theoretically complete coverage of residences. The frame suffers undercoverage if the listing of residences within selected area units misses some units. The frame suffers duplication when one person has more than one residence. The frame suffers clustering when it is used to sample persons because multiple persons live in the final frame. These are the frame issues for the NCVS and NSDUH.

Increasingly, lists of addresses are being used to replace or complement the listing of addresses in the final chosen areas (often similar to city blocks). Many of these are based on the U.S. Postal Delivery Route list, which contains listings of units to which mail is delivered. These lists have the advantage of high coverage of addresses, but the disadvantage that units receiving mail in other ways are not covered. The lists are also sorted by the order of mail delivery, which often poses challenges to link an address on the list to a given sample area. Some commercial firms match the delivery route list to enhance the variables on the list, and this can produce enhanced stratification possibilities for selection of housing units within the areas. Active evaluation of these lists is ongoing (see, for example, Innachione, Staab, and Redden, 2003).

3.4.2 Telephone Number Frames for Households and Persons

Another household population frame is that associated with telephone numbers for landline (or "fixed line") telephones in housing units. This fails to cover about 20% of U.S. households. A minority of households has more than one landline number and are "overcovered." There are many nonresidential numbers on the frame that have to be screened out when it is used for person-level samples. These are the frame issues for BRFSS and SOC, both of which are random digit dialed surveys, using the landline telephone number frame.

The frame of listed residential telephone numbers in the United States is smaller than the landline telephone number frame. Commercial firms derive the frame from electronic and printed telephone directories. They sell samples from the frame to mass mailers and survey researchers. For uses in household surveys, it is much more efficient because most of the nonworking and nonresidential numbers are absent. However, a large portion of residential numbers, disproportionately urban residents and transient persons, is not listed in the directory. There are duplication problems because the same number can be listed under two different names, typically names of different members of the same household.

Mobile or cell phones are rapidly replacing landline service in many countries. As early as the mid-1990s in Finland, for example, fixed-line telephone subscriptions began to decline while mobile phone subscribers rapidly increased (Kuusela, Callegaro, and Vehovar, 2008). This shift represented a loss of fixed-line telephone coverage because cell phone numbers were not included in the existing frames. The coverage loss was greatest among younger persons and those just forming households independent of their parents.

Figure 3.4 shows how rapidly the U.S. adult population is switching from line telephones to wireless or mobile phone subscription. In just four years (2004–2007) the percentage of adults having mobile phone subscriptions but not line phone subscriptions has grown from about 4% to 16.1% in 2008. Blumberg and Luke (2008) show that those with only wireless phones tend to live in households with unrelated persons, to rent their homes, to receive lower incomes, and to be younger.

In addition, mobile phones differ from line phones in that they are often associated with one person, not an entire household (as with line phones). In that sense, any sampling frame built on cell phone numbers is closer to a frame of persons than a frame of households (clusters of persons). (We note that early studies of wireless

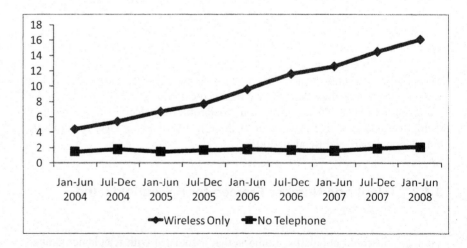

Figure 3.4 Percentage of U.S. adults with wireless telephone service only and percentage without telephones, January, 2004–June, 2008. (Source: Blumberg and Luke, 2008.)

phones show that there is some sharing of telephone numbers across persons in the same household, complicating this generalization somewhat.)

Another feature of frames of wireless telephone numbers in the U.S. is that the geographical stratification variables are weak. That is, it is easy for cell phone subscribers to acquire a number that is usually assigned to a different location than their residence. This means that using such a frame for geographically restricted samples leads to undercoverage of such persons.

Telephone surveys that sample mobile phone numbers require movement away from the household as a frame and sampling unit. At the present time, though, there are a number of frame problems associated with a mix of clustering and duplication occurring in fixed-line and cell service telephone numbers that are unsolved. In the United States, the majority of cell phone numbers are isolated in designated number series. Given this, it is common to consider an approach in which separate samples are taken from the cell phone number series and other number series (those covering line phones). Using the frame this way, there are some persons who have more than one element in the frame (i.e., their cell phone number and their landline phone number). Some designs attempt to account for such duplication through weighting; others attempt to associate the person with one and only one number and screen out the other, if sampled. There is no consensus at this writing about the relative benefits of the two approaches.

There are many other issues involving cell phone surveys, reviewed at www.aapor.org/uploads/Final_AAPOR_Cell_Phone_TF_report_041208.pdf. These include issues of nonresponse (mentioned in Chapter 6) and measurement (discussed in Chapter 5)

3.4.3 Frames for Web Surveys of General Populations

With the growing interest in Web surveys, there is much attention paid to the possibility of developing a frame of e-mail addresses for the household population. The e-mail frame, however, fails to cover large portions of the household population. It has duplication problems because one person can have many different e-mail addresses, and it has clustering problems because more than one person can share an e-mail address (e.g., smithfamily@aol.com). The nonstandardized format of e-mail addresses also precludes the generation of RDD-like methods for Web surveys. E-mail addresses are not the only form of sampling frame for Web surveys; address frames can be used (see Section 3.4.1), with sample persons invited by mail to visit the URL to complete the Web survey.

opt-in internet panel

In the absence of a universal frame of e-mail addresses, opt-in Internet panels (or access panels) often attempt to assemble large sets of e-mail addresses and later send Web survey requests to samples of them. They often claim to have an inferential population of the full adult population of a country. It is important to distinguish these approaches from the attempt to build a complete frame of e-mail addresses connected to persons.

First, a common criterion of the Internet panel firm is not full coverage but the collection of a set of persons with sufficient diversity on attributes related to the type of surveys it conducts. For an Internet panel firm concerned with household products, this might include diversity on income, household size, ethnic subcultural identity, age, and various other lifestyle attributes. For an Internet panel focused on public opinion, the diversity of interest might be income, age, gender, ethnicity, political identification, and social engagement. E-mail addresses can be acquired from organizations that assemble such lists as part of their business (e.g., websites of retailers and membership organizations). E-mail addresses so identified are solicited for their willingness to do surveys. When the firm is satisfied with the diversity and size of list it has achieved, then it stops assembling cases, but may need to replenish the list because of attrition within specific subgroups.

Second, often the entire notion of a sample frame is skipped in an access or Internet panel. Instead, the method jumps immediately to the sampling and recruitment steps. Persons with Internet access are directly solicited through banner ads (e.g., "Make money doing surveys!") or through pop-up solicitations during a visit to a website. The task of the Internet panel is to gain consent from as diverse and large a group as possible.

Third, Internet panels often repeatedly sample from the assembled willing survey participants. They often will continue to send survey solicitations to a member as long as the member responds. There is little attention to systematically reflecting dynamic change in the full population of person-level e-mail addresses. Market research organizations are developing guidelines for purging inactive members from opt-in panels and identifying inattentive or fraudulent respondents (see www.esomar.org/index-php/26-questions.html).

In short, without a universal frame of e-mail addresses with known links to individual population elements, some survey practices have begun to ignore the frame development step. Without a well-defined sampling frame, the coverage error of resulting estimates is completely unknowable.

3.5 FRAME ISSUES FOR OTHER COMMON TARGET POPULATIONS

This section gives a brief overview of frame issues for surveys of customers, employees, organizations, events, and the handling of rare populations.

3.5.1 Customers, Employees, or Members of an Organization

Most surveys that study populations of customers, employers, or members of organizations use a list frame. Sometimes, the list frame is an electronic file of person records; other times it can be a physical set of records. Such record systems have predictable coverage issues. Undercoverage issues stem from the out-of-date files. New employees or customers tend to be missed if the files require several administrative steps before they are updated.

ineligible elements   Similarly, the lists can contain ineligible elements, especially if persons leaving the organization are not purged from the list quickly. For example, in a file of customers, some of the customers may have experienced their last transaction so long ago that in their minds they may not perceive themselves as customers. In addition, there may be some ambiguity about whether a person should be counted as a member of the organization. In surveys of employees of a business, how should "contract" employees be treated? Although they may work day to day at the business, they are employees of another company that has a contract to supply services to the business.

Duplication of elements within a frame of employees or members can occur, but is rarer than with the household frames. In a frame of customers, duplication is inherent if the record is at the level of a transaction between the customer and the organization. Every person who is a customer has as many records as transactions they had with the organization. Hence, the survey researcher needs to think carefully about whether the target population is one of people (i.e., the customers) or transactions or both.

Increasingly, sampling frames of work and membership organizations take advantage of the use of the Web to communicate within the organization. When all employees or members have one and only one e-mail address on the network, such a list can be a near-perfect frame. There *are* sometimes problems of addresses that are groups of individuals, addresses that describe a role (e.g., personnelmanager@ABC.COM) when the person holding the role also has a personal e-mail address. However, samples from these frames avoid all the problems of the absence of web survey frames mentioned in the last section.

Survey researchers attempt to learn how and why the list was developed as part of the evaluation of alternative frames. For example, payroll lists of employees of a hospital may fail to cover volunteer and contract staff, but records from a security system producing identification cards may do so. Learning how and why the frame is updated and corrected is important. For example, payroll records for employees on monthly pay periods may be updated less frequently than those for daily or weekly pay periods. Updating procedures for temporary absences may complicate coverage issues of the frame. If an employee is on extended medical leave, is his/her record still in the frame? Should the target population include such persons? All of these issues require special examination for each survey conducted.

3.5.2 Organizations

Organizational populations are diverse. They include churches, businesses, farms, hospitals, medical clinics, prisons, schools, charities, governmental units, and civic groups. Sampling frames for these populations are often lists of units. Of all the types of organizational populations, perhaps businesses are the most frequent target population for surveys.

Business populations have distinct frame problems. First, a very prominent feature of business populations is their quite large variation in size. If the target population of software vendors is chosen, both Microsoft (with revenues over $20 billion per year) and a corner retail outlet that may sell $5,000 in software per year should be in the frame. Many business surveys measure variables for which size is related to the variables of interest (e.g., CES estimates of total employment in the industry). Hence, coverage issues of business frames often place more emphasis on including the largest businesses than the smallest businesses.

Second, the business population is highly dynamic. Small businesses are born and die very rapidly. Larger businesses purchase others, merging two units into one (e.g., Hewlett-Packard buys Compaq and becomes a single corporation). A single business splits into multiple businesses (e.g., Ford Motor Company splits off its parts division, creating Visteon, an independent company). This means that frame populations need to be constantly updated to maintain good coverage of new businesses and to purge former businesses from the lists.

Third, the business population demonstrates a distinction between a legally defined entity and physical locations. Multiunit, multilocation companies are common (e.g., McDonald's has over 30,000 locations in the world, but only one corporate headquarters). Hence, surveys of businesses can study "enterprises," the legally defined entities, or "establishments," the physical locations. Some legally defined businesses may even not have a physical location (e.g., a consulting business with each employee working on his/her own). Some locations are the site of many businesses owned by the same person.

Besides the business population, other organizational populations exhibit similar features, to larger or smaller extents. These characteristics demand that the survey researcher think carefully through the issues of variation in size of organization, the dynamic nature of the population, and legal and physical definitions of elements.

3.5.3 Events

Sometimes, a survey targets a population of events. There are many types of events sampled in surveys: purchases of a service or product, marriages, pregnancies, births, periods of unemployment, episodes of depression, automobiles passing over a road segment, or criminal victimizations (like those measured in the NCVS).

Often, surveys of events begin with a frame of persons. Each person has either experienced the event or has not. Some have experienced multiple events (e.g., made many purchases) and are in essence clusters of event "elements." This is how the NCVS studies victimizations as events. It first assembles a frame of persons, each of whom is potentially a cluster of victimization events. NCVS measures each event occurring during the prior six months and then produces statistics on victimization characteristics.

Another logical frame population for event sampling is the frame of time units. For example, imagine wanting to sample the visits to a zoo over a one-year period. The purpose of the survey might be to ask about the purpose of the visit, how long it lasted, what were the most enjoyable parts of the visit, and what were the least enjoyable parts. One way to develop a frame of visits is to first conceptually assign each visit to a time point, say, the time of the exit from the zoo. With this frame, all visits are assigned to one and only one point in time. If the study involves a sample, then the research can select a subset of time points (say, 5-minute blocks) and attempt to question people about their visit as they leave during those 5-minute sample blocks.

Some time-use surveys (which attempt to learn what population members are doing over time) use electronic beepers that emit a tone at randomly chosen moments. When the tone occurs, the protocol specifies that the respondent report what they were doing at the moment (e.g., working at the office, watching television, shopping) (see Csikszentmihalyi and Csikszentmihalyi, 1988; Larson and Richards, 1994).

Surveys that study events may involve multiple populations simultaneously. They are interested in statistics about the event population, but also statistics about the persons experiencing the event. Whenever these dual purposes are involved, various clustering and duplication issues come to the fore. In a study of car purchases, for the event element of a purchase by a family, which persons experienced the event—the legal owner(s), all family members, or just the poten-

COVERAGE ERROR

tial drivers of the car? The NCVS produces statistics like the percentage of households experiencing a crime (based on the household units) and the percentage of household break-ins occurring while the residents were at home (based on the incident population). Careful thinking about key survey estimates is important in choosing the target and frame populations of events.

3.5.4 Rare Populations

"Rare populations" is a term that is used to describe small target groups of interest to researchers. Sometimes, what makes a population rare is not its absolute size but its size relative to available frames that cover it. For example, consider the population of welfare recipients in the United States. If there were 7.5 million persons receiving welfare benefits in a population of 305 million, the population would be rare principally because it forms less than 3% of the total population. When chosen as target populations, rare populations pose considerable problems for identifying suitable sampling frames.

rare population

There are two basic approaches to building sampling frames for rare populations. First, lists of rare population elements themselves can be made. For example, one can attempt to acquire lists of welfare recipients directly (although these might be kept confidential) through records in welfare disbursement offices. Sometimes, no single list has good coverage and multiple lists are assembled (see the discussion of multiple frame designs in Section 3.7.3). Second, and more commonly, a frame that includes the rare population as a subset of elements can be screened. For example, the household population can be screened to locate families that receive welfare payments. If all elements of the rare population are members of the larger frame population, complete coverage of the rare population is possible (albeit at the expense of screening to locate the rare population among sample elements).

3.6 COVERAGE ERROR

There are remedies for many of the sampling frame problems discussed in Section 3.3, but the remedies do not always eliminate coverage error. Undercoverage is a difficult problem, and may be an important source of coverage error in surveys. It is important to note, though, that coverage error is a property of sample statistics and estimates made from surveys. One statistic in a survey may be subject to large coverage errors; another from the same survey can be unaffected by the same coverage issues. In the jargon of survey methodology, undercoverage, duplication, clustering, and other issues are problems of a sampling frame. Coverage error is the effect of those problems on a survey statistic.

The nature of coverage error in a simple statistic like the sample mean was presented in Section 2.3.4. Recall that if a mean is being estimated, the coverage bias was given as

$$\bar{Y}_C - \bar{Y} = \frac{U}{N}(\bar{Y}_C - \bar{Y}_U)$$

where $\bar{Y}$ denotes the mean for the total population; $\bar{Y}_C$ and $\bar{Y}_U$ are the means in the population of the eligible units on the frame (covered) and not in the frame (not covered), respectively; U is the total number of target population elements off the frame; and N is the full target population. Thus, the error due to not covering the $N - C$ units left out of the frame is a function of the proportion "not covered" and the difference between means for the covered and the not covered.

The survey (regardless of its sample size) can only estimate the mean of the covered, $\bar{Y}_C$. The extent to which the population of U noncovered units is large, or there is a substantial difference between covered and undercovered, determines the size of the bias, or coverage error. The proportion not covered will vary across subclasses of the eligible persons. That is, undercoverage could be higher for the total sample than for a particular subgroup. In addition, since coverage error depends on the difference of estimates between covered and undercovered, coverage error can vary from one statistic to another, even if they are based on the same subclass of eligible units.

3.7 REDUCING UNDERCOVERAGE

Remedies for common frame problems such as duplication, clustering, many-to-many mappings, undercoverage, and foreign units in frames have been examined in Section 3.3. However, specific remedies for undercoverage, and consequent coverage error, were not addressed in detail in that section. There is a general class of coverage improvement procedures that involve frame supplementation designed to reduce coverage error more specifically.

3.7.1 The Half-Open Interval

Frames that are slightly out of date, or that provide reasonably good coverage except for some kinds of units, may be brought up to date through additions in update listing operations during or shortly before data collection. If there is a logical order to the list, it may be possible to repair the frame by finding missing units between two listed units.

Consider, for example, address or housing unit lists used in household surveys (see Figure 3.5). These lists may become out of date and miss units quickly. They may also have missed housing units that upon closer inspection could be added to the list. Since address lists are typically in a particular geographic order, it is possible to add units to the frame only for selected frame elements, rather than updating the entire list. That is, frames can be updated after selection and during data collection.

half-open interval

One such tool is called the "half-open interval." Consider the example of one block with the address list shown in Figure 3.5. The geographic distribution of addresses is available for the block in a sketch map shown in Figure 3.6. Suppose that an address from this frame has been selected: 107 Elm Street. From an area frame perspective, the address of 107 Elm Street will be viewed not as a physical structure but a geographic area bounded by "property lines" from 107 Elm Street up to but not including the next listed address, 111 Elm Street. List order defines what mathematicians would define from set theory as a half-open interval for each address appearing on the list. The interval begins with 107 Elm Street (the closed end of the interval) and extends up to but does not include 111 Elm Street (the open end of the interval).

REDUCING UNDERCOVERAGE

No.	Address	Selection?
1	101 Elm Street	
2	103 Elm Street, Apt. 1	
3	103 Elm Street, Apt. 2	
4	107 Elm Street	Yes
5	111 Elm Street	
6	302 Oak Street	
7	308 Oak Street	
---	---	---

Figure 3.5 Address list for area household survey block.

When an interviewer arrives at the selected address, he inspects the "half-open interval" associated with 107 Elm Street to determine if there are any newly constructed units or any missed units in the interval. If a new or missed unit is discovered, the interviewer adds it to the list, selects it as a sample unit, and attempts to conduct an interview at all addresses in the interval. All addresses in the half-open interval thus have the same probability of selection (that of the selected address), and missed or newly constructed units are automatically added to the frame during data collection.

On occasion, the number of new addresses found in a half-open interval is too large for the interviewer to be able to sustain the added workload. For instance, if a new apartment building with 12 apartments had been constructed between 107 and 111 Elm Street, the interviewer would be faced with conducting

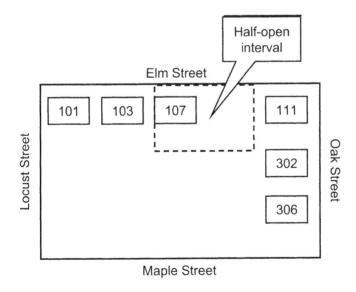

Figure 3.6 Sketch map for area household survey block.

13 interviews instead of an expected single household interview. In such instances, the additional addresses may be subsampled to reduce the effect of this clustering on sample size and interviewer workload. Subsampling, as for clustering in sample frames, introduces unequal probabilities of selection that must be compensated for by weighting (see Chapter 10 for discussion of weighting and weighted estimates). In continuing survey operations, it is also possible to set aside such added units into a separate "surprise" stratum (see Kish and Hess, 1959) from which a sample of missed or newly constructed units are drawn with varying probabilities across additional samples selected from the frame.

Similar kinds of linking rules can be created for coverage checks of other logically ordered lists that serve as frames. For example, a list of children attending public schools ordered by address and age within address could be updated when the household of a selected child is visited and additional missed or recently born children are discovered in the household.

3.7.2 Multiplicity Sampling

The half-open interval concept supplements an existing frame through information collected during the selection process. Some frame supplementation methods add elements to a population through network sampling. This is commonly termed "multiplicity sampling." A sample of units can be selected, and then all members of a well-defined network of units identified for the selected units.

multiplicity sampling

For instance, a sample of adults may be selected through a household survey, and asked about all of their living adult siblings. The list of living adult siblings defines a network of units for which information may be collected. Of course, the network members have multiple chances of being selected, through a duplication in the frame, since they each could be selected into the sample as a sample adult. The size of the network determines the number of "duplicate chances" of selection. If an adult sample person reports two living adult siblings, the network is of size three, and a weight of one-third can be applied that decreases the relative contribution of the data from the network to the overall estimates. This "multiplicity" sampling and weighting method (Sirken, 1970) has been used to collect data about networks to increase sample sizes for screening for rare conditions, such as a disease. The method does have to be balanced against privacy concerns of individuals in the network. In addition, response error (see Chapter 7), such as failing to report a sibling, including someone who is not a sibling, or incorrectly reporting a characteristic for a member of the network, may contribute to errors in the network definition and coverage as well as in the reported levels of a characteristic.

"Snowball sampling" describes a closely related, although generally nonprobability, method to supplement a frame. Suppose an individual has been found in survey data collection who has a rare condition, say blindness, and the condition is such that persons who have the condition will know others who also have the condition. The sample person is asked to identify others with the condition, and they are added to the sample. Snowball sampling cumulates sample persons by using network information reported by sample persons. Errors in reports, isolated individuals who are not connected to any network, and poorly defined networks make snowball sampling difficult to apply in practice. It generally does not yield a probability sample.

REDUCING UNDERCOVERAGE

Although multiplicity sampling offers theoretical attraction to solve frame problems, there is much research left to be conducted on how to implement practical designs. These include problems of measurement error in reports about networks, nonresponse error arising from incomplete measurement of networks, and variance inflation of multiplicity estimators.

3.7.3 Multiple Frame Designs

Coverage error can sometimes be reduced by the use of multiple frames, in several ways. A principal frame that provides nearly complete coverage of the target population may be supplemented by a frame that provides better or unique coverage for population elements absent or poorly covered in the principal frame. For example, an out-of-date set of listings of housing units can be supplemented by a frame of newly constructed housing units obtained from planning departments in governmental units responsible for zoning where sample addresses are located. Another example concerns mobile homes that may be present on an address list but poorly covered. A supplemental frame of mobile home parks may be added to the principal address list to provide better coverage of the population residing in mobile homes.

multiple frame sampling

At times, the supplemental frame may cover a completely separate portion of the population. In most cases, though, supplemental frames overlap with the principal frame. In such cases, multiple frame sampling and estimation procedures are employed to correct for unequal probabilities of selection and possibly to yield improved precision for survey estimates.

Suppose, for example, that in a household survey random digit dialing (RDD) from the full telephone frame is used to reach U.S. telephone households. RDD will, in principle, cover all telephone households in the country, but it fails to cover approximately 2% of households that do not have telephones. Figure 3.7

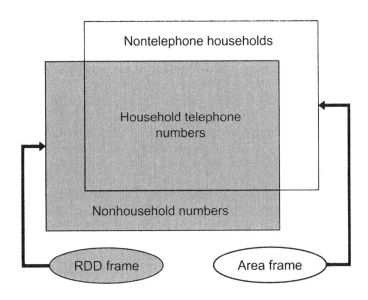

Figure 3.7 Dual frame sample design.

shows the telephone frame as a shaded subset of the area frame of housing units. A remedy for undercoverage is a supplementary area frame of households. Under such a dual frame design, a sample of households is drawn from an area frame, but it will require visits to households, which is considerably more expensive than contact by telephone. Together, the two frames provide complete coverage of households. For the NCVS, Lepkowski and Groves (1986) studied the costs and error differences of an RDD and an area frame. They found that for a fixed budget, most statistics achieve a lower mean square error when the majority of the sample cases are drawn from the telephone frame (based on a simulation including estimates of coverage, nonresponse, and some measurement error differences).

These two frames overlap, each containing telephone households. The dataset from this dual frame survey combines data from both frames. Clearly, telephone households are overrepresented under such a design since they can be selected from both frames.

There are several solutions to the overlap and overrepresentation problem. One is to screen the area household frame. At the doorstep, before an interview is conducted, the interviewer determines whether the household has a fixed-line telephone that would allow it to be reached by telephone. If so, the unit is not selected and no interview is attempted. With this procedure, the overlap is eliminated, and the dual frame sample design has complete coverage of households.

A second solution is to attempt interviews at all sample households in both frames, but to determine the chance of selection for each household. Households from the nonoverlap portion of the sample, the nontelephone households, can only be selected from the area frame and, thus, have one chance of selection. Telephone households have two chances, one from the telephone frame and the other from the area household frame. Thus, their chance of selection is $p_{RDD} + p_{area} - p_{RDD} \times p_{area}$, where p_{RDD} and p_{area} denote the chances of selection for the RDD and area sample households. A compensatory weight can be computed as the inverse of the probability of selection: $1/p_{area}$ for nontelephone households and $1/(p_{RDD} + p_{area} - p_{RDD} \times p_{area})$ for telephone households, regardless of which frame was used.

A third solution was proposed by Hartley (1962) and others. They suggested that the overlap of the frames be used to obtain a more efficient estimator. They proposed that a dual frame (in their case, multiple frame) design be examined as a set of nonoverlapping domains, and results from each domain combined to obtain a target population estimate. In the illustration in Figure 3.7, there would be three domains: nontelephone households (*Non-tel*), RDD telephone households (*RDD-tel*), and area sample telephone households (*area-tel*). The *RDD-tel* and *area-tel* households are combined with a mixing parameter chosen to maximize mathematically the precision of an estimate (say, a mean). The telephone and nontelephone domains are combined using a weight that is the proportion of the telephone households in the target population, say W_{tel}. The dual frame estimator for this particular example is

$$\bar{y} = (1 - W_{tel}) p_{non-tel} + W_{tel} \left[\theta p_{RDD-tel} + (1 - \theta) p_{area-tel} \right]$$

where θ is the mixing parameter chosen to maximize precision.

The dual frame illustration in Figure 3.7 is a special case of the multiple frame estimation approach. The method can be applied to more complex situations involving three or more frames, where the overlap creates more domains. Even in dual frame sampling, there are at least four domains: frame 1 only, frame 2 only, frame 1 sample overlapping with frame 2, and frame 2 sample overlapping with frame 1. (Although the last two are intersections of the two frames, which frame they are actually sampled from may affect the survey statistics; hence, they are kept separate.) These kinds of designs are found in agriculture surveys. Suppose that a sample is to be drawn in a state of farm holdings that have a particular kind of livestock, say, dairy cows. Suppose also that there is a list of dairy farmers available from the state department of agriculture, but it is known to be out of date, having some dairy farms listed that no longer have dairy cows and not listing small farms with dairy cows. A second area frame is used to draw a sample of all farms. There will be four domains: list frame only, area frame only, list frame also found on area frame, and area frame also found on list frame. Again, screening, weighting, or multiple frame estimation may be used to address the overlap problem.

One last example of more recent interest concerns Web survey design. Suppose that a list of e-mail addresses is available from a commercial firm. It is inexpensive to use for self-administered surveys, but it has foreign elements (addresses that are no longer being used, persons who are not eligible) and lacks complete coverage of the eligible population. A second supplementary RDD frame can be used to provide more expensive and complete coverage of eligible persons in all telephone households. Samples would be drawn from each frame, interviews conducted with sample eligible persons in both, and a dual frame estimation procedure used to combine results from these overlapping frames. There is much methodological research to be conducted in these mixes of frames and modes (see more in Section 5.4 regarding these issues).

foreign element

3.7.4 Increasing Coverage While Including More Ineligible Elements

The final coverage repair arises most clearly in coverage of persons within sample households, as part of a survey to produce person-level statistics. When a housing unit is identified for measurement, the interviewer attempts to make a list of all persons who live in the household. There appear to be consistent tendencies of underreporting of persons who are inconsistent residents of the household or who fail to fit the household informant's native definition of a "household."

Both Tourangeau, Shapiro, Kearney, and Ernst (1997) and Martin (1999) investigated what happens when questions about who lives in the unit are altered to be more inclusive. The typical question that generates the frame of persons within a household is "Who lives here?" Additional questions included who slept or ate there the previous evening, who has a room in the unit, who has a key to the unit, who receives mail at the unit, who usually is at the home but was away temporarily, and so on (see box on page 94). The questions appeared to increase the number of different people mentioned as attached to the household.

The next step in the process is the asking of questions that determine whether each person mentioned did indeed fit the definition of a household

> **Tourangeau, Shapiro, Kearney, and Ernst (1997) and Martin (1999) on Household Rosters**
>
> Two studies on why persons are omitted from listings of household members inform survey practice.
>
> *Study designs*: The Tourangeau et al. study mounted a randomized experimental design of three different rostering procedures: asking for names of all living at the unit, asking for names of all who spent the prior night at the unit, and asking for initials or nicknames of all those who spent the prior night. Follow-up questions asked whether all persons listed fulfilled the definition of living in the unit. Face-to-face interviews were conducted in 644 units on 49 blocks in three urban areas. The Martin study used an area probability sample of 999 units. Roster questions asked for all persons with any attachment to the household during a two-month period. Reinterviews were conducted on a subsample with further questions about residency. In both studies, follow-up questions after the roster asked whether all persons listed fulfilled the definition of living in the unit.
>
> *Findings*: The Tourangeau et al. study found that after probes to identify usual residence, only the technique of asking for initials produced more than the standard procedure. They conclude that concealment of the identity of some residents contributes to undercoverage. The Martin study found inconsistent reporting for unrelated persons, persons away from the home for more than a week, and persons not contributing to the financial arrangements of the unit. Martin concluded that informant's definitions of households do not match the survey's definition, causing underreporting.
>
> *Limitations of the studies*: There was no way to know the true household composition in either study. Both assumed that follow-up questions producing larger counts of persons were more accurate reports.
>
> *Impact of the studies*: They helped to document the size of household listing errors. They demonstrated that both comprehension of the household definition and reluctance to report unusual composition produce the undercoverage of household listings.

member according to the survey protocol. After such questions, those mentioned who have households elsewhere are deleted.

In essence this repair strategy "widens the net" of the frame and then trims out those in the net who were erroneously included. The burden of the strategy is that it requires more time and questions to assemble the frame of eligible persons in the household. Many times, this questioning is one of the first acts of the interviewer, at which point continued cooperation of the household informant is most tenuous. Hence, at this writing, adoption of the new approach is limited.

3.8 SUMMARY

Target populations, sampling frames, and coverage are important topics in survey design because they affect the nature of the inference that can be made directly from survey data. The problems that arise when comparing a frame to its target populations have remedies, many of which are standard approaches in survey research. They are also not necessarily complete corrections to the coverage error that may arise.

Coverage errors exist independent of the sampling steps in surveys. The sample selection begins with the frame materials. Samples can be no better than the frames from which they are drawn. We examine in the next chapter how samples for surveys are drawn. The discussion assumes that the kind of coverage errors and frame problems examined here are considered separately from the issue of how to draw a sample that will yield precise estimates for population parameters.

EXERCISES

KEYWORDS

area frame
area probability sample
clustering
coverage
duplication
de facto residence rule
de jure residence rule
elements
foreign element
foreign unit
half-open interval
household
housing unit

ineligible element
inelibible unit
multiple mappings
multiplicity sampling
multiple frame sampling
opt-in internet panel
random digit dialing (RDD)
rare population
sampling frame
survey population
target population
undercoverage
usual residence

FOR MORE IN-DEPTH READING

Kish, L. (1965), *Survey Sampling*, Chapter 11, Section 13.3, New York: Wiley.

Lessler, J. and Kalsbeek, W. (1992), *Nonsampling Error in Surveys*, Chapters 3–5, New York: Wiley.

Levy, P. and Lemeshow, S. (2008), *Sampling of Populations: Methods and Applications*, 4th Edition, New York: Wiley.

EXERCISES

1) Using one of the six example surveys (see Chapter 1), describe a situation through which altering the definition of the target population or the definition of the sampling frame could eliminate a coverage error in a resulting survey estimate. (Mention the survey estimate you have in mind, the target population, and the sampling frame.)

2) Name three concerns you would have in transforming an area probability face-to-face survey for the target population of U.S. adult household members into a Web survey attempting to estimate the same statistics.

3) Name two conditions (whether or not they are realistic) under which there would be no coverage error in an estimate (say, a sample mean) from a telephone survey attempting to describe the target population of all adults in U.S. households.

4) You are interested in the target population of farm operators in a three-county area encompassing 360 square miles. You lack a list of the farm operations

and instead plan on using a grid placed on a map, with 360 square-mile segments. You plan to draw a sample of farm operations by drawing a sample of square-mile segments from the grid. Identify three problems with using the frame of 360 square-mile grids as a sampling frame for the target population of farm operations in the three-county area.

5) Five years after the last census, you mount a household survey using a telephone number frame. If a selected telephone number is a household number, interviewers ask to speak to the person most knowledgeable about the health of the household members. After the survey is over, someone suggests evaluating your survey by comparing the demographic distributions (i.e., age, sex, race/ethnicity, and gender) of your "most knowledgeable" health informants to the demographic distributions of adults from the last census. Comment on the wisdom of this suggestion.

6) You are working for the political campaign of a presidential candidate. She gives you the result of a poll conducted on a volunteer Internet panel, showing that she leads her chief opponent. The Internet panel consists of volunteers recruited from among users of a credit card that has large annual fees and unusually generous benefits (e.g., discounts on products and services). This poll result is more favorable to your candidate than those using telephone sampling and centralized telephone interviewing methods. She asks you to evaluate the two polls and give advice on what questions she should ask of the pollsters. What do you say to her? (Identify 2–4 distinct points of evaluation.)

7) In most countries of the world, cell (or mobile) telephones have numbers in series distinct from those of line phones. Given the discussion in this chapter, invent a set of estimates that might have large coverage errors if a telephone survey omitted these cases from the sampling frame.

8) Imagine that you are considering sampling from a population of patients who visited a health clinic in the prior two years. The sampling frame available to you consists of paper records in a filing cabinet, with multiple pieces of paper from each visit of the person and their family to the clinic, supplemented by one sheet per person with background information. For each of the frame weaknesses that were reviewed in the chapter, give an example of how the weakness might be exhibited in this situation.

CHAPTER FOUR

SAMPLE DESIGN AND SAMPLING ERROR

4.1 Introduction

The selection of the elements from a sampling frame to include in a survey is a critical part of the survey process. A survey needs a well-developed questionnaire, highly trained and motivated interviewers, excellent field supervision and management, a data collection mode matched to the type of information to be collected, and a well-conceived editing plan. However, if the sample is chosen haphazardly or with subjective selection of elements, there may be little hope of making inferences to the population that was the target of the investigation.

In the fall of 1998, the U.S. National Geographic Society (NGS) launched "Survey 2000," a Web-based survey, designed to measure, among other things, the frequency of the population's movie attendance, museum visits, and book reading. Invitations on the NGS website and in the NGS magazines urged people to complete the survey. There were over 80,000 visitors to the survey Web page and over 50,000 completions of the questionnaire (Witte, Amoroso, and Howard, 2000).

A little earlier, in 1997, the U.S. National Endowment for the Arts sponsored the Survey of Public Participation in the Arts (SPPA) (National Endowment for the Arts, 1998). The SPPA was a telephone survey with sample telephone households selected based on randomly generated telephone numbers. Its response rate was approximately 55%, and it contained interview data from over 12,000 adults. It measured many of the same behaviors as did the NGS survey.

The results of the SPPA survey based on more rigorous sampling methods, however, were dramatically different than those of the NGS self-selected Web survey. For example, about 60% of the NGS respondents reported seeing live theater in the last 12 months, compared to about 25% for musical theater and 16% for nonmusical theater in the SPPA. Similarly, about 77% of the NGS respondents but only 35% of the SPPA respondents reported visiting an art museum or gallery.

How the sample of a survey is obtained can make a difference. The self-selected nature of the NGS survey, coupled with its placement on the National Geographic Society's website, probably yielded respondents more interested and active in cultural events (Couper, 2000).

Contrast the NGS survey with one of our example surveys, say, the Survey of Consumers (SOC). The SOC selects households for participation in the survey at random from among all telephone households in the country, except those in Alaska and Hawaii. "Random" or "chance" selection means that all human influence, both known and unknown, is removed from the selection process. Random

> random selection

Survey Methodology, Second Edition. By Groves, Fowler, Couper, Lepkowski, Singer, and Tourangeau
Copyright © 2009 John Wiley & Sons, Inc.

numbers, used to identify the selected elements, are such that they display no sequential order for single, double, triple, and larger digits. The selection also controls the sample distribution geographically, making sure it is spread over the spatial dimension like the target population. The random sampling mechanism and geographic controls are designed to avoid the selection of a sample that has higher incomes, or fewer members of ethnic or racial monorities, or more females, or any of a number of other distortions when compared to the entire U.S. population.

Essentially, in its simplest form, the kind of sample selection used in carefully designed surveys has three basic features:

1) A list, or combinations of lists, of elements in the population (the sampling frame described in Chapter 3).
2) Chance or random selection of elements from the list(s).
3) Some mechanism that assures that key subgroups of the population are represented in the sample.

It is important to note that random selection alone does not guarantee that a given sample will be representative. For example, there are many random samples of telephone households from a list of all such households in the 48 coterminous states that, by chance, consist only of households from the urban areas of the country. They are randomly selected, but they do not represent the entire country. So both randomized selection and techniques for assuring representation across population subgroups need to be used in well-designed samples.

probability sample

When chance methods, such as tables of random numbers, are applied to all elements of the sampling frame, the samples are referred to as "probability samples." Probability samples assign each element in the frame a known and nonzero chance to be chosen. These probabilities do not need to be equal. For example, the survey designers may need to overrepresent a small group in the population, such as persons age 70 years and older, in order to have enough of them in the sample to prepare separate estimates for the group. Overrepresentation essentially means that the members of the small group of interest have higher chances of selection than everyone else. That is, probability selection is used, but the probabilities are not equal across all members of the sample.

sampling bias

Section 2.3 makes the distinction between fixed errors or biases and variable errors or variance. Both errors can arise through sampling. The basis of both sampling bias and sampling variance is the fact that not all elements of the sampling frame are measured. If there is a systematic failure to observe some elements because of the sample design, this can lead to "sampling bias" in the survey statistics. For example, if some persons in the sampling frame have a zero chance of selection and they have different characteristics on the survey variable, then all samples selected with that design can produce estimates that are too high or too low. Even when all frame elements have the same chance of selection, the same sample design can yield many different samples. They will produce estimates that vary; this variation is the basis of "sampling variance" of the sample statistics. We will use from time to time the word "precision" to denote the levels of variance. That is, if the sampling variance of a survey statistic is low, we will say it has high precision.

sampling variance

precision

4.2 SAMPLES AND ESTIMATES

Not all samples for surveys are selected using chance methods. Many surveys use haphazardly or purposefully selected samples. For example, mall intercept surveys approach individuals entering a mall and ask each one for an interview, then continue sampling until a desired number of interviews are completed. These haphazard or convenience selection methods have a shared weakness: there is no direct theoretical support for using them to describe the characteristics of the larger frame population. Probability sampling provides the payoff of statistical inference, the ability to make statements about the population with known levels of confidence using data from a single sample.

Understanding sampling requires understanding some important concepts about phenomena not observed when doing a survey. Figure 4.1 displays the distribution of values of a variable Y in a frame population. Using the example of the Survey of Consumers, this frame population consists of all adults in households with telephone service. We never observe this full distribution. We never know it perfectly. We conduct the survey to learn about it. The distribution has a mean, which we will describe in capital letters as $\bar{Y}$. This mean is unknown to us and estimating it is the purpose of the survey. The spread of the distribution of the individual values of Y, called Y_i, is labeled as S^2, the population element variance.

Remember that the capital letters in this chapter will always refer to characteristics of the frame population, and will be unknown to us.

Figure 4.2 repeats a figure from Chapter 2, illustrating the process of estimation or inference that we perform in a sample survey. When probability selection methods are used, there is no one single sample that can be selected, but many.

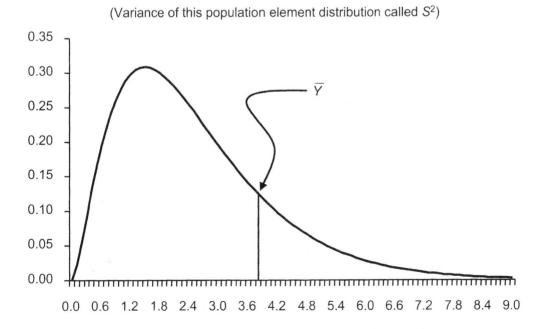

Figure 4.1 Unknown distribution for variable *Y* in frame population.

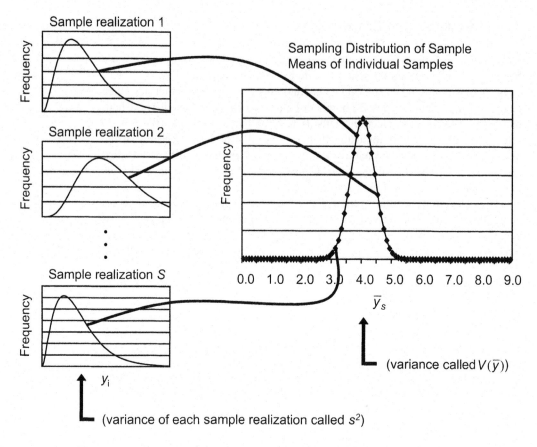

Figure 4.2 Distributions of *y* variable from sample realizations samples and the sampling distribution of the mean.

sample realization

These alternatives, called "realizations," are portrayed on the left of Figure 4.2. A "sample realization" is one set of frame elements selected using a specific sample design. Each sample design can yield many possible sample realizations. The number of possible realizations is a function of the number of units in the sample, the number in the frame, and the sample design. (Keep in mind that in survey research, we do a survey based on only one of these realizations.) In order to understand this idea, it is necessary to consider the variability in sample results from one possible sample realization to another.

With Figure 4.2 we can describe other key terms in sampling theory. In some sense, you should think of each survey as one realization of a probability sample design among many that could have occurred. It is conventional to describe attributes of the sample realizations using lower case letters. So each sample realization has a mean and a variance of the distribution of values of *y* in the sample realization. The sample mean is called $\bar{y}$, which summarizes all of the different values of *y* on the sample elements, and the variance of the distribution of y_i's in the sample realization is labeled as s^2. We will describe sample designs in which values of $\bar{y}$ from one sample realization will be used to estimate $\bar{Y}$ in the frame population. We will describe sample designs where the values of s^2 (the sample ele-

sample element variance

sampling variance of the mean

SAMPLES AND ESTIMATES

ment variance) in the sample realization will be used as estimates of S^2 in the frame population.

Finally, we will estimate variance of the sampling distribution of the sample mean [look at the $V(\bar{y})$ in Figure 4.3] using computations on one sample realization. Another term for the variance of the sampling distribution of the sample mean is the "sampling variance of the mean," the square root of which is the "standard error of the mean." If this notation is new to you, it is worth spending a moment committing it to memory. Figure 4.3 might help. Make sure you commit this notation to memory; it will help you understand what follows.

standard error of the mean

When analysts want to estimate the frame population mean using the sample mean from a given sample survey (one realization), they employ the standard error to develop confidence limits for a given estimate. "Confidence limits" describe a level of confidence that the means from all the different sample realizations will lie within a certain distance of the full frame population mean (ignoring coverage, nonresponse, and all the other errors of surveys). For example, suppose that we estimate the average age of the sample of adults in the Survey of Consumers as $\bar{y}$ = 42 years. We have estimated that the standard error of this estimate is 2.0. Confidence limits can be formed for the actual mean in the population by adding to and subtracting from the mean the standard error, or some multiple of it. For example, a 68% confidence interval for the mean age is based on one standard error on each side of the estimate, (42 – 2.0, 42 + 2.0) = (40, 44). The interpretation of these limits is that for 68% of the samples drawn from the same population using the same sampling method, similarly

confidence limits

	Type of Distribution for Y Variable		
	Distribution within Sample Realizations	Distribution within Frame Population	Sampling Distribution of Mean
Status	*Known for one sample realization*	Unknown	Unknown
Number of elements	$i = 1,2,...,n$	$i = 1,2,...,N$	$s = 1,2,...,S$
Values for an element	y_i	Y_i	$\bar{y}_s$
Mean	$\bar{y}$	$\bar{Y}$	$E(\bar{y}_s)$
Variance of distribution	s^2	S^2	$V(\bar{y})$
Standard deviation of distribution	s	S	$Se(\bar{y})$

Figure 4.3 Key notation for sample realization, frame population, and sampling distribution of sample means.

> **Warning**
>
> Do not confuse the different "variances" discussed. Each frame population has its own distribution of Y values, S^2, the population element variance, estimated by s^2, the sample element variance from any given sample realization. These are each different from the sampling variance of the sample mean, labeled $V(\bar{y})$, estimated by using sample realization data, using $v(\bar{y})$. Capital letters denote frame population quantities; lower case letters generally denote sample quantities.

computed intervals will include the actual population mean 68% of the time. For a commonly used 95% limit, we multiply the standard error by 1.96 (from the *Normal* distribution), or $(42 - 1.96 \times 2.0, 42 + 1.96 \times 2.0) = (38, 46)$. Again, in 95% of the samples drawn in exactly the same way from this population, this interval will include the actual mean population age.

The standard error of a sample statistic is a measure of the spread, or variability, of the statistic across the possible sample realizations from the same design. The notation used for the standard error is $se(\bar{y})$. It is the square root of the sampling variance $v(\bar{y})$, that is,

$$se(\bar{y}) = \sqrt{v(\bar{y})}$$

The standard error and sampling variance depend on a number of features of the sample design or method used to select the sample. Typically, larger samples chosen using a given sampling technique have smaller variances of the mean. Samples selected using clusters of elements have larger standard errors. Samples that divide the population into groups and select from the groups independently can have smaller standard errors. Larger standard errors mean wider confidence limits, and vice versa.

sampling fraction

In sampling for surveys, one additional measure appears in the consideration of sampling variability: the sampling fraction $f = n/N$, which is merely the fraction of units selected into the sample (of size n) from among all the N elements in the frame population. When we want to make inferences or statements about the population using the sample data, the inverse of this sampling fraction, $F = (1/f) = (N/n)$, reverses the sampling operation and projects the n sample elements to the N population elements.

probability sample

Summarizing the discussion thus far, we do surveys to discover properties of unknown distributions of variables that describe something in the frame population. Probability samples assign known, nonzero chances of selection to each frame element. Their benefits are that we can use one sample realization to estimate characteristics of the frame population within known levels of confidence. If the statistic of interest on the frame population is the sample mean, we can use the mean computed on the sample realization, estimate the standard error of the mean, and construct confidence limits on the mean.

As we noted in Section 2.3.5, the extent of the error due to sampling is a function of four basic principles of the design:

1) How large a sample of elements is selected.
2) What chances of selection into the sample different frame population elements have.
3) Whether individual elements are drawn directly and independently or in groups (called "element" or "cluster" samples).
4) Whether the sample is designed to control the representation of key subpopulations in the sample (called "stratification").

SIMPLE RANDOM SAMPLING

This chapter is organized around these features. The chapter begins with the simplest sample design, in order to document the basics of probability sampling.

4.3 SIMPLE RANDOM SAMPLING

Simple random sampling, or SRS, is often used as a basic design to which the sampling variance of statistics using other sample designs are compared. Simple random samples assign an equal probability of selection to each frame element, equal probability to all pairs of frame elements, equal probability to all triplets of frame elements, and so on. One way to think about a simple random sample is that we would first identify every possible sample of size n distinct elements within a frame population, and then select one of them at random. Sample designs assigning equal probabilities to all individual elements of a frame population are called "epsem" for *E*qual *P*robability *SE*lection *M*ethod.

simple random sample

epsem

In practice, one never writes down all possible samples of size n. It is just too time-consuming for a large population. For example, the number of possible samples of size 300 different adults for the Survey of Consumers drawn from the population of over 228 million adults in the United States is enormous. To select an SRS, random numbers can be applied directly to the elements in the list. SRS uses list frames numbered from 1 to N. Random numbers from 1 to N are selected from a table of random numbers, and the corresponding population elements are chosen from the list. If by chance the same population element is chosen more than once, we do not select it into the sample more than once. Instead, we keep selecting until we have n distinct elements from the population. (This is called sampling "without replacement.")

sampling without replacement

From an SRS, we compute the mean of the variable y, as

$$\bar{y} = \left(\frac{1}{n}\right)\sum_{i=1}^{n} y_i$$

which is merely the sum of all the sample y_i values divided by the total number of elements in the sample. The sampling variance of the mean can be estimated directly from one sample realization as

$$v(\bar{y}) = \frac{(1-f)}{n} s^2$$

the element variance of y_i's in the sample realization divided by the sample size, then multiplied by $(1-f)$. The term $(1-f)$ is referred to as the finite population correction, or *fpc*. The finite population correction factor equals the proportion of frame elements *not* sampled or $(1-f)$, where f is the sampling fraction. As a factor in all samples selected without replacement, it acts to reduce the sampling variance of statistics. If we have a situation in which the sample is a large fraction of the population, where f is closer to 1, then the *fpc* acts to decrease the sampling variance. Often, the frame population is so large relative to the sample that the *fpc* is ignored.

finite population correction

When f is small, the *fpc* is close to 1, and

$$v(\bar{y}) \doteq \frac{s^2}{n}$$

which is merely the element variance for y_i divided by the sample size. The sampling variance, the standard error of the mean, and the width of the confidence interval, depend on just two factors: how variable the values of the y_i's are across the sample elements and how large the sample is. If the y_i's are quite variable (i.e., S^2 is large), then the sampling variance of the mean will be large, and the confidence limits wide. A larger n reduces the sampling variance, improving the quality of the estimate.

Lastly, it is useful to note that $v(\bar{y})$ is only a sample based estimate of the sampling variance of the sample mean. There is an actual population value $V(\bar{y})$ that is being estimated here, and $v(\bar{y})$ is, in SRS, an unbiased estimate for $V(\bar{y})$. Confidence limits, such as the 95% limits, for the actual population mean $\bar{Y}$ are computed as $\bar{y} \pm 1.96[se(\bar{y})]$, adding to and subtracting from the mean twice the standard error.

There is a closely related result for proportions that yields a well-known formula. Suppose that in the Survey of Consumers the characteristic of interest y_i is whether or not an adult thinks the economy is at present better than it was a year ago. Then, y_i is a binary variable, taking on two values: 1 if the adult thinks the economy better, and 0, otherwise. The mean $\bar{y}$ of these values is called a "proportion," often using the notation p, and the proportion always lies between 0.0 and 1.0.

For this type of variable, the sampling variance of the mean for an SRS has a simpler form:

$$v(p) = \frac{(1-f)}{n-1} p(1-p)$$

> **Comment**
>
> The estimated sampling variance of a proportion leads to the common approximation for the sampling variance of a proportion of pq/n, where $q = (1 - p)$, and the finite population correction is ignored.

Here, the sampling variance of p depends not only on the *fpc* and the sample size but also on the value of p itself. This is a great convenience in estimating $v(p)$, since one does not need to have all the individual y_i values, but only the proportion itself.

How large a sample should be taken to achieve the goals of a survey? Sample sizes are determined in several different ways. One way is to try to find a sample size such that the confidence limits obtained from the subsequent sample will not exceed some value. For example, in the Survey of Consumers, suppose the sponsor wants to have a large enough sample such that the 95% confidence limits for the proportion thinking the economy is better are from 0.28 to 0.32, an interval that contains 4 points. What standard error does this imply?

Given an expected proportion of 0.30, these limits imply that the "half width" of the confidence interval, or half of $1.96[se(\bar{y})]$, must be equal to 0.01, or the standard error should be 0.01. Since the standard error of the mean is $\sqrt{(1-f)s^2/n}$, we have to solve for n. We will need an estimate of s^2, however, in order to get a value for the sample size n.

SIMPLE RANDOM SAMPLING

We might get a value to approximate s^2 for our survey from a calculation on data from a prior survey on the same population, from another survey measuring the same characteristic in a slightly different population, or from a small pilot survey in which we compute s^2. We will never have an exact value of S^2 to use instead of one of these approximations, and a choice must be made to use a value that is the best one we can manage to get with reasonable cost.

> **Comment**
>
> The vast majority of U.S. surveys do *not* use simple random samples. They implement some stratification (Section 4.5) and/or some clustering (Section 4.4).

In the case of proportions, though, the problem is much easier. Since, approximately, $s^2 = p(1 - p)$, if we guess, even with some degree of error, a value of p in advance, we can estimate a s^2 to use in sample size calculations. For example, in the BRFSS, we might expect a proportion of obese individuals in the low income population of $p = 0.3$. Then, $s^2 = p(1 - p) = 0.3 (1 - 0.3) = 0.21$. That is, the element variance would be 0.21. If we want a 95% confidence interval for p of (0.28, 0.32) (which implies a standard error of p of 0.01), then

$$n = \frac{s^2}{(0.01)^2} = \frac{0.21}{(0.01)^2} = 2100$$

In some instances, the population size N is small enough that we suspect that the *fpc* may not, for reasonable sample sizes, be close to 1.0. In these cases, it is customary to adjust the needed sample size to account for the *fpc*. For example, suppose we mounted the BRFSS estimating obesity in a population of $N = 10,000$ adults in a medium-sized town. What would be the effect of the *fpc* be on the needed sample size of 2100, before adjusting for the *fpc*? We should anticipate some benefits of a small population, reducing the needed sample size. We adjust this SRS sample size as

$$n = \frac{n'}{1 + (n'/N)}$$

or, in this case, as

$$n = \frac{2,100}{1 + (2,100/10,000)} = 1,736$$

Because the frame population is so small, we need only 1736 adults rather than the 2100 to achieve a standard error of 0.01.

There is a popular misconception that the proper way to compute sample size is based on some fraction of the population. It is assumed that if you select the same fraction of two populations, regardless of their size, the samples will yield equally precise results. As the above calculations show, that is simply not true. Once the population size is large, the sample size needed for a given standard error does not change, even if another population even larger is to be studied. For example, the same sample size is needed for the United States as is needed for the People's Republic of China, despite the fact that China has five times the population of the United States.

The example above chooses a sample size to provide a predetermined standard error for a key statistic. This does not usually happen. Instead, an investigator might have allocated a certain amount of money to conduct the survey. The question then becomes, "How large a sample size can I afford?" When this is the question asked, the investigator tries to estimate what standard error is possible, given guesses about what S^2 might be for a given statistic.

It is also important to keep in mind that surveys often ask respondents many questions, and many statistics can be computed from each survey. For a given sample size, the levels of precision of different statistics will vary. The single sample size chosen is often a compromise among issues such as the how much the investigator is willing to spend for the survey, which statistics are most important for the investigator, and what levels of precision are obtained for them with different budget allocations.

4.4 Cluster Sampling

cluster sample

Often, the cost of a simple random sample is exorbitant. Cluster samples do not select frame elements directly, but groups of frame elements jointly.

For example, even if there were a population register for the United States listing every person in the population, face-to-face surveys would still not use simple random samples because of the costs of sending interviewers all over the country. This occurs when populations are geographically widespread, such as the population of adults that is the target of the NCVS, or the population of 4th, 8th, and 12th graders in the NAEP. Similarly, if there is no frame listing directly all the elements of the population, the cost of developing a frame of elements could be very expensive. A widespread population often does not appear on a single list but many lists (as in lists of 4th graders in each elementary school in the country) or there may be no list of the elements at all (as for individuals aged 12 years and older for the NCVS).

One way to construct frames at reasonable cost is to sample clusters of elements and then collect a list of elements only for the selected clusters. And when a sample cluster is selected and visited, it makes sense to interview or collect data from more than one element in each cluster also to save costs. To simplify the presentation, consider first the case in which we have a widespread population, and we can obtain a list of clusters for them. Further, suppose that each of the clusters is of exactly the same size. This would never be the case in practice, but it is a useful assumption in order to start with less-complicated results for standard errors of a mean or proportion.

Figure 4.4 illustrates a simple clustered frame. Imagine that we have a small village of six city blocks pictured on the left of the figure. On each block there are 10 houses, represented either as crosses or circles. There is a total of 60 households in the village. Half of the village consists of rich households (labeled as crosses in the figure) and half of poor households (labeled as circles in the figure). Notice that the poor persons tend to live next to other poor persons, and the rich persons, next to other rich persons. This is the typical pattern of residential segregation by income that is seen around the world.

If we plan a sample of size 20 households from this village using a clustered design of city blocks, we would sample two blocks for each sample realization. If we merely drew two random numbers from 1 to 6, all of the sample realizations

CLUSTER SAMPLING

Frame Population

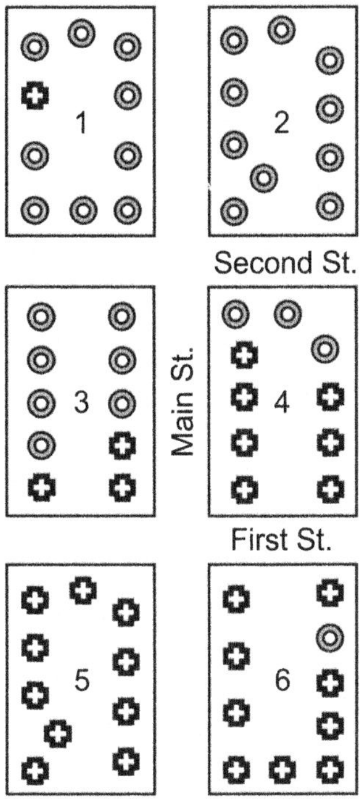

Samples of Two Distinct Blocks
(Proportion "◎" Households in Realization)

Block Numbers	Proportion "◎"
1,2	19/20 = .95 *
2,3	17/20 = .85 *
1,3	16/20 = .80 *
2,4	13/20 = .65
1,4	12/20 = .60
2,6	11/20 = .55
1,6	10/20 = .50
2,5	10/20 = .50
3,4	10/20 = .50
1,5	9/20 = .45
3,6	8/20 = .40
3,5	7/20 = .35
4,6	4/20 = .20 *
4,5	3/20 = .15 *
5,6	1/20 = .05 *

Figure 4.4 A bird's-eye view of a population of 30 "✧" and 30 "◎" households clustered into six city blocks, from which two blocks are selected.

possible are displayed on the right of the figure. Remember that the proportion of poor households (the circles) is 0.50. Across the 15 sample realizations, the average proportion is 0.50; that is, the sample design yields an unbiased estimate of the proportion of poor households. However, look at the large variability in the estimated proportion of poor households across the 15 different sample realizations! Six of the fifteen different samples yield estimates of 0.80 or more, or 0.20 or less. The true proportion is 0.50; these are extremely poor estimates. The sampling variance of the estimated mean (proportion in this case) is very high. This reflects the damaging effect of clustering when the survey statistic itself is highly variable across the clusters.

How do we compute statistics for clustered samples? Let us use NAEP as an illustration. For example, suppose that for the NAEP, we could obtain a list of all $A = 40,000$ 4th grade classrooms in the United States, and that each of those classrooms has exactly $B = 25$ students. We do not have a list of the $A \times B = N = 40,000 \times 25 = 1,000,000$ students, but we know that when we get to a classroom, we can easily obtain a list of the 25 students there.

The sampling procedure is simple: chose a sample of a classrooms using SRS, visit each classroom, and collect data from all students in each sample classroom. If we select $a = 8$ classrooms, our sample size is $n = 8 \times 25 = 200$ students. It should be clear that this is not the same kind of sampling as an SRS of elements. There are many SRS samples of size 200 students that cannot possibly be chosen

> **Comment**
>
> Analyzing clustered survey data with typical software packages will produce estimated standard errors that are too low. See Chapter 10 to learn how to get proper estimates.

in this kind of a design. For instance, some SRS sample realizations consist of exactly one student in each of 200 classrooms. None of these are possible cluster samples of 200 students from 8 classrooms.

Statistical computations get a little more complex with cluster samples because we need to think of the survey dataset as consisting of student records, grouped into classrooms. Hence, we need some notation to reflect both students and classrooms. Suppose that we obtain for each of our 200 sample students a test score of $y_{\alpha\beta}$ for the β th student in αth classroom, and we compute the mean test score

$$\bar{y} = \frac{\sum_{\alpha=1}^{a}\sum_{b=1}^{B} y_{\alpha\beta}}{aB}$$

Thus, there are $\alpha = 1, 2, \ldots, a$ classrooms in the sample; and $\beta = 1, 2, \ldots, B$ students in each sample classroom. This mean is computed exactly as before, except that we first add all the test scores within a classroom, then classroom sums, and finally divide by the sample size.

The sampling variance of this mean is different from the SRS sampling variance. The randomization in the selection is applied only to the classrooms. They are the sampling units, and depending on which classrooms are chosen, the value of $\bar{y}$ will vary. In one sense, everything remains the same as SRS, but we treat the clusters as elements in the sample.

In this case,

$$v(\bar{y}) = \left(\frac{1-f}{a}\right)s_a^2$$

where s_a^2 is variability of the mean test scores across the a classrooms. That is,

$$s_a^2 = \left(\frac{1}{a-1}\right)\sum_{\alpha=1}^{a}(\bar{y}_\alpha - \bar{y})^2$$

where $\bar{y}_\alpha$ is the mean test score in classroom α. We use "between cluster variance" in the case of cluster sampling, instead of the element variance s^2.

Suppose that a sample of eight classrooms yielded the following mean test scores for the 4th graders: 370, 370, 375, 375, 380, 380, 390, and 390, for a mean of 378.75. Then

$$s_a^2 = \left(\frac{1}{a-1}\right)\sum_{\alpha=1}^{a}(\bar{y}_\alpha - \bar{y})^2 = 62.5 \text{ and } v(\bar{y}) = \left(\frac{1-f}{a}\right)s_a^2 = 7.81$$

We have noted above that clustering tends to increase the standard error of the mean over that expected from a simple random sample of the same size. Suppose an SRS of size $n = 200$ were chosen from this population, and we

CLUSTER SAMPLING

obtained the same mean (i.e., 378.75). Suppose the 200 student test scores have an element variance of $s^2 = 500$. Then

$$v(\bar{y}) = \left(\frac{1-f}{n}\right)s^2 = 2.50$$

instead of 7.81. We note immediately that the sampling variance of the cluster sample mean is larger. It is in fact larger by the ratio

$$d^2 = \frac{v(\bar{y})}{v_{srs}(\bar{y})} = \frac{7.81}{2.50} = 3.13$$

We have more than tripled the sampling variance of the mean by selecting classrooms of students instead of students directly! This implies an increase in the standard error and the confidence limits of 77% (because $\sqrt{3.13} = 1.77$) due to the clustering effect.

This statistic d^2 is referred to as the design effect. It is a widely used tool in survey sampling in the determination of sample size, and in summarizing the effect of having sampled clusters instead of elements. It is defined to be the ratio of the sampling variance for a statistic computed under the sample design [in this case, $v(\bar{y})$] divided by the sampling variance that would have been obtained from an SRS of exactly the same size [$v_{srs}(\bar{y})$]. Every statistic in a survey has its own design effect. Different statistics in the same survey will have different magnitudes of the design effect.

design effect

If this result showing a big increase in sampling variance of the mean were true in general, why would anyone select clusters instead of elements? The reason is that the cost of an element or nonclustered design may be too high to pay. In order to obtain the frame of all students to draw the SRS, we may have to visit all 40,000 classrooms to collect individual lists. Further, the sample of 200 may lie in that many classrooms, requiring many more recruitment efforts than with merely eight sample classrooms. In this trade-off of costs and precision, the larger standard error of the mean from the classroom sample will be acceptable in order to make the survey costs fit the budget.

4.4.1 The Design Effect and Within-Cluster Homogeneity

Cluster samples generally have increased sampling variance relative to element samples, the result of mean test scores varying across the classrooms. With test score differences across classrooms, it also means that we have a tendency to have similarity or homogeneity of test scores within classrooms. The more variability between cluster means, the more relatively homogeneous the elements are within the clusters.

Another way of thinking about clustering effects asks, "What new information about the population do we obtain by adding to the sample one more element from the same cluster?" If in the extreme, all the students in a classroom have the same test score, once we know one test score for a student in a classroom, the other 24 measurements in the room are unnecessary. We could have saved a lot of money by only getting a test score from a single child in each classroom.

intracluster homogeneity

roh

One way to measure this clustering effect is with an intraclass correlation among element values within a cluster, averaged across clusters. This correlation measures the tendency for values of a variable within a cluster to be correlated among themselves, relative to values outside the cluster. The intracluster homogeneity, denoted *roh* for "rate of homogeneity," is almost always positive (greater than zero). In addition, we can link this *roh* to the design effect that summarizes the change in sampling variance as we go from an SRS of size n to a cluster sample of size n: $d^2 = 1+(b-1)roh$. That is, the increase in sampling variance we observed for the school classroom example actually depends on the amount of homogeneity among students in classrooms on test scores, and on the size of the sample taken from each classroom, b. In the illustration, $b = B = 25$, or all children are taken in a sample cluster. One could also draw only a subsample of students into the sample in each classroom. We will turn to subsampling in the next section.

For now, we note that d^2 will increase as *roh* increases. The *roh* will vary across types of variables. Variables with larger rates of homogeneity within clusters will have larger design effects for their means. For example, Table 4.1 presents some *roh* values from area probability samples in different countries, using clusters that were similar to counties in the United States. In general, this table shows higher *roh* values for socioeconomic variables and lower ones for attitudes and fertility experiences of women. The finding of higher *roh* values for socioeconomic variables follows from the residential segregation of rich and poor households common to most countries. This means that design effects for means on socioeconomic variables are higher on average than those on attitudes or fertility experiences.

The larger the subsample size, the larger the design effect. The subsample size b magnifies the contribution of the homogeneity to the sampling variance. The illustration above yields a moderately large value of d^2 because we elected to take all students in each sample classroom. If we select only $b = 10$ students in a classroom, d^2 will drop. Of course, we would have to increase the number of clusters from 8 to

Table 4.1. Mean *roh* Values for Area Probability Surveys about Female Fertility Experiences in Five Countries by Type of Variable

Type of Variable	Country and Year of Survey				
	Peru 1969	U.S. 1970	S. Korea 1973	Malaysia 1969	Taiwan 1973
Socio-economic	0.126	—	0.081	0.045	0.016
Demographic	0.024	0.105	0.025	0.010	0.025
Attitudinal	0.094	0.051	0.026	0.017	0.145
Fertility experiences	0.034	0.019	0.009	0.025	0.014

Source: Kish, Groves, and Krotki, 1976.

20 in order to maintain the overall sample size and, thereby, increase the cost of data collection. The dependence of the design effect on b, the number of sample cases per cluster, also means that means computed for subclasses of a sample will tend to have lower design effects than means computed on the full sample.

The worst case for our sample would be if we had $roh = 1$ and we selected $b = B = 25$ because d^2 would be at its maximum value. Alternatively, if $roh = 0$, then d^2 is equal to one; we cannot do any better than an unclustered sample of students directly, as long as roh is positive.

But how do we get an estimate of roh to understand the homogeneity for a given variable and cluster definition? A simple way to get an approximate value for roh is to extract it directly from d^2. For our example of a sample of classrooms, we already know that $d^2 = 3.13 = 1 + roh(25 \text{ students} - 1)$. We can estimate roh as

$$roh = \frac{(d^2 - 1)}{(b - 1)}$$

In this case, we have

$$roh = \frac{(3.13 - 1)}{(25 - 1)} = 0.0885$$

That is, there is a modest level of homogeneity present (the value of roh is closer to zero than one), and it is magnified by the subsample size b.

In survey sampling estimation systems, roh is often estimated routinely. The process is always the same. Compute $v(\bar{y})$ for the sample design actually used to select sample elements. Then treat the cluster sample elements as though they had been selected by SRS and compute

$$v(\bar{y}) = \left(\frac{1-f}{n}\right)s^2$$

using the cluster sample data. Finally, compute d^2, and then compute

$$roh = \frac{(d^2 - 1)}{(b - 1)}$$

Kish and Frankel (1974) on Design Effects for Regression Coefficients

Kish and Frankel (1974) discovered that many of the sample design impacts on sample means had analogues for analytical statistics, like correlation coefficients, multiple correlation coefficients, and regression coefficients

Study design: In a simulation study, 200 to 300 repeated clustered subsamples were drawn from a Current Population Survey dataset of 45,737 households in 3240 primary sampling units. The CPS is a multistage area probability sample, and the subsamples used the same stages as the base design. The subsample designs varied, with 6, 12, or 30 strata, and produced on average sample household counts of 170, 340, and 847, respectively. Two multiple regression equations with eight coefficients each, correlation, and multiple correlation coefficients were estimated. Sampling distributions for the statistics were estimated by studying the average values and dispersion of values from the 200 to 300 samples.

Findings: The design effects for the correlation and regression coefficients tended to be greater than 1.0, but less than the average design effect for sample means. The multiple correlation coefficient had larger design effects than the other statistics.

Limitations of the study: The study simulated a sampling distribution of complex statistics by drawing repeated subsamples of identical design from a larger sample survey. The study itself cannot address how the design of the base survey affected the empirical results.

Impact of the study: The study led to the widespread caution about design effects on analytic statistics, which had previously been largely ignored in practical analysis of complex survey data.

How can one use estimates of *roh* in practice? Say we have computed a value for *roh* from a prior survey on the same or similar topic from virtually the same population. Then, to estimate the sampling variance under a new design, first compute a new design effect $d^2_{new} = 1 + (b_{new} - 1)roh_{old}$, where b_{new} is the new number of sample elements per cluster and roh_{old} is the *roh* value from the prior survey. Now take this new manufactured design effect and multiply it by an estimate of the SRS variance for the mean of the new sample, say

$$v(\bar{y}) = \left(\frac{1-f}{n}\right)s^2$$

with s^2 estimated from a prior survey, and n determined from the new design.

There is one other way to view d^2. It is a measure of the loss in precision we incur by sampling clusters. Suppose we think about this loss in terms of sample size. In the illustration, the number of sample students in the sample is 200. But with the large increase in sampling variance reflected in $d^2 = 3.13$, we really do not have the equivalent of an SRS of size 200. Instead, we really have effectively the much smaller sample size of an SRS that achieves the same $v(\bar{y}) = 7.81$. In this case, the effective sample size is $n_{eff} = 200/3.13 = 64$. The "effective sample size" for a statistic is the SRS sample size that would yield the same sampling variance as achieved by the actual design.

effective sample size

4.4.2 Subsampling within Selected Clusters

Above, we mentioned that one could reduce the effect of clustering on the sampling variance of the mean test score by reducing the sample size within selected clusters (e.g., the number of students selected per classroom). This attempts some compromise, a dampening of the harmful effects of clustering on the precision of results. Our example design chose eight sample classrooms of 25 students each to yield 200 selected students. We could randomly select 10 of the 25 students from a selected classroom. With that subsampling, a sample of 200 students would have to be spread over 20 classrooms. The b, the number of selected students in a classroom, moves from 25 to 10, while a, the number of selected classrooms, moves from 8 to 20.

What do we gain from subsampling? The design effect for the sample mean goes down and the precision of the sample mean increases. The design effect is now $d^2 = 1 + (10 - 1)(0.0885) = 1.80$ (instead of 3.13 with 25 sample students per sample classroom), and the effective sample size increases to $n_{eff} = 200/1.80 = 111$ (instead of 64). The sample mean is more precise even though the total number of students in the sample has not changed.

Spreading a cluster sample over more clusters while keeping the overall number of sample elements fixed usually increases total costs. For our example, the negotiations to gain cooperation of each sample school to perform the tests on its students would be extensive. In contrast, however, the cost of testing one more student within the sample school would be trivial. We have increased the cost of the study, but also reduced the sampling variance of the mean from the cluster sample of classrooms.

In summary, changing our unit of sample selection from an element, as in SRS, to a cluster of elements reduces costs, but the cost reduction comes at the

STRATIFICATION AND STRATIFIED SAMPLING

price of higher sampling variance for the mean, or larger sample sizes to obtain the same sampling variance. The increase in sampling variance is measured by a design effect. The sampling variance for a cluster sample mean is driven by the rate of homogeneity and how many elements are selected in each cluster.

There are other sampling techniques that can have the opposite effect of cluster sampling. We turn now to one of those—stratification.

4.5 STRATIFICATION AND STRATIFIED SAMPLING

Probability sample designs can be made better with features to assure representation of population subgroups in the sample. Stratification is one such feature. The technique is, at first view, straightforward to use. On a frame of population elements, assume that there is information for every element on the list that can be used to divide elements into separate groups, or "strata." Strata are mutually exclusive groups of elements on a sampling frame. Each frame element can be placed in a single "stratum" (the singular of "strata"). In stratified sampling, independent selections are made from each stratum, one by one. Separate samples are drawn from each such group, using the same selection procedure (such as SRS in each stratum, when the frame lists elements) or using different selection procedures (such as SRS of elements in some strata, and cluster sampling in others).

Figures 4.5 and 4.6 illustrate how stratification works. Figure 4.5 lists in alphabetical order a population of 20 employers eligible for the CES. Imagine that the frame is a list of employers in one industrial sector of a state. In addition to the name of the employer, the list contains a grouping variable—Low, Medium, High, Highest—which indicates the number of employees in the establishment. For example, Bradburn Corp. is in the High size category, with several employees; Cochran Inc. is in the Highest group, with many employees. Each of the four size groups contains five frame elements (five employers). In CES, we plan to measure payroll sizes, and expect that they will vary across the four size groups. If we ignore the frame information about size, we might draw an SRS. For simplicity of illustration, consider an SRS of size $n = 4$. Figure 4.5 presents one such realization, based on the random numbers 9, 13, 14, and 18. Notice that the size distribution in that sample realization is one in the Low group, two in the Medium group, and one in the High group. The realization misses any elements of the Highest group, fully 20% of the frame population. Another realization of the design might have four Highest group members or four Low group members. Using SRS, there is little control over the representation of size group members in each sample realization.

Figure 4.6 shows the same frame population, prepared for a stratified sample. First, we sort the frame by group (notice in the figure that the Highest size group members are listed first, then the High experience members, and so on). Then an SRS is selected within each group. To keep the total sample size the same, we draw an SRS of size 1 from each size group, attaining the same sample size of 4 with the same sampling fraction, 1/5. Figure 4.6 shows one realization of that design. Note that each group is represented by one sample establishment. Indeed, *every* realization of the stratified design would have each of the four groups (strata) represented by one and only one sample establishment. The realizations that include only those Low or Highest in size are eliminated from occur-

stratification

stratum

strata

Record	Name	Group	
1	Bradburn Corp.	High	
2	Cochran Inc.	Highest	
3	Deming Design	High	
4	Fuller & Fuller	Medium	One SRS of Size 4
5	Habermann AG	Medium	
6	Hansen PLC	Low	
7	Hu Electronics	Highest	
8	HydeBev	High	
9	Kalton Group	Medium	→ Kalton Group
10	Kish Consulting	Low	
11	Madow USA	Highest	
12	M.P.H. Bank	Highest	
13	Norwood LC	Medium	→ Norwood LLC
14	Rubin Inc.	Low	→ Rubin Inc.
15	Sheatsley Co.	Low	
16	Steinberg Ltd.	Low	
17	Sudman Inc.	High	
18	Wallman AG	High	→ Wallman AG
19	Wolfe & Enix	Highest	
20	WXM Ventures	Medium	

Figure 4.5 Frame population of 20 establishments sorted alphabetically, with SRS sample realization of size n = 4.

ring. To the extent that the key survey variable, payroll, varies across the groups, this yields smaller standard errors on the overall sample mean.

To estimate population values, such as a mean or proportion, the strata results must be combined. An estimate of the population mean or proportion can be computed in each stratum and combined across strata to give an estimate for the population. The combining method will depend on how the sample is divided up across the strata.

4.5.1 Proportionate Allocation to Strata

Proportionate allocation is identical to selecting the sample in each stratum with the same probabilities of selection, an equal probability selection method, or *epsem*. That is, if we let $f_h = n_h/N_h$ denote the sampling fraction for stratum h, then proportionate allocation, as described in the last paragraph, is the same thing as selecting samples independent from each stratum with the same sampling fractions. In other words, $f_h = f$ for all strata. For example, one can choose a sample size n_h for stratum h such that the proportion of elements from a given stratum in the sample, n_h/n, is the same as the proportion of elements in the population, N_h/N (here N_h is the number of population elements in stratum h). We let $W_h = N_h/N$ denote this population proportion for each stratum.

STRATIFICATION AND STRATIFIED SAMPLING

Record	Name	Group	
2	Cochran Inc.	Highest	
7	Hu Electronics	Highest	One Stratified Random Sample of Total Size 4
11	Madow USA	Highest	
12	M.P.H. Bank	Highest	
19	Wolfe & Enix	Highest	→ Wolfe & Enix
1	Bradburn Corp.	High	→ Bradburn Corp.
3	Deming Design	High	
8	HydeBev	High	
17	Sudman Inc.	High	
18	Wallman AG	High	
4	Fuller & Fuller	Medium	→ Fuller & Fuller
5	Habermann AG	Medium	
9	Kalton Group	Medium	
13	Norwood LC	Medium	
20	WXM Venture	Medium	
6	Hansen PLC	Low	
10	Kish Consulting	Low	
14	Rubin Inc.	Low	→ Rubin Inc.
15	Sheatsley Co.	Low	
16	Steinberg Ltd.	Low	

Figure 4.6 Frame population of 20 establishments sorted by group, with stratified element sample of size $n_h = 1$ from each stratum.

To obtain estimates for the whole population, results must be combined across strata. One method weights stratum results by the population proportions W_h. Suppose we are interested in estimating a mean for the population, and we have computed means $\bar{y}_h$ for each stratum. The stratified estimate of the population mean is called $\bar{y}_{st}$ with the subscript st denoting "stratified." It is computed by

$$\bar{y}_{st} = \sum_{h=1}^{H} W_h \bar{y}_h = \text{weighted sum of strata means}$$

In this formulation, each stratum mean contributes to the total an amount that is proportionate to the size of its stratum.

The sampling variance of $\bar{y}_{st}$ is also a combination of values across strata, but it is more complicated. Let us consider a particular type of stratified sampling in which we have elements in each stratum, and we select an SRS of size n_h in stratum h. Thus, the sampling variance for the hth stratum mean $\bar{y}_h$ can be given as

$$v(\bar{y}_h) = \left(\frac{1-f_h}{n_h}\right) s_h^2$$

the sampling variance computed for a simple random sample of size n_h. The element variances within strata, the s_h^2's, are estimated separately for each stratum and are the within stratum element variances around the mean $\bar{y}_h$:

$$s_h^2 = \left(\frac{1}{n_h - 1}\right) \sum_{i=1}^{n_h} (y_{hi} - \bar{y}_h)^2$$

Thus, in stratified random sampling, we must compute not one element variance as in SRS, but one for each stratum.

For $\bar{y}_{st}$, we combine these SRS sampling variance estimates as

$$v(\bar{y}_{st}) = \sum_{h=1}^{H} W_h^2 \left(\frac{1-f_h}{n_h}\right) s_h^2$$

where we "weight" each stratum's SRS variance by the square of the population proportion W_h.

How does the precision of a stratified sample mean compare to those from an SRS of equal size? As we did for cluster sampling, we will compare the sampling variance for a stratified sample (in this case, a stratified random sample with SRS in each stratum) to the sampling variance for an SRS of the same size. Thus, the design effect for stratified random sampling is given as

$$d^2 = \frac{v(\bar{y}_{st})}{v_{srs}(\bar{y})} = \frac{\sum_{h=1}^{H} W_h^2 \left(\frac{1-f_h}{n_h}\right) s_h^2}{\left(\frac{1-f}{n}\right) s^2}$$

This design effect can be less than one, or equal to one, or even greater than one. The value of the design effect depends a great deal on how large a sample is drawn from each stratum, which is known as the sample allocation across strata.

Of course, the estimation procedure for proportions is similar to the one for means, and actually can use the same formula. However, the estimation for proportions is often written in terms of proportions, as

$$p_{st} = \sum_{h=1}^{H} W_h p_h$$

and

$$v(p_{st}) = \sum_{h=1}^{H} W_h^2 \left(\frac{1-f_h}{n_h - 1}\right) p_h (1 - p_h)$$

Consider an example from NAEP's School Characteristics and Policy Survey, which samples schools. To simplify somewhat, imagine a frame separated by urban location. Suppose there are $N = 8,000$ schools in the frame, and a stratified sample of $n = 480$ is to be selected. Table 4.2 presents the frame population divided into three strata. Further, assume we decide to allocate the sample proportionally across strata, so that a sampling fraction of 0.06 is used in each stratum. Once the

Table 4.2. Proportionate Stratified Random Sample Results from a School Population Divided Into Three Urbanicity Strata

Stratum	N_h schools on frame	W_h proportion of population	n_h stratum sample size	f_h stratum sampling fraction	y_h sample stratum mean	s_h^2 sample stratum element variance
Central city schools	3200	0.4	192	0.06	6	5
Other urban schools	4000	0.5	240	0.06	5	4
Rural schools	800	0.1	48	0.06	8	7
Total	8000	1.0	480	0.06		

> **Cochran (1961) on How Many Strata to Use**
>
> Cochran (1961) addressed the question of how many different categories should be used for a stratifying variable. For example, if two strata are better than one, are seven strata better than six?
>
> *Study design*: Cochran analyzed different conditions involving y, the survey variable, and x, the stratifying variable, assuming strata of equal sizes. He examined the stratification effects on the sampling variance of the mean given different levels of correlation between y and x. Cochran also displayed some real examples.
>
> *Findings*: Most of the stratification gains are obtained with six or fewer categories (strata) on a stratifying variable. The table shows reduced sampling variance from additional strata on the same variable, but at a diminishing rate of return for each additional stratum. With larger correlations between the stratifying variable (x) and the survey variable (y), creating more strata pays off. The real data follow the simulation results.
>
> **Design Effects for the Stratified Mean**
>
> | No. of Strata | Simulated Correlation between x and y | | Real Data | |
> |---|---|---|---|---|
> | | 0.99 | 0.85 | College Enrollments | City Sizes |
> | 2 | 0.265 | 0.458 | 0.197 | 0.295 |
> | 3 | 0.129 | 0.358 | 0.108 | 0.178 |
> | 4 | 0.081 | 0.323 | 0.075 | 0.142 |
> | 6 | 0.047 | 0.298 | 0.050 | 0.104 |
> | Infinity | 0.020 | 0.277 | | |
>
> *Limitations of the study*: The study used a linear model to describe the x–y relationship; other results apply with more complex relationships.
>
> *Impact of the study*: The study guided practice to a relatively small number of strata on each stratifying variable.

sample is selected, the survey asks principals of the 480 schools how many noncollege preparatory or "vocational" classes are offered in the fall term. We compute a mean number of vocational classes for each stratum, and we compute an element variance s_h^2 as well. See the results of this stratified sampling in Table 4.2.

Although the sampling fractions are equal in all strata, the strata sample sizes differ. The means and element variances also vary across the strata, as we might expect. The rural schools have a higher mean number of vocational classes. Element variances increase as the mean number of vocational classes increase.

If we simply computed an unweighted mean across the 480 sample cases, we would have a biased mean because it would give too high a weight to rural schools. In this case, the unweighted mean would be simply

$$\bar{y} = (6+5+8)/3 = 6.3$$

We need to use a mean that weights each stratum by its proportion in the frame population, which in this case would be $\bar{y}_{st}$:

$$(0.4 \times 6) + (0.5 \times 5) + (0.1 \times 8) = 5.7$$

This yields a lower mean because the unweighted mean overrepresented the rural schools.

One of the problems with weighted estimation like this is that standard statistical software does not automatically compute weighted averages of strata means. Instead, it computes weighted means for which there is a weight for each individual case. It is possible to translate this weighted sum of stratum means into the kind of weighted sum of individual elements used in standard software. In particular, suppose we look at the stratified mean estimator $\bar{y}_{st}$ from a different perspective:

$$\bar{y}_{st} = \sum_{h=1}^{H} W_h \bar{y}_h = \sum_{h=1}^{H} \left(\frac{N_h}{N} \right) \bar{y}_h$$

STRATIFICATION AND STRATIFIED SAMPLING

This form of the computation does not display how it could be a function of values on individual data records. With a little algebra, you can show that this is equivalent to

$$\bar{y}_{st} = \sum_{h=1}^{H}\sum_{i=1}^{n_h} w_{hi} y_{hi} \bigg/ \sum_{h=1}^{H}\sum_{i=1}^{n_h} w_{hi}$$

where w_{hi} is a weight variable on the dataset, equal to $w_{hi} = (N_h/n_h)$ for all of the elements, i, in the hth stratum. In short, the weighted mean is the ratio of the weighted total to the sum of the weights. This is the computing formula used by SPSS, STATA, or SAS in computing the weighted mean. Notice that the value of the weight for each element is merely the inverse of its sampling probability.

The sampling variance of $\bar{y}_{st}$ is most easily represented as a weighted sum of variances across the strata. If simple random sampling has been used in each stratum, then

$$v(\bar{y}_{st}) = \sum_{h=1}^{H} W_h^2 \text{ (variance of } h\text{th stratum mean)}$$

$$= W_1^2 \left(\frac{1-f_1}{n_1}\right)s_1^2 + W_2^2 \left(\frac{1-f_2}{n_2}\right)s_2^2 + W_3^2 \left(\frac{1-f_3}{n_3}\right)s_3^2$$

$$= (0.4)^2 \left(\frac{1-0.06}{192}\right)(5) + (0.5)^2 \left(\frac{1-0.06}{240}\right)(4) + (0.1)^2 \left(\frac{1-0.06}{48}\right)(7)$$

$$= 0.00920$$

That is, the estimation of the variance is a stratum-by-stratum calculation, followed by a combining of stratum results. This underscores that the key to stratification gains is finding groups that have low internal variation on the survey variable.

The 95% confidence interval for the estimated mean of 5.7 visits uses the estimated mean and standard error. First, compute the standard error of the mean:

$$se(\bar{y}) = \sqrt{v(\bar{y})} = \sqrt{0.00920} = 0.096$$

Then, construct the 95% confidence interval around the mean as

$$\bar{y} \pm z_{1-\alpha/2} \times se(\bar{y}) = 5.7 \pm (1.96)(0.096) \quad \text{or} \quad (5.5, 5.9)$$

It is also useful to examine the design effect of the mean for this sample. The estimated sampling variance for the stratified random sample has been computed. The design effect requires that the simple random sampling variance for a sample of the same size be estimated. For the sample shown in Table 4.2, a separate calculation shows that $s^2 = 5.51$ for the sample of 480. With a SRS of 480 sample schools, the sampling variance of the mean is given as

$$v_{srs}(\bar{y}) = \left(\frac{1-f}{n}\right)s^2 = \left(\frac{1-0.06}{480}\right)(5.51) = 0.0108$$

Hence, the design effect is estimated as

$$d^2 = \frac{v(\bar{y}_{st})}{v_{srs}(\bar{y})} = \frac{0.00920}{0.0108} = 0.85$$

That is, the proportionate stratified random sample has a sampling variance that is about 85% of what a simple random sample of the same size would have obtained. This translates to a decrease in the standard error (and the confidence interval widths) of about

$$100 \times (1 - \sqrt{0.85}) = 8\%$$

These kinds of gains from proportionate allocation are almost always guaranteed, provided that the stratification itself is effective. Stratification almost always leads to gains in precision when (1) the frame is divided into groups that are substantively meaningful—groups that differ from one another on the study variables, and (2) proportionate allocation is used. Stratification, then, is a process that first seeks frame variables that are likely to be correlated with the kinds of variables to be measured in the survey. Even variables with modest levels of correlation with respect to survey variables can lead to good stratification.

What can we do when there is no frame information to use in stratification? Fortunately, this "no information" case is quite rare in practice. There is almost always some kind of information for each sample case that can be used to group population elements into groups that are going to be different with respect to the survey variables. For example, when a list of names is used, rough gender and ethnicity strata can be formed based only on the names. Perfection in classifying cases into strata is not necessary to achieve some of the precision gains of stratification.

4.5.2 Disproportionate Allocation to Strata

In addition to proportionate allocation, there are other allocations that can lead to smaller sampling variances than for simple random samples of the same size. There is one allocation in every problem that produces the smallest sampling variance for the sample mean that can be obtained for any allocation. This allocation is named after its inventor, Jerzy Neyman. The allocation is more complicated than the proportionate or equal sample size allocations, but a brief examination will uncover an important design principle.

The Neyman allocation for a stratified sample is one of the many disproportionate allocations available. It just happens to be one that has a smaller sampling variance than any other element sample of the same size. The Neyman allocation requires that we know not only the W_h for each stratum, but also either the standard deviations

$$S_h = \sqrt{S_h^2}$$

or some values that are themselves proportionate to them.

Compute the product $W_h S_h$ for each stratum, and sum these products across the strata. Then the Neyman allocation sample sizes are

$$n_h = n \frac{W_h S_h}{\sum_{h=1}^{H} W_h S_h}$$

That is, the sample sizes are made proportionate not to the W_h, but to the $W_h S_h$. Thus, if the stratum is large, allocate more of the sample to it, just as in proportionate allocation. However, if the stratum has more variability among the elements, as when

$$S_h = \sqrt{S_h^2}$$

is larger, then also allocate more of the sample to the stratum. In other words, when there is higher element variance in a stratum, increase the size of the sample taken from the stratum, relative to other strata that have lower element variance. This has great intuitive appeal. If some parts of the population have large variability, we need larger samples from them to achieve more stable sample survey statistics.

The Neyman allocation can lead to very large gains in precision over simple random sampling, and even over proportionate allocation. However, there are several weaknesses to the method. For one, it does not necessarily work well for proportions. When working with proportions, very large differences in the proportions between strata are needed, and it may be difficult to find variables that will achieve the large differences that are needed. Second, Neyman allocation works well for one variable at a time. If the survey is collecting data on more than just that one variable, the Neyman allocation for the other variables may differ from the one for the first. And it is possible that the Neyman allocation for primary variable of interest will not work well for the other variables, so much so that the design effects for the other variables are actually greater than one.

Disproportionate allocation to strata without good information about the within-stratum variances (S_h^2) can be dangerous, leading to increases in overall standard errors. For exam-

---

Neyman (1934) on Stratified Random Sampling

Neyman (1934) addressed the simple problem of how to choose a sample at a time when there was great controversy about purposive selection versus random selection.

Study design: Neyman presented the work of other researchers, mostly about social problems like poverty, birth rates, and so on. Some had argued that separating the sampling frame into separate groups that were different from one another and then picking some units was important to "represent" the population. Others were arguing that random sampling giving each person an equal chance of selection through randomization was important.

Findings: Neyman observed that the two designs each had merits and could be combined. Through examples and some mathematical reasoning, he showed that separating the population into "strata," as he called them, which had different values on the key statistic of interest, could be done prior to random selection. By sampling from each stratum using random sampling, confidence limits could be derived. Further, by using higher sampling fractions in strata with large internal variability on the key variables, the confidence limits of the statistic from the stratified sample could be minimized for any given sample size.

Limitations of the study: Optimal allocations could vary for different statistics from the same survey. Implementing the allocation required knowledge of population variances within strata. The findings ignore nonresponse, coverage, and measurement errors in the survey.

Impact of the study: The paper was a fundamental breakthrough and encouraged the use of stratified probability sampling throughout the world.

ple, one naïve allocation is to assign equal numbers of sample cases to each stratum, even though the strata are of different sizes. If such a design were used in the illustration of Table 4.2, each stratum might be assigned 160 sample schools, totaling the same 480 as with proportionate sampling. Such a design, however, would have achieved a sampling variance of the stratified mean of 0.0111 instead of the 0.0108 of a simple random sample (or the 0.0092 of the proportionate stratified design).

Neyman allocation

There is much research left to be done on disproportionate allocation in sample design. The Neyman allocation addresses a survey with a single purpose in mind: the estimation of the population mean. However, most surveys are done to estimate scores, hundreds, or thousands of statistics simultaneously. Optimizing the design for just one statistic is generally not desirable, so the practicing survey researcher cannot rely on Neyman allocation for just one. There have been some serious studies of how to improve sample designs for multipurpose samples (Kish, 1988), but practical tools have yet to be developed. Further, Neyman allocation ignores coverage, nonresponse, and measurement errors that might vary over strata in the design.

Despite these complications, stratification and careful allocation of the sample are important tools in sample design and selection. Stratified sampling can achieve the goal of obtaining a sample that is representative of the population for every possible sample that can be drawn, a goal that cannot be obtained through purely random selection. Practical sampling invariably uses some kind of stratification, and yields sampling variances that are smaller than those obtained if no stratification had been used.

4.6 Systematic Selection

systematic selection

Systematic selection is a simpler way to implement stratified sampling. The basic idea of systematic sampling is to select a sample by taking every kth element in the population. Determine the population and sample sizes and compute the sampling interval k as the ratio of population to sample size. Then choose a random number from 1 to k, take it, and select every kth element thereafter.

Figure 4.7 displays the same example as Figure 4.6, illustrating a stratified element sample. It shows a frame population of 20 employers, sorted into four size strata based on number of employees (Highest to Low). In stratified element sampling, there was a sample of size one employer drawn from each of four strata, implementing a sampling fraction of 1/5 in each stratum. A systematic selection on this same sort order can provide many of the same advantages of stratified element sampling. The sampling fraction of 1/5 translates to a systematic selection interval of $k = 1/f = 5$. To implement a systematic selection, first choose a "random start" (in this case, from 1 through 5); in Figure 4.7 this is $RS = 2$. The first selected element is the second on the list, the next selected element is the $(2 + 5) = $ 7th on the list, the third is the $(7 + 5) = $ 12th on the list, and the fourth is the $(12 + 5) = $ 17th on the list. (If we kept going like this, the next number would be 22, beyond the total number of 20 frame elements.) The figure illustrates how this process always achieves representation of cases from each successive group of five elements in the listing order. In this case, every five elements is one of the strata in the original stratified element sample, so the systematic selection attains the same representational properties of the stratified element design.

SYSTEMATIC SELECTION

Record	Name	Group
1	Cochran Inc.	Highest
2	Hu Electronics	Highest
3	Madow USA	Highest
4	M.P.H. Bank	Highest
5	Wolfe & Enix	Highest
6	Bradburn Corp.	High
7	Deming Design	High
8	HydeBev	High
9	Sudman Inc.	High
10	Wallman AG	High
11	Fuller & Fuller	Medium
12	Habermann AG	Medium
13	Kalton Group	Medium
14	Norwood LC	Medium
15	WXM Venture	Medium
16	Hansen PLC	Low
17	Kish Consulting	Low
18	Rubin Inc.	Low
19	Sheatsley Co.	Low
20	Steinberg Ltd.	Low

Systematic Selection, RS = 2

→ Hu Electronics
→ Deming Design
→ Habermann AG
→ Kish Consulting

Figure 4.7 Frame population of 20 establishments sorted by group with systematic selection; selection interval = 5 and random start = 2.

The systematic selection process is thus much easier to apply than simple random or stratified random selection. The selection process does have a couple of drawbacks. For one, the interval k is not always an integer. What should be done if k has a decimal part to it? There are a number of methods to handle "fractional intervals" in systematic selection. One is to round the interval to the nearest integer. For example, suppose a sample of 1000 is to be selected from a population of 12,500, requiring an interval of 12.5. The interval could be rounded up to 13. A problem arises, though: the sample size will no longer be 1000. After a random start from 1 to 13, the sample would either be 961 or 962. If the interval were rounded down to 12, the sample size would be 1041 or 1042. Rounding the interval gets back to an integer, but the price is a departure from the desired sample size.

A second approach is to treat the list as circular. Calculate the interval and round it up or down (it actually doesn't matter which is done). Suppose we choose the interval of 12 for this illustration. Choose a starting place anywhere on the list by choosing, say, a random number from one to 12,500. Suppose the start is 12,475. Start there, and begin counting every 12th element: 12,475, 12,487, and 12,499 are the first three selections. Since the end of the list has been reached, continue counting from the end of the list back around to the beginning—12,500,

> **Comment**
>
> Look again at Figure 4.7. Notice that there are only five possible sample realizations from the systematic selection. Random start 1 selects elements (1,6,11,16); random start 2 selects (2,7,12,17); random start 3 selects (3,8,13,18); random start 4 selects (4,9,14,19); and random start 5 selects (5,10,15,20). Only these five realizations of the sample design are possible. The sampling variance of the sample mean is a function of the variation across these five realizations. (One by-product of selecting just one of them in a survey is that an unbiased estimate of the sampling variance is not possible. Generally, variance estimates assuming SRS or a stratified element sample are used in practice for systematic samples.)

1, 2, ... , 11. The next selection is the 11th element from the beginning of the list. Continue counting and selecting until the needed 1000 elements are chosen.

A third method directly using the fractional intervals is also available. It includes the fraction in the interval, but rounds after the fractional selection numbers have been calculated. This turns out to be an easy type of selection method when the sampler has a calculator handy. Start by choosing a random number that includes a decimal value. If the interval is 12.5, choose a random number from 0.1 to 12.5 by choosing a three digit random number from 001 to 125, and then inserting a decimal before the last digit. This random number, with inserted decimal, is the starting selection number. Then continue by counting every 12.5 elements; add 12.5 to the start, and add 12.5 to that number, and 12.5 to the next, and so on, until the end of the list is reached. Then go back and drop the decimal portion of each of these numbers. The sample will be of exactly the needed size.

For example, suppose the random start were 3.4. Then the selection numbers would be 3.4, 15.9, 28.4, 40.9, 53.4, and so on. "Rounded" down by dropping the decimal portion of the selection number, the sample is

fractional interval   the 3rd, the 15th, the 28th, the 40th, the 53rd, and so on, elements. Notice that the "interval" between the selections is not constant. First it is of size 12, and then 13, and then 12, and so on. The interval between the actual selections varies, and it averages 12.5.

Systematic sampling from an ordered list is sometimes termed "implicitly stratified sampling" because it gives approximately the equivalent of a stratified proportionately allocated sample. Thus, for systematic samples from ordered lists, there will be gains in precision compared to simple random selection for every survey variable that is correlated with the variables used to do the sorting.

An important type of sorting of some kinds of elements is by geography. For many kinds of units, geographic location is correlated with characteristics of the units. For example, suppose a survey were conducted to estimate average number of employees per firm in a state with 12,500 firms, and the firms were sorted geographically by their location from the southeast corner of the state to the northwest in some fashion. Those firms in or near large metropolitan areas would have larger numbers of employees than those firms in rural areas. Sometimes, firms would tend to be grouped together geographically by size. A systematic sample drawn from the geographical list would implicitly stratify firms by size, and lead to gains in precision compared to simple random selection.

4.7 COMPLICATIONS IN PRACTICE

In the discussion of cluster sampling in Section 4.4, only the rather limited case of clusters of exactly the same size was considered. In practice, clusters seldom

COMPLICATIONS IN PRACTICE

contain the same number of elements. There is a selection technique that is often used to select clusters of unequal size. It sets up a design that selects the sample in multiple stages. This is a very common design in large surveys and, hence, important to understand. We will illustrate the technique using a two-stage example.

Suppose that a sample of 21 housing units is to be selected from a frame population of nine blocks that have a total of 315 total housing units. And suppose that the number of housing units in the nine blocks are, respectively, 20, 100, 50, 15, 18, 45, 20, 35, and 12. The sample of 21 housing units from 315 housing units produces a sampling fraction of $f = 21/315 = 1/15$.

Imagine that we decide to select the sample by first randomly choosing three blocks, at a sampling rate of $f_{blocks} = 3/9 = 1/3$, and then subsampling housing units from each block at random with a sampling rate for housing units of $f_{hu} = 1/5$. The overall probability of selection for a housing unit is the product of the probability of selecting its block and the probability of selecting the housing unit, given that its block was selected. Across the two stages of selection, the sampling probability for selecting housing units is thus

$$f = f_{blocks} \times f_{hu} = (1/3)(1/5) = 1/15$$

two-stage design

the desired fraction. One would expect, on average, a sample of size $(1/15) \times 315 = 21$ to be obtained. This is a two-stage sample design; the first stage selects blocks; the second, housing units on the selected blocks. Multistage designs always have this nested property. First, "primary" selections are made; then, within the selected primary units, "secondary" units, and so on. Some ongoing sample surveys like the NCVS can have four or more stages in parts of their designs.

However, when this procedure is applied to the housing units in the nine blocks, the sample size of 21 will only rarely be achieved. For example, suppose the blocks of size 100, 50, and 45 are selected. Then the number of selected housing units would be the second-stage fraction applied to the three blocks,

$$x = \left(\frac{1}{5}\right)(100 + 50 + 45) = 39$$

selected housing units, much larger than the expected 21 selected housing units. On the other hand, if the block sample chose the 4th, 5th, and 9th blocks, the sample would be of size

$$x = \left(\frac{1}{5}\right)(15 + 18 + 12) = 9$$

selected housing units. This variation in the achieved number of selected housing units would cause logistical problems in completing the survey, but also causes statistical inefficiencies. The logistical problems arise in anticipating how many questionnaires need to be prepared or interviewers hired. The statistical problems arise through the loss of precision that results from a lack of control by design over the total number of sample households. Hence, there are methods to reduce this lack of control over sample size.

4.7.1 Two-Stage Cluster Designs with Probabilities Proportionate to Size (PPS)

What is needed with a two-stage sample from blocks of unequal size is a method to select housing units (or whatever the elements are) with equal chances, and yet control the sample size so it is the same across possible samples. One way to do this would be to select three blocks at random, but then take exactly seven housing units in each. That would achieve the same sample size each time, no matter which blocks were chosen. Unfortunately, the selection does not give equal chances of selection to all housing units. Housing units from small blocks would tend to be overrepresented in such a sample, because once their block was chosen they would have a higher chance of selection than housing units in a much larger block.

probability proportionate to size

The method referred to as "probability proportionate to size," or PPS, addresses this problem by changing the first- and second-stage selection chances in such a way that when multiplied together the probability is equal for every element, and the sample size is the same from one sample to the next.

Consider, for example, Table 4.3, in which blocks are listed in order, with their sizes in terms of housing units, and the cumulative number of housing units given for each block across the list. Assume that we are attempting the same overall probability of selection of housing units, 1/15, and that the same number of sample blocks, 3, as used in the example above. Draw three random numbers from 1 to 315, say, 039, 144, and 249. Find on the cumulative list the first block with a cumulative size that first includes the given random number. Blocks 2 (with cumulative measure 120, the first larger than 039), 3 (the first with cumulative measure larger than 144), and 7 (the first with cumulative measure greater than or equal to 249) are chosen.

These blocks have now been chosen with probabilities that are not all the same. For example, the first chosen block, with 100 housing units, had a much greater chance of being selected for this sample than did the first block on the list with only 20 units. For each of the three random selections, it had a chance

Table 4.3. Block Housing Unit Counts and Cumulative Counts for a Population of Nine Blocks

Block	Housing units on Block α	Cumulative	Selection Numbers for the Block
1	20	20	001–020
2	100	120	021–120 ← 039
3	50	170	121–170 ← 144
4	15	185	171–185
5	18	203	186–203
6	45	248	204–248
7	20	268	249–268 ← 249
8	35	303	269–303
9	12	315	304–315

COMPLICATIONS IN PRACTICE

of being selected of 100 out of 315 or 3 × 100/315 = 300/315. The second chosen block had a chance of selection of (number of random selections) × (probability of block being selected once) = 3 × 50/315 = 150/315. The third chosen block had a much smaller chance of selection, 3 × 20/315 = 60/315. To compensate for these unequal selection probabilities, probability proportionate to size selection chooses housing units within blocks with chances that are inversely proportional to the block selection rates. For example, choose a sampling rate in the first chosen block for housing units such that the product of the first-stage rate,

$$f'_{block} = 3 \times 100/315 = 300/315$$

is "balanced" by this within-block rate to obtain the overall desired sampling rate of $f = 1/15$. In this case, if housing units were sampled within the block at a rate of

$$f_{hu} = \frac{(1/15)}{(300/315)} = \frac{21}{300} = \frac{7}{100}$$

the overall rate would be achieved, since for the first block

$$f = f'_{block} \times f_{hu} = \left(\frac{300}{315}\right)\left(\frac{7}{100}\right) = \frac{1}{15}$$

That is, once the first chosen block has been selected, subsample housing units at the rate of 7/100 or select at random 7% of them. Similarly for the second chosen block, select housing units within it with the rate 7/50, or an overall rate of

$$f = f'_{block} \times f_{hu} = \left(\frac{150}{315}\right)\left(\frac{7}{50}\right) = \frac{1}{15}$$

Take exactly seven housing units from the second chosen block. The same would be done in the third chosen block.

This kind of a *pps* procedure yields an equal chance selection for elements, and it ends up with the same number of elements being chosen from each cluster. Thus, it achieves *epsem* and fixed sample size.

There are several problems that arise with the application of this method in practice. First, the choice of random numbers in the first stage may choose the same block more than once. For example, the random numbers could have been 039, 069, and 110, choosing the second block three times. This "with replacement" selection of the blocks is acceptable, but often a sample of three distinct blocks is preferred. There are several methods to achieve a sample of distinct blocks, and not all of them are *epsem* for elements. One involves systematic selection of block with *pps*, but it is beyond the scope of this volume to consider this method here (see Kish, 1965).

A second problem arises when some of the blocks do not have sufficient numbers of elements to yield the required subsample size. Suppose, for example, that the third chosen block in the example had only six housing units in it. Then it would be impossible to get a sample of seven housing units, at least distinct ones as in simple random sampling, from the block. In cases like this, where there are undersized blocks that can be identified in advance of selection, a simple solution is to link blocks to form units that are of sufficient size. So the seventh block in the list, if too small, could be linked to the sixth, reducing the total number of units from nine to eight. The *pps* selection could be applied to these eight units, cumulating size across them and proceeding as above.

Finally, it is also possible to have blocks that are too large. For example, suppose the second block had a size of 120 instead of 100. Then its first-stage sampling fraction would exceed 1.0 (3 × 120/315 = 360/315). In other words, it would always be selected, regardless of the random start. One common remedy for this kind of a block is to take it out of the list, and choose housing units from within it directly. So, in this case, a sample would be drawn from block 2 with probability 1/15, or a sample of size

$$x_2 = \left(\frac{1}{15}\right)(120) = 8$$

housing units. In addition, a sample would be drawn from the remaining eight blocks with *pps*, choosing two blocks and housing units within them at the overall rate $f = 1/15$.

4.7.2 Multistage and Other Complex Designs

Many surveys based on probability methods use sample designs that are combinations of techniques discussed in previous sections. They employ stratification of clusters, selection of clusters, and subsampling of clusters. The subsampling may extend to three or four stages. There may also be unequal probabilities of selection and other complexities introduced into selection to make it practical to select a sample with available resources.

Subsampling through multiple stages of selection is often used in situations in which a list of population elements is completely lacking. It is cost-effective, as discussed in Section 4.4, to select a sample of clusters of elements, and then obtain or create a list of elements for each of the sample clusters. However, if the clusters selected in the first stage are large, having many elements, creating a list is still not feasible. Multistage sampling is used to further reduce the size of the listing task by identifying and choosing another set of clusters within the selected first-stage clusters. It may be the case that even the second-stage units are too large to create lists of elements, and a third stage of selection may be needed. The multistage selection continues until a unit of small enough size to create a list at reasonable cost is obtained.

This kind of multistage selection arises most often in the context of a widely used procedure called "area sampling." Area sampling chooses areas, such as counties, blocks, enumeration areas, or some other geographic unit defined by government agencies, as sampling units.

There is much methodological research that is needed to make multistage cluster designs more efficient. Sampling theory provides some tools for guiding

area sampling

COMPLICATIONS IN PRACTICE   129

decisions on how many first-stage, second-stage, third-stage, etc., selections to make for different precision targets. However, optimization requires knowledge of both variation in key statistics across the different stages of selection and relative costs of selections at different stages. Cost models in complex data collections (e.g., knowing the costs of recruiting interviewers who are residents of the primary areas, of traveling to second-stage areas, of contact attempts, etc.) need to be seriously studied and made more useful at the design phase. The use of computer-assisted methods has given survey researchers more administrative data useful for such design decisions, and these should be used in future research.

More research is needed also for cluster designs when the purpose of the survey is both to describe the population of clusters and the population of elements within clusters. For example, NAEP selects schools, then students within sample schools. The data could be useful for studying properties of schools and properties of students. In recent years, there have been important new developments in statistical models that measure effects on some phenomena that stem from multiple levels (e.g., impact on assessment scores from the student's, her parent's, her teacher's, and her school's attributes). Sample designs to maximize the precision of such models need further development.

4.7.3 How Complex Sample Designs Are Described: The Sample Design for the NCVS

We can combine all the lessons of sample design above to examine one of our example surveys in detail, the National Crime Victimization Survey (NCVS). Recall that the target population of the NCVS is the civilian noninstitutionalized household population age 12 years old and older. We note that in the United States there is no list of names and addresses of such persons, so one problem in the sample design is frame development. The design is based on a multistage list and area probability design.

Below appears the description of the sample design that is provided to users of the National Crime Victimization Survey when the survey data are requested (Interuniversity Consortium for Political and Social Research, 2001). We present the description and then offer comments on each section of it.

> The NCVS sample consists of approximately 50,000 sample housing units selected with a stratified, multistage cluster design. Currently, the sample is about 51,000 households per month. [Note: these are assigned addresses, before ineligible (i.e. vacant, nonresidential ones) are removed.] The number of eligible households is the 42,000 cited above, which results in about 38,000 household interviews (91% of 42,000). The Primary Sampling Units (PSUs) composing the first stage of the sample were counties, groups of counties, or large metropolitan areas. PSUs are further grouped into strata. Large PSUs were included in the sample automatically and each is assigned its own stratum. These PSUs are considered to be self-representing (SR) since all of them were selected. The remaining PSUs, called nonself-representing (NSR) because only a subset

of them was selected, were combined into strata by grouping PSUs with similar geographic and demographic characteristics, as determined by the Census used to design the sample.

Comment: The first-stage selection units are area-based units—counties and groups of counties. Some are so large (like the problem treated in Section 4.7.1) that they are in all realizations of the design. These are called self-representing units (e.g., New York, Los Angeles, Chicago, and other big urban counties).

Prior to 1995, the sample was drawn from the 1980 decennial Census. From January, 1995 until December, 1997, the sample drawn from the 1990 Census was phased in. Since January, 1998, the complete NCVS sample has been drawn from the 1990 Census.

Comment: This means that the probabilities of selection of the primary sampling units are based on population counts from the 1990 Census for these data files. Ideally, the probabilities of selection would be based on current data. If the measure is out of date, the result is variation in realized sample size across the sample units.

The current design consists of 84 SR PSUs and 153 NSR strata, with one PSU per stratum selected with probability proportionate to population size.

Comment: The primary sampling units are stratified before selection takes place. There are 153 strata used in the design, based on region, size of PSU, and other variables.

segments
The sample of housing units within a PSU was selected in two stages. These stages were designed to ensure a self-weighting probability sample of housing units and group quarters dwellings within each of the selected areas. (That is, prior to any weighting adjustments, each sample housing unit had the same overall probability of being selected.) The first stage consisted of selecting a sample of Enumeration Districts (EDs) from designated PSUs. (EDs are established for each decennial Census and are geographic areas ranging in size from a city block to several hundred square miles, and usually encompassing a population of 750 to 1500 persons.) EDs were systematically selected proportionate to their 1980 or 1990 population size. In the second stage, each selected ED was divided into segments (clusters of about four housing units each), and a sample of segments was selected.

Comment: The nested stages of selection are (1) PSUs (counties or county groups), (2) enumeration districts (census units of about 1000 persons), (3) groups of adjacent housing units (about 4 housing units) called "segments." Then every housing unit in a selected segment is designated for interviewing.

COMPLICATIONS IN PRACTICE											131

> The segments were formed from the list of addresses compiled during the 1980 and 1990 Censuses. However, procedures allow for the inclusion of housing constructed after each decennial Census enumeration. A sample of permits issued for the construction of new residential housing is drawn, and for jurisdictions that do not issue building permits, small land area segments are sampled. These supplementary procedures, though yielding a relatively small portion of the total sample, enable persons living in housing units built after each decennial Census to be properly represented in the survey. In addition, units in group quarters known as special places were also selected in special place segments. These units, such as boarding houses and dormitories, constitute a small portion of the total sample. The sample design results in the selection of about 50,000 housing units and other living quarters.

Comment: There are problems with using lists of addresses covered by the last decennial census. There may have been new construction of units since the last census. Multiple frames are used to update the list of addresses on the census frame. Thus, the NCVS is a multiple frame design (see Section 3.7.3). For other survey organizations, the address lists from the last decennial census are not available. These organizations will send interviewers to sample blocks in order to have them list all of the addresses, or housing units, in the sample blocks. These are returned to a central facility where they are keyed, and then a sample selected for the last household listing operation to identify eligible persons.

> Because of the continuing nature of the National Crime Victimization Survey, a rotation scheme was devised to avoid interviewing the same household indefinitely. The sample of housing units is divided into six rotation groups, and each group is interviewed every six months for a period of three-and-a-half years. Within each of the six rotation groups, six panels are designated. A different panel is interviewed each month during the six month period.

rotation group

Comment: After the entire sample is defined, six random subgroups are defined and enter the sample over consecutive months. These "rotation groups" are part of the rotating panel nature of the NCVS. In any given month, some sample members are providing their first interview; some, their second; some, their third; and so on. Those providing their first interview are "rotating into" the panel for the first time. Those providing their seventh interview will "rotate out" of the NCVS.

An interviewer visits each selected housing unit, makes a listing of all persons who are household members, and then attempts to collect NCVS data on all those 12 years old and older.

What results from all of this is a sample of persons, clustered into housing units, that are clustered into segments, that are clustered into enumeration districts, that are clustered into primary sampling units. Stratification takes place before selection at the segment level (through spatial groups), at the enumeration district level (again, through spatial ordering), and at the primary sampling unit level. Probabilities proportionate to size are introduced at all stages, but the overall probability of selection of each person is equal within each panel used in the sample.

4.8 SAMPLING U.S. TELEPHONE HOUSEHOLDS

A large number of surveys in the United States employ the landline telephone as a medium of interviewing. When telephone interviewing is used, the sampling frame most often used is one containing landline telephone numbers, each of which might be assigned to a household. Persons in the household of a selected telephone number are then listed and one or more selected. Because telephone samples are prevalent in the United States, there is well-developed information about the telephone number sampling frame.

Telephone numbers in the United States are ten-digit numbers with three parts:

$$734 - 764 - 8365$$
$$\text{Area Code} - \text{Prefix} - \text{Suffix}$$

At this writing, there are over 300 area codes in the United States that contain working household numbers. Prefixes are restricted in range from 2XX to 9XX, where the Xs can be any digit from 0 to 9. This means that there are about 800 possible prefixes in each area code. For each prefix, there are 10,000 possible individual phone numbers (i.e., 0000–9999), and this level has the largest change over time, as subscribers move in and out of the population. From this we can observe that there are over 2.4 billion possible telephone numbers in that frame. In 2000, there were approximately 110 million households, with about 89 million having at least one line telephone subscription. If every household had one and only one telephone number, the percentage of household numbers among all possible numbers would be less than 4%. That is, most of the frame would contain ineligible numbers for a household survey. This is the simplest "random digit dialed" sample design. It takes all working area code-prefix combinations that are working and appends to them at random four-digit random numbers (from 0000 to 9999). The resulting ten-digit numbers are the same numbers, among which we would expect that a very small percentage would be working household numbers.

Such designs are so inefficient that they are never used. At the current time, if the frame were restricted to prefixes that are activated, a higher percentage of the possible numbers would be working household numbers. This offers a large boost in efficiency. But even this efficiency can be improved by separating the prefixes into smaller sets of numbers.

The use of the numbering plan is such that household numbers cluster in certain series. A common series is the 100 block of numbers, representing 100 consecutive numbers within an area code–prefix combination (e.g., 734-764-8365 lies in the 100 block defined by 734-764-8300 to 734-764-8399). As it turns out, only a small portion of blocks have one or more listed telephone numbers. In those blocks, the rate of working household numbers is much higher. Thus, many practical uses of random digit dialing, like those used in BRFSS and SOC, limit the sample to 100 blocks with some minimum number of listed numbers. By limiting the frame, there is more undercoverage of telephone households, but very large increases in the proportion of numbers that are household numbers.

In addition to the random digit dialed frame, there are alternative telephone number frames, but they suffer from problems of undercoverage of the telephone household population. The most obvious alternative is the set of numbers listed in the residential sections of telephone directories across the country. There are commercial companies that assemble these numbers into electronic files. A large portion of numbers are not listed in these directories, either because they are newly activated (and thus missed the printing date of the directory) or the subscriber requested that the number remain unlisted. In large urban areas it is common that less than half of the households have their numbers listed in the directory. Hence, using telephone directories as a frame would severely limit the coverage of the frame.

The frame based on all numbers in working area codes and prefixes is becoming less efficient over time. In the early 1990s, two changes occurred to the U.S. telephone system that added more area codes and prefixes. First, there was a large increase in the use of telephone numbers to facilitate fax transmissions and computer electronic communication via modems. This led to the creation of new telephone numbers, some of which are installed in homes, which are not primarily used for voice communication. When dialed, they are often answered by an electronic device or never answered at all.

Second, to reduce the cost of telephone service, local area competition was introduced. Prior to the early 1990s, one company was authorized to provide telephone service for a given area (the so-called telephone exchange). The system was changed so that more than one company could offer service in the same area. In addition, the new company could offer service to customers that would allow them to retain their previous telephone number, a service known as telephone number portability. However, the telephone numbering system is also a billing system. The area code–prefix combination is assigned to a single company for billing purposes for its customers. If the new company allowed its customers to retain their old number, the billing system would be confused. To allow correct billing, the new company was given an area code–prefix combination to use for billing that contained "shadow" or "ghost" numbers. When the customer places a call from their retained telephone number, the bill for the call is transferred to the "shadow" number. But this addition of new area code–prefix combinations enlarged the pool of available telephone numbers much faster than the number of telephone households increased. The result was a drop in the proportion of telephone numbers that were actually assigned to households. This change in the telephone system led to an increase in the number of activated area codes and prefixes in the United States. Figure 4.8 shows how the telephone system changed between 1986 and 2008. The figure presents on the x axis the number of phone numbers within a block of 100 consecutive phone numbers that are listed in a phone directory. It is based on dividing all the active area code–prefix combinations into such 100 blocks of numbers (i.e., there are about 2.5 million such 100 blocks). In 1986, the modal number of listed numbers in the population of 100 blocks in the United States was about 55; in 2008 it was less than 25, a dramatic decline in the density of listed household numbers.

In short, telephone surveys face a trade-off between one frame—listed numbers—which fails to cover large portions of telephone households, but contains a

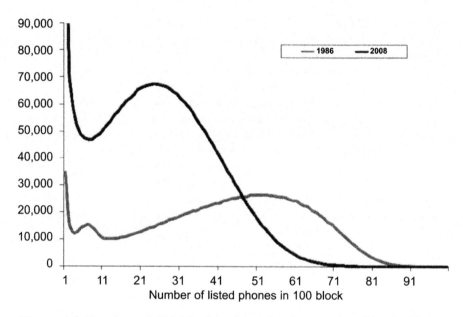

Figure 4.8 Number of 100 blocks of number by number listed within the block, 1986 and 2008. (Source: Survey Sampling, Inc.)

very high percentage of eligible numbers, and the RDD frame, which covers all numbers but also contains large number of ineligible numbers. With such a situation, various more complex design alternatives have been developed, usually implementing some stratification connected to oversampling of portions of the RDD frame (e.g., list assisted sampling; see Tucker, Lepkowski, and Pierkarski, 2002), cluster samples that oversample 100 blocks densely filled with household numbers, dual-frame designs that combine RDD frame and the listed frame (e.g., Traugott, Groves, and Lepkowski, 1987), and other schemes that enjoy the benefits of the RDD frame but attempt to reduce the costs of screening through nonresidential numbers. There is much methodological research that needs to be conducted on U.S. telephone sample designs. As mobile telephones become more prevalent, there will be many research questions about how to blend frames of the line and mobile telephones, how to optimize the allocation between frames, how to measure overlaps of frames, and how to minimize nonresponse errors of such mixed mode designs.

4.9 SELECTING PERSONS WITHIN HOUSEHOLDS

In many household surveys estimating person-level statistics, the last step of sampling involves the selection of one person within a selected household. There are several methods in practice for this step. They include: (a) making a sampling frame of all eligible household members and selecting one of them at random, (b) selecting at random a person with given age and gender attributes, (c) selecting an eligible household member whose month and day of birth is closest to the date of interview (either before or after), and (d) selecting persons in priority order based on their likelihood of being interviewed.

Kish (1949) first described a method of sampling adults in a household by listing all persons in the household, first males from oldest to youngest, then females oldest to youngest. The interviewer numbered sequentially those householders eligible for the survey (e.g., those 18 years or older). In paper questionnaire surveys, one of eight different patterns of selections was assigned to each sample household. This yielded nearly an equal probability selection scheme within households of the same number of eligibles. That is, both the eligible persons in two-person households had the same probability of selection, all eligible persons in three-person households had the same probability, and so on. (In very large households, there were some problems of giving all persons equal probabilities with the set of eight patterns; this was repaired with computer-assisted software by using a random number between 1 and the total eligible persons.) There is research that evaluates the listings of persons within households, principally motivated by within-household coverage problems of censuses. It appears that when the household has complicated membership, informants tend to omit persons with tenuous ties to the household (Martin, 1999). There is also some evidence of the same when a household member violates some legal or normative rule (male partners to single mothers receiving child support payments dependent on her living without a partner) (Tourangeau, Shapiro, Kearney, and Ernst, 1997). These measurement errors produce frame errors on these household listings.

The Kish method requires the informant to identify all eligible persons in the household prior to selecting one at random. Rizzo, Brick, and Park (2004) describe a scheme whereby only one question is initially asked. For example, if the survey was limited to adults, "How many persons 18 years or older live here?" If there is only one eligible, that person is immediately selected. If there are two, then, randomly, one of the two is selected; if the "other" is selected, the interviewer says, "The other adult has been selected." If there are more than two in the household, the software selected a number between 1 and the number of eligibles. If "1" is selected, then the interviewer says, "You have been selected." If another number is randomly generated, the interviewer completes a household listing and selects the person corresponding to the random number. This limits the full listing to those households with three or more eligibles (in the United States this is only about 15% of households).

Another approach can be effective in household populations with relatively small number of eligible persons in each household. It is most often implemented in a computer-assisted format, using a random number selection to identify the sex and relative age of the sample person. To illustrate this, imagine a household with two eligible males and three eligible females. To prepare for the selection, the interviewer first asks how many eligible males live in the household; then how many females live in the household. Given the example of two males and three females, the software then chooses a random number between 1 and 5. If "1" is selected, the software tells the interviewer to identify the "older male" as the sample person. If "2" is selected, the software tells the interviewer to identify the "younger male." If "3" is selected the software specifies to select the "oldest female"; "4," the "second oldest female"; if "5," the "youngest female." The probability of selection of each eligible person is 1/5 in this case. In general, the probability of selection is a function of the number of eligible persons listed.

The third method does not yield a probability sample (the chances of selection of each eligible person cannot be determined). It attempts to avoid intrusive questions required to build a frame of eligible persons. After the introduction of the survey, the interviewer says, "I'd like to interview the adult in the household with the next birthday." (Alternatively, the interviewer asks, "I'd like to interview the adult in the household with the last birthday.") If the month and year of birth in households were randomly assigned, then this procedure gives all persons an equal chance of selection. If the month and day of birth is correlated to the survey variables in some way, then the procedure is biased. In some circumstances (e.g., a survey of attitudes toward aging), the person with the last birthday may have distinctive attributes relative to others.

The final method worth noting is a nonprobability selection approach that is popular in U.S. commercial surveys with short data collection periods limiting callbacks on sample households. The method attempts to select persons who tend to be rarely encountered among respondents (either because they are difficult to contact or because they tend to refuse survey requests). An extreme variant of this limits selection to those who are at home at the time the interviewer calls (Hill, Donelan, and Frankel, 1999). The interviewer asks first to speak to "the youngest male, 18 years or age or older, who is now at home." If no adult male is at home, the interviewer then asks to speak to the "oldest female, 18 years of age or older, who is now at home." This obvious limits respondents to those at home when the interviewer calls; thus, the sample properties are a function of calling times. If no weekend calls are made, those only at home on the weekends cannot be in the respondent pool. If people with different at-home patterns vary on the survey attributes (e.g., a survey of leisure time use), then the method can yield biased results.

Gaziano (2005) presents a summary of the methodological evaluations about within-household selection and points out the key issues with the various techniques. Many arise because of concerns that full probability methods tend to produce higher nonresponse rates. Some of the full probability methods require more questions of the informant before the selection of the respondent can be made. Gaziano notes that many studies speculate that an important determinant of nonresponse is how the interviewers are trained to administer the methods. Thus, there is lack of replication in the evaluations of alternative methods.

All of the methods require some knowledge of all household members; some, the age and gender; others, who is at home at a specific time. When the informant does not have perfect knowledge (common among households with unrelated persons), the methods can break down. For example, Rizzo, Brick, and Park (2004) find that informants lack birthday information in 16% of households with more than four adults.

At this point, different survey organizations have made different decisions about whether the questions required to select a respondent at random are too intrusive to obtain high response rates. There is no debate that probability sampling requires knowledge of the total number of eligible persons in the household and a way to identify uniquely one chosen at random. Practice differs because of the desire to minimize selection biases and nonresponse rates simultaneously, and the belief that full probability methods depress response rates.

With probability designs that select one person per household, the selected person has a probability of selection that is inversely proportional to the number of eligible persons in the household. This number must be recorded in the data in order to compute a selection weight for the case, which is simply the number of eligible persons in the household. You can remember this by considering that the one sample person in some sense represents all persons in the household and must be weighted up by this number to correctly represent their contribution to the overall population estimate. This weight is then multiplied by the selection weight of the household, as well as any other adjustment weights (see Chapter 10), to produce the final weight for the respondent case.

4.10 SUMMARY

Sample design is the most fully developed component of survey research. Most of the practices of sample design are deduced from theorems in probability sampling theory. Simple random sampling is a simple base design that is used for the comparison of all complex sample designs. There are four features of sample design on which samples vary:

1) The number of selected units on which sample statistics are computed (other things being equal, the larger the number of units, the lower the sampling variance of statistics).
2) The use of stratification, which sorts the frame population into different groups that are then sampled separately (other things being equal, the use of stratification decreases the sampling variance of estimates).
3) The use of clustering, which samples groups of elements into the sample simultaneously (other things being equal clustering increases the sampling variance as a function of homogeneity within the clusters on the survey statistics).
4) The assignment of variable selection probabilities to different elements on the frame (if the higher selection probabilities are assigned to groups that exhibit larger variation on the survey variables, this reduces sampling variance; if not, sampling variances can be increased with variable selection probabilities).

Much of sample design is dictated by properties of the sampling frame. If there is no frame of elements, clustered samples are often used. If there are auxiliary variables on the frame (in addition to the identifying variables for the element), these are often used as stratifying variables.

KEYWORDS

area sampling
cluster sample
confidence limits
design effect

rotation group
sample element variance
sample realization
sampling bias

effective sample size
epsem
finite population correction
fractional interval
intracluster homogeneity
Neyman allocation
precision
probability proportional to size (PPS)
probability sample
random selection
roh
sampling fraction
sampling variance
sampling variance of the mean
sampling without replacement
segments
simple random sample
standard error of the mean
strata
stratification
stratum
systematic selection
two-stage design

For More In-Depth Reading

Kish, L. (1965), *Survey Sampling*, New York: Wiley.

Lohr, S. (1999), *Sampling: Design and Analysis*, Pacific Grove, CA: Duxbury Press.

Exercises

1) Consider a small population consisting of $N = 8$ students with the following exam grades:

$$Y_1 = 72, Y_2 = 74, Y_3 = 76, Y_4 = 77, Y_5 = 81, Y_6 = 84, Y_7 = 85, Y_8 = 91$$

a) Compute the population mean $\bar{Y} = \frac{1}{N}\sum_{i=1}^{8} Y_i$.

b) Compute the population variance $S^2 = \frac{1}{N-1}\sum_{i=1}^{8}(Y_i - \bar{Y})^2$.

c) Identify all possible simple random samples of size 2. (There should be 28 possible samples, ignoring the order of selection.) Compute the mean of each possible sample $\bar{y} = \frac{1}{2}(y_1 + y_2)$ and make a histogram of the sample means.

d) Compute the sampling variance of the sample means obtained in c) by computing $V(\bar{y}) = \frac{1}{S}\sum_{s=1}^{S}(\bar{y}_s - \bar{Y})^2$, where S is the number of samples of size $n = 2$ selected from the $N = 8$ elements in the population, and the s subscript represents the different samples. Compare this result to the computed sampling variance $V_{srs}(\bar{y}) = \frac{1-f}{n}S^2$.

e) Identify all the possible simple random samples of size $n = 6$. Compute their means and make a histogram of the sample means. (Hint: It is eas-

EXERCISES

iest to do this using the results of part c. How does the distribution differ from parts b and c?)

2) Using the same population of $N = 8$ students and exam scores from Exercise 1, consider the stratified population in two groups:

Low scores: $\quad Y_{11} = 72,\ Y_{12} = 74,\ Y_{13} = 76,\ Y_{14} = 77$

High scores: $\quad Y_{21} = 81,\ Y_{22} = 84,\ Y_{23} = 85,\ Y_{24} = 91$

a) Identify all possible stratified random samples of size $n = 6$ obtained by selecting $n_h = 3$ from each stratum.
b) Compute the means from each of the samples identified in part a and make a histogram of the sample means. How does the distribution of these sample means compare with the unstratified version from Exercise 1?
c) Calculate the sampling variance of the mean from the stratified sample means obtained in part b by computing $V(\bar{y}) = \dfrac{1}{S}\sum_{s=1}^{S}(\bar{y}_s - \bar{Y})^2$, where S is the number of samples of size $n = 6$ selected from the $N = 8$ elements in the population.
d) Compare the result obtained in part c to the computed sampling variance
$$V(\bar{y}_{st}) = \sum_{h=1}^{2} w_h^2 \dfrac{(1 - f_h)}{n_h}.$$
(Hint: calculate $S_1^2,\ S_2^2$ from the population data.)

3) Design effects are used to evaluate the precision of statistics for different sample designs.

a) Is the design effect of a clustered element sample likely to be larger or smaller than one?
b) Is the design effect of a stratified element sample likely to be bigger or smaller than one?
c) In a single stage clustered sample, if within a cluster a variable has nearly the same value for all elements within the cluster, what value will the intraclass correlation be close to?
d) For a single-stage clustered sample, the intraclass correlation for a key variable is 0.016 and the cluster size is 10. Calculate the design effect for the mean of that key variable.
e) What does the design effect in part d mean?

4) A survey is to be conducted to study work absence due to acute illness in a factory with 1200 workers. Suppose the mean number of days lost per year is 4.6 and that the standard deviation is 2.7 days lost per year, and the sample is to be selected by simple random sampling.

a) What sample size is needed to produce an estimate of the mean number of days lost with a standard error of $se(\bar{y}) = 0.15$?

b) Now suppose that a proportionate stratified sample with $H = 6$ strata, formed by the cross-classification of sex and age (in three age groups) is to be drawn instead. Consider the sample size necessary to achieve a standard error for the mean of 0.15. Will the sample size for the proportionate stratified design be smaller, the same, or larger than that computed in part a? Briefly explain your answer.

5) The following is a list of $A = 10$ blocks. Draw a PPS systematic sample, using $X\alpha$ as the measure of size. Use a random start of 6 and an interval of 41.

Block	X_α	Cumulative X_α	Selection
1	32		
2	18		
3	48		
4	15		
5	37		
6	26		
7	12		
8	45		
9	46		
10	21		

6) The frame for an education survey includes $A = 2000$ high schools, each containing $B = 1000$ students. An epsem sample of $n = 3000$ students is selected in two stages. At the first stage, $a = 100$ schools are selected randomly, and at the second stage $b = 30$ students are selected in each sampled school. Of the selected students, 30% reported having access to computers at home. A published estimate gives the standard error of this percentage as 1.4%. Ignoring finite population corrections and approximating $(n - 1)$ by n, estimate the following:

a) The design effect d^2 for the sample percentage.

b) The within-school intraclass correlation *roh* for percentage of student who have access to a computer at home.

c) The standard error for the sample percentage for a sample design that selected $a = 300$ schools and $b = 10$ students per school. (Hint: calculate a new design effect $d_{new}^2 = 1 + (b_{new} - 1)roh$ and multiply it by the simple random sampling variance of the proportion $\dfrac{p(1-p)}{n}$, where p is the sample proportion, and not the percentage.)

7) A sample of $n = 10$ was selected by simple random sampling from a list of 12,000 registered voters in a city. Each household where the sample registered voters lived was visited and information collected on whether the household had central air conditioning.

Case Number	Number of Registered Voters	Air Conditioning Indicator
1	1	Present
2	1	Absent
3	1	Present
4	2	Present
6	2	Absent
7	3	Absent
8	3	Absent
9	3	Present
10	4	Present

a) Estimate the percentage of registered voters living in households with central air conditioning.
b) Estimate the percentage of households with central air conditioning.

8) A stratified sample was selected to estimate annual visits to a doctor. The stratum size (N_h), the proportion of the population (W_h), the sample size (n_h), the sampling fraction (f_h), the total number of doctor visits (y_h), the mean number of doctor visits $(\bar{y}_h)$, and the sample element variance (s_h^2) in each of three strata are as follows:

Stratum	N_h	W_h	n_h	f_h	Y_h	$\bar{y}_h$	S_h^2
Young	3200	0.40	192	0.06	1152	6	5
Middle aged	4000	0.50	240	0.06	1200	5	4
Old	800	0.10	48	0.06	384	8	7
Total	8000	1	480	0.06	2736		

a) Compute the unweighted mean of the strata means.
b) Compute the weighted stratified mean.
c) Compute the sampling variance of the mean from part b.
d) Compute a 95% confidence interval for the mean computed in part b.

9) A medical practice has records for $N = 900$ patients. A simple random sample of $n = 300$ was selected, and 210 of the sample patients had private health insurance.

a) Estimate the percentage of patients with private health insurance and the standard error of this estimate.
b) Calculate a 95% confidence interval for the population percentage.
c) The study is to be repeated in another medical practice that has $N = 1000$ patients. A standard error of 2.5 percentage points for the sample percentage of patients with private health insurance is required. What sample size is needed for a simple random sample to achieve this level of precision? For planning purposes, assume that the population percentage is 50%.

10) In a sample of $a = 10$ clusters containing $b = 10$ completed household interviews each, the total number of persons and the total number of cell phone users among those in the table below.

Cluster	Total # of people	# of Cell Phone Users	Proportion of cell phone users
1	40	10	0.25
2	38	8	0.21
3	25	10	0.40
4	13	5	0.38
5	22	13	0.59
6	28	12	0.43
7	34	10	0.29
8	42	14	0.33
9	20	8	0.40
10	30	10	0.33
Total	292	100	

EXERCISES

a) Under an assumption of simple random selection of households, estimate the proportion p with cell phones and its standard error. (Assume that the finite population correction is close to 1.)
b) Since the data were collected using cluster sampling, estimate the standard error of the prevalence of cell phone users p using a method for cluster samples.
c) Compare the results in parts a and b, discussing the differences.
d) Compute the design effect d^2 for the estimated prevalence of cell phone users under the clustered sample design.
e) Estimate roh.
f) Calculate the effective sample size under the cluster sample design.

11) A population of $N = 10,000$ farms has a variance for annual crop yield of $S^2 = 1,000,000$. Calculate the simple random sample size needed to obtain a standard error for the mean annual crop yield of $\sqrt{1,000}$.

12) A college has $N = 150$ faculty members. The dean wants to do a faculty salary survey selecting a systematic sample of $n = 20$ faculty. The list of faculty members appears in the table at the end of this chapter's exercises.

a) What should be the sampling interval for this design?
b) What are the possibilities for the first random start?
c) Draw a sample of 20 faculty members.
d) Estimate the mean salary for the faculty using the data in the sample obtained in part c.

13) There are two entrances to a convention center that delegates must pass through to attend a conference. Systematic samples of subjects are selected at each entrance with intervals 25 and 75, respectively, where entrance 2 admits three times as many delegates as entrance 1. Sample delegates are asked the distance they have traveled to attend the conference. The following results are obtained for 10 sample delegates:

Entrance	1	1	1	1	1	2	2	2	2	2
Distance	12	34	450	75	240	470	455	24	16	200

a) Estimate the mean distance traveled and its standard error. Provide a justification for any assumptions you make about the delegates in order to estimate the standard error.
b) Compute 95% confidence interval for the estimated mean.
c) A sample is to be drawn at another convention center with three entrances, where it is known that $S_1 = 100$, $S_2 = 200$, and $S_3 = 400$. Twice as many delegates pass through entrance 1 as entrance 2, and three times as many delegates pass through entrance 2 as entrance 3. That is, $W_1 = 0.6$, $W_2 = 0.3$, and $W_3 = 0.1$. Suppose we select from Entrance 1 with the

interval 25. What sampling intervals should be used for each the other two entrances to have an *epsem* sample design?

14. A two-stage cluster sample of $n = 1200$ in $a = 60$ clusters is selected from a large population with $S^2 = 500$. In a published report, $v(\bar{y}) = 9$ is reported. A colleague says there is an error in the calculation of the variance of the mean. Do you agree? Show a numeric justification for your answer. (Hint: What is the value of *roh* implied by this reported variance?)

15) Answer the following statements true or false. Give a brief justification for each of your answers:

   a) All probability samples are measureable. (If false, give an example of a probability sample that is not measureable.)
   b) All epsem samples are measureable.
   c) Cluster samples are always less precise than simple random samples of the same size.
   d) To increase precision, stratification seeks heterogeneity between strata, while clustering seeks heterogeneity among clusters. (If false, give an explanation of what stratification and clustering do seek to do.)
   e) Element variances S^2 will vary depending on what sample design is used.

16) For a PPS sample of 100,000 employees of businesses with sample size 1000, and 10 sample employees per business, Chamber of Commerce counts of employees in each business in 2001 are to be used as measures of size.

   a) What would be the probability of selection of a business with a 2001 count of 40?
   b) Describe how you would select two businesses from one stratum that had the following measures of size in it: 56, 14, 84, 92, 16, 8, 30.

17) The following are $n = 20$ simple random selections from a population of $N = 270$ blocks. For each selection, the number of rented dwellings is given.

i	1	2	3	4	5	6	7	8	9	10
y_i	31	21	2	19	35	0	17	0	27	27
i	11	12	13	14	15	16	17	18	19	20
y_i	1	8	31	11	59	11	5	47	0	54

   a) Compute the mean number of rented dwellings per block,
   $$\bar{y} = y/n = \sum_{i=1}^{n} y_i / n.$$

b) Compute the element variance $s_y^2 = \left(\sum_{i=1}^{n} y_i^2 - (y^2/n)\right)/(n-1)$, the sampling variance $\text{var}(\bar{y}) = (1 - f)s^2/n$, and standard error of the mean $se(\bar{y}) = \sqrt{\text{var}(\bar{y})}$.

c) Estimate the total number of rented dwellings in all blocks, $N\bar{y}$, and its standard error $se(N\bar{y}) = N \times se(\bar{y})$.

d) Compute a 95% confidence interval for the estimated mean, $\bar{y} \pm t_{(n-1, 1-\alpha/2)} \cdot se(\bar{y})$.

e) If the sample size were increased from $n = 20$ to $n = 50$, what would be the estimated standard error?

f) What sample size n is needed to make the standard error 3.5?

18) Consider the population of faculty in the list of $N = 150$ at the end of the problem set. Select an SRS of $n = 15$ faculty by drawing three digit random numbers from 001 to 150 until 15 different faculty have been selected. Disregard random numbers that are blank (there is no corresponding faculty on the frame). Compute the following quantities for your sample:

a) The sample mean: $\bar{y} = y/n = \sum_{i=1}^{15} y_i / n$.

b) The sample element variance:
$$s^2 = \sum_{i=1}^{n}(y_i - \bar{y})^2 / (n-1) = \left(\sum_{i=1}^{n} y_i^2 - \frac{y^2}{n}\right)/(n-1).$$

c) The sampling variance of the mean: $\text{var}(\bar{y}) = (1 - f)s^2/n$.

d) The standard error of the mean: $se(\bar{y}) = \sqrt{\text{var}(\bar{y})}$.

e) A 95% confidence interval for the sample mean: $\bar{y} \pm t_{(1-\alpha/2; n-1)} \times se(\bar{y})$.

19) The following table gives population information for a stratified population. You want to estimate the proportion of units in the population that have the characteristic. The total sample size is $n = 30$.

Stratum h	Population count Nh	Proportion with a characteristic Ph
1	100	0.9
2	200	0.5
3	300	0.1

a) Compute the stratum sample sizes for proportional and equal allocation. (Do not round your answers to integers.)

b) Compute the standard errors for the estimated proportion for each of the allocations in part a. What are the degrees of freedom for each estimated standard error? (Retain 4 decimal places for calculations. Finite population correction factors can be omitted.)
c) The Neyman allocation will result in the smallest standard error of any of the allocations, but survey designers do not necessarily use it. Give two reasons why we do not always use the Neyman allocation.
d) If you wanted to publish an estimate for each stratum what allocation would you use? Explain your reasoning.

EXERCISES

Faculty Salaries (in $1,000)

#	Division	Sex	Rank	Salary	#	Division	Sex	Rank	Salary	#	Division	Sex	Rank	Salary
1	Eng&Prof	m	3	$88	51	Eng&Prof	m	3	$55	101	Lit&SocSci	m	2	$55
2	Medicine	f	3	$45	52	Biol&Sci	m	1	$49	102	Medicine	m	3	$80
3	Medicine	m	3	$57	53	Eng&Prof	m	3	$57	103	Eng&Prof	m	1	$114
4	Medicine	m	1	$133	54	Medicine	m	1	$118	104	Lit&SocSci	m	1	$63
5	Eng&Prof	f	2	$71	55	Medicine	m	3	$84	105	Medicine	m	1	$112
6	Lit&SocSci	m	1	$113	56	Eng&Prof	m	3	$52	106	Medicine	m	1	$93
7	Medicine	f	3	$65	57	Medicine	m	3	$64	107	Lit&SocSci	m	2	$47
8	Biol&Sci	m	3	$47	58	Eng&Prof	m	1	$75	108	Biol&Sci	m	1	$127
9	Lit&SocSci	f	3	$39	59	Medicine	f	1	$87	109	Eng&Prof	m	2	$121
10	Biol&Sci	m	1	$74	60	Eng&Prof	m	3	$58	110	Medicine	m	3	$58
11	Medicine	m	1	$88	61	Medicine	f	3	$39	111	Biol&Sci	f	3	$97
12	Lit&SocSci	m	1	$62	62	Medicine	m	3	$69	112	Lit&SocSci	m	1	$71
13	Lit&SocSci	m	1	$49	63	Medicine	f	2	$46	113	Eng&Prof	m	1	$72
14	Medicine	m	3	$88	64	Eng&Prof	f	1	$86	114	Lit&SocSci	m	3	$29
15	Medicine	m	1	$181	65	Medicine	m	3	$87	115	Medicine	m	2	$167
16	Eng&Prof	m	3	$63	66	Medicine	m	3	$59	116	Lit&SocSci	m	3	$36
17	Medicine	m	2	$94	67	Eng&Prof	f	3	$44	117	Medicine	m	1	$57
18	Eng&Prof	m	1	$91	68	Medicine	m	2	$123	118	Biol&Sci	m	1	$107
19	Medicine	m	1	$60	69	Lit&SocSci	f	3	$37	119	Medicine	m	2	$88
20	Eng&Prof	m	3	$55	70	Lit&SocSci	m	1	$106	120	Medicine	m	2	$87
21	Biol&Sci	m	2	$55	71	Lit&SocSci	m	1	$91	121	Lit&SocSci	f	2	$43
22	Medicine	f	1	$106	72	Lit&SocSci	m	1	$78	122	Lit&SocSci	m	1	$79
23	Medicine	m	1	$116	73	Biol&Sci	m	1	$77	123	Medicine	m	2	$113
24	Medicine	m	3	$79	74	Medicine	m	1	$90	124	Medicine	m	3	$55
25	Lit&SocSci	m	1	$61	75	Eng&Prof	m	2	$71	125	Medicine	m	3	$57
26	Lit&SocSci	f	3	$37	76	Medicine	f	3	$42	126	Eng&Prof	m	3	$56
27	Medicine	m	2	$72	77	Medicine	f	2	$59	127	Eng&Prof	m	2	$65
28	Eng&Prof	m	1	$105	78	Eng&Prof	m	2	$49	128	Medicine	m	2	$42
29	Medicine	m	2	$79	79	Biol&Sci	m	1	$83	129	Medicine	m	1	$102
30	Medicine	m	1	$61	80	Lit&SocSci	m	1	$34	130	Medicine	f	3	$40
31	Medicine	m	1	$86	81	Medicine	f	3	$42	131	Eng&Prof	m	3	$53
32	Biol&Sci	m	1	$103	82	Medicine	m	2	$97	132	Medicine	m	3	$82
33	Lit&SocSci	m	1	$48	83	Medicine	m	1	$109	133	Medicine	m	2	$64
34	Eng&Prof	m	2	$64	84	Lit&SocSci	f	2	$48	134	Eng&Prof	m	1	$72
35	Eng&Prof	m	1	$78	85	Medicine	m	1	$47	135	Biol&Sci	f	3	$36
36	Medicine	f	2	$53	86	Eng&Prof	m	2	$45	136	Lit&SocSci	f	1	$66
37	Biol&Sci	m	1	$85	87	Medicine	m	3	$83	137	Medicine	f	3	$66
38	Eng&Prof	m	1	$61	88	Medicine	m	2	$51	138	Medicine	m	2	$102
39	Medicine	m	1	$106	89	Biol&Sci	m	1	$78	139	Biol&Sci	m	1	$103
40	Lit&SocSci	m	2	$60	90	Lit&SocSci	m	1	$70	140	Medicine	m	1	$148
41	Biol&Sci	f	1	$73	91	Eng&Prof	f	2	$46	141	Lit&SocSci	f	1	$60
42	Medicine	m	1	$70	92	Eng&Prof	m	1	$85	142	Lit&SocSci	f	3	$46
43	Medicine	f	3	$32	93	Lit&SocSci	m	1	$53	143	Lit&SocSci	f	1	$57
44	Lit&SocSci	m	2	$49	94	Medicine	f	3	$40	144	Medicine	f	2	$50
45	Eng&Prof	m	3	$43	95	Eng&Prof	m	1	$87	145	Lit&SocSci	m	1	$90
46	Medicine	m	1	$75	96	Lit&SocSci	m	1	$71	146	Eng&Prof	m	3	$63
47	Lit&SocSci	m	1	$92	97	Medicine	m	1	$75	147	Eng&Prof	m	1	$80
48	Medicine	m	2	$107	98	Biol&Sci	m	1	$85	148	Medicine	m	3	$56
49	Biol&Sci	m	2	$57	99	Lit&SocSci	m	2	$50	149	Medicine	m	1	$72
50	Medicine	m	2	$114	100	Medicine	m	3	$118	150	Eng&Prof	m	1	$96

CHAPTER FIVE

METHODS OF DATA COLLECTION

No matter how good its sample design, a survey will yield biased results if the rest of the survey design is not appropriate to its purpose. This includes writing and testing a measurement instrument, typically a questionnaire (see Chapters 7 and 8). In addition, if interviewers are used, they must be recruited, trained, and supervised (see Chapter 9). This chapter focuses on another key set of decisions that affects the survey, that is, the method of data collection used.

The term "data collection" is a little misleading, since it implies that the data already exist and merely need to be gathered up (see Presser, 1990). Survey data are usually produced or created at the time of the interview or completion of the questionnaire. In other words, they are a product of the data collection process, a notion that underscores the importance of this step. Despite this, we will use the usual terminology of "data collection" here. As Figure 5.1 shows, all design aspects flow through the data collection process to produce survey estimates. The two inferential steps described in Chapter 2 come together in the data collection process. Although often viewed as an operational step following design, data collection is an essential element in the production of useful data for analysis and is subject to empirical research informed by theory. This chapter focuses on decisions related to the choice of a data collection method and the implications of such decisions on costs and errors in surveys.

Surveys have traditionally relied on three basic data collection methods: mailing paper questionnaires to respondents,

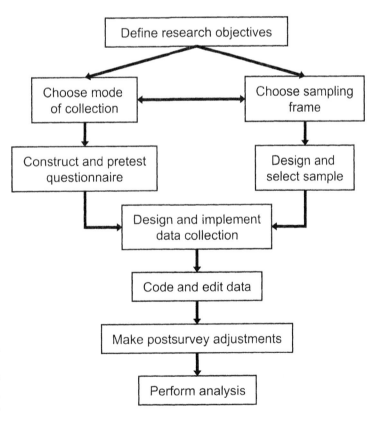

Figure 5.1 A survey from a process perspective.

Survey Methodology, Second Edition. By Groves, Fowler, Couper, Lepkowski, Singer, and Tourangeau
Copyright © 2009 John Wiley & Sons, Inc.

who fill them out and mail them back; having interviewers call respondents on the telephone and asking them the questions in a telephone interview; and sending interviewers to the respondents' home or office to administer the questions in face-to-face (FTF) interviews. The computer revolution has altered each of these traditional methods and added new ones to the mix. Not only has there been a proliferation of methods, but researchers are increasingly combining modes to minimize costs and/or errors. This has greatly expanded the design choices that need to be made during this phase, and the need for research evidence to inform such choices. This chapter focuses on the decisions related to the choice of a data collection method and the implications of such decisions on costs and errors in surveys. There are two basic issues that affect the choice of a data collection method:

1) What is the most appropriate method to choose for a particular research question?
2) What is the impact of a particular method of data collection on survey errors and costs?

To answer these questions, both the survey designers and the analysts of the data must understand the data collection process. From the design perspective, the decisions made at the outset will often determine how the data can be collected, with implications for costs, data quality, nonresponse, coverage, and so on. From the analytic perspective, understanding how the data were collected is critical to evaluating the quality of the resulting estimates. Thus, whether one is faced with choosing the mode or that decision has already been made, understanding the details of the particular method(s) used is an important part of understanding the quality of the data obtained.

5.1 Alternative Methods of Data Collection

Over the past 25 years or so, survey methodologists have invented many new methods of collecting survey data. The possibilities are even more numerous since surveys sometimes use multiple methods simultaneously. For instance, the method used to contact sample units may not be the same method used to collect the data. Different modes of data collection may be used for different parts of the interview, or one method may be used initially and another method used during later follow-up. For example, in past years, the National Survey on Drug Use and Health (NSDUH) used interviewer administration for some items but self-administration for others. In a longitudinal survey (which collects data from the same respondents several times), one mode may be used in the first wave of the survey, but another mode may be used for later waves. For example, the National Crime Victimization Survey (NCVS) interviews respondents up to seven times; it uses face-to-face interviews in the initial round but attempts to conduct telephone interviews in later rounds. The initial contact is still in person. Subsequently, the telephone is used if possible. Face-to-face interviews are used for first-time interviews because they have been found to boost response rates and to motivate the respondents to provide accurate information in later waves. Later rounds use telephone interviews to save money over face-to-face methods.

In the first few decades of the survey research enterprise, "mode" typically meant either face-to-face (or personal-visit) interviews or mail surveys. In the late 1960s, telephone surveys became more common, and their use grew dramatically in the following decades. Early mode-comparison studies contrasted these three methods. Much of the mode effects literature has its roots in the face-to-face versus telephone comparisons of the 1970s, and later expanded to include other methods. For example, Groves and Kahn's (1979) study compared a national sample (clustered in 74 counties and metropolitan areas) interviewed face-to-face with two national telephone samples, one clustered in the same primary areas and one selected via random digit dialing throughout the United States. Their study compared response rates, costs, coverage, and measurement error by mode of administration. On the whole, the telephone and face-to-face interviews yielded similar results, providing further impetus to the use of telephone interviews to gather survey data.

With the proliferation of new data collection methods in recent years, largely associated with the introduction of computers to the survey process, survey mode now encompasses a much wider variety of methods and approaches, including combinations of different approaches or mixed-mode designs. Some of the most common methods of data collection in use now are:

1) Computer-assisted personal interviewing (CAPI), in which the computer displays the questions on screen, the interviewer reads them to the respondent, and then enters the respondent's answers.
2) Audio computer-assisted self-interviewing (audio-CASI or ACASI), in which the respondent operates a computer, the computer displays the question on its screen and plays recordings of the questions to the respondent, who then enters his/her answers.
3) Computer-assisted telephone interviewing (CATI), the telephone counterpart to CAPI.
4) Interactive voice response (IVR), the telephone counterpart to ACASI (and, therefore, also called telephone ACASI, or T-ACASI), in which the computer plays recordings of the questions to respondents over the telephone who then respond by using the keypad of the telephone or saying their answers aloud.
5) Web surveys, in which a computer administers the questions online.

Figure 5.2 presents a more comprehensive list of current methods of data collection and traces the relations among them. All of the methods have their roots in the traditional trio of mail surveys, telephone interviews, and face-to-face interviews. The best way to read the figure is to move from left to right, the dimension that reflects both the passage of time and the increased use of computer assistance. The original mailed paper questionnaire survey was first enhanced with the use of optical character recognition (OCR) for printed responses and later with intelligent character recognition (ICR) for handwriting, using machines to read and code answers from completed questionnaires; this has evolved to the use of fax delivery of self-administered paper questionnaires, most often used in business surveys. This branch of development used computers in the processing stage of a survey. The other branch of evolution from mail used computer assistance in capturing the data. For populations with personal computers (often business pop-

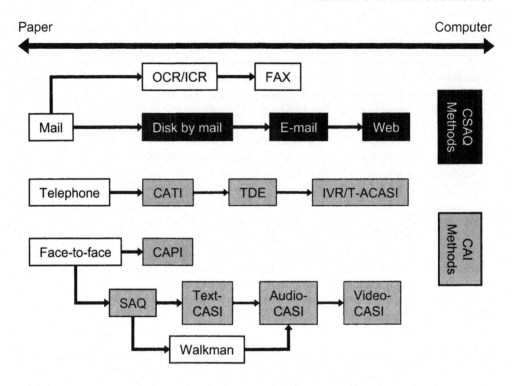

Figure 5.2 The evolution of survey technology.

disk by mail

CSAQ

TDE

IVR/ T-ACASI

ulations), survey researchers sent computer disks containing software presenting the survey questions and capturing data (disk-by-mail method). The respondent returned the disk upon completion of the survey. With the onset of the Internet, the questionnaire could be sent by e-mail to the respondent (e-mail method), and then later via Web pages (Web method). These methods are collectively known as computerized self-administered questionnaires (CSAQ). Web surveys have themselves expanded into a wide array of methods for selecting and inviting participants (see Couper, 2000) and ways of implementing survey questionnaires (see Couper, 2008b).

The evolution of methods using the telephone network first led to telephone interviewers' terminals connected to mainframe computers, then terminals connected to minicomputers, then individual networked personal computers. CATI software presented questions to the interviewers in the interview, the interviewers read the questions, and the software received and checked answers entered on the keyboard. Some business surveys, with limited data needs, used touchtone data entry (TDE). TDE had respondents call a toll-free number, hear recorded voice requests for data items, and enter the data using the keypad of their telephone. An advance in this area was interactive voice response (IVR or T-ACASI), in which the interviewer initiates the call before switching the respondent over to the automated system (see Steiger and Conroy, 2008; Tourangeau, Steiger, and Wilson, 2001). The distinction between recruit-and-switch and inbound IVR has implications for survey errors, particularly errors of nonobservation.

The evolution of the face-to-face mode followed two paths: one was replacement technologies for the paper questionnaire that interviewers used; the other was the self-administered questionnaires interviewers sometimes gave to respondents as part of a face-to-face protocol. Computer-assisted personal interviewing (CAPI) replaced the paper questionnaire, using laptop computers to display questions and receive data input. For the self-administered questionnaires (SAQ) that were sometimes handed to respondents to complete, computer assistance took various forms. The first was a short-lived transition technology—use of the Sony Walkman cassette tape player to provide audio delivery of questions that were answered by the respondent using a paper data form. The second was various computer-assisted self-interviewing (CASI) approaches, including laptops presenting text questions and accepting responses (text-CASI), an enhancement that delivers the questions in audio form also (ACASI), and a further enhancement that presents graphical stimuli as part of the measurement (video-CASI).

SAQ

CASI

text-CASI

video-CASI

Note that Figure 5.2 charts the progression from older (paper-based) technologies to newer, computer-based methods (see Couper and Nicholls, 1998, and Couper, 2008a, for a discussion of these different technologies). This list is by no means exhaustive, and new technologies or variations of existing methods are continually evolving. Clearly, it is no longer enough simply to refer to "mode effects" in surveys. We need to be explicit about what methods are being compared and what features of the design are likely to account for any effects we may observe. The different methods of data collection differ along a variety of dimensions: the degree of interviewer involvement, the level of interaction with the respondent, the degree of privacy for the respondent, which channels of communication are used, and the degree of technology use. The sections below discuss these.

5.1.1 Degree of Interviewer Involvement

Some survey designs involve direct face-to-face interactions between an interviewer, on the one hand, and the respondent or informant, on the other. For example, in a face-to-face interview, the interviewer reads the survey questions to the respondent. Other survey designs (e.g., a mail survey) involve no interviewers at all. Between these extremes, there are a variety of approaches, such as self-administered questionnaires (SAQs) completed as part of an interviewer-administered survey, with the interviewer present during completion, like the NSDUH use of ACASI for sensitive drug use measurement in the context of a personal-visit survey. Other variants involve interviewer recruitment and persuasion over the telephone before transfer to an automated system, using recruit-and-switch IVR or T-ACASI.

The continuum of interviewer involvement has many implications for survey quality and cost. Interviewer-administered surveys use a corps of trained, equipped, and motivated persons, requiring supervision, monitoring, and support. It is common that interviewer costs form a large portion of total costs in interviewer-assisted surveys. Use of interviewers affects the entire organization of the data collection effort.

Interviewers, however, can be effective recruiters of the sample persons, potentially affecting the nonresponse error features of the survey statistics. Section 6.7 shows how the interviewers' ability to address the sample persons' questions and concerns about the survey can lead to higher cooperation rates. Callbacks from interviewers appear to be effective reminders of the survey request to respondents who have not taken the time to respond. Interviewers are also effective in implementing the sampling design, from locating and identifying eligible sample units to carrying out within-household selection.

Interviewers can also assist in clarifying, probing, and motivating respondents to provide complete and accurate responses. This appears to be directly related to the percentage of questions unanswered in a survey questionnaire. However, in addition, the presence of an interviewer may harmfully affect the answers provided, especially to sensitive questions (see Section 5.3.5). With interviewer-administered surveys, such attributes of the interviewers as their race and sex and such task-related behaviors as the way they ask the questions can affect the answers (see Section 9.2.2). Indeed, almost all interviewer effects on respondent answers appear to be related to their important role in defining the social context of the measurement. For example, a question about racial attitudes may be interpreted differently depending on the race of the interviewer. Thus, interviewers can both increase and decrease the quality of survey statistics.

5.1.2 Degree of Interaction with the Respondent

Face-to-face or personal-visit surveys have a high degree of interaction with the respondent. The interview is often conducted in the respondent's home, with the answers being provided by the respondent to the interviewer. Telephone interviews, like the BRFSS, involve less direct contact with the respondent. Other surveys, such as agricultural surveys that measure crop yield, involve minimal interaction with the respondent, essentially seeking the farmer's permission to do the direct measurement on the crop. Still other surveys, especially those involving administrative records, involve no contact at all with the sample unit; field staff simply extract the necessary information from the records. For example, part of NAEP involves periodic examination of high school course transcripts from samples of students, based on the administrative record systems of the sample schools. There was a transcript study in 2005, and another is planned for 2009.

The more the data emerge from the interaction between the respondent and an interviewer, the more control the researcher has, at least in principle, over the measurement process. If a survey is based solely on administrative records, then the quality and content of the records are determined before the survey begins. Similarly, the coverage of the target population by the administrative records is out of the control of the researcher. On the other hand, the more that the data emerge during an interaction with a human respondent, the more they are susceptible to the vagaries of the particular data collection situation.

When administrative records are used, the nature of the questionnaire construction process should be informed by the nature of the record-keeping systems (Edwards and Cantor, 1991), not merely by the nature of the respondents' ability to comprehend the questions and access their memories to get the answers. Further, not all respondents will have access to the records needed to respond to all questions. Thus, with modes that cannot be easily used by multiple respondents, the completeness of the survey data depends on the sample person contacting others in the sample unit (e.g., household, business, or farm).

The researcher typically has less control over this dimension. The nature of the measurement being sought and the availability of administrative records may determine whether it is necessary to interact with a respondent or a less intrusive approach can be taken. Whether it makes sense to ask a farmer about crop yield or to measure this more objectively in the field may be determined by the particular research questions being asked. There is a growing tendency for surveys to combine objective measures (administrative records, direct observation, transaction data, etc.) with direct questioning of survey respondents. There is increasing recognition that respondents and records provide complementary information and that the combination may help overcome the weaknesses of each source of information. However, we should acknowledge that the error properties of the various sources of information might differ.

5.1.3 Degree of Privacy

Surveys can be conducted in a variety of settings, in part dictated by the mode of data collection, and these settings can differ in how much privacy they offer the respondent while completing the survey. Both the presence of the interviewer and the presence of other persons may affect respondents' behavior. To the extent that others may overhear the questions or see the answers provided by the respondents, their privacy is compromised.

In survey measurement, the principles underlying privacy effects are often similar to those involving issues of confidentiality of data. The presence of an interviewer implies that at least that person, in addition to the researcher, will be aware that the respondent was asked a question. With oral response by the respondent, the interviewer knows the answer. If there are other persons present during the interview or questionnaire filling process (e.g., other household members, a colleague in a business setting, passers-by in interviews taken in public settings), then others can overhear the interviewer-respondent interaction or see the questionnaire being completed. The loss of privacy implies a loss of control by the respondents over who knows that they were survey respondents. In some situations, it implies that the respondents' answers will be known by those who have nothing to do with the survey.

The impact of the privacy of the setting is likely to increase when the information sought is sensitive or potentially embarrassing to the respondent. For example, as its name implies, the National Survey on Drug Use and Health focuses on an extremely sensitive topic: illicit drug use. At one extreme, in face-to-face surveys conducted in the respondent's home, there is often little control over the presence of other family members. Exit surveys and group-administered surveys (e.g., in a classroom setting) also have low privacy. For example, the NAEP assessments are conducted in sample classrooms, in which all students in the classroom are simultaneously administered the examination.

The degree of privacy may depend on the proximity of respondents to each other (see Beebe, Harrison, McRae, Anderson, and Fulkerson, 1998). At the other extreme, a self-administered survey completed in a private room in a clinic or in a laboratory setting offers considerable privacy to the respondent. Threats to privacy come both from other members of the household (e.g., admitting in the presence of one's parents that one has used illicit drugs) and from the mere presence of the interviewer. For example, it may be difficult to admit negative attitudes toward a target group when the interviewer is a member of that group.

A variety of methods have been used to improve reporting by increasing the privacy of the responses. For example, paper-based, self-administered questionnaires (SAQs) have long been used as part of a face-to-face survey to elicit information of a sensitive nature. This is the original rationale of NSDUH using self-administration for measures of drug and alcohol use. This method has evolved along with data collection technology into various forms of computer-assisted self-interviewing (CASI) in which the respondents interact directly with the laptop computer for a portion of the interview. In the case of audio-CASI, the questions need not even be visible on the screen; the respondent hears the questions through headphones and enters the answer into the computer. Similar techniques, variously called interactive voice response (IVR) or telephone audio-CASI, have evolved for telephone surveys, in part to increase privacy for the respondent while answering sensitive questions.

5.1.4 Channels of Communication

channels of communication

"Channels of communication" is a phrase that denotes the various sensory modalities that humans use to obtain information from their external world. Various combinations of sight, sound, and touch can be used to communicate; each combination may yield different issues of comprehension, memory stimulation, social influence affecting judgment, and response hurdles. Survey modes vary in how they communicate the questions to the respondents, and in how the respondents communicate their answers to the survey researchers. Interviewer-administered surveys are principally aural; the interviewer reads the questions out loud and the respondent answers in like fashion. Mail surveys are visual; the respondent reads the questions and writes the answers on paper. Some surveys offer a mix of different communication channels. For example, face-to-face surveys can employ visual aids (such as show cards that list the response options for a question). CASI can be text only (visual), audio plus text, or audio only. Further, words and pictures are distinct tools. Hence, another distinction between data collection approaches may be the use of verbal stimulus and response material only (as in the telephone mode) versus the use of other visual stimuli (such as pictures, videos, and so on). The introduction of computers to the data collection process (CAPI and CASI) and the use of the World Wide Web have greatly expanded the type of material that can be presented to respondents in surveys (see Couper, 2001).

social presence

Early research in the social psychology of telecommunications studied the effect of different combinations of visual and aural channels of communication. For example, Short, Williams, and Christie (1976) identified a greater sense of "social presence" in the video channel of communication. "Social presence" is the psychological sense that a subject is aware of and in direct contact with the full person of the other actor, including his or her emotional state, indicating a single focus on the interaction (versus some multitasking). Social presence thus is a function of the salience to the subject of the other actor in an interaction. Short, Williams, and Christie found in a set of experiments that the audio channel complicates joint tasks that depend on reading the emotional states of actors engaged in the task. From the survey perspective, this might suggest that interviewers would tend to have smaller effects on the comprehension of questions in the telephone mode than in the face-to-face mode, in which nonverbal cues from the interviewer could be read by the respondent.

In the same vein, an interviewer in a face-to-face survey is more likely to pick up nonverbal indicators of reluctance or confusion, leading the interviewer to offer encouragement or clarification, even without explicit requests from the respondent. In self-administered surveys where no such multichannel communication occurs, such intervention is not possible.

Much of the earlier research on survey mode effects, especially telephone versus face-to-face comparisons, focused on the issue of the channels of communication. More recently, attention is being focused on the visual/aural differences between mail or Web surveys, on the one hand, and interviewer-administered surveys or telephone self-administered surveys, on the other. For example, primacy effects are more prevalent in visual modes of data collection, whereas recency effects are more common in auditory modes. With a primacy effect, presenting an option first (or at least near the beginning of the list) increases the chances that respondents will choose that option. With a recency effect, the opposite happens; putting an option at or near the end of the list increases its popularity (see Section 7.3.6).

primacy effects

recency effects

Research on both mail and Web surveys is also showing that visual syntax—the layout of the question and answer elements on the page or screen—affects the answers provided. Thus, principles of visual design and visual communication are germane to understanding how measurement error can be reduced in self-administered surveys.

5.1.5 Technology Use

Finally, survey data collection methods can vary in the degree and type of technology used (see Fuchs, Couper, and Hansen, 2000). Mail surveys, for example, use paper-based methods for data collection. Other than literacy and a pencil, no specialized skills or equipment are needed. In Web surveys, the respondent interacts with the survey instrument via the Internet, using their own hardware (computer, modem, etc.) and software (ISP, browser). In computer-assisted interviewing, the interviewer uses technology provided by the survey organization. The technology used and how it is used may have implications for standardization (who controls the equipment), degree and type of training required (e.g., CAPI interviewers require more training in the care and handling of the laptop computer and in data transmission issues than CATI interviewers in a centralized facility), coverage (Web surveys require access to the Internet; mail surveys require only the ability to read and write), and the costs of data collection. Technology can be used to reduce measurement error, through automated routing in complex questionnaires, edit checks, and other tools. On the other hand, the increased complexity of computerized instruments may introduce other types of errors (such as programming errors).

One issue in the use of technology is the degree of control exercised over the respondent. At one extreme, a paper SAQ places few constraints on the respondent. At the other extreme, CASI instruments may impose such constraints as restricting the range of possible answers, requiring an answer before proceeding, and limiting navigation around the instrument. Although cleaner data may result (see Chapter 10), these constraints may affect the way respondents answer the survey questions. For example, in a meta-analysis of social desirability distortions in various modes of test administration, Richman, Kiesler, Weisband, and

Drasgow (1999) found that giving respondents the ability to backtrack to review or change previous answers reduced social desirability effects.

Related to this, there is an increasing recognition that the design of a computerized instrument, whether it be CAPI, the Web, or some other method, can affect interviewer and respondent behavior and, hence, influence measurement error. Increasingly, principles of human–computer interaction and user-centered design are being applied to improve the design of such instruments.

5.1.6 Implications of these Dimensions

There are several implications of this broadened array of data collection tools at the survey researcher's disposal. One implication is that when we talk about a method of data collection, we need to be explicit about the particular method used. It is no longer sufficient to say that one type of survey (e.g., a personal-visit survey or an interviewer-administered survey) is better or worse than another type (e.g., a telephone survey or a self-administered survey), without being explicit about the particular implementation of each survey in terms of each of the five dimensions described here. For example, the strength of Web surveys lies in reduced costs, increased timeliness, and improvements in measurement; however, they suffer from problems of coverage and nonresponse, and challenges of constructing suitable frames for probability samples.

A related implication is that it is harder to make broad generalizations about the results of mode comparisons. The effect of a particular data collection method on a particular source of error may depend on the specific combination of methods used. The research literature does not yet cover all variations. As new methods are being developed, studies comparing them to the methods they may replace must be done. For this reason, theory is important to inform our expectations about the likely effect of a particular approach. Such theory is informed by past mode-effects literature as well as by an understanding of the features or elements of a particular design.

Furthermore, surveys often combine different methods, in mixed-mode or hybrid designs. As we already noted, the NCVS interviews part of the sample face to face (e.g., those completing their first interview) and the rest over the telephone. Although this combination may reduce overall costs, it may also affect nonresponse, coverage, and measurement error. Similarly, like NSDUH, many surveys now use a self-administered component (whether on paper or computer) to supplement an interviewer-administered survey. The choice of when to do this, what questions to include in the self-administered component, and so on, must be made in light of the survey error and cost implications of the various options. We return to the issue of mixed-mode designs later in this chapter (see Section 5.4).

Mode choices explicitly involve trade-offs and compromises. Survey designs are increasingly complex and often serve a variety of different needs. What might be best for one part of the survey, or for minimizing one source of error, may have implications for other parts of the survey or other error sources. For example, direct interaction with the computer, as in audio-CASI, may minimize some types of measurement error, but have an impact on nonresponse. Older or less-educated respondents may be reluctant to respond via computer (Couper and Rowe, 1996). No one data collection method is best for all circumstances. The choice of a par-

5.2 Choosing the Appropriate Method

If there is no one ideal mode for all survey applications, how does one go about choosing the appropriate method for a particular study? There are many considerations that need to be weighed in reaching a decision. These include the various sources of survey error as well as cost considerations (which could be viewed broadly to include logistical, personnel, time and other organizational issues). At times, the choice is fairly evident. For example, it would not make sense to do a survey to assess literacy by mail, since those low in literacy would be unable to comprehend the questions. Similarly, it would be hard to assess literacy over the telephone. How would one measure literacy without giving the respondent something to read and evaluating their ability to do so? These considerations imply a method involving face-to-face interaction. The NAEP measures knowledge of reading and mathematics as a way to assess the performance of schools. Situating the measurement in the school environment is appropriate because that is where the learning is focused.

At other times, some alternatives can easily be eliminated from consideration. For example, if one is estimating the prevalence of Internet use, doing so in a Web survey makes little sense. Estimating how many persons are having trouble getting medical care (as the BRFSS does) just from using medical record surveys would similarly miss the point.

But for a vast array of other survey designs, there are genuine trade-offs that can be made between alternative approaches, and the survey researcher must weigh the relative importance of a variety of factors in reaching a decision. Depending on the importance of coverage error, say, relative to measurement error or costs, one may reach different decisions on choice of method.

Despite the wide range of choices, there are obviously logical groupings of some methods. For example, telephone surveys are often considered as alternatives to face-to-face surveys, probably because both use interviewers and frame coverage issues are similar. Hence, the BRFSS, a telephone survey, could indeed be conducted by personal visit (with improved coverage of the household population), but would be much more expensive. The NCVS conducts first interviews by face-to-face methods and then switches to the telephone for most interviews in later waves.

Mail surveys are possible alternatives to telephone surveys of list frames containing both addresses and telephone numbers. Web surveys are being proffered as a replacement technology for mail surveys, intercept-based methods, and even telephone surveys, but few studies have contrasted Web surveys with face-to-face surveys, for example. The CES, given its longitudinal nature and target population of businesses, can offer many modes to its sample units because after the initial contact using a list of employer names and addresses, it can acquire identifications for other modes (e.g., telephone numbers and e-mail addresses).

If one were to array the different methods in a multidimensional space based on the dimensions we distinguished earlier, some methods would be closer to one another, others more distant. Typically, the more proximate methods are considered as reasonable alternatives to each other, and these comparisons dominate the mode comparison literature.

5.3 Effects of Different Data Collection Methods on Survey Errors

Much of what we know about the relative strengths and weaknesses of different modes of data collection comes from mode comparison studies. Typically, these are field experiments that assign a random subset of sample units to one mode and the rest to a second mode. Several meta-analyses have also advanced our understanding of the effect on mode (e.g., Goyder, 1985; de Leeuw and van der Zouwen, 1988). It is not easy to carry out such mode comparison studies, and the information they provide about alternative designs is often limited. It is important to understand these issues when drawing conclusions about the relative merits of alternative approaches.

We do not attempt an exhaustive review of all possible mode comparisons here. We focus on a subset of key comparisons, including face-to-face versus telephone, telephone versus mail, and mail versus Web.

5.3.1 Measuring the Marginal Effect of Mode

Survey estimates from different modes can differ for any number of reasons. For example, different sampling frames could be used (e.g., an RDD frame for a telephone sample, an area probability frame for face-to-face). The frame populations could vary (e.g., in mail versus Web comparisons, not all frame elements may have access to the Web). Nonresponse may differ across mode. For these reasons, identifying specific sources of the differences between modes is often difficult. For some comparisons, this can be done, for example, by excluding nonphone households in comparisons of face-to-face versus telephone surveys, any effects of differential coverage of the household population is removed. Similarly, to remove the confounding of frame population differences, comparisons of mail

Table 5.1. Design Issues in Research Comparing Face-to-Face and Telephone Surveys

Design feature	Important questions
Sampling frame	Same frame or frames with equivalent coverage of the telephone population?
Interviewers	Same interviewers? Same hiring and training? Same experience?
Supervision	Is telephone interviewing centralized? Is supervisory contact equivalent?
Respondent rule	Same respondent selection procedures?
Questionnaire	Identical questionnaires? Visual aids?
Callback rules	Same rules? Equivalently enforced?
Refusal conversion	Equivalent efforts?
Computer assistance	Use of CATI/CAPI?

and Web surveys could be restricted to those persons with Web access in each mode. Table 5.1, reproduced from Groves (1989, Figure 11.1), illustrates the design issues that arise in a comparison of two modes: telephone versus face-to-face interviewing. Similar issues arise for other mode comparisons.

One strategy for mode comparisons is to treat each mode as a package of features and to look for the net differences between them. This strategy addresses the practical question of whether different conclusions would be derived from a survey in one mode versus another. The best methods are chosen for each mode (e.g., face-to-face versus telephone) without trying to make every aspect strictly comparable. This may require, for example, different interviewing corps, different sampling frames, or somewhat different instruments designed to be optimal for each mode. This approach is often used when a survey is considering replacing one mode with another. The focus is on whether the resulting estimates are similar or different, regardless of the particular reasons behind any differences. When the Current Population Survey switched from paper to computer administration, this strategy was used to assess both the overall impact of the change in mode and of the implementation of a revised CPS questionnaire (Cohany, Polivka, and Rothgeb, 1994).

A different strategy for mode comparisons focuses on understanding the causes or mechanisms underlying any differences between two modes. It often uses a single frame. This approach attempts to isolate one particular factor (e.g., the channel of communication) by randomly assigning respondents to one mode or the other after an initial screening step (to eliminate coverage and nonresponse differences). Another example is the evaluation of the impact of technology, by converting a paper telephone questionnaire to CATI without any corresponding enhancements to the instrument

---

Hochstim (1967) on Personal Interviews Versus Telephone Interviews Versus Mail Questionnaires

Hochstim (1967) reported one of the earliest and most influential comparisons of survey modes.

Study design: Face-to-face interviewers visited a sample of 350 blocks in Alameda County, California, and enumerated households in 97% of 2148 housing units. The study then assigned at random face-to-face, telephone interviewing, or mail questionnaire modes to each sample unit. Two different surveys were then conducted, using the same design. If the originally assigned mode generated a nonresponse, the case was assigned to the other modes. The first study concerned general medical, familial, and demographic characteristics; all household members in a unit were included. The second concerned use of the Pap smear, with women 20 years or older as respondents. For the latter, visits to medical providers elicited record information about the survey reports.

Findings: The face-to-face mode generated the highest initial response rate before reassignment of modes for uncompleted cases. Costs per completed case were highest for the face-to-face interview, with the telephone and the mail questionnaire survey being within 12% of one another. There were no differences among modes on demographic variables, and few substantive differences. There was more item-missing data and more mentions of socially undesirable behaviors (e.g., drinking alcoholic beverages) in the mail mode. When survey reports were compared to medical records for Pap smear tests and pelvic examinations, there were no differences among modes in agreement rates.

Limitations of the study: The experimental groups were not single modes but mixes of a dominant mode and secondary modes, weakening inference about mode differences. The use of one county as a target population and focus on health and medical conditions limited use of findings to other surveys.

Impact of the study: The similarity of results across modes forced a reconsideration of "preferred" modes. The costs and social desirability sensitivities across modes were replicated in later studies.

> **The de Leeuw and van der Zouwen (1998) Meta-Analysis of Data Quality in Telephone and Face-to-Face Surveys**
>
> In 1988, de Leeuw and van der Zouwen reported a meta-analysis (i.e., a statistical summary and synthesis of documented research studies) of 31 comparisons of face-to-face and telephone surveys.
>
> *Study design*: A total of 28 studies between 1952 and 1986 compared telephone and face-to-face surveys in published journals and unpublished papers. Quality measures included comparisons of survey answers to record data, absence of social desirability bias, item-missing data, number of responses (to open questions), similarity across modes, and unit nonresponse rates.
>
> *Findings*: The average response rate for face-to-face surveys was 75%; for telephone, 69%. Differences between modes in detectable social desirability bias and rates of item-missing data were small and inconsistently found, but when present, tended to favor the face-to-face interview over the telephone. There was evidence that the mode differences themselves declined over time, with later studies showing smaller effects.
>
> *Limitations of the study*: As with all meta-analyses, the findings are limited to a set of studies that could be found at the time. No studies included the effects of computer assistance. Most studies confounded nonresponse and measurement error differences. Some studies included the nontelephone households in the face-to-face data, confounding coverage differences with measurement differences. Hence, the "data quality" includes errors of observation and nonobservation. All studies were conducted before 1986, a period when telephone surveys were less common.
>
> *Impact of the study*: The study added support for the growing use of telephone surveys in a wide variety of substantive fields.

or use of advanced CATI features. This approach often requires compromises in design, that is, not redesigning the survey to take advantage of the unique strengths of each alternative. Sometimes, these studies are conducted within the context of a longitudinal survey. Typically, the first wave receives a face-to-face interview, and later waves some other mode. Method comparisons are thus confounded with other differences between the waves.

Many of the mode comparison studies in the literature can be arrayed along a continuum from the more practically oriented designs (that compare bundles of features) to the more theoretically oriented approaches (that attempt to isolate the effects of specific design features). The relative merits of alternative approaches vary with the population under study, the topic of the survey, and a host of other design features, making broad generalizations difficult. Many of the common packages of features (e.g., regarding interviewer behavior, question format, and populations covered) have been studied, however. Combining the results from several such studies, we can build a picture of the relative advantages and disadvantages of different modes in terms of various sources of survey error and costs.

The next several sections focus on empirical evidence of differences and similarities between modes in terms of various sources of survey error and cost.

5.3.2 Sampling Frame and Sample Design Implications of Mode Selection

The available sampling frames often affect the choice of method. For example, mail surveys make the most sense when there is a list frame that includes a mailing address for each element. Web surveys do best when e-mail addresses are available on the frame, but mailed invitations to Web surveys are also used. Conversely, the desired mode of data collection may dictate the sampling frame. If the design calls for telephone interviews with a sample of the general population, then almost certainly a telephone number frame will be used. The choice of a mode of data collection and the basic sampling strategy are often made simultaneously.

Almost all surveys that use area probability frames start out conducting face-to-face interviews, though they may switch to less-expensive methods of data collection in subsequent rounds (assuming they interview the respondents more than once). Telephone number frames are used almost exclusively with telephone interviewing or IVR.

Method choice often has indirect sample design implications. Because of the expense of face-to-face interviews, this mode of data collection is almost always used with clustered samples, even though clustering reduces sample efficiency. In contrast, mail and Web surveys gain no cost savings through clustered sample designs. (See Chapters 3 and 4 for a discussion of alternative sampling designs and frames.)

A final consideration involves mode implications for designs that sample a single respondent from within a household. Within-household selection is often best done by interviewers rather than left to the respondent. Surveys that require a listing of all household members with a subsequent random selection of a sample person tend to produce fewer errors in coverage when done face to face than over the phone. This has led to the development of several alternative procedures to minimize the intrusiveness of this approach (see Chapters 3 and 4). Attempting an equivalent strategy in a self-administered survey is unlikely to yield much success (see Battaglia et al., 2008 for research on this issue). Similarly, surveys that require screening (e.g., for age eligibility) are best done by trained interviewers. Hence, an unsolved weakness of most self-administered methods is researcher control over who actually answers the questions. Much research remains to be done on this issue.

5.3.3 Coverage Implications of Mode Selection

Because mode and sampling frame choices are often linked, so too are mode choice and coverage of the target population. For example, an RDD frame generates a list of telephone numbers, which makes initial contact by telephone the most reasonable choice for studying the household population. The coverage properties of telephone directory frames are typically much worse, but they do usually yield mailing addresses. Using reverse directory lookups for RDD-generated numbers typically yields addresses only for about half of the numbers. List frames of e-mail addresses lend themselves to Web surveys, unless other information is available. As these examples demonstrate, some modes become more attractive or feasible depending on the frame information available. Still, one should not lose sight of the coverage properties associated with various sampling frames.

In terms of coverage of the household population, the combination of area probability frames and face-to-face interviews is viewed as the gold standard by which other modes are compared. But even for such personal-visit surveys, we typically restrict the population of interest in some way for a variety of cost, efficiency, or coverage reasons. For example, frames are often limited to the civilian (military personnel excluded), noninstitutionalized (prisons, hospitals, and other institutions excluded), and household (homeless persons and transients excluded) population in the contiguous United States (Alaska, Hawaii, and territories excluded for cost reasons). In short, some subgroups of the population are harder to include in sample surveys than others, and these are often excluded for cost or efficiency reasons. Most of our example surveys are restricted to the civilian household population.

The telephone coverage rate for the United States (defined as the proportion of households with landline telephones) has been above 90% for a couple of decades (92.9% in 1980, 94.8% in 1990, 94.5% in 2000). However, with the rapid rise in cell-phone-only households (see Section 3.4.2), the landline coverage has recently declined to 80% in early 2008, but with 17.5% of all households being wireless only, the overall telephone coverage rate is around 97.5% (Blumberg and Luke, 2008). Those not covered by telephone differ from those covered on several key dimensions, especially sociodemographic variables (Blumberg, Luke, Cynamon, and Frankel, 2008). For some surveys, for example, those focusing on crime victimization, drug use, unemployment, and welfare receipt, this may be an issue.

For mail surveys, no general population list exists in the United States. Some proxies exist, such as voter registration lists, and are often used for statewide mail surveys on particular topics. Obviously, not every adult resident in a state is registered to vote; approximately 72% of eligible U.S. adult citizens were registered to vote in November, 2006 (Bureau of the Census, 2006). Thus, mail surveys are more often used for surveys of specialized populations, for which a frame or list exists. The coverage properties (and, hence, noncoverage error) are highly dependent on the particular list used. (In countries that maintain population registers, this absence of a national mailing list is less of a problem, as mailing addresses are theoretically available for all residents. But just because a list exists does not mean it is complete and accurate.)

Turning to Web or Internet surveys, the last few years have seen a rapid rise in the proportion of U.S. adults with access to the Internet, from less than 15% in 1995 to 50% in 2000 and 75% in 2008 (see Figure 5.3). But unlike other modes of contact (e.g., mail or telephone), access to or use of the technology does not mean

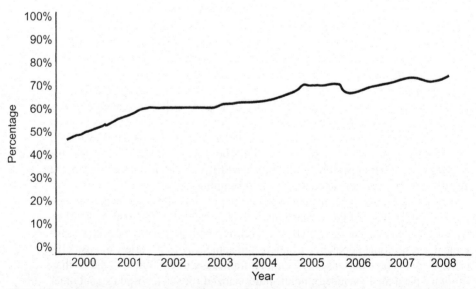

Figure 5.3 Percentage of U.S. adults who ever use the Internet, quarter 1, 2000, to quarter 3, 2008. (Source: Pew Internet and American Life Project Surveys, 2008.)

that the sample person can actually be reached with the technology; most Web surveys start by contacting the sample person via e-mail. No good frame has been developed for sampling the Internet population (see Couper, 2000, for a review; see also Section 3.4.3). In addition to a noncoverage rate of about 25%, the differences between those with access and those without are large enough to be of concern to those interested in making inference to the full household population. Evidence on demographic differences can be found in the series of reports on the digital divide by the U.S. National Telecommunications Information Administration (NTIA). For example, data from the Pew Internet Project Report for May 2008 reveals that whereas 90% of persons 18–25 years old were online, only 35% of those 65 or older were online. Similarly, although 91% of those with a college degree were online, only 44% of those with less than a high school education were online. The differences between those online and those not, the so-called "digital divide," are not limited to demographic characteristics (see Robinson, Neustadtl, and Kestnbaum, 2002; Couper, Kapteyn, Schonlau, and Winter, 2007).

For establishment surveys, like the CES, the U.S. government agencies use frames that they control, which have the name and address of the firm, but generally no reliable telephone number. For that reason, many establishment surveys begin with a mail contact attempt. Those choosing the telephone method are generally forced to match telephone numbers to the frame of company addresses. Survey researchers outside the Federal government mounting business surveys have access to commercial frames, some of which offer both telephone and mailing address identifications, thus permitting both telephone and mail modes.

Thus, the coverage properties of different modes need to be considered when choosing or evaluating a mode for a particular purpose. The fact that those not accessible by telephone in the household population are more likely to be poor and unemployed may make this an unsuitable mode for an unemployment survey or a survey of welfare recipients, whereas coverage may be a smaller concern for surveys on a variety of political issues. In the latter case, several factors outweigh concerns about coverage in the choice of mode:

1) Speed is sometimes of the essence (the half-life of the estimates produced by political polls is extremely short).
2) High levels of precision may not be most important to the users of the survey.
3) The inference is often to those likely to participate in an election, a variable correlated with telephone ownership.

Because the coverage of common target populations by different methods of data collection is constantly changing, the survey methodologist needs ongoing assessment. New frames for the new methods (e.g., Web surveys) are likely to develop over time. Coverage research will be necessary to give practical guidance to survey researchers.

5.3.4 Nonresponse Implications of Mode Selection

The choice of mode can also affect nonresponse rates and nonresponse bias, and much of the research on mode differences has focused on response rates. In addition to several two-mode and three-mode comparisons, there are several meta-analyses of response rates within (e.g., Heberlein and Baumgartner, 1978; Yu and Cooper, 1983; Edwards, Roberts, Clarke, DiGuiseppi, Pratap, Wentz, and Kwan, 2002; de Leeuw and de Heer, 2002) and across (Goyder, 1985; Hox and de Leeuw, 1994) modes. For example, Goyder (1985) assembled response rate and other information for 112 face-to-face surveys and 386 mail surveys for a meta-analysis. He found that, controlling for such variables as number of contacts, type of sample, and sponsorship, "response on questionnaires and interviews is essentially reducible to a common model, but that 'inputs' into that model have traditionally been greater on interviews." (Goyder, 1985, p. 246). Put differently, face-to-face surveys have higher response rates on the average than mail surveys, but some of that difference reflects the number of contact attempts and other variables. Hox and de Leeuw (1994) undertook a meta-analysis of 45 studies that explicitly compare response rates obtained from mail, telephone, or face-to-face surveys. They conclude that, on the average, face-to-face surveys have the highest response rates, followed by telephone, then mail. Similarly, Groves and Lyberg (1988) noted that telephone surveys tend to get lower response rates than face-to-face surveys.

What principles might underlie these differences? Section 6.7 notes the role of the interviewer in overcoming obstacles to delivering the survey request, answering questions, and addressing concerns of the sample persons. The effectiveness of this role appears to vary by the sponsor of the survey and the experience of the interviewer. Some of the higher response rates in face-to-face surveys probably arise because of greater credibility that the interviewers are who they say they are. The telephone mode, dependent only on the audio channel, strips away the ability to present physical evidence (e.g., identification cards, signed official letters). In the absence of this, the interviewers' abilities to recruit sample persons decline. In self-administered surveys that have no personal recruitment effort, sample persons may assign even less legitimacy and importance to the request. The ability to tailor persuasion strategies to address a sample person's concerns is highest with face-to-face contact, and is minimal in self-administered surveys.

For interviewer-administered surveys, there is no evidence that the technology used (for example, CATI or CAPI) affects response rates. In reviewing studies that compared computer-assisted methods with their paper counterparts, Nicholls, Baker, and Martin (1997) found no significant differences in either noncontact or refusal rates. Likewise, CASI methods have little effect on nonresponse rates; this is not surprising since they are typically part of an interviewer-administered survey.

However, for self-administered surveys, paper-based methods tend to obtain higher response rates than their electronic equivalents (e-mail and Web). Although there are some exceptions, two recent meta-analyses (Lozar Manfreda, Bosnjak, Haas, and Vehovar, 2008; Shih and Fan, 2008) show a clear advantage for mail over Web surveys. Whether this is due to fundamental differences between modes or it simply reflects the absence of well-tested strategies for inducing respondents to complete Web surveys remains to be seen.

In general, though, it is not very clear whether there are inherent differences across modes that directly affect response rates or whether the different methods as typically implemented vary in the number of contacts, the methods used to elicit cooperation, the delivery of incentives and other persuasive or legitimizing materials, and so on. For example, Cook, Heath, and Thompson (2000) undertook a meta-analysis of 68 Internet (Web or e-mail) surveys, and found that number of contacts, personalized contacts, and precontacts were associated with higher response rates (see also Shih and Fan, 2008; Couper, 2008b). These same variables are known to affect response rates in mail surveys and similar variables affect response rates in telephone and face-to-face surveys as well.

Different methods of data collection also vary in how much information is available to determine whether a sample case is in fact a nonrespondent and, if so, why (see, e.g., Dillman, Eltinge, Groves, and Little, 2002). For example, in mail surveys, it can be hard to distinguish ineligible units (e.g., bad addresses) from nonresponding units. Some questionnaires are explicitly returned as undeliverable (PMR or postmaster returns) and others come back completed, but for the remainder there is no easy way to tell whether the questionnaire ever actually reached the intended target. Similarly, in telephone surveys, a repeated "ring, no answer" outcome may indicate a nonworking number or someone who is away during the field period of the survey. In face-to-face surveys, the interviewers can usually determine the eligibility of a selected housing unit by observation.

The richness of information available about the nonresponse process also typically varies by mode. A noncontact in a telephone survey yields little information about the possible causes or potential strategies to overcome the problem, other than making repeated call attempts. Initial interactions in telephone surveys are typically also much shorter than those in face-to-face surveys (Couper and Groves, 2002), again yielding less information on causes or correlates. Mail surveys typically yield very little information on the process of nonresponse. However, Web surveys provide somewhat more detail (Bosnjak and Tuten, 2001; Vehovar, Batagelj, Lozar Manfreda, and Zaletel, 2002) when prespecified samples are used. Web surveys allow the researcher to distinguish, for example, between someone who does not access the survey at all, someone who logs onto the survey but fails to complete any questions, and someone who completes a large proportion of the questions but then decides not to continue.

The strengths of a method in one area of nonresponse are often counterbalanced by weaknesses in others. For example, mail surveys have fewer access problems and are a cheaper method of making contact with a person, household, or establishment than interviewer-administered methods. But the ability to obtain cooperation from the sample person is reduced by the lack of personal contact. The cost of repeated telephone calls is much lower than repeated visits in personal-visit surveys, but the increasing use of a variety of access controls (answering machines, caller ID) is reducing the rate of contact in telephone surveys. The per-call cost for a personal-visit survey is much higher, but the likelihood of gaining cooperation given contact may also be higher than for other methods, given the presence of the interviewer and the ability to tailor persuasive strategies to the particular concerns raised by the sample person (Groves and Couper, 1998). Groves and Lyberg (1988) note that the lion's share of nonresponse in telephone surveys is associated with refusals. Thus, the ability to persuade reluctant persons may depend on the richness of the media (e.g., in mail, motivational messages are limited to written materials) and the feedback provided by the potential respondent (permitting varying levels of tailoring).

In summary, not only do nonresponse rates differ across modes, but the reasons for nonresponse may differ, too. The latter may have more serious implications for nonresponse bias. For example, mail surveys may have greater nonresponse bias than interviewer-administered surveys because the content of the survey is revealed to the sample person when he or she decides whether to take part. The decision whether to take part may be determined in part by the sample person's reaction to the survey content, which may well depend on his or her values on the variables of interest. In interviewer-administered surveys, the content of the survey is often disguised behind vague euphemisms (such as "health and social life" for drug use or sex surveys) to avoid response decisions based on the topic of the survey. It is clear that ongoing methodological research is needed to discover the mechanisms that underlie nonresponse differences across methods.

5.3.5 Measurement Quality Implications of Mode Selection

Mode can also affect the quality of the data collected. We focus on three aspects of data quality here: the completeness of the data, the extent that answers are distorted by social desirability bias, and the extent the data show other response effects. "Social desirability bias" refers to the tendency to present oneself in a favorable light. Survey respondents exhibit this bias when they overreport socially approved behaviors (like voting) and underreport socially disapproved behaviors (like using illicit drugs). The mode of data collection may affect the level of social desirability bias. "Response effects" refer to measurement problems in surveys due to such features of the questions as their exact wording, the order in which they list the answer categories, or the order of the questions. For example, the answers may change when slightly different wording is used in the question or when the order of two related items is varied.

social desirability bias

response effects

With respect to data completeness, more questions go unanswered in self-administered (mail) questionnaires than in those surveys administered by interviewers (Tourangeau, Rips, and Rasinski, 2000). There are three possible reasons for this: (1) respondents do not understand the question (and interviewers are not there to help), (2) respondents do not follow the instructions in the questionnaire, or (3) respondents are not willing to give an answer (and interviewers are not there to encourage them to do so). Although there are exceptions (e.g., de Leeuw, 1992), this general trend is supported by the mode comparison literature (e.g., Brøgger et al., 2002; Van Campen, Sixma, Kerssens, and Peters, 1998; O'Toole, Battistutta, Long, and Crouch, 1986). The findings on missing data in Web surveys are less clear, in part because of the many different ways Web surveys can be designed, for example, to accept missing data (as in mail surveys), to prompt respondents to provide the missing responses (as in interviewer-administered surveys), or to require responses before proceeding. Thus, missing data in Web surveys may be more a function of design than of mode.

Less conclusive evidence exists for differences in rates of missing data in telephone versus face-to-face surveys. Groves and Kahn (1979) found higher overall rates of missing data on the telephone, as did Jordan, Marcus, and Reeder (1980), and Körmendi and Noordhoek (1989) for several income questions. Béland and St-Pierre (2008) report slightly higher rates in CATI than CAPI across a variety of health items. Several other studies report equivalent rates (Aquilino, 1992; Aneshensel, Frerichs, Clark, and Yokopenic, 1982; Dillman, 1978; Hochstim, 1967). The increased impersonality of the telephone (which may encourage respondents to provide sensitive information) may be offset by the ability of interviewers in the face-to-face mode to reassure respondents about the legitimacy of the survey and the confidentiality of their responses. As noted above, the technology used may also affect item-missing data. Computer-assisted methods also tend to produce lower missing data rates than paper-based surveys, primarily through eliminating items that are inadvertently skipped, to the extent that the instrument is programmed correctly. As Martin, O'Muircheartaigh, and Curtice (1993) found, for example, CAPI produced significantly lower item-missing data rates than similar paper surveys, but the rate of "do not know" and "refused" responses did not differ by method.

Another measure of completeness of response is the length of responses to open-ended questions. "Open-ended questions" allow respondents to formulate an answer in their own words. For example, many surveys ask respondents to describe their occupation. The answers are then classified by trained coders. By contrast, closed questions require respondents to select an answer from among the options listed as part of the question. In one of the first comparisons between face-to-face and telephone surveys, Groves and Kahn (1979) found significantly higher numbers of codable responses to open-ended questions in the personal-visit survey. They attributed this to the faster pace of the telephone interview and the absence of nonverbal cues that more information may be desired (see also Groves, 1979). Körmendi and Noordhoek (1989) obtained similar results. Few comparisons exist between self- and interviewer-administered surveys on this dimension. On the one hand, we may expect longer responses in mail surveys, as there are fewer time pressures on respondents, and they can write as much (or as little) as they desire. On the other hand, an interviewer can be quite successful in eliciting further information from the respondent through careful probing. De Leeuw (1992) compared the responses to four open-ended items, and found no differences between mail and interviewer-administered methods for two of them, longer responses for one, and shorter responses for another. Bishop, Hippler, Schwarz, and Strack (1988) found that respondents were more likely to give multiple answers to an open question on

---

The Tourangeau and Smith (1996) Study of Mode Effects on Answers to Sensitive Questions

In 1996, Tourangeau and Smith compared CAPI, CASI, and ACASI on reports about sexual behavior.

Study design: A randomized experiment was embedded in an area probability sample survey of 643 respondents 18–45 years old from 32 segments in the Chicago area, with a response rate of 56.8%. Within each segment, the interview used either CAPI, CASI, or ACASI for the entire sample household. Key outcomes were levels of reporting illicit drug use and sexual behavior and unit nonresponse rates.

Findings: ACASI and CASI generally elicited higher levels of reporting of drug use and sexual behavior. For example, the proportion of respondents reporting ever using marijuana was 48% higher using ACASI and 29% higher using CASI than using CAPI. The proportion reporting experiencing anal sex was 421% higher in ACASI, and 204% higher in CASI. Men reported fewer sexual partners and women more sexual partners in ACASI and CASI than in CAPI (suggesting fewer social desirability effects with the self-administered modes). There were no important nonresponse differences among the three modes.

Limitations of the study: There was no way to validate the results of different modes. However, most questions used have been shown to produce underreports of socially undesirable behaviors. The sample consisted of relatively young, well-educated urban residents who may have used CASI and ACASI with greater ease than others.

Impact of the study: The study added empirical support to the theory that the increased privacy offered respondents by self-administered modes leads to more accurate reporting of socially undesirable attributes. For that reason, many surveys measuring sensitive attributes use self-administration.

what they would most prefer in a job in a self-administered form than in a telephone interview. They attributed this to the inability to probe and clarify such responses in the self-administered condition. Comparisons of paper-based versus computerized surveys suggest no impact of computer-assisted interviewing on the responses to open-ended questions. In comparisons of CAPI (Bernard, 1989) and CATI (Catlin and Ingram, 1988) with equivalent paper-and-pencil methods, no differences were found in the length or quality (codability) of responses to questions on industry and occupation. Research by Denscombe (2008) and Deutskens, De Ruyter, and Wetzels (2006) suggests that Web and mail surveys produce equivalent responses to narrative-type open-ended questions.

In summary, then, interviewer administration and computer assistance seem to reduce the rate of missing data relative to self-administration and paper-based data collection. Face-to-face interviews also seem to yield fuller responses to open-ended questions than telephone interviews.

Some of the clearest differences in the literature on mode effects involve socially desirable responding. This refers to the tendency of survey respondents to present themselves in a favorable light; for example, nonvoters may claim to have voted, smokers may deny smoking, and those who are intolerant of minorities may give misleading answers to questions assessing their attitudes toward minorities. Respondents are often reluctant to reveal embarrassing information about themselves in a survey, and the mode of data collection appears to affect their willingness to admit to undesirable attitudes or behaviors. In general, the presence of an interviewer increases social desirability effects: the overreporting of socially desirable behaviors such as voting, church attendance, exercise and healthy eating, and so on, and the underreporting of socially undesirable behaviors such as drug and alcohol use, sexual behavior, and the like. Self-administered methods thus usually produce fewer social desirability effects than interviewer-administered methods.

Figure 5.4 shows the results of two large studies that compared the data on illicit drug use obtained in face-to-face interviews and self-administered questionnaires. Both studies asked about the use of several drugs, including marijuana and cocaine, during the past month and past year and over the respondent's lifetime. The figure plots the ratio between the proportions of respondents admitting they had used each drug in each time period under self- and interviewer-administration of the questions. For example, Turner, Lessler, and Devore (1992) found that respondents were 2.46 times more likely to report that they had used cocaine in the past month when the questions were self-administered than when they were administered by an interviewer. The results are similar (though less striking) in the study by Schober, Caces, Pergamit, and Branden (1992).

Within interviewer-administered methods, telephone interviews appear to be less effective than personal interviewing in eliciting sensitive information, and the data typically show a higher social desirability bias (e.g., Groves and Kahn, 1979; Henson, Roth, and Cannell, 1977; Johnson, Hoagland and Clayton, 1989; de Leeuw and van der Zouwen, 1988; Aquilino, 1992). However, several studies have found the opposite effect (e.g., Sykes and Collins, 1988; Hochstim, 1967) or no differences (e.g., Mangione, Hingson, and Barrett, 1982) between face-to-face and telephone interviews. The de Leeuw and van der Zouwen study is particularly telling since it is a meta-analysis reviewing a large number of mode comparisons.

A number of studies have explored mode differences in response effects. Response effects refer to a number of measurement problems in surveys, such as

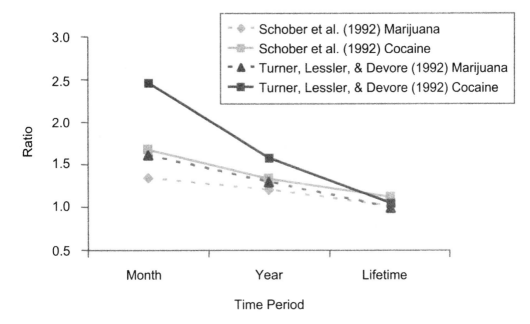

Figure 5.4 Ratio of proportion of respondents reporting illicit drug use in self-administered versus interviewer-administered questionnaires, by time period by drug.

question order effects (in which the answers change depending on the order of the questions) and response order effects. "Response order effects" refer to changes in the distribution of the answers as a result of changes in the order in which the possible answers are presented. In some cases, respondents tend to select the first or second answer category presented to them (a primacy effect). In other cases, they favor the options presented at the end of the list (a recency effect). Bishop, Hippler, Schwarz, and Strack (1988) concluded that question order and response order effects are significantly less likely to occur in a self-administered survey than in a telephone survey, whereas question form and wording effects are probably just as likely to occur with one mode of data collection as another. These differences appear to arise from the ability of respondents in self-administered surveys to see entire groups of questions before they answer them.

However, the direction of response order effects seems to depend, in part, on the method of data collection. When the questions are delivered aurally, respondents are more likely to select the last options presented; when the questions are delivered visually, they are more likely to select the one presented at the beginning of the list (Schwarz, Hippler, Deutsch, and Strack, 1985; Tourangeau, Rips, and Rasinski, 2000). In terms of question order, although mail surveys do not deliver the items in as strictly controlled a sequence as interviewer-administered surveys (thereby reducing question order effects), the placement of questions on the same or different page, and visually linked or separated (Schwarz, Hippler, Deutsch, and Strack, 1991) may increase or decrease such context effects, respectively. In similar fashion, we expect a scrollable Web survey, where all of the questions are visible to respondents, to have fewer such order effects than a Web

response order effects

survey in which the items are presented sequentially to the respondent by the software. Again, these examples point to the importance of clarifying what is meant by mode, in particular, how an instrument is designed and delivered in a particular mode.

Other response styles that have been investigated with regard to mode include acquiescence (the tendency to answer affirmatively regardless of the content of the question) and extremeness (the tendency to choose scale endpoints). It is often hard to disentangle these effects from social desirability and response order effects. Even so, the results are generally mixed, with some studies finding more acquiescence and more extremeness in telephone than in face-to-face interviews (e.g., Jordan, Marcus and Reeder, 1980; Groves, 1979) and in telephone than mail (e.g., Tarnai and Dillman, 1992), but others (e.g., de Leeuw, 1992) finding no differences among modes.

To this point, we have briefly reviewed a variety of different measurement effects and contrasted several modes of data collection on these dimensions. But how does one generalize from these comparisons to newer or emerging forms of data collection? In general, the aural modes (i.e., interviewer-administered, IVR, and audio-CASI) deliver the questions in sequence to a respondent. Upon answering one question, the respondent gets the next question in a predetermined order. In predominantly visual modes (i.e., mail, paper SAQ), the respondent is not constrained by the order in which questions are presented in the questionnaire, but can answer the questions in any order. Therefore, we expect fewer context effects (e.g., question order or response order) in visual modes, and this is borne out by the research evidence. However, Web surveys can be designed to resemble either paper questionnaires (e.g., by using a single scrollable HTML form) or interviewer-administered surveys (by successively revealing one or more questions at a time using separate forms). The scrollable designs are likely to produce fewer context effects. Thus, the context effects may not be a property of the mode itself, but rather how the instrument is designed for that mode.

Thus far, our focus has been on measurement error differences between methods. This requires the collection of comparable data across modes. However, we can also consider alternative modes in terms of the measurement opportunities they provide. For example, show cards (listing response alternatives) or other visual aids (e.g., pictures) are a useful adjunct to face-to-face surveys, but are more difficult to employ in telephone surveys. Show cards may alter some response effects, and they can be used to improve recall and recognition of a variety of subjects (e.g., magazine readership, ad testing, pill cards, etc.). Other additions to the traditional methods of survey measurement may be possible depending on the data collection method. For example, personal-visit surveys can include observations (e.g., of the house, neighborhood, etc.) or physical measures (height, weight, etc.). The collection of physical samples (hair, urine, blood, etc.) from the respondents or their environment (radon, water quality, etc.) is facilitated by face-to-face data collection, although is not impossible in other modes of data collection (e.g., Etter, Perneger, and Ronchi, 1998; Boyle et al., 2007). In addition, computer-assisted data collection methods permit the measurement of response latencies (i.e., the amount of time passing between delivery of a question and the respondent's delivery of an answer) and other data that would be difficult to obtain using paper-based methods. Similarly, the ability to randomize questions and response options (to both measure and reduce question and response order effects) is facilitated by computer-assisted interviewing (CAI).

This is particularly true of Web surveys, where we have seen a surge in design and measurement experiments.

5.3.6 Cost Implications

Survey costs are commonly divided into two types: fixed and variable. The fixed costs refer to costs that are incurred regardless of the sample size. For example, the costs of developing, pretesting, and programming the questionnaire are fixed costs; they do not depend on the size of the sample. The variable or per-case costs refer to the costs incurred in contacting the sample cases, interviewing them, following up with nonrespondents, and so on. They vary depending on the number of sample cases that are fielded. We would be remiss in our review of alternative methods of data collection if we did not discuss the relative costs of alternative approaches. The cost of different methods of data collection depends on a number of operational details. For example, the relative impact of interviewer travel costs is reduced in a local survey versus a national survey; this may affect comparisons of the relative costs of face-to-face and telephone interviews. Similarly, the number of callback attempts or follow-up mailings may affect the relative costs of mail versus telephone. The cost differential between mail and Web surveys may depend on the size of the sample, with Web having proportionately larger fixed costs and mail having larger variable costs.

fixed costs
variable costs

Despite these caveats, and the fact that relatively few studies provide a detailed reporting of costs, we can offer some generalizations. Personal-visit surveys are typically much more expensive than telephone surveys. The ratio of costs, when reported, cluster around 2; that is, personal-visit surveys cost twice as much per case as telephone surveys (see Warner, Berman, Weyant, and Ciarlo, 1983; Weeks, Kulka, Lessler, and Whitmore, 1983; Van Campen, Sixma, Kerssens, and Peters, 1998). Our experience, however, suggests that the typical ratio for national surveys is higher than this, ranging from 5 or 10 to 1. A key component of the cost of face-to-face interviewing is the time taken by an interviewer to travel to selected segments. Each successive callback adds appreciably to the cost of the survey. In addition, because they are dispersed, field interviewers typically need to be more highly trained and experienced than those in centralized telephone facilities, where the task is less complex and centralized supervision is possible. When computer-assisted interviewing is used, the ratio between the cost of face-to-face and telephone interviews is likely to increase because of the added equipment costs. Each interviewer in a field interviewing setting must be provided with a laptop computer, and it is put to less use than the desktop computers in a centralized telephone facility. On the other hand, the fixed costs of developing the CAI instrument are not affected by mode, so that the overall cost differential may depend on the sample size.

The cost differences between telephone and mail are generally smaller, yielding ratios of between 1.2 (e.g., Hochstim, 1967; Walker and Restuccia, 1984) to around 1.7 (e.g., McHorney, Kosinski, and Ware, 1994; O'Toole, Battistutta, Long, and Crouch, 1986; Warner, Berman, Weyant, and Ciarlo, 1983). Again, this may depend on the number of mailings or callbacks, as well as a variety of other factors (sample size, use of computers, etc.).

Although there are many claims that Web surveys are significantly cheaper than mail surveys, the relative costs of the two methods may depend on what is included in the cost estimate and on the volume of work. Web surveys typically have larger fixed costs than mail surveys, reflecting both general infrastructure costs and the costs of developing and testing the particular survey questionnaire. On the other hand, if the survey is fully electronic (i.e., e-mail is used for invitations and reminders), the per-case costs of a Web survey are close to zero. By contrast, the fixed costs of a mail survey are usually smaller, but the variable costs (printing, postage, keying, or scanning, etc.) are larger than those of Web surveys. The relative cost of these two methods depends in large part on the number of sample cases over which the fixed costs are amortized.

There are obviously many other components that factor into the cost structure of a particular survey. For example, the inclusion of an audio-CASI component in a face-to-face survey increases the costs relative to interviewer administration, but may increase data quality. Similarly, adding a mailout to a telephone survey (to increase legitimacy or to deliver an incentive) increases costs, but may improve response rates. It should be clear that broad generalizations about relative costs of different modes should be treated with caution, but that cost factors, like other elements of survey quality, involve trade-offs among many different design elements.

If, as expected, future survey designs will involve multiple methods of data collection, research into cost models for different methods will be needed. If survey researchers desire to use their resources in an optimal fashion (i.e., to maximize the quality of statistics within a given budget constraint), then both the error properties of different methods and their relative costs must be studied.

5.3.7 Summary on the Choice of Method

For a new cross-sectional survey, the choice of method of data collection can be made by weighing the relative impact of the various sources of error on the overall estimates, and with consideration of other factors such as cost, timeliness, availability of staff and other resources, and so on. Because of its expense, face-to-face interviewing is generally only used in large-scale, federally funded surveys that place a premium on high rates of coverage of the target population, high response rates, and high levels of data quality. Three of our example surveys (the NCVS, NSDUH, and NAEP) rely primarily on in-person data collection, at least in the first interview with respondents, in order to achieve the required coverage, response rates, and data quality. Two of the others (the SOC and BRFSS) rely on data collection by telephone, although both are exploring mixed-mode alternatives. The CES is a classic mixed-mode survey.

For ongoing data collection efforts with an existing time series, the calculus regarding the switch to a new method becomes more complex. The advantages of the new mode also have to be weighed against possible perturbations in the time series. In such cases, care must be taken in the transition from one method to another, often with the use of split-sample experiments to assess the impact of the mode switch. Longitudinal or panel surveys considering a mode switch face similar challenges. The decision to switch from a single mode of data collection to multiple modes (see Section 5.4) also requires similar care. Many of the mode comparison studies reviewed above were designed and conducted to inform decisions about switching particular studies from one method of data collection to another.

In general, the measurement error properties of alternative modes of data collection are fairly well known. Where equivalent modes are compared (e.g., face-to-face versus telephone and mail versus Web), the resulting estimates are often not very different. The notable exceptions appear to involve highly sensitive behaviors, such as drug use, sexual behavior, and the like (see Figure 5.4), where self-administered methods appear to have a consistent advantage. But for a large array of nonthreatening survey items, interviewer-administered surveys yield similar results whether administered by telephone or face-to-face, and self-administered methods yield similar results whether they are done on paper or via the Web. The differences, and hence the key trade-offs, are often with respect to errors of nonobservation (coverage and nonresponse) and efficiency considerations (time, costs, etc.).

5.4 USING MULTIPLE MODES OF DATA COLLECTION

Many surveys use multiple modes of data collection. Although there are many reasons for using more than one mode in a survey, three reasons are especially common. The first reason is that using a blend of methods may reduce costs. Typically, this entails attempting to collect data from each case with the least expensive mode (mail) first, then moving to the next most expensive mode (telephone) to collect data from nonrespondents to the initial mail attempt, and, finally, resorting to face-to-face interviewing for the remaining cases. In such designs, costlier methods are applied to successively fewer cases. The U.S. Census 2000 employed such a design, beginning with mail questionnaires and moving to face-to-face interviewing for follow-up with nonrespondents. (CATI, IVR, and Web data collection were also used in some circumstances.)

A second reason for mixing modes is to maximize response rates. Establishment surveys such as the Current Employment Statistics (CES) program employ multiple methods of data collection, including Web, fax, touchtone data entry (inbound IVR), telephone, and mail. Concerns about nonresponse (and the importance of timeliness) outweigh concerns about any mode effects that may be introduced. Offering a mix of modes allows respondents to select the one that is most convenient for them.

A third reason for employing mixed modes is to save money in a longitudinal survey. As we noted, the CPS and NCVS use face-to-face interviews for respondents in their first wave of data collection. This maximizes the response rate (typically, losses to nonresponse are worst in the initial round of a longitudinal survey) and also allows the interviewers to obtain telephone numbers for later rounds. The bulk of the data thereafter can be collected by telephone, reducing the data collection costs. The BRFSS is experimenting with mixed-mode data collection (see, for example, Link and Mokdad, 2005, 2006), in response to coverage and nonresponse threats to traditional RDD telephone designs.

But the method mixes described above are only some of the possibilities. The mode may be respondent-specific (e.g., some respondents are interviewed face-to-face, others by telephone; or some respond by mail, others by Web), stage-specific (e.g., recruitment done by telephone, survey completed using IVR; or initial response by mail, with telephone reminder calls), or even question-specific (e.g., some items answered using audio-CASI, others administered by an interviewer). Figure 5.5 lists various types of mixed-mode approaches (see also de Leeuw, 2005).

1. One mode for some respondents, another for others, for example:
   - telephone survey, with FTF component for those without telephones
   - mail survey with Web response option
2. One mode for recruitment, another for survey administration, for example:
   - mail invitation for Web survey
   - telephone recruitment for IVR survey
3. One mode for data collection, another for reminders or follow-up, for example:
   - telephone reminders for mail or Web survey
4. One mode for one wave of a panel survey, another for others, for example:
   - first wave of panel is FTF, subsequent waves by telephone or mail
5. One mode for main part of the interview, another mode for some subset of items for example:
   - audio-CASI for sensitive items

Figure 5.5 Five different types of mixed mode designs.

The rise of mixed-mode data collection is due in large part to the research findings summarized above, indicating that alternative modes generally yield similar results, but often have different coverage, nonresponse, and cost properties. Once telephone surveys were found to perform similarly to face-to-face surveys in terms of measurement error, for example, they were increasingly used not only to replace face-to-face surveys (i.e., mode switches) but also to supplement them (mixed-mode designs). A similar trend is evident with Web surveys supplementing data collection by mail. The logic of mixed-mode designs is to exploit the advantages of one mode (e.g., the reduced costs of telephone surveys) while neutralizing the disadvantages (e.g., coverage) through a combination of methods. Mixed-mode design thus involves explicit trade-offs of one source of error for another. Combining modes does not always produce gains. For example, offering respondents a Web option to a mail survey has not been found to increase response rates (see, e.g., Griffin, Fisher, and Morgan, 2001). However, careful sequencing of the invitations can maximize the proportion using the Web, thereby reducing costs (see Holmberg, Lorenc, and Werner, 2008).

A key distinction can be made between those designs that obtain all the data for one set of questions using the same mode, and those that obtain answers to the same questions from different respondents using different modes. When data are obtained from different subsets of the sample using different methods, it is critical to ensure that any effects of the mode can be disentangled from other characteristics of the sample. This is especially true when respondents choose which mode they prefer, or when access issues determine the choice of mode. This is either done by conducting or reviewing carefully designed mode comparisons prior to the study (as in the case of telephone and face-to-face surveys) or by embedding mode comparisons into the study in question, with a small subsample being randomly assigned to the mode. This helps ensure that differences between the two datasets reflect true differences in the population rather than the mode of data collection.

The design of a mixed-mode survey may reflect different considerations than the design of a single-mode survey. Instead of optimizing the design features for one particular mode, the instruments and procedures need to be designed to ensure equivalence across modes (see Dillman, Smyth, and Christian, 2009). For example, this may mean designing a Web survey instrument to resemble as much as possible the paper version, rather than building in the complex skips, randomization, edit checks, and the like that are possible with Web surveys. Similarly, decisions must be made whether the use of show cards in a face-to-face survey changes the measurement in a mixed-mode design that includes telephone interviewing. The goal is to keep the essential survey conditions as similar as possible across modes.

Mixed-mode designs add other operational complexities to the data collection process. Case management becomes more critical to avoid duplication, that is, to avoid collecting data from a single respondent via two modes. The timing of case transfers from one mode to another must also be carefully considered. Given that different types of error may occur in different modes, particularly when self-administered and interviewer-administered modes are mixed, the data cleaning and merging process must take note of this in creating a single analytic file. However, on balance, mixed-mode designs are increasingly preferred for a number of reasons, most notably reducing costs, increasing the speed of data collection, and increasing response rates. We are likely to see even more creative use of mixed-mode designs in the future in pursuit of these goals.

5.5 SUMMARY

Although mail, telephone, and face-to-face methods were the three dominant modes in past decades of survey research, there are now a multitude of data collection methods. The methods vary on whether and how they use interviewers, how much they require direct interaction with the sample persons, the extent of privacy they provide the respondent during measurement, what channels of communication they employ, and whether and how they use computer assistance. Each of these features can have impacts on the statistical error properties of survey results.

Choosing the method of a survey requires considering coverage and contact information on the sampling frame for the method, the appropriateness of the topic to the method, cost constraints of the survey, and the value of timely results.

Survey methodology contains many randomized experimental studies of the effects of data collection method. Some attempt to measure the pure effect of channels of communications; others compare different typical packages of design features common to a method. Face-to-face and telephone modes generally use sampling frames that offer better coverage of the household population than mail and e-mail frames. The typical ranking of methods by response rates is as follows: face-to-face surveys, telephone surveys, mail surveys, and Web surveys. Most of these results are based on household surveys of adult target populations; other populations might produce different rankings. Interviewers appear to act as effective agents to improve response rates. Self-administered modes achieve lower response errors due to social desirability effects. Interviewer-administered modes tend to achieve lower item-missing data rates. Methods using aural presentation of questions tend to generate recency effects in response selection on closed-ended questions. Methods using visual presentation of all response options tend to produce primacy effects.

The data collection methods vary greatly in their required costs. The typical ranking of costs from highest to lowest is face-to-face surveys, telephone surveys, mail surveys, and Web surveys. The cost comparisons are sensitive to the sample size because the methods vary on the relative sizes of fixed costs (independent of sample size) and variable costs (increasing with sample size).

Mixed-mode designs are increasing in frequency as a way to gain better balancing of costs and errors in survey results. Longitudinal surveys often begin with a face-to-face interview and move to cheaper methods in later waves.

The proliferation of modes is likely to continue as new technologies are exploited to reduce costs and improve the quality of survey data. It will thus be increasingly important to understand how the methods differ from each other, and the characteristics or dimensions underlying such differences. Understanding these properties is essential for making informed decisions of which mode or modes to use for a particular study, for evaluating and adopting new and possibly currently untested modes, and for combining different methods in innovative ways to minimize survey costs and survey errors.

KEYWORDS

ACASI
acquiescence
CAPI
CASI
CATI
channels of communication
CSAQ
disk by mail
extremeness
fixed costs
ICR
IVR
mode
open-ended question

OCR
primacy
recency
response effects
SAQ
show cards
social desirability bias
social presence
T-ACASI
text-CASI
TDE
variable costs
video-CASI
Web

FOR MORE IN-DEPTH READING

Couper, M., Baker, R., Bethlehem, J., Clark, C., Martin, J., Nicholls, W., and O'Reilly, J. (1998), *Computer Assisted Survey Information Collection*, New York: Wiley.

Dillman, D., Smyth, J., and Christian, L. (2009), *Internet, Mail, and Mixed Mode Surveys: The Tailored Design Method*, 3rd Edition, New York: Wiley.

Gwartney, P. (2007), *The Telephone Interviewer's Handbook: How to Conduct Standardized Conversations*, New York: Wiley.

Lepkowski, J., Tucker, C., Brick, J.M., de Leeuw, E., Japec, L., Lavrakas, P., Link, M., and Sangster, R. (2008), *Advances in Telephone Survey Methodology*, New York: Wiley.

EXERCISES

1) A nonprofit public policy "think tank" organization is planning to conduct a survey examining the condition of low-income families nationwide. The questionnaire covers many topics including family income, participation in state-run welfare programs and health insurance programs for low-income families, children's education and health status, and employment status of parents. Low-income families in rural and inner city areas are of special interest to the researchers. The researchers are considering two modes: a CATI telephone survey or CAPI face-to-face interviews. Discuss the advantages and disadvantages of each mode choice for this project.

2) Briefly describe two advantages and two disadvantages of using interviewers for survey data collection.

3) Name one advantage and one disadvantage of using a supplemental self-administered questionnaire to collect information regarding sexual behavior/history during a face-to-face interview.

4) Which of the following modes of data collection is least likely to produce question order effects: face-to-face, telephone, or mail? Provide a reason for your answer.

5) Identify which mode (telephone, face-to-face, or mail) would be most desirable for household surveys, if each criterion below were of highest priority. Explain your choice of mode in each case.

   a) The survey is completed quickly.
   b) The survey costs are low for a given sample size.
   c) The response rate is high.
   d) Populations speaking a language different from the majority are well measured.
   e) Sampling error is minimized for a given sample size.

6) Indicate whether you agree or disagree with each of the following statements. In each case, provide a reason for your answer.

   a) Coverage error is a major concern in CAPI surveys of the U.S. household population.
   b) Coverage error is a major concern in Web-based surveys of the U.S. household population.

c) The large number of respondents in Web-based surveys for describing the general adult household population implies that we are able to obtain better survey estimates (i.e., estimates that describe the population).
d) It is relatively easy to assess nonresponse rates in opt-in Web surveys (e.g., where recruitment is done through banner links, etc.).

7) Your market research firm has been asked to conduct a Web survey to see if adult consumers would purchase a new beverage product. The client has requested that estimates and "margins of error" be available in two weeks and, apparently, has ample money to finance the project. Your boss is not sure she wants to take the job, and has asked you to do a point–counterpoint presentation that outlines the benefits and drawbacks of conducting a survey on the Web. First, list three advantages of this mode of data collection. Second, list at least one limitation related to each source of error (sampling, coverage, nonresponse, and measurement) that could undermine the usefulness of the Internet as a tool for survey research. Third, briefly (2–4 sentences) make a recommendation about whether or not your boss should take the job, justifying your recommendation on the practical and empirical implications.

8) What if, instead of a Web-based survey, the client in Exercise 7 wanted to conduct an invound IVR with a mailed invitation? Mention two limitations of IVR for survey data collection.

9) The governor of your state has asked you to design a survey to assess the views of residents of the state toward public transportation and to evaluate how likely they would be to use alternative modes of transportation. Which mode of data collection would you recommend be used for this study, and why?

10) The American Statistical Association wishes to do a survey of its members in the United States on their views toward the certification of statisticians. Which mode of data collection would you recommend for the survey? Give three reasons for your choice.

11) In comparing statistics from a telephone survey to similar ones from a face-to-face survey, what three differences between the two designs can be due to the medium of communication?

12) Give two reasons why face-to-face household surveys often obtain response rates higher than comparable telephone household surveys.

13) A survey has been designed to study differential access to health care among persons with different income levels and different health insurance coverage. The funding for the study is suddenly cut by about 25%. In a discussion of how to redesign the study in light of reduced funding, the client suggests a shift from a face-to-face interview based on an area probability sample of households to a telephone interview of a random-digit dialed sample. The

EXERCISES

sample size in terms of interviewed individuals could be maintained if we follow the client's suggestion. Identify two sources of error that need to be addressed in considering the desirability of the client's suggestion. Give an example of how the proposed design change might affect each of two errors.

14) Give two attractions and two detractions of a mixed-mode design that moves from a mail survey to phone on the remaining nonrespondents, then to face-to-face interviews on those who are still nonrespondent.

15) Give one advantage and one disadvantage of a mixed-mode design that moves from an initial mail survey, to phone on the remaining nonrespondents, then to face-to-face interviews on those still nonrespondent.

16) Briefly mention what factor(s) account for the big cost differential between face-to-face surveys and telephone surveys of a national sample.

17) Briefly describe the challenge(s) cell-phone-only users pose for the following sources of error in a telephone survey:

   a) Coverage error
   b) Nonresponse error
   c) Measurement error

18) Briefly describe the difference between CASI and CSAQ methods.

19) Discuss briefly the differences between in-bound and out-bound recruit-and-switch IVR with respect to two types of survey error.

20) Discuss briefly the advantages of centralized versus decentralized CATI.

CHAPTER SIX

NONRESPONSE IN SAMPLE SURVEYS

6.1 INTRODUCTION

Chapters 1 and 2 discussed how survey researchers use the term "nonresponse" to describe the failure to obtain measurements on sampled units. Sometimes, the failure is complete—the person chosen for the sample refuses to cooperate with the survey request entirely (e.g., Sample Person: "I never participate in surveys; please don't call me again"). Sometimes, the failure affects only one item in the survey measurement (e.g., Interviewer: "What was your total family income last year?" Respondent: "I don't know that; my wife keeps those records"). The total failure is termed "unit nonresponse"; the partial failure is called "item nonresponse."

unit nonresponse

item nonresponse

Nonresponse can affect the quality of survey estimates. If the nonrespondents have different values on variables that are components of the estimate, its value based on the respondents may differ from that based on the total sample. When the departure is a systematic one, endemic to all implementations of the survey design, it is called "nonresponse bias." On some simple estimates (like the sample mean), nonresponse bias is a function of the correlation between the survey variable and the likelihood of participating in the survey (see Section 2.3.6).

Nonresponse rates (the percentage of eligible sample cases that are nonrespondent) in most household surveys have increased over time in the United States and Western Europe. There are three principal sources of nonresponse that appear to have different causes: failure to deliver the survey request to the sample person, failure to gain the cooperation of a contacted sample person, and inability of the sample person to provide the requested information. They appear to affect different types of estimates produced by surveys.

This chapter provides the reader with the essential concepts and practices involving nonresponse. It first presents information on response rates and their trend over time; then it discusses the link between nonresponse rates and bias. Next, it dissects the phenomenon of nonresponse into distinct types. It ends with a discussion of how survey design features can affect nonresponse error in different estimates.

6.2 RESPONSE RATES

In simple terms, a survey's response rate is merely the percentage of eligible sample cases that were measured. The nonresponse rate is its complement. Before we examine some typical values of response rates, we need to note that this simple definition belies the complexity of computation of response and nonresponse rates.

6.2.1 Computing Response Rates

Since nonresponse rates have traditionally been viewed as direct measures of survey quality, there is a checkered history of unethically presenting deflated estimates of nonresponse to give the appearance of high-quality statistics. This problem has been the focus of several committees of professional associations (Frankel, 1983; American Association for Public Opinion Research, 2000). The guidelines of the American Association for Public Opinion Research (AAPOR) can be found at www.aapor.org. That website provides several different response rate computations, depending on the survey design used. In general, there are three complications in calculating response rates:

1) Some sampling frames contain units that are not target population members, requiring a screening step to determine eligibility (e.g., a telephone surveys of households when the frame contains business numbers). With such designs, there is uncertainty about the eligibility of nonrespondent cases and, hence, what the denominator of the response rate should be.
2) Some sample frames consist of clusters of sample elements in which the number of elements is unknown at the time of sampling (e.g., a survey of school children chosen from a sample of schools). When a full cluster is nonrespondent, it is unclear how many sample elements are nonrespondent.
3) Unequal probabilities of selection are assigned to different elements in the sampling frame (e.g., oversamples of minority ethnic groups). In this case, it is unclear whether selection weights should be used in the computation of response rates (see Groves, 1989).

One way to approach the first two problems is to estimate the value of the denominator, using either external information or information from other cases. Thus, the response rate might be

$$\frac{I}{I+R+NC+O+e(UH+UO)}$$

where

$I =$ Complete interview
$R =$ Refusal and breakoff
$NC =$ Noncontact
$O =$ Other eligible
$UH =$ Unknown if household/occupied household unit
$UO =$ Unknown eligibility, other
$e =$ Estimated proportion of cases of unknown eligibility that are eligible

An estimate of e can be obtained from the current survey, e.g., $(I + R + NC + O)/(I + R + NC + O +$ ineligibles chosen into sample). Alternatively, some special methodological studies might be mounted to learn e, by studying a sample of cases with initially unknown eligibility to learn more about them. Finally, if no estimate of e can be obtained, it is recommended that two response rates are presented: one including $(UH + UO)$ in the denominator and the other excluding the term. This produces a range of response rates within which the true response rate must lie. More sophisticated modeling approaches to estimating e in an RDD telephone survey are presented in Brick, Montaquila, and Scheuren (2002).

The third problem in estimating response rates arises when (as described in Chapter 4) unequal probabilities of selection are used. For example, if a community survey about city services oversampled areas where poor persons lived (stratum 1), using twice the sampling fraction as other areas (stratum 2), how should the response rate be computed? There are often two response rate issues in such a design. One is a comparison of the response rates in the poor areas (stratum 1) with the response rates in the nonpoor areas (stratum 2). These would be relevant to the analysis that compares means of the two strata. In this case, the two response rates would use the same approach as above. If, however, the interest was in the overall sample mean, which, as noted in Chapter 4, Section 4.5, must use selection weights, w_i, then the response rate for the overall sample might use the same weights, adjusting each sample element for the selection probability it received.

A variety of other rates are used for different purposes. For example, refusal rates [e.g., $R/(I + R)$] and refusal conversion rates (initial refusals that were subsequently interviewed) are used to evaluate interviewer performance. For establishment surveys, coverage rates (the proportion of the total being measured that is accounted for by the respondent units) are used to reflect that, when estimating things like output or number of employees, missing Walmart is not the same as missing the corner convenience store. For surveys such as NAEP, in which selection occurs over several levels (e.g., schools and pupils), compound rates are calculated to reflect nonresponse at each level.

Merkle and Edelman (2002) on How Nonresponse Rates Affect Nonresponse Error

Merkle and Edelman (2002) present an observational study that finds no relationship between nonresponse rates and nonresponse error.

Study design: Exit polls use a probability sample of voting places, with interviewers selecting a systematic random sample of voters exiting the voter station. For each sample voting place, the difference between the Democratic and Republican vote percentages was compared between respondents and the public vote totals.

Findings: The response rates at the sample places varied from 10 to 90%, mostly falling in the 45–75% range. The plot below has a point for each voting place; the x axis is the response rate for the sample place; the y axis is the error of the difference between Democratic and Republican vote percentages. There is no apparent relationship between response rate and total error of the estimated difference. A good predictor of cooperation was how far away from the exit door the interviewer was asked to stand.

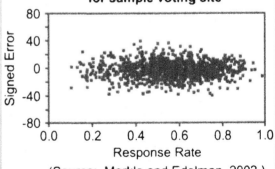

(Source: Merkle and Edelman, 2002.)

Limitations of the study: Because of lack of randomization of interviewers to sample place, there could be a confounding of interviewer effects and true differences in cooperation likelihoods. Further, the exit poll is a unique setting, unlike most other surveys.

Impact of the study: The study offers an example of the absence of nonresponse error when nonresponse causes are unrelated to the statistic.

6.2.2 Trends in Response Rates Over Time

Some notion of variation in response rates can be attained by examining the trends in response rate over time for ongoing surveys. For example, Figure 6.1 shows the response rate for the NCVS. The NCVS attempts interviews with all household members 12 years old and older. The NCVS reports response rates at the household level and the person level. The household nonresponse rate measures the percentage of households in which no household member was successfully interviewed. This is the top (dashed) line in Figure 6.1. It shows relatively stable values, ranging between 3 and 4 percentage points (a household response rate of 96–97%). The lowest line (shadow line) is the household refusal rate; it similarly shows a relatively consistent level over the years. The solid black line is the person-level refusal rate, which is the ratio of all sample household members enumerated who refused the interview request to all those enumerated. In short, it is an estimate of the percentage of persons providing an interview, once access to the household has been achieved. This nonresponse component is rising over the years, nearly doubling in the years covered by the chart.

Figure 6.2 shows the response rate trend for the Current Population Survey conducted for the Bureau of Labor Statistics by the US Census Bureau, which produces the monthly unemployment rates for the United States. A very different trend is displayed—for many years an overall consistent nonresponse rate of about 4 to 5%, but a continually increasing refusal rate. This is an example of a survey that kept the overall nonresponse rate low by reducing the noncontact rate over the years, in the face of higher refusals. What happened in 1994? There was a large change in the interviewing protocol in that year, moving to a computer-assisted personal interviewing design and a new questionnaire. One possible explanation for the impact is that the short questionnaire (8–12 minutes) was often taken on the

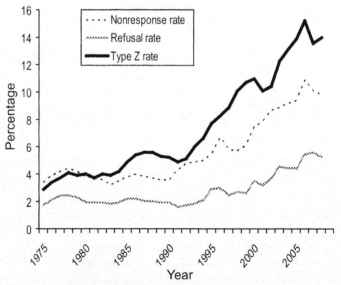

Figure 6.1 Household nonresponse rate, household refusal rate, and person refusal rate for the National Crime Victimization Survey by year. (Source: U.S. Census Bureau, 2007.)

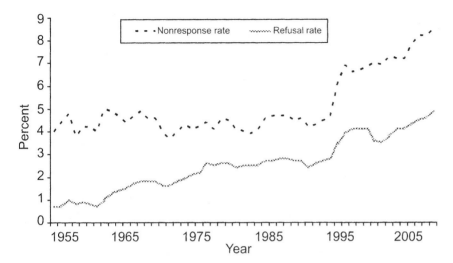

Figure 6.2 Nonresponse and refusal rates for the Current Population Survey by year. (Source: U.S. Census Bureau.)

doorstep with a paper questionnaire, but could not be done that way with a laptop computer and, hence, refusal rates increased (see Couper, 1996).

Both of the nonresponse rates above are quite low but the increasing trends are common. Both studies are well-funded face-to-face surveys; both are conducted by the Federal government. In general, response rates for other surveys are lower, with academic surveys somewhat lower and private sector surveys considerably lower.

Figure 6.3 shows overall nonresponse and refusal rates for the Survey of Consumers. The survey has much higher nonresponse rates than NCVS or CPS.

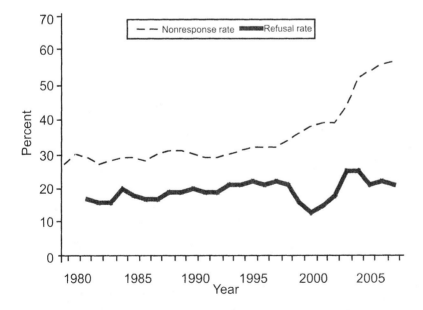

Figure 6.3 Nonresponse rate and refusal rate for the Survey of Consumers by year. (Source: Survey of Consumers.)

The overall nonresponse rate shows a steadily increasing trend, from percentages in the 30s in the 1980s to percentages near 60 more recently. The refusal rate shows a much smaller rise. At the same time, the percentage of the interviews that were taken only after an initial refusal climbed from about 7% to about 15% of the interviews. Again, the lesson is clear: response rates have declined over time.

Figure 6.4 presents nonresponse rates for the Behavioral Risk Factor Surveillance System (BRFSS). Because there are individual nonresponse rates for each state conducting the BRFSS, the graph shows the median nonresponse rate among all reporting states. The figure shows nonresponse rates that are more like the SOC than the NCVS. This probably results both from the telephone mode (versus the face-to-face mode of the NCVS) and the fact that the U.S. government is only an indirect sponsor of the surveys. The overall nonresponse rate increases from percentages in the 30s to percentages near 50% in later years.

Increasing nonresponse rates are not only a phenomenon of the United States. A study of nonresponse trends in 16 European countries over a 20-year period ending in the 1990s found that the noncontact rate increased on average 0.2% per year and the refusal rate 0.3% per year (de Leeuw and de Heer, 2002).

These are results from household surveys. Surveys of business establishments are most often conducted using mail questionnaires or the telephone (or some combination of both). There is less evidence of increasing nonresponse in business surveys (Atrostic and Burt, 1999).

6.3 Impact of Nonresponse on the Quality of Survey Estimates

Perspectives on the role of nonresponse rates in survey quality are undergoing rapid change. For some decades, the dominant goal of survey researchers was to

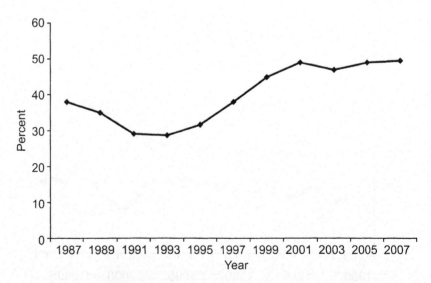

Figure 6.4 Median nonresponse rate across states, Behavioral Risk Factor Surveillance System, 1987–2007. (Source: BRFSS.)

minimize nonresponse rates. For example, Alreck and Settle (1995, p. 184) say, "It's obviously important to do as much as possible to reduce nonresponse and encourage an adequate response rate." Babbie (2004) is bold enough to say "I believe that a response rate of at least 50 percent is adequate for analysis and reporting. A response rate of 60 percent is good; a response rate of 70 percent is very good" (p. 261). Finally, Singleton and Straits (2005) note, "Therefore, it is very important to pay attention to response rates. For interview surveys, a response rate of 85 percent is minimally adequate; below 70 percent there is a serious chance of bias" (p.145). All of these quotations come from books used to teach students about survey methods.

The goal of minimizing nonresponse rates as the sole method of reducing nonresponse error resulted from an interpretation of the expression

$$\bar{Y}_r = \bar{Y}_n + \left(\frac{M}{N}\right)(\bar{Y}_r - \bar{Y}_m)$$

where $\bar{Y}_r$ is the unadjusted respondent mean, $\bar{Y}_n$ is the full sample mean, M/N is the proportion of the population that is nonrespondent, and $\bar{Y}_m$ is the nonrespondent mean (a value unknown in most surveys). Many researchers examined the two-factor term that is the bias of the respondent mean, inferred that $(\bar{Y}_r - \bar{Y}_m)$ was fixed in the population and concluded the only way to reduce the nonresponse bias was to reduce the nonresponse rate.

nonresponse bias

The formula above is sometimes labeled the "deterministic" view of survey nonresponse. It is compatible with the notion that there is a fixed set of respondents and nonrespondents in the population. For all sample persons, posing the survey request reveals to the researcher who are respondents and who are the nonrespondents. Instead of this viewpoint, a more modern view of nonresponse is that each person is potentially a respondent and potentially a nonrespondent—the decision to participate is the realization of a stochastic process (i.e., a process subject to random variability). In this perspective the expression above changes to

$$\bar{Y}_r = \bar{Y}_m + \frac{\sigma_{yp}}{\bar{p}}$$

where σ_{yp} is the covariance between, y, the variable of interest in the survey, and p, the propensity to respond, among units of the population; and $\bar{p}$, the mean propensity in the population. The covariance measures the joint variation of two variables,

$$\sum_{i=1}^{N}(y_i - \bar{y})(p_i - \bar{p})/(N-1)$$

so the expression above implies that when the likelihood of responding is strongly related to the variable of interest in the survey, then nonresponse bias for the respondent mean will be large. In short, nonresponse bias should vary over different estimates in a survey (as a function of the covariance of y and p). Instead of focusing on the response rate solely, the researcher has to focus on whether response propensity and the survey variable are correlated.

response propensity

Is there empirical support for large nonresponse error variation within surveys across different estimates? Yes. A "meta-analysis" synthesizes the findings of many different scientific studies of the same phenomenon. There are a set of surveys specially designed to measure the nonresponse bias of estimates from the survey. Sometimes these compare respondents and nonrespondents on variables contained on the sampling frame, sometimes on supplementary data matched to the sample cases, sometimes on data from a prior screening interview, and sometimes on data from a followup effort to measure nonrespondents. We note that across all the estimates, the nonresponse rates of the studies range from 14% to 72%, with a mean nonresponse rate of 36%. Most of the estimates come from studies using nonsurvey records (24% from the sampling frame, 32% from a supplementary dataset); 28% come from studies using follow-up of nonrespondents with some extraordinary effort. Most of the remaining studies use screener interview data (14%). A very small percentage (2%) use reports of intentions about responding to a future survey request.

Figure 6.5 presents a scatterplot of 959 estimates of the absolute value of the relative nonresponse bias:

$$\left| \frac{100 \times (\bar{y}_r - \bar{y}_n)}{\bar{y}_n} \right|$$

where the numerator contains the difference between respondent and full sample means, and the denominator is the full sample mean. This figure contains a point for each of the means reported in the 59 studies, with complementary percentages for binary variables. For each binary variable, two percentages can be computed. The smaller of the two tends to generate higher relative nonresponse bias. Hence, the figure presents the nonresponse bias of both complementary percentages. The plot displays vertical sequences of points, representing different estimates com-

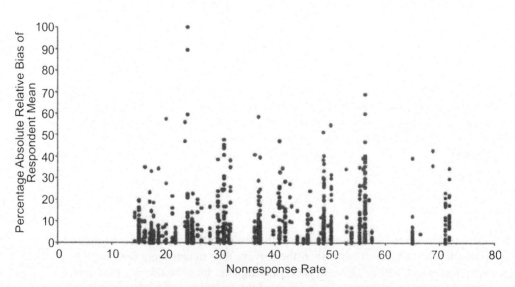

Figure 6.5 Estimates of the absolute value of the relative nonresponse bias for 959 estimates by nonresponse rate of survey. (Source: Groves and Peytcheva, 2008.)

puted from the same survey. The figure clearly shows: (a) large relative nonresponse biases exist in the studies, (b) most of the variation in nonresponse lies across estimates within the same survey, and, as implied by that observation, (c) the nonresponse rate of a survey, by itself, is a poor predictor of the absolute relative nonresponse bias. If a naïve OLS regression line were fit to the scatterplot, the R^2 would be 0.04. In short, insight into the linkage between nonresponse rates and nonresponse bias requires more information about the circumstances of each survey measurement.

There are two practical implications for survey researchers from the figure above. First, often there are very small σ_{yp} terms; that is, the variable of interest is not correlated to response propensity, so there is little nonresponse bias (regardless of the response rate of the survey). Second, finding out how response propensity is related to the important survey variables is a new task for researchers attempting to reduce nonresponse error. This requires thinking about nonresponse error in a causal way. There is an important note of caution about the figure. Since the scatterplot comes from different surveys, it does not provide an answer to the question, "For my specific survey, will increasing the response rate reduce my nonresponse error?" Sometimes, the answer will be "yes"; sometimes, "no." There are even examples in surveys in which increasing the response rate *increased* the nonresponse error (see Merkle, Edelman, Dykeman, and Brogan, 1998). The answer depends on whether the types of nonrespondents brought into the respondent pool as the response rate is increased act to increase σ_{yp} or decrease it.

It is important to note that nonresponse can affect both descriptive statistics and analytic statistics (like a regression coefficient). The expressions for the effect of nonresponse error on such statistics are more complex than those above, often functions of covariances of the variables involved. A large set of modeling techniques have been developed to address nonresponse error in such estimates, most drawing on developments in econometrics (Heckman, 1979; Berk, 1983).

In the next section, we argue that, to become more sophisticated in our understanding of when nonresponse rates produce nonresponse error in an estimate, we need to think causally about nonresponse mechanisms.

6.4 Thinking Causally About Survey Nonresponse Error

Three alternative causal models underlie the relationship between response propensity and nonresponse bias (Groves, 2006). As graphically shown in Figure 6.6, the "separate cause" model asserts that the vector of causes of the Y variable is independent of the causes of response propensity, P. In this case, expected values of Y among respondents would be unbiased estimates of those among all sample persons and corresponds to the "missing completely at random" case (Rubin, 1987). The "common cause" model asserts that there are shared causes (Z) of response propensity and the Y variable; this model corresponds to the "missing at random" case. The "survey variable cause" model asserts that Y itself is a cause of response propensity; this is the "nonignorable" condition of nonresponse.

nonignorable nonresponse

In each model nonresponse bias in the simple respondent mean can be portrayed as $\sigma_{yp}/\bar{p}$, where σ_{yp} is the covariance between a given survey variable, y,

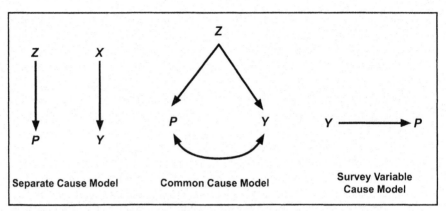

Figure 6.6 Alternative models for relationship between response propensity (*P*) and survey variable (*Y*), involving auxilliary variables (*S*, *Z*). (Source: Groves, 2006.)

and the response propensity, p; and $\bar{p}$ is the expected propensity over the sample members to be measured (Bethlehem, 2002). The separate cause model would produce a zero covariance, the common cause model would produce a nonzero covariance (but a zero covariance controlling for *Z*), and the survey variable cause model would have a nonzero covariance.

The expression above reminds us that nonresponse bias varies over different estimates within a survey, as a function of whether the likelihood of survey participation is related to the variable underlying the estimate. The scientific question associated with this expression is "what causes a correlation between *Y* and *P*" or "what causes a survey variable to be correlated to the likelihood to respond?"

6.5 Dissecting The Nonresponse Phenomenon

Methodological research has found that three types of unit nonresponse have distinctive causes and, for many surveys, distinctive effects on the quality of survey statistics. They are:

1) The failure to deliver the survey request (e.g., "noncontacts," failure to locate the sample unit, or postmaster returns mail surveys).
2) The refusal to participate (e.g., a contacted person declines the request).
3) The inability to participate [e.g., a contacted person cannot understand the language(s) of the questionnaire].

6.5.1 Unit Nonresponse Due to Failure to Deliver the Survey Request

Nonresponse due to noncontact or failure to deliver the survey request misses sample persons whose activities make them unavailable in the specific mode of data collection. The key concept is the "contactability" of sample units—whether

DISSECTING THE NONRESPONSE PHENOMENON

they are accessible to the survey researcher. Figure 6.7 shows a basic diagram of the influences acting on the contactability of sample units.

In household surveys, if we knew when people were at home and accessible to us, we could make a successful contact in one attempt. However, the accessible times of sample units are generally unknown; hence, interviewers are asked to make multiple calls on a unit until a first contact is made. Some sample units have "access impediments" that prevent strangers from contacting them (e.g., locked apartment buildings or telephone answering machines). People who throw away mail from unfamiliar sources often are missed in mail questionnaire surveys. People who are rarely at home often remain uncontacted even after repeated call attempts by interviewers. People who have call blocking services on their telephones often are not aware of the attempts of telephone interviewers to reach them.

For example, about 2% of sample households in the NSDUH are never contacted at the screening stage. The same statistic for the SOC is much higher. In the NSDUH, these noncontacted units are likely to be disproportionately in multi-unit structures and other structures with access impediments. Because the SOC is a telephone survey, such structures do not necessarily pose contact problems; instead, houses with caller ID and call blocking devices tend to be disproportionately noncontacted.

In practice, the percentage of successful calls declines with each successive call. For example, Figure 6.8 shows the percentage of sample U.S. households contacted by call number, among those never yet contacted over five different household surveys, some telephone, some face-to-face. About half of the contacted households are reached in the first call. With each succeeding call, a smaller and smaller percentage is reached. Variation across the surveys in the figure probably reflect effects of sample design variation and calling rules after the first call.

What predicts how many calls are required to gain first contact in household surveys? There are two principal answers:

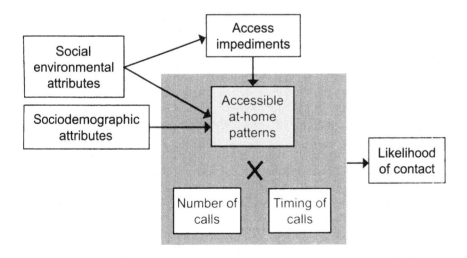

Figure 6.7 Causal influences on contact with sample household.

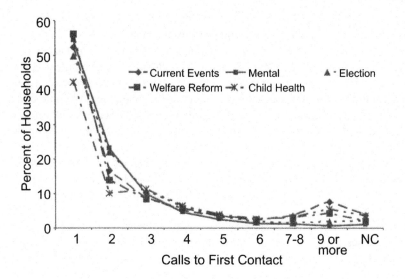

Figure 6.8 Percentage of eligible sample households by calls to first contact for five surveys. (Source: Groves, Wissoker, Greene, McNeeley, and Montemarano, 2001.)

1) Calls in the evening and on weekends are more productive than calls at other times.
2) Some populations have different accessibility likelihoods than others.

Sample persons tend to be more accessible to interviewers when they are at home. When are people at home? Most households have very predictable schedules. For those who are employed out of the home, most are away from home at set times, often the same periods each week. Most employed persons in the United States are away from home in the period of 8:00 AM to 6:00 PM Monday through Friday. If interviewers call at those times, proportionately fewer persons are reached. As Figure 6.9 shows, the best times appears to be Sunday through Thursday evenings from 6–9 PM local time, no matter what call attempt number is considered. Those evenings share the feature that the next day is a work day. Friday and Saturday evenings are different, with lower rates of contact in general. Times during the day on the weekends are better than times during the day during the work week. As it turns out, there are very few households in the United States where no one is ever at home in the evening.

There are systematic differences in noncontact rates across subpopulations in the United States. The easiest households to contact tend to be those in which someone is almost always at home. These include households with persons who are not employed outside the house, either because they are retired, they care for young children not yet in school, or some other reason.

One measure of difficulty of accessing different types of persons is the mean number of calls required to achieve the first contact with someone in the household. Persons in households that have some access impediments require more calls to first contact. These include apartment buildings with locked central

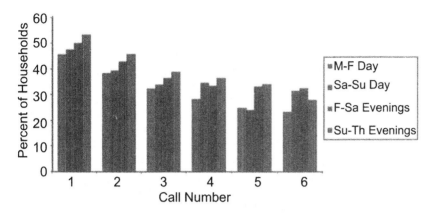

Figure 6.9 Percentage household contacted among those previously uncontacted by call number by time of day. (National Survey of Family Growth, Cycle 6.)

entrances, gated communities, or rural residences with locked gates. In telephone surveys, numbers connected to caller ID or other screening devices require more calls to first contact. Persons who live in urban areas tend to require more calls to first contact (partly because they tend to be single-person households in units with access impediments).

Mail, e-mail, and Web surveys (those modes that do not use interviewers to make contact) make the survey request accessible to the sample person continuously after it is sent. That is, in a mail survey, once the mail questionnaire is received by the household, it remains there until some household member does something with it. This can be at any time of the day or day of the week. Thus, the number of attempts required to gain contact with the sample unit has a different pattern. Chapter 5 notes that one attribute of these self-administered modes is that the researcher cannot easily differentiate a questionnaire that has not been seen by the sample household from one that has been seen and rejected.

In short, in interviewer-assisted surveys some of the noncontact arises because of at-home patterns of different groups and the number of different persons in the household. Other noncontacts arise from access impediments. The former affects all modes of data collection. The latter is specific to individual modes. For example, locked apartment buildings affect the productivity of face-to-face interviewers but not necessarily telephone interviewers. Thus, the composition of noncontacts can vary by mode. A corollary of this is that surveys that combine multiple modes can reduce the noncontact rate.

There are unanswered research issues about noncontact nonresponse that need to be tackled in the future. Most noncontact nonresponse is likely to be independent of the purpose of the survey. That is, the sample unit is not difficult to contact because of the topic of the survey but rather because of a set of influences that would be present for any survey request. This implies that nonresponse error would arise only for statistics related to those influences. Research into why sample units are difficult to contact may yield practical guidance to researchers on when nonresponse error might arise from noncontact and when it would not.

Estimates Biased by Noncontact Nonresponse. Bias flows from nonresponse when the causes of the nonresponse are linked to the survey statistics measured. For example, if one is mounting a survey whose key statistic is the percentage of persons who are employed outside the home and live by themselves, as Figure 6.10 shows, the value of the statistic is heavily affected by noncontact nonresponse. The figure shows estimated nonresponse bias for four different surveys. Imagine that we were interested in estimating what percentage of households contained only one person. For statistics like that, there are external sources of estimates from the Census Bureau, which have very high accuracy. If one would make only one call attempt on the households, the survey would disproportionately miss households with one person because they tend to be at home less frequently. The value of the statistic is underestimated by 27 to 35% if the design calls for a maximum of one call. A two-call survey overestimates the statistic by 20–32%. This statistic is sensitive to noncontact error because noncontact is partially caused by the amount of hours the sample person spends at home, accessible to the interviewer. In contrast, it is quite likely that another statistic, say the percentage of persons who report an interest in politics, is not as heavily affected by the noncontact rate.

6.5.2 Unit Nonresponse Due to Refusals

It could well be argued that the essential societal ingredients for surveys to gain cooperation of sample persons are rare in human history (Groves and Kahn, 1979, p. 227). Success requires the willingness of persons to respond to a complete stranger who calls them on the telephone, mails them a request, or visits their home. The persons must have little fear of physical or financial harm from the

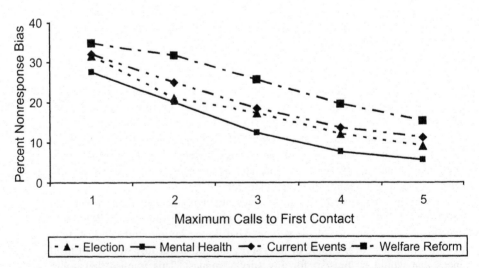

Figure 6.10 Percentage nonresponse bias for estimated proportion of single person households, by number of calls required to reach the house hold, for four surveys. (Source: Groves, Wissoker, Greene, McNeeley, and Montemarano, 2001.)

stranger, of reputational damage from the interaction, or of psychological distress caused by the interview. The sample persons must believe the pledge of confidentiality that the interviewer proffers; they must believe that they can speak their minds and report intimate details without recrimination or harm. Think, for a moment, about whether there are societies right now in the world that miss one or more of those ingredients.

Despite the ubiquity of surveys, in most countries they are not daily (or even frequent) experiences of individual persons. But the frequencies of surveys have increased over time. For some years, a survey was conducted that asked persons whether they had participated in any other survey in the past few years. In 1980, the percentage reporting participating in the last year was about 20%; in 2001, about 60%. (These are likely to be overestimates because they do not represent the answers of nonrespondents to the survey that asked about past participation) (Sheppard, 2001). In the United States, two types of survey efforts are most common. First, many service-providing organizations conduct customer satisfaction surveys, which collect ratings of the quality of the service. Second, every two years legislative elections produce large numbers of surveys, especially in congressional districts with close races. Outside of these circumstances, the frequency of survey requests to most people is low.

Groves and Couper (1998) put forward the hypothesis that when persons are approached with a survey request, they attempt quickly to discern the purpose of the encounter with an interviewer. Since the encounter was not initiated by the householders, they value returning to their original pursuits. The fact that there are common reasons for requests, purposes of requests, and institutions that are making requests, may lead to standardized reactions to requests. These might be default reactions that are shaped by experiences over the years with such requests. Survey requests, because they are rare relative to the other requests, might easily be confused by householders and misclassified as a sales call, for example. When this occurs, the householder may react for reasons other than those pertinent to the survey request. The fact that surveys often use repeated callbacks is probably an effective tool to distinguish them from sales calls. When surveys are conducted by well-known institutions that have no sales mission, interviewers can emphasize the sponsorship as a means to distinguishing themselves from a salesperson. There are several findings from survey methodological research that support the hypothesis that some sample persons may misidentify the intent of the survey interviewer:

1) Decisions to decline a request are made quickly (on the telephone most refusals take place in less than 30 seconds).
2) Persons who refuse at one point often accept when recontacted (so-called refusal conversion rates often lie between 25–40%).
3) Some survey interviewers are trained to attempt to avoid the misattribution (e.g., they say "I'm not selling anything" early in the call).

The speed of decision making suggests that persons focus on a small number of features of the request that are salient to them. Interviewer introductions on telephone and face-to-face surveys are often quite brief (see box). Some persons might choose the sponsorship of the survey as most salient; others, the vocal attributes of the interviewer; still others, the topic of the survey. In some self-administered surveys (e.g., mailed paper questionnaires), seeing the questions makes more salient the purpose of the measurement.

Why Unit Nonresponse Occurs. The causes of nonresponse are of growing interest to survey methodologists. Many theoretical frameworks have been applied to the survey participation problem, but they all involve influences arising from four different levels:

1) The social environment [e.g., large urban areas tend to generate more refusals in household surveys; households with more than one member generate fewer refusals than single-person households (Groves and Couper, 1998)].
2) The person level [e.g., males tend to generate more refusals than females (Smith, 1983)].
3) The interviewer level [e.g., more-experienced interviewers obtain higher cooperation rates than less-experienced interviewers (Groves and Couper, 1998)].
4) The survey design level (e.g., incentives offered to sample persons tend to increase cooperation).

The first two influences are out of the control of the researcher. For example, events that have nothing to do with a survey request affect how people react to the request (e.g., the Tuskegee experiment on syphilis among African American men is often cited as an influence on lowered response rates in that population, see Section 11.5.2). The last two influences, the interviewer level and the survey design level, are features that the researcher can manipulate to increase response rates (this is discussed in Section 6.6).

The theoretical perspectives that are most commonly applied to survey participation include "opportunity cost" hypotheses, based on the notion that busy persons disproportionately refuse to be interviewed because the cost of spending time away from other pursuits is more burdensome than for others. They include notions of "social isolation," which influence persons at the high and low ends of the socioeconomic spectrum to refuse survey requests from the major institutions of the society; notions of "topic interest," which fuel hypotheses about how the "interested" disproportionately responding may induce nonresponse error in key statistics; and notions of "oversurveying" that suggest fatigue from survey requests. There are many variants of these concepts that arise in different disciplines. Unfortunately, there seems to be spotty support for any one theory explaining the phenomenon. Most of the theories offer explanations that fit into the "person level" or the "social environment level" mentioned above.

Most theories do not inform how the diverse influences on participation to cooperate manifest themselves

---

The "I'm Not Selling Anything" Phenomenon

Evidence that sample persons misjudge the intent of interviewers has led to what some interviewers are sometimes trained to say in their introduction: "Hello, I'm Mary Smith from the RCDF Research Services. This is not a sales call. I'm conducting a survey on your recent experiences with telephone service." This is clearly an attempt to correct a feared misinterpretation of why the interviewer is calling.

Does it work? Experiments in using the technique show mixed results (van Leeuwen and de Leeuw, 1999). The effects are probably dependent on whether other information provided in the introduction underscores the credibility of the interviewer when she says, "I'm not selling anything."

at the moment the decision is made. One theoretical perspective that attempts to describe the underpinnings of these behaviors is called "leverage-salience theory" (Groves, Singer, and Corning, 2000). Under the theory, different persons place different importance on features of the survey request (e.g., the topic of the survey, how long the interview might take, the sponsor of the survey, or what the data will be used for). Some persons may positively value some attribute; others, negatively. Of course, these differences are generally unknown to the survey researcher. When the survey approach is made to the sample person, one or more of these attributes are made salient in the interaction with the interviewer or the survey materials provided to the sample person. Depending on what is made salient and how much the person negatively or positively values the attribute, the result could be a refusal or an acceptance.

In Figure 6.11, two different sample persons are represented by a scale, which if tilted to the right implies an acceptance, and to the left, a refusal. Prior to the exposure to the survey request, Person 1 most positively values the topic of the survey (the position of the "topic" hook is to the right of the fulcrum); Person 2 is quite uninterested in the topic. Person 1 has limited free time and is very sensitive to the time demands of a request; Person 2 has no sensitivity toward the burden of the interview. Person 1 is only mildly sensitive to an incentive offer; Person 2 is quite positively disposed to receiving a cash incentive. When the interviewer makes contact, she emphasizes the sponsor of the survey and the incentive (this emphasis is represented in the figure by the size of the ball weights placed on the scale hooks). The result of the request is that Person 1 is more likely to accept the request than Person 2. Using the metaphor of a scale, the value the sample person places on a specific attribute of the request is called the "leverage" of the request. How important the attribute becomes in the description of the request is called its "salience." There are several implications of this theory:

> **What Interviewers Say**
>
> Interviewers are trained to provide various pieces of information quickly in their introduction:
>
> On the telephone, "Hello, my name is Mary Smith. I am calling from the University of Michigan in Ann Arbor. Here at the University we are currently working on a nationwide research project. First, I would like to make sure I dialed the right number. Is this 301-555-2222?"
>
> OR
>
> In face-to-face encounters, "Hello, my name is Mary Smith, and I work for the University of Michigan's Survey Research Center. Let me show you my identification. The University of Michigan is conducting a nationwide survey, and we are interested in talking to people about their feelings on a variety of topics, including their feelings about the economy, the upcoming presidential elections, and some important issues facing the country these days. You should have received a letter from the University of Michigan telling you about this survey."

1) People have many different reasons for acceding to, or declining, a survey request, and these are not known at the outset by the requestor.
2) No one introduction is suitable to address the concerns of diverse sample persons.
3) Interviewers must have ways of quickly discerning the concerns to make salient those attributes given positive leverage by the sample person.

We discuss this in more detail in Section 6.6.

Surveys that make their requests without using interviewers make more or less salient various attributes of the request, generally through words or symbols

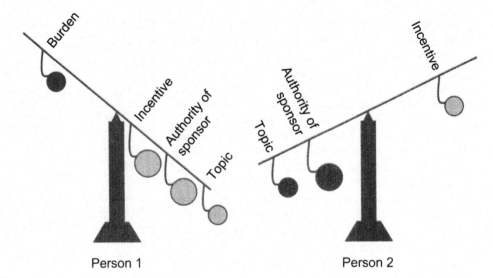

Figure 6.11 Two sample persons with different leverages for attributes of a survey request.

in the survey materials. These might include prominent display of the sponsor of the survey through envelopes and stationery, or muted display of the sponsorship. They might include a cash incentive affixed to the cover letter in a prominent position. It might involve placing sensitive questions toward the end of the questionnaire, attempting to mute their saliency prior to the decision of the sample person.

Estimates Biased from Nonresponse Due to Refusals. Similar logic underlies how refusal nonresponse error is generated. If the cause of the refusal is related to key statistics of the survey, the refusal rates portend nonresponse error in those statistics. Evidence for this is not plentiful but one example involves a comparison of a survey design that used an incentive for respondents and one that did not. The incentive acted to increase the cooperation of sample persons who were not interested in the survey about physician-assisted suicide, a topic then being debated in the community. One statistic was whether the respondent was involved in the community (e.g., a member of a civic organization or an attendee at political events). In the survey without the incentive, about 70% of the respondents reported such activity. In the survey design without the incentive, about 80% reported such activity (Groves, Singer, and Corning, 2000). Levels of community involvement are overestimated in the design without the incentive.

There are some ubiquitous correlates of the tendency to refuse a survey request. That is, many persons place "hooks" on the left of the scale above because they find these relevant to a survey request. For example, sponsorship effects are common, with central government surveys generating higher response rates than academic than commercial surveys (Groves and Couper, 1998). Burden of the survey as measured by pages in a self-administered questionnaire produces lower response rates (Goyder, 1985; Heberlein and Baumgartner, 1978); burden as measured by length of telephone and face to face interviews shows less clear effects (Bogen, 1996). Males tend to refuse more than females (Smith, 1983). Urbanicity is a powerful indicator of response rates in all modes (de Leeuw and de Heer, 2002). Adults who live alone tend to be refusers (Groves and Couper, 1998); households with young children show higher response rates than others (Lievesley, 1988). When the key variables of the survey are related to these attributes, we can anticipate nonresponse biases in the respondent-based estimates.

6.5.3 Unit Nonresponse Due to the Inability to Provide the Requested Data

Sometimes, sample persons are successfully contacted and would be willing to be respondents, but cannot. Their inability stems from a myriad of sources. Sometimes, they cannot understand any of the languages that are used in the survey. Sometimes, they are mentally incapable of understanding the questions or retrieving from memory the information requested. Sometimes, they have physical health problems that prevent them from participating. Sometimes, because of literacy limitations, they are unable to read or understand the materials in a mail survey. Sometimes, in business surveys, establishments do not have the necessary information available in the format or time frame required by the survey.

Since the reasons for their inability to comply with the survey request are diverse, it is also the case that what statistics are affected by the nonresponse are diverse. For example, in a survey measuring the health characteristics of a population, the inability to respond for health reasons portends nonresponse biases. Estimates of the well-being of the population would be overestimated because of the systematic failure to measure the unhealthy. On the other hand, estimates of the political attitudes of the population in the same survey may not be as severely affected by the same causes of nonresponse.

There has been relatively little methodological research into the causes and effects of unit nonresponse of this type. In many household surveys, the relative amount of such nonresponse is small. However, in studies of the elderly or immigrant populations, the rates can rise to substantial levels. For such designs, research discoveries have yet to be made regarding the role played by the interviewer, the method of data collection, features of the languages used, and characteristics of the sample persons.

6.6 DESIGN FEATURES TO REDUCE UNIT NONRESPONSE

At this point in the chapter, we have noted that nonresponse error can harm the quality of survey statistics, that the extent of harm is a function of how the influences toward nonparticipation relate to the statistic in question, and that the larger the nonresponse rate, the larger the risk of nonresponse error (other things being equal). In general, the only visible part of the quality impacts of nonresponse is the nonresponse rate. Hence, much effort of survey researchers has been focused on reducing nonresponse rates.

Figure 6.12 decomposes survey participation into three steps: contact, the initial decision regarding participation, and the final decision regarding participation. The last step arises because many survey designs use repeated efforts to address the concerns of reluctant respondents.

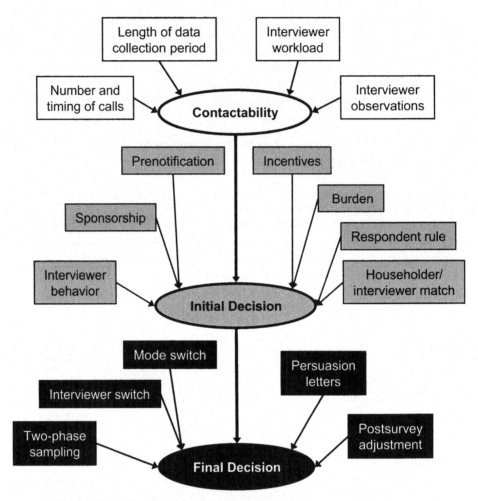

Figure 6.12 Tools for reducing unit nonresponse rates.

We have learned in Chapter 5 that the different modes of data collection tend to have different average response rates. The typical finding is that face-to-face surveys have higher response rates than telephone surveys. Telephone surveys have higher response rates than self-administered paper and Web surveys, other things being equal. It is a common finding that the use of interviewers in face-to-face and telephone surveys increases response rates, both because of higher success at delivering the survey request and because of their effectiveness in addressing any concerns about participation that sample persons may have.

There are several features in Figure 6.12 that address interviewer actions. First, leverage-salience theory of survey participation offers several deductions about interviewer behavior. Recall that different sample persons are likely to vary in how they evaluate the survey request (assigning different "leverages" to different attributes). Since these are unknown to the interviewer, the interviewer must somehow discern them in order to gain their cooperation.

One further deduction from leverage-salience theory is that training interviewers to recite the same introductory description to each sample person will not be effective. This appears to have empirical support. When Morton-Williams (1993) compared response rates for interviewers instructed to recite a standard script with those trained to address the sample persons in a freer manner, the unscripted interviewers had higher response rates (see box on the Morton-Williams study on p. 206). Groves and Couper (1998) propose two principles of interviewer behavior that may underlie the Morton-Williams experimental findings: maintaining interaction and tailoring. Expert interviewers appear to engage the sample persons in extended conversations (whether or not they are pertinent to the survey request). The interviewers "maintaining interaction" in this way attempt to gain information about the concerns of the sample person. Effective interviewers then "tailor" their remarks to the perceived concerns of the sample person. This tailoring appears to explain some of the tendency for experienced interviewers to achieve higher cooperation rates than novice interviewers. They carefully observe the verbal and nonverbal behavior of the sample persons in order to discern their concerns. When they form hypotheses about those concerns, the interviewers "tailor" their behavior to the concerns. They customize their description of the survey to those concerns.

maintaining interaction

tailoring

Figure 6.12 also indicates that if the initial decision of the sample person does not yield an interview, surveys seeking high response rates often make further efforts to bring the person into the respondent pool. These can involve switching interviewers, changing to a different mode, or sending persuasion letters. Finally, when the refusal decision appears final (or when all efforts to contact the sample unit have failed), then a variety of postsurvey features can be used. First, two-phase designs can be introduced whereby a sample of nonrespondents are followed up using a different recruitment protocol. Second, statistical adjustments can be made at the analysis stage (these are discussed in Chapter 10).

The remainder of this section briefly reviews the literature on methods to increase response rates, whenever possible making comments about when the method might affect the nonresponse bias of different kinds of estimates. The literature is a large one, consisting of several books (Goyder, 1987; Brehm, 1993; Groves and Couper, 1998; and Groves, Dillman, Eltinge, and Little, 2002), as well as hundreds of articles in scientific journals. The research that produced these methods often used randomized experimental designs to compare two ways of designing a survey protocol. One sample of persons received one protocol; another, the other. That protocol that produced the highest response rate was judged to be preferable. Sometimes, the researcher had available external indicators of the characteristics of sample person and, thus, could estimate whether increasing the response rate increased the quality of the survey statistics. Usually, however, the purpose of the studies is to demonstrate how to achieve higher response rates, whether or not they improved the survey statistics. We use Figure 6.12 to organize the discussion.

What Interviewers Say about Approaching Sample Households

In focus groups, interviewers describe how they prepare for a visit to a sample household:

"I use different techniques depending on the age of the respondent, my initial impression of him or her, the neighborhood, etc."

"I try to use a 'gimmick' in my attire when visiting HUs. Bright colors, interesting pins, jewelry, nothing somber or overly 'professional' or cold looking—fun items of attire like ceramic jewelry, scarves tied in odd ways. If my [initial drive through of the neighborhood] spots cats or dogs in windows or doors, I make a note and wear something like a cat pin on my coat, etc."

Number and timing of attempts to access the sample person. In both self-administered questionnaires and interviewer-assisted surveys it has been repeatedly found that the larger the number of sequential attempts to deliver the survey request to the sample person, the higher the likelihood of successfully contacting them. Goyder (1985) and Heberlein and Baumgartner (1978) show that repeated efforts to gain access to the sample units tend to reduce nonresponse. For U.S. telephone and face-to-face surveys, contact with household samples is easier on Sunday–Thursday evenings (the nights preceding work days) and during the day on weekends. Only a tiny minority of households can be reached only during the weekday day hours. As Figure 6.10 showed, these efforts radically reduce the nonresponse bias of estimates of the prevalence of single-person households. Estimates based on other strong correlates of "at-homeness" would show similar effects.

Data collection period. The longer the data collection period, the higher the likelihood that all persons are made aware of the survey request. How short is too short? It is noteworthy that the Current Population Survey is conducted in about 10 days and achieves near 100% contact rates. Thus, with proper staffing of interviewers such surveys can make initial contact with the vast majority of persons in a relatively short time. Mail surveys need longer periods because of the time requirements of the postal service. Choosing the data collection period probably affects the bias of estimates in similar ways to choosing the number of calls to make on sample cases.

Interviewer workload. Each sample case assigned to an interviewer requires time to contact. In household surveys, only about 50% of the first contact attempts are successful using the usual calling rules. If too many cases are assigned to an interviewer, insufficient effort will be given to some cases. Botman and Thornberry (1992) note when there is insufficient time or too many cases to work, nonresponse can increase, both noncontacts (because of insufficient effort to access the sample units) and refusals (because of inadequate refusal conversion effort). Since this feature also limits calls, its biasing effects should be focused on similar estimates to those of callback rules (e.g., single person households of employed persons tend to be nonrespondent).

Interviewer observations. The face-to-face mode has a distinct advantage in making contact with sample households—the ability of an interviewer to observe characteristics of the sample unit. Sometimes, these are visual observations (e.g., toys in the yard imply that children are resident); sometimes they are the result of neighbors providing information about the sample unit. Interviewers documenting what informants say in intermediate contacts can be useful to survey managers. When informants ask questions about the survey, it portends higher odds of their eventually consenting (Groves and Couper, 1998). To the extent that interviewers can observed attributes of the sample that are correlates of the key variables, the researcher can attempt to reduce bias in estimates based on those variables.

Sponsorship. In most countries of the world, the central government survey organizations attain higher cooperation rates than academic or private sector survey organizations. When the sponsor of the survey has some connection to the target population (e.g., a membership organization) the strength of the connection is related to the response propensities. For example, random half samples in a face-to-face survey were assigned to U.S. Census Bureau and University of Michigan Survey Research Center interviewers. The cases with survey requests from the Census Bureau refused at a 6% rate; those with survey requests from the University of Michigan refused at a 13% rate (National Research Council, 1979). Many speculate that the higher response rates of governments arise from (a) the widespread understanding that democratic governments legitimately need information from the residents to benefit them, or (b) a belief that participation is mandatory.

Prenotification. An advance letter mailed to a sample household can generate higher rates of cooperation than no such prenotification (e.g., see Traugott, Groves, and Lepkowski, 1987; for a contrary finding, see Singer, Van Hoewyk, and Maher, 2000). If the letter explicitly notes the sensitive content of the interview, however, it can depress response rates [as in ACSF (1992) cited in deLeeuw, Hox, Korendijk, Lensvelt-Mulders, and Callegaro, 2007]. Further, its effect may be dependent on the agency of the letter author, perhaps as evidenced by the stationery. For example, in a randomized experiment the same letter written on a market research letterhead received a lower response rate than one on university letterhead. In fact, the market research company letter generated a lower response rate than no letter at all (Brunner and Carroll, 1969). Interviewers appear to value such letters, and for that reason alone they are standard practice in most scientific surveys.

What kinds of estimates might have their nonresponse bias affected by advance letters? Unfortunately, there are few studies showing differential effects of letters across subgroups. One could speculate that nonresponse biases sensitive to advance letters would be located among correlates of likelihoods of receipt or readership of the letters. For example, since advance letters are a device requiring literacy for some of their

---

Berlin, Mohadjer, Waksberg, Kolstad, Kirsch, Rock, and Yamamoto (1992) on Incentives and Interviewer Productivity

Berlin et al. (1992) discovered that offering incentive can *reduce* total survey costs.

Study design: Three randomized treatment groups promised no incentive, a $20 incentive, or a $35 incentive upon the completion of literacy assessments in a pretest of the National Survey of Adult Literacy. The interviewer began with background questions and then gave the respondent the assessments for self-completion. The study used an area probability sample of over 300 segments of census blocks with about 2800 housing units, with each segment assigned to one of the treatments.

Findings: The incentives brought into the respondent pool less educated persons who in the past tended to be nonrespondent. As shown below, the $35 incentive achieved the highest response rates; the no-incentive condition, the lowest.

	Incentive Level		
	None	$20	$35
Response Rate	64%	71%	74%
Interviewer costs	$130	$ 99	$ 94
Incentive costs	0	20	35
Total costs	$130	$ 119	$129

With fewer interviewer hours per case, the $20 incentive more than paid for itself, relative to no incentive. The $35 incentive group was not as cost-effective, even though its response rate was higher.

Limitations of the study: Applying the results to other studies is limited by the fact that the survey task (i.e., self-completion of literacy assessment packages) is unusual, the survey was sponsored by a government agency, and the number of primary areas was small (16). This produced standard errors that themselves were rather unstable.

Impact of the study: The finding that giving incentives could save money was counterintuitive and important.

> **Morton-Williams (1993) on Tailoring Behavior by Interviewers**
>
> Morton-Williams (1993) presented the results of a randomized experiment in Britain that supports the value of interviewers' tailoring their introductory remarks to the concerns of the sample person.
>
> *Study design*: Fourteen of 30 interviewers were asked to use a scripted introduction to a face-to-face survey; the complement were asked to use their own judgment about the content of their remarks. All interviewers were asked to give their own names, identify their organization, show an identity card, mention the independence of their organization, describe the survey, and note how the address was selected. The scripted introduction specified the exact words the interviewers were to use.
>
> *Findings*: The scripted interviewers obtained a 59% response rate; the unscripted, a 76% response rate.
>
> *Limitations of the study*: The description does not specify how interviewers or housing units were assigned to the treatment groups. Hence, there may be some confounding with true differences among units that cooperate and the treatment groups. The statistical analysis does not appear to account for the interviewer component of variability; it is not clear that the response rate differences are beyond those expected by chance, given the small numbers of interviewers.
>
> *Impact of the study*: The experiment was the first to support the growing evidence that scripting interviewers in the recruitment phase of a survey may harm cooperation rates.

effects, one would expect that letters would have lower effects in semiliterate populations.

Incentives. Offering an extrinsic benefit to participation increases cooperation rates (Singer, 2002). Cash incentives tend to be more powerful than in-kind incentives of similar value. Incentives paid prior to the survey request are more powerful than those offered contingent on the survey being completed. Higher incentives are more powerful, but some studies report diminishing returns as the amount rises. If the incentive effect is powerful enough, the total cost of the survey can actually decline because of lower interviewer or follow-up costs (see the box on p. 205).

There is some evidence that when an incentive is used, it disproportionately affects the participation of persons less interested in the topic (Groves, Couper, Presser, Singer, Tourangeau, Piani Acosta, and Nelson, 2002). Without an incentive, the respondents consist of those interested in the topic (and often reporting different attributes on the key variables). With an incentive, the respondent pool better reflects the full population.

Burden. The evidence for how survey length or cognitive burden affects cooperation is mixed (Singer and Presser, 2007). However, there is evidence that perceived length of time or complexity of self-administered instruments reduces cooperation. For example, in examining a large number of surveys, both Heberlein and Baumgartner (1978) and Goyder (1985) found that each additional page of a self-administered questionnaire reduced the response rate by 0.4 percentage points.

Respondent rules. Surveys that permit several alternative persons in a household to provide the survey information tend to have higher cooperation rates than those selecting a random adult respondent. Similarly, surveys that permit proxy reporting have higher response rates than those requiring self-report. There is some indication of bias effects of such rules. For example, in health surveys the prevalence of impairments and illnesses is often higher in surveys accepting proxy reports (since illness is both a cause of nonresponse and a survey variable).

Interviewer introductory behavior. Especially in telephone surveys, the first few seconds of interviewer interaction affect cooperation rates. There is little empirical research on this, but what exists suggests that variation in inflection and speed of communication are related to higher cooperation rates (Oksenberg, Coleman, and Cannell, 1986). Morton-Williams (1993) shows that rigid scripting of interviewers' introductory remarks can increase refusals (see the box on p. 206). Given the causal approach to nonresponse bias reviewed above, one would worry about biasing effects of these behaviors if interviewers tended to use the survey content as a persuasive tool.

Interviewer/household matching. There is no research but much speculation that matching interviewers with sample persons in a way that improves the likelihood of trust and acceptance of the interviewer improves cooperation. For example, Nealon (1983) shows that female interviewers obtained higher cooperation rates than male interviewers for a sample of spouses of farm operators. Similar practices of using "indigenous" interviewers are common in studies in anthropology. When thinking about nonresponse bias and this practice, the researcher has to consider whether the same interviewer attribute that improves participation may affect the answers of the respondents. (We note in Chapter 9 that interviewer gender and race have been found to affect answers for survey topics related to gender and race.)

Interviewer switches. It is common for data collection supervisors to replace the interviewer who receives an initial refusal with another interviewer who has characteristics that might be more acceptable to the sample person (e.g., often attempting to match attributes as the prior paragraph noted). The new interviewer reviews the history of the case, including any documentation on the initial refusal, then approaches the sample unit again. This can be viewed as a naïve form of tailoring, by manipulating some characteristics of the interviewer to be more amenable to the concerns of the sample person. For example, a male face-to-face interviewer who appears to frighten an elderly woman living alone can be replaced with an older female interviewer.

Mode switches. There are many designs that begin with a cheaper mode of data collection (e.g., mail questionnaires and Web surveys) and then use more expensive modes (e.g., telephone and face-to-face) for nonrespondents from the first mode. In single-mode surveys, the clear pattern is that face-to-face surveys can achieve higher response rates than telephone surveys or mailed self-administered surveys. Mixing the modes allows one to optimize resources to improve cooperation. Estimates related to literacy might have nonresponse bias properties sensitive to such switches.

Persuasion letters. It is common to send persons who initially refuse to be interviewed a letter that reinforces the serious purpose of the survey and notes that an interviewer will again call on the household to address any concerns. Common practice attempts to tailor the letters to expressed concerns of the sample unit (e.g., sending letters emphasizing confidentiality to those persons concerned about their privacy).

This brief review suggests that there are a number of methods employed to reduce nonresponse in surveys. Some of these are based on firm research evidence; others are based on practical experience. Not all techniques work in all circumstances. Examples of failed replications of experiments can be found. The challenge for the practitioner is to determine the strategies likely to be most successful given a particular design and target population.

The final two boxes in Figure 6.12 are tools that require different statistical analysis techniques. **Two-phase sampling** draws a probability sample of nonrespondents, for whom new methods of contact and requests for participation are used. Those sampled who are thereby measured are used to estimate the characteristics of all the nonrespondents existent at time of the sampling. Postsurvey adjustment (see Chapter 9) is a technique using the existing respondents in ways that compensate for those nonrespondents who were missed (e.g., giving more weight in the analysis to urban respondents than rural ones when the response rates in urban areas are lower).

two-phase sampling

postsurvey adjustment

Despite the well-developed methodological literature on ways to increase response rates, there are many unanswered research questions. These include:

1) When efforts to interview reluctant respondents succeed, do they provide responses more tainted by measurement errors?
2) When do efforts to increase response rates affect nonresponse errors and when do they not?
3) How should efforts to reduce noncontact versus refusal rates be balanced against one another?
4) Considering both sampling error and nonresponse error, given a fixed budget, when can the researcher justify less than full effort to reduce nonresponse rates?

The research in this arena is likely to be very important for the future of the field, because survey designs are now devoting large portions of their research budgets to nonresponse reduction, with little scientific basis for this practice.

6.7 ITEM NONRESPONSE

The discussion above centers on unit nonresponse (the failure to obtain any measures on a sample unit). Item nonresponse is a severe problem in some kinds of surveys. Item nonresponse occurs when a response to a single question is missing. For example, a respondent to the Survey of Consumers may willingly participate and start answering questions, but when the interviewer asks about family income in the last year, they may refuse to give an answer to that question.

The impacts of item nonresponse on a statistic are exactly the same as that for unit nonresponse, but the damage is limited to statistics produced using data from the affected items. Thus, the expression on page 189 for a sample mean applies to the combined effects of unit and item nonresponse.

It often appears that the causes of item nonresponse are different from those of unit nonresponse. Whereas unit nonresponse arises from a decision based on a brief description of the survey, item nonresponse occurs after the measurement has been fully revealed. The causes of item nonresponse that have been studied in methodological research include: (a) inadequate comprehension of the intent of the question, (b) judged failure to retrieve adequate information, and (c) lack of willingness or motivation to disclose the information (see Beatty and Herrmann, 2002; Krosnick, 2002). However, research in this area is in its infancy. Most of the methodological research on question wording (see Chapter 7) has focused on properties of the questions that change the substantive answers. A research program in the causes of item-missing data would be of great utility to practitioners.

There is evidence for item-missing data arising from respondents' judging that they do not have adequately precise answers. Some experiments show persons failing to give a specific income value but willingly providing an estimate within an income range (e.g., between $50,000 and $75,000; Juster and Smith, 1997). Evidence for the effect of motivation for the respondent role is that open questions (requiring the respondents to write in answers they invent) tend to have larger missing data than closed questions (requiring the respondents to choose from a list of answers).

Figure 6.13 is a model of the response process posited by Beatty and Herrmann, which distinguishes four levels of cognitive states regarding the information sought by the survey question: available, accessible, generatable, and inestimable. The four states are ordered by level of retrieved knowledge suitable for a question response. They posit both errors of commission (reporting an answer without sufficient knowledge) and errors of omission (failing to report an answer when the knowledge exists). Social influence to give an answer may produce data with measurement errors. Item-missing data can arise legitimately (for those in an "inestimable" cognitive state) or as a response error (for those with the knowledge available). The latter situation might arise when social desirability

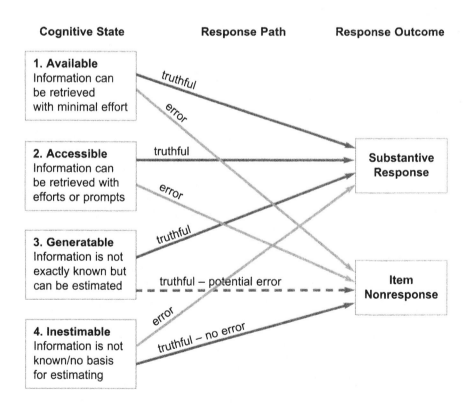

Figure 6.13 Beatty-Herrmann model of response process for item-missing data.

influences a respondent to refuse to answer a question (or answer, "don't know") instead of revealing a socially unacceptable attribute.

The tools used to reduce item nonresponse are reduction of the burden of any single question, the reduction of psychological threat or increase in privacy (e.g., self-administration), and interviewer actions to clarify or probe responses. The strategies used to compensate for item nonresponse are often quite different from those for unit nonresponse, as in the former case the analyst usually has a rich vector of other responses with which to adjust. Thus, imputation is most often used for item-missing data, whereas weighting class adjustments are more common for unit nonresponse (see Chapter 10).

6.8 ARE NONRESPONSE PROPENSITIES RELATED TO OTHER ERROR SOURCES?

Before we close out this chapter, the reader should be alerted to some survey methodological research that shows that nonresponse may be linked to other errors, especially coverage and measurement errors. From the literatures on censuses, we see that young, single adult males, who often have tenuous ties to several households (and thus tend to be missed in address based sampling frames), also tend to be nonrespondent. Similarly, there is evidence that those most reluctant to respond sometimes provide answers to questions that are less thoughtful and subject to greater measurement error (see Olson, 2006). Although this research is just now emerging, it is another important reminder that merely increasing response rates is an overly simple reaction to fears of nonresponse bias in survey estimates.

6.9 SUMMARY

Surveys produce numbers that attempt to describe large populations by measuring and estimating only a sample of those populations. When the designated sample cannot be completely measured and estimates are based only on responding cases, the quality of survey statistics can be threatened. There are two types of nonresponse: unit nonresponse and item nonresponse.

In their simplest form, nonresponse rates are the ratios of eligible sample units that were not measured to the total number of eligible units in the sample. In practice, nonresponse rates are sometimes difficult to compute because the eligibility of nonrespondents remains unknown and the sample design assigns varying probabilities of selection to different frame elements.

Not all nonresponse hurts the quality of survey estimates. Nonresponse produced by causes that are related to key survey statistics is the most harmful kind (e.g., failure to contact persons who are rarely at home in a survey of how they use their time). Such nonresponse is termed "nonignorable" nonresponse. Nonresponse can harm the quality both of descriptive and analytic statistics. Different statistics in the same survey can vary in their magnitudes of nonresponse error.

There are three classes of unit nonresponse that have different causes and, hence, affect the quality of survey statistics in different ways: inability to gain access to the sample unit, failure to gain the cooperation upon delivery of the survey request, and the sample unit's inability to provide the data sought.

Nonresponse rates by themselves do not predict the nonresponse error of individual estimates in a survey. Indeed, there is evidence of large variation of nonresponse error among estimates within the same survey. Despite continuing ignorance about when nonresponse matters and when it does not, there is a strong set of professional norms to increase the response rates of surveys. Guidelines of most survey professional organizations describe efforts to obtain high response rates.

There are many tools that survey researchers have to increase the response rates in surveys. These include repeated callbacks, long data collection periods, small interviewer workloads, using interviewer observations to guide their behaviors, advance letters, using trusted sponsors, short questionnaires, use of proxy respondents, tailoring of interviewer behavior to the concerns of the sample person, matching interviewer and sample person characteristics, persuasion letters for initial refusals, mode and interviewer switches for reluctant respondents, and two-phase samples for nonresponse. Almost all of the methods require spending more time or effort contacting or interacting with the sample units. This generally increases the costs of surveys.

An important remaining challenge to survey researchers regarding nonresponse is determining when it hurts the quality of survey statistics and when it does not. More research is needed on this issue. Without it, there is no guarantee that efforts to increase response rates are wise. Without it, there is no way to justify being satisfied with low response rates.

KEYWORDS

access impediments
contactability
ignorable nonresponse
inability to participate
item nonresponse
leverage-salience theory
maintaining interaction
noncontact
nonignorable nonresponse
nonresponse bias

opportunity cost
oversurveying
postsurvey adjustment
refusal
response propensity
social isolation
tailoring
topic interest
two-phase sampling
unit nonresponse

FOR MORE IN-DEPTH READING

Groves, R., and Couper, M. (1998), *Nonresponse in Household Interview Surveys*, New York: Wiley.

Groves, R., Dillman, D., Eltinge, J., and Little, R. (eds.) (2002), *Survey Nonresponse*, New York: Wiley.

Särndal, C., and Lundström, S. (2005), *Estimation in Surveys with Nonresponse*, New York: Wiley.

EXERCISES

1) Shown below is the distribution of final case results for an RDD telephone survey conducted on a sample of 2127 randomly generated telephone numbers covering the contiguous 48 United States (excluding Alaska and Hawaii) plus the District of Columbia. The target population consists of households in the 48 states and the District of Columbia. The topic of the survey is the household's recycling activity: availability of recycling pickups in the household's city/town, and the household's use thereof. All phone numbers were dialed a maximum of 20 times, covering weekdays and weekends, daytime and evenings, until one of these final results was achieved.

Completed interviews	614
Refusals	224
Answering machine with residential message on every call	180
Never answered on every call	302
Eligible household contacted, but no interview for other than refusal reasons	127
Businesses/nonresidences	194
Nonworking numbers	486
Total	

Compute the response rate in three different ways, as follows (You might want to consult the "Resources for Researchers" section of www.aapor.org).

a) Assume all unanswered numbers are eligible.
b) Assume all unanswered numbers are ineligible.
c) Estimate eligibility among the unanswered numbers
d) Estimate the response rate among the unanswered numbers.

2) You conducted a survey of 1500 randomly selected county school systems nationwide. Your study aims to examine factors related to whether or not school systems offers sex education programs/classes, comparing school systems located in areas where the majority of the local population belongs to a conservative religious group (CRG) to those in areas where the members of conservative religious groups represent a minority of the local population. The sample yielded the following results, and from external sources you already know which school systems offer sex education programs/classes.

Population Within School System	Sample Size	Response Rate	Responding School Systems	Percent Offering Sex Education	
				Respondents	Non-respondents
Majority CRG	500	50%	250	5%	0%
Minority CRG	1000	60%	600	50%	35%

EXERCISES

Assume that the sample sizes represent the proportions in the population (i.e., an equal probability sample). The estimate of the overall percent of school systems offering sex education, as computed based on responding school system reports, is thus 36.8%, computed as follows:

Majority CRG: 5% × 250 = 12.5
Minority CRG: 50% × 600 = 300
12.5 + 300 = 312.5   312.5/850 = 36.8%

Estimate the nonresponse bias in the estimate of "percent of school systems offering sex education" of the full sample.

3) You have studied the effects of incentives on cooperation in surveys.

   a) Describe the reasoning behind the common effect that prepaid incentives have larger net effects on cooperation rates than promised incentives.
   b) Describe why incentives can sometimes reduce the cost of a survey.

4) You are completing a telephone survey of members of a professional organization, regarding their level of activity in support of the organization. You face the decision about whether to use some funds to increase the response rate over that currently achieved, which is 80%. A key indicator to be estimated from the survey is the percentage of members who attend every monthly meeting of the local chapter of the organization. With the 80% response rate, the current estimate is that 42% of the membership attends monthly meetings. (For purposes of your answer, ignore sampling error differences in the estimates.)

   a) Is it possible that, in truth, over half of the membership attends monthly meetings?
   b) What arguments would you make regarding the likely characteristics of the nonrespondents on the attendance measure?

5) Briefly explain in words the relationship between the nonresponse rate and the bias of nonresponse for an estimate of the population mean.

6) You are planning a survey about health care practices in a population of participants in the past year's U.S. masters' swimming competitions. The survey interview consists of questions about what diet and exercise regimens the respondent pursues. The same questionnaire was used in a general population survey recently. You discover that a randomized experiment regarding a $10 incentive produced a 20 percentage point increase in response rates and a decline in the estimated proportion of respondents reporting they exercised. Identify one reason to expect the incentive to have similar effects and one reason why the incentive may have different effects for the survey you are planning.

7) Given what you have learned about the effect of advance letters on survey cooperation rates, make a choice of which organization's stationery to use for the letter in the following three surveys.

a) Assume that your goal is to maximize cooperation rates. Choose whether to use the sponsoring organization's letterhead (the organization paying for the survey) or the data collection organization's letterhead (the organization collecting the data, which you can assume to be relatively unknown among sample persons); then explain the reason for your decision.

Sponsor	Data Collection Organization	Target Population	Which organization's stationery for advance letter?	Reason for answer
Commercial credit company	Academic survey center	U.S. households		
Federal government	Commercial market research firm	Low-income households		
Highway Construction lobbying organization	Nonprofit research organization	Presidents of highway construction companies		

b) Now assume that you are interested in minimizing the nonresponse bias of the following estimates for each survey.

Sponsor	Data Collection Organization	Target Population	Which organization's stationery for advance letter?	Reason for answer
Commercial credit company	Academic survey center	U.S. households	Mean amount of credit card debt	
Federal government	Commercial market research firm	Low-income households	Percent with favorable attitudes about food stamp program	
Highway Construction lobbying organization	Nonprofit research organization	Presidents of highway construction companies	Percentage judging lobbying effort effective	

8) What subpopulations within the household population in your country tend to be noncontact nonrespondents in telephone surveys with restricted time periods for the data collection? Why?

9) Using leverage-salience theory, describe one hypothesis of why estimates of support for a given political candidate may be affected by the sponsorhip of the survey.

10) In studying the effect of incentives, an experiment was conducted among new college graduates. The sample was randomized into two groups, one group receiving questions related to prescription drug coverage by govern-

EXERCISES

ment health plans and the other group receiving questions related to environmental problems. Each of these two groups was further divided into two: one receiving no incentives and another receiving $5 incentive, thus yielding a 2 × 2 experimental design.

Briefly describe one possible finding from this experiment in relation to incentives and survey topic.

11) Give three reasons why face-to-face surveys typically achieve higher response rates than telephone surveys.

12) This chapter described alternative ways to increase response rates in surveys.

   a) Identify three methods used to increase response rates.
   b) Thinking of the adult household population, identify a subgroup for whom one of the methods you listed in (a) does not appear to be as effective as for other subgroups, and briefly explain why.

13) The chapter shows that reducing nonresponse rates can sometimes increase nonresponse error on simple statistics like the respondent mean. Identify a survey design feature than might create such an outcome in a survey of members of a social organization (a local dance group) when attempting to measure the mean number of years that a member has been active in the organization.

CHAPTER SEVEN

QUESTIONS AND ANSWERS IN SURVEYS

As we pointed out in Chapter 5, surveys use a variety of methods to collect information about their respondents. Perhaps the most common is the use of the questionnaire, a standardized set of questions administered to the respondents in a survey. The questions are typically administered in a fixed order and often with fixed answer options. Usually, interviewers administer a questionnaire to the respondents, but many surveys have the respondents complete the questionnaires themselves. Among our six example surveys, three (the NCVS, the SOC, and the BRFSS) rely almost exclusively on interviewer administration of the questions, and the other three (the NSDUH, the NAEP, and the CES) mix interviewer administration with self-completion of the questions by the respondents. Over the last 30 years or so, survey questionnaires have increasingly taken electronic form, with computer programs displaying the questions either to the interviewer or directly to the respondent. But whether an interviewer is involved or not, and whether the questionnaire is a paper document or a computer program, most surveys rely heavily on respondents to interpret a preestablished set of questions and to supply the information these questions seek.

questionnaire

7.1 ALTERNATIVES METHODS OF SURVEY MEASUREMENT

Surveys do not always force respondents to construct answers during the interview. For example, many surveys collect information from businesses or other establishments, and such surveys often draw information from company records. For these, the questionnaire may be more like a data recording form than like a script for the interview, and the interviewer may interact with the records rather than with a respondent. (The CES survey sometimes comes close to this model.) Similarly, education surveys may supplement data collected via questionnaires with data from student transcripts; and health surveys may extract information from medical records instead of relying completely on respondents' reports about their medical treatment or diagnosis. Even when the records are the primary source of the information, the respondent may still play a key role in gaining access to the records and in helping the data collection staff extract the necessary information from them. Some surveys collect the data during interviews but ask respondents to assemble the relevant records ahead of time to help them supply accurate answers. For instance, the National Medical Expenditure Survey and its successor, the Medical Expenditure Panel Survey, encouraged the respondents to keep doctor bills and other records handy to help answer questions about doctor visits and medical costs during the interviews. Sometimes, records might be extremely helpful to the respondents as they answer the survey questions, but the necessary records just do not exist. For example, few households keep detailed records of their everyday expenses; if they did, they would be in a much better position to provide the information sought by the U.S. Bureau of Labor Statistics in its Consumer Expenditure Survey (CES), which tracks household spending. These and other surveys attempt to persuade the respondents to create contemporaneous record-like data by keeping diaries of the relevant events. Like surveys that rely on existing records, diary surveys shift the burden from the respondents' memories to their record keeping.

Another type of measurement used in surveys involves standardized psychological assessments. Many educational surveys attempt to relate educational outcomes to characteristics of the schools, teachers, or parents of the students; cognitive tests are used in such studies to provide comparable measurements across sample students. One of our example surveys, the National Assessment of Educational Progress (NAEP), leans heavily on standardized tests of academic achievement for its data.

This chapter focuses on the issues raised by survey questionnaires. Virtually all surveys use questionnaires, and even when surveys do not use questionnaires, they still rely on standardized data collection instruments, such as record abstraction forms or diaries. Many of the principles involved in creating and testing questionnaires apply to these other types of standardized instruments as well.

7.2 Cognitive Processes in Answering Questions

Almost all surveys involve respondents answering questions put to them by an interviewer or a self-administered questionnaire. Several researchers have attempted to spell out the mental processes set in motion by survey questions. Most of the resulting models of the response process include four groups of processes: "comprehension" (in which respondents interpret the questions), "retrieval" (in which they recall the information needed to answer them), "judgment" (in which they combine or summarize the information they recall), and "reporting" (in which they formulate their response and put it in the required format). See Figure 7.1.

encoding   In some cases, it is also important to take into account the cognitive processes that take place before the interview, when the respondent first experiences the

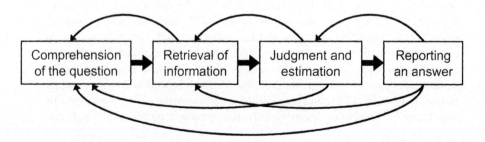

Figure 7.1 A simple model of the survey response process.

events in question. "Encoding" is the process of forming memories from experiences. Generally, survey designers have little impact on these processes, but there is good evidence that survey questions can be improved if they take into account how respondents have encoded the information the survey questions seek to tap.

Finally, with self-administered questions, respondents also have to figure out how to work their way through the questionnaire, determining which item comes next and digesting any other instructions about how to complete the questions. We treat this process of interpreting navigational cues and instructions as part of the comprehension process. (For early examples of models of the survey response process, see Cannell, Miller, and Oksenberg, 1981; and Tourangeau, 1984. For more up-to-date treatments, see Sudman, Bradburn, and Schwarz, 1996; and Tourangeau, Rips, and Rasinski, 2000.)

It is easy to give a misleading impression of the complexity and rigor of the cognitive processes involved in answering survey questions. Much of the evidence reviewed by Krosnick (1999), Tourangeau, Rips, and Rasinski (2000), and others who have examined the survey response process in detail indicates that respondents often skip some of the processes listed in Figure 7.1 and that they carry out others sloppily. The questions in surveys are often difficult and place heavy demands on memory or require complicated judgments. Consider these items drawn from our example surveys below. We follow the typographical conventions of the surveys themselves. For example, the blank slot in the NCVS item (used until 2000) is filled in at the time of the interview with an exact date.

> On average, during the last 6 months, that is, since _____,
> how often have YOU gone shopping? For example, at drug,
> clothing, grocery, hardware and convenience stores. [NCVS]

> Now turning to business conditions in the country as a whole,
> do you think that during the next 12 months we'll have good
> times financially, or bad times, or what? [SOC]

Imagine how difficult it would be to come up with an exact and accurate answer to the question on shopping posed on the NCVS. Fortunately, the NCVS is looking only for broad categories, like "once a week," rather than a specific number. Even so, the memory challenges this item presents are likely to be daunting and most respondents probably provide only rough estimates for their answers. Similarly, most of us are not really in a position to come up with thoughtful forecasts about "business conditions" (whatever that means!) during the next year. There is no reason to think that the typical respondents in surveys have either the time or the inclination to work hard to answer such questions, and there is ample evidence that they take lots of shortcuts to simplify their task.

We also do not mean to give the impression that respondents necessarily carry out the relevant cognitive processes in a fixed sequence of steps beginning with comprehension of the question and ending with reporting an answer. Some degree of backtracking and overlap between these processes is probably the rule rather than the exception. In addition, although most of us have some experience with surveys and recognize the special conventions that surveys employ (such as five-point response scales), we also have a lifetime of experience in dealing with a wide range of questions in everyday life. Inevitably, the habits and the strategies

7.2.1 Comprehension

comprehension

"Comprehension" includes such processes as attending to the question and accompanying instructions, assigning a meaning to the surface form of the question, and inferring the question's point (i.e., identifying the information sought). For concreteness, let us focus our discussion of how people answer survey questions on a specific item from the NSDUH:

> Now think about the past 12 months, from [DATE] through today. We want to know how many days you've used any prescription tranquilizer that was not prescribed to you or that you took only for the experience or feeling it caused during the past 12 months. [NSDUH]

It is natural to suppose that the respondent's first task is to interpret this question. (Although this item is not in the interrogative form, it nonetheless conveys a request for information and will generally be treated by respondents in the same way a grammatical question would be.) Like most survey items, this one not only requests information about a particular topic, in this case, the illicit use of prescription tranquillizers, it also requests that the information be provided in a specific form, ideally the number (between one and 365) corresponding to the number of days the respondent has used a certain type of prescription drugs in a certain way. (Respondents who reported that they had not used prescription tranquilizers illicitly during the past year do not get this item, so the answer should be at least one day.) The next part of the question (which is not shown) goes further in specifying the intended format of the answer, offering the respondents the choice of reporting the average days per week, the average days per month, or the total days over the whole year. Thus, one key component of interpreting a question (some would argue *the* key component) is determining the set of permissible answers; surveys often offer the respondents help in this task by providing them with a fixed set of answer categories or other clear guidelines about the form the responses should take.

Interpretation is likely to entail such component processes as parsing the question (identifying its components and their relations to each other), assigning meanings to the key substantive elements (terms like "used" and "prescription tranquilizers"), inferring the purpose behind the question, and determining the boundaries and potential overlap among the permissible answers. In carrying out these interpretive tasks, respondents may go beyond the information provided by the question itself, looking to the preceding items in the questionnaire or to the additional cues available from the interviewer and the setting. And, although it is obvious that the NSDUH item has been carefully framed to avoid confusion, it is easy to see how respondents might still have trouble with the exact drugs that fall under the heading of "prescription tranquilizers" or how they might differ in implementing the definitional requirements for using such drugs illicitly. Even everyday terms like "you" in the NCVS example above can cause problems—does "you" mean you personally or you and other members of your household?

(The use of all capital letters in the NCVS is probably intended to convey to the interviewer that the item covers only the respondent and not the other persons in the household.)

7.2.2 Retrieval

"Retrieval" is the process of recalling information relevant to answering the question from long-term memory. Long-term memory is the memory system that stores autobiographical memories, as well as general knowledge. It has a vast capacity and can store information for a lifetime.

retrieval

To determine which of the possible answers to the NSDUH item is actually true, the respondent will typically draw on his or her memory for the type of event in question. The study of memory for autobiographical events is not much older than the study of how people answer survey questions, and there is not complete agreement on how memory for autobiographical information is organized (see Barsalou, 1988, and Conway, 1996, for two attempts to spell out the relevant structures). Nonetheless, there is some consensus about how people retrieve information about the events they have experienced. Survey respondents might begin with a set of cues that includes an abstract of the question itself (e.g., "Hmm, I've used tranquilizers without a prescription a number of times over the past year or so"). A retrieval cue is a prompt to memory. It provides some clue that helps trigger the recall of information from long-term memory. This initial probe to memory may call up useful information from long-term memory—in the best case, an exact answer to the question itself. In many situations, however, memory will not deliver the exact answer but will provide relevant information, in the form of further cues that help lead to an answer (e.g., "It seems like I took some pills that Lance had one time, even though I didn't really need them"). Inferences and other memories based on these self-generated cues (e.g., "If Lance had a prescription, it probably contained about fifty tablets or so") help constrain the answer and offer further cues to probe memory. This cycle of generating cues and retrieving information continues until the respondent finds the necessary information or gives up.

retrieval cue

Several things affect how successful retrieval is. In part, success depends on the nature of the events in question. Some things are generally harder to remember than others; when the events we are trying to remember are not very distinctive, when there are a lot of them, and when they did not make much of an impression in the first place, we are hard put to remember them. If you abused prescription tranquilizers often, over a long period of time, and in a routine way, chances are you will find it difficult to remember the exact number of times you did so or the details of the specific occasions when you used tranquilizers illicitly. On the other hand, if you have abused tranquilizers just once or twice, and you did so within the last few days, retrieval is more likely to produce the exact number and circumstances of the incidents.

Of course, another major factor determining how difficult or easy it is to remember something is how long ago it took place. Psychologists who have studied memory have known for more than 100 years that memory gets worse as the events to be remembered get older. Our example from the NSDUH uses a relatively long time frame (a year) and that may make it difficult for some respondents to remember all the relevant incidents, particularly if there are a lot of them.

Another factor likely to affect the outcome of retrieval is the number and richness of the "cues" that initiate the process. The NCVS item on shopping tries to offer respondents help in remembering by listing various kinds of stores they may have visited ("drug, clothing, grocery, hardware and convenience stores"). These examples are likely to cue different memories. The best cues are the ones that offer the most detail, provided that the specifics in the cue match the encoding of the events. If, for example, in the NSDUH question a respondent does not think of Valium as a "prescription tranquilizer," then it may not tap the right memories. Whenever the cues provided by the question do not match the information actually stored in memory, retrieval may fail.

7.2.3 Estimation and Judgment

estimation

judgment

"Estimation" and "judgment" are the processes of combining or supplementing what the respondent has retrieved. Judgments may be based on the process of retrieval (e.g., whether it was hard or easy). In addition, judgments may fill in gaps in what is recalled, combine the products of the retrieval, or adjust for omissions in retrieval.

Though the NSDUH question seems to ask for a specific number, the follow-up instructions tacitly acknowledge that the information the respondent is able to recall may take a different form, such as a typical rate. People do not usually keep a running tally of the number of times they have experienced a given type of event, so they are generally unable to retrieve some preexisting answer to questions like the one posed in the NSDUH item or the question on shopping from the NCVS. By contrast, a company might well keep a running tally of its current employees, which is the key piece of information sought in the CES. The CES also demonstrates that in some cases retrieval may involve an external search of physical records rather than a mental search of memory. The respondent could try to construct a tally on the spot by recalling and counting individual incidents, but if the number of incidents is large, recalling them all is likely to be difficult or impossible. Instead, respondents are more likely estimate the number based on typical rates. Which strategy a particular respondent adopts—recalling a tally, constructing one by remembering the specific events, giving an estimate based on a rate, or simply taking a guess—will depend on the number of incidents, the length of the period covered by the survey, the memorability of the information about specific incidents, and the regularity of the incidents, all of which will affect what information the respondent is likely to have stored in memory (e.g., Blair and Burton, 1987; and Conrad, Brown, and Cashman, 1998).

It may seem that answers to attitudinal items, like the example from the Survey of Consumers on business conditions, would require a completely different set of response processes from those required to answer the more factual NSDUH item on illicit use of prescription tranquilizers. But a number of researchers have argued that answers to attitude items also are not generally preformed, waiting for the respondent to retrieve them from memory (see, for example, Wilson and Hodges, 1992). How many respondents keep track of their views about the likely business climate in the upcoming year, updating these views as conditions change or they receive new information? Respondents are more likely to deal with issues like the one raised in the SOC as they come up, basing their answers on whatever considerations come to mind and seem relevant at the time the question is asked (e.g., trends in unemployment and inflation, recent news about the world markets, or how the stock market has done lately). The same sorts of judgment strategies used for answering questions about behaviors have their counterparts for questions about attitudes. For example, respondents may recall specific incidents in answering the NSDUH item about tranquilizers just as they may try to remember specific facts about the economy in deciding what business conditions might bring over the next year. Or they may base their answers on more general information, like typical rates in the case of the NSDUH item or long-term economic trends in the case of the Survey of Consumers question.

7.2.4 Reporting

"Reporting" is the process of selecting and communicating an answer. It includes mapping the answer onto the question's response options and altering the answer for consistency with prior answers, perceived acceptability, or other criteria. As we already noted, our example NSDUH item not only specifies the topic of the question, but also the format for the answer. An acceptable answer could take the form of an exact number of days or a rate (days per week or per month). There are two major types of questions based on their formats for responding. A "closed" question presents the respondents with a list of acceptable answers. "Open" questions allow respondents to provide the answers in their own terms, although, typically, the answer is nonetheless quite circumscribed. Roughly speaking, open questions are like in fill-in-the-blank questions on a test; closed questions are like multiple choice questions. Attitude questions almost always use a closed format, with the answer categories forming a scale.

reporting

How respondents choose to report their answers will depend in part on the fit between the information they retrieve (or estimate) and the constraints imposed by the question. For questions that require a numerical answer, like the NSDUH item, they may have to adjust their internal judgment to the range and distribution of the response categories given. For example, if most of the response categories provided involve low frequencies, the reports are likely to be skewed in that direction. Or when no response options are given, respondents may have to decide on how exact their answer should be and round their answers accordingly. Respondents may also give more weight to certain response options, depending on the order in which they are presented (first or last in the list of permissible answers) and the mode of presentation (visual or auditory). When the topic is a sensitive one (like drug use), respondents may shade their answer up or down or refuse to answer entirely. This response censoring is more likely to happen when an interviewer administers the questions (see Section 5.3.5).

7.2.5 Other Models of the Response Process

It is worth mentioning that the simple model of the survey response process depicted in Figure 7.1 is not the only model that researchers have proposed. Cannell, Miller, and Oksenberg (1981) proposed an earlier model distinguishing

> **Comments on Response Strategies**
>
> Respondents may adopt broad strategies for answering a group of survey questions. Several of these strategies—selecting the "don't know" or "no opinion" answer category or choosing the same answer for every question—can greatly reduce the amount of thought needed to complete the questions. Such strategies are examples of survey "satisficing," in which respondents do the minimum they need to do to satisfy the demands of the questions.

acquiescence

social desirability

satisficing

two main routes that respondents may follow in formulating their answers. One track that includes most of the same processes that we have described here—comprehension, retrieval, judgment, and reporting—leads to accurate or at least adequate responses. The other track is for respondents who take shortcuts to get through the interview more quickly or who have motives that override their desire to provide accurate information. Such respondents may give an answer based on relatively superficial cues available in the interview situation, cues like the interviewer's appearance or the implied direction of the question. Responses based on such cues are likely to be biased by "acquiescence" (the tendency to agree) or "social desirability" (the tendency to present oneself in a favorable light by underreporting undesirable attributes and overreporting desirable ones).

A more recent model of the survey response process shares the Cannell model's assumption of dual paths to a survey response—a high road taken by careful respondents and a low road taken by respondents who answer more superficially. This is the "satisficing" model proposed by Krosnick and Alwin (1987) (see also Krosnick, 1991). According to this model, some respondents try to "satisfice" (the low road), whereas others try to "optimize" (the high road) in answering survey questions. Satisficing respondents do not seek to understand the question completely, just well enough to provide a reasonable answer; they do not try to recall everything that is relevant, but just enough material on which to base an answer; and so on. Satisficing thus resembles the more superficial branch of Cannell's two-track model. Similarly, optimizing respondents would seem to follow the more careful branch. In his later work, Krosnick has distinguished some specific response strategies that satisficing respondents may use to get through questions quickly. For example, they may agree with all attitude items that call for agree–disagree responses, a response strategy called "acquiescence."

Like the Cannell model, Krosnick's satisficing theory makes a sharp distinction between processes that probably vary continuously. Respondents may process different questions with differing levels of care, and they may not give the same effort to each component of the response process. Just because a respondent was inattentive in listening to the question, he or she will not necessarily do a poor job at retrieval. For a variety of reasons, respondents may carry out each cognitive operation carefully or sloppily. We prefer to think of the two tracks distinguished by Cannell and Krosnick as the two ends of a continuum that varies in the depth and the quality of thought respondents give in formulating their answers.

There is more research that could be done on appropriate models of the interview. One challenge of great importance is discovering the role of the computer as an intermediary in the interaction. Is a CAPI or ACASI device more like another actor in the interaction or is it more like a static paper questionnaire? What are the independent effects of computer assistance on the interviewer and the respondent in face-to-face surveys? How can software design change respondent behavior to improve the quality of survey data?

7.3 PROBLEMS IN ANSWERING SURVEY QUESTIONS

One of the great advantages of having a model of the response process, even a relatively simple one like the one in Figure 7.1, is that it helps us to think systematically about the different things that can go wrong, producing inaccurate answers to the questions. As we note elsewhere, the goal of surveys is to reduce error and one major form of error is measurement error—discrepancies between the true answer to a question and the answer that finds its way into the final database. (Although this definition of measurement error does not apply so neatly to attitude measures, we still want the responses to attitude questions to relate systematically to the underlying attitudes they are trying to tap. As a result, we prefer attitude measures that have a stronger relation to the respondents' attitudes.)

The major assumption of the cognitive analysis of survey responding is that flaws in the cognitive operations involved in producing an answer are responsible for errors in the responses. Here, we distinguish seven problems in the response process that can give rise to errors in survey reports:

1) Failure to encode the information sought.
2) Misinterpretation of the questions.
3) Forgetting and other memory problems.
4) Flawed judgment or estimation strategies.
5) Problems in formatting an answer.
6) More or less deliberate misreporting.
7) Failure to follow instructions.

There are several book-length examinations of the response process that offer longer, more detailed lists of response problems (e.g., Sudman, Bradburn, and Schwarz, 1996; Tourangeau, Rips, and Rasinski, 2000). All of these approaches share the assumption that measurement errors can generally be traced to some problem in the response process (e.g., the respondents never had the necessary information or they forgot it, they misunderstand the questions, they make inappropriate judgments, and so on).

7.3.1 Encoding Problems

The mere fact that someone has lived through an event does not necessarily mean that he or she absorbed much information about it. Studies of eyewitness testimony suggest that eye witnesses often miss key details of the unusual and involving events they are testifying about (e.g., Wells, 1993). With the more routine experiences that are the stuff of surveys, respondents may take in even less information as they experience the events. As a result, their after-the-fact accounts may be based largely on what usually happens. A study by A. F. Smith illustrates the problem. He examined survey respondents' reports about what they ate and compared these reports to detailed food diaries the respondents kept. There was such a poor match between the survey reports and the diary entries that Smith (1991, p. 11) concluded, "dietary reports ... consist in large part of individuals' guesses about what they probably ate." The problem with asking people about what they eat is that most people do not pay that much attention; the result is they cannot report about it with much accuracy.

There is a practical lesson to be drawn from this example. People cannot provide information they do not have; if people never encoded the information in the first place, then no question, no matter how cleverly framed, is going to elicit accurate responses. A key issue for pretests is making sure respondents have the information the survey seeks from them.

7.3.2 Misinterpreting the Questions

Even if the respondents know the answers to the questions, they are unlikely to report them if they misunderstand the questions. Although it is very difficult to say how often respondents misunderstand survey questions, there are several indications that it happens quite frequently.

One source of evidence is a widely cited study by Belson (1981; see also Belson, 1986), who asked respondents to report what key terms in survey questions meant to them. He found that respondents assigned a range of meanings even to seemingly straightforward terms like "you" (does this cover you personally, you and your spouse, or you and your family?) or "weekend" (is Friday a weekday or part of the weekend?). Belson also studied an item whose problems probably seem a lot more obvious in retrospect:

> Do you think that children suffer any ill effects from watching programmes with violence in them, other than ordinary Westerns?

Fowler (1992) on Unclear Terms in Questions

In 1992, Fowler reported a study showing that removing unclear terms during pretesting affects answers to survey questions.

Study design: About 100 pretest interviews of a 60-item questionnaire were tape-recorded. Behavior coding documented what interviewers and respondents said for each question–answer sequence. Seven questions generated calls for clarification or inadequate answers in 15% or more of the interviews. Revisions of the questions attempted to remove ambiguous words. A second round of pretesting interviewed 150 persons. Response distributions and behavior-coding data were compared across the two pretests. For example, the first pretest included the question, "What is the average number of days each week you have butter?" The second addressed the ambiguity of the word, "butter," with the change, "The next question is just about butter. Not including margarine, what is the average number of days each week that you have butter?"

Findings: The number of calls for clarification and inadequate answers declined from pretest 1 to pretest 2. Response distributions changed; for example, on the question about having butter:

	% Never Having Butter
Pretest 1	33%
Pretest 2	55%

The authors conclude that the exclusion of margarine increased those who reported never having butter.

Limitations of the study: There was no external criterion available for the true values of answers. The results provide no way of identifying what level of behavior-coding problems demand changes to question wording. The work assumes that the same question wording should be used for all.

Impact of the study: The study demonstrated that pretesting with behavior coding can identify problem questions. It showed how changes in wording of questions can improve interaction in interviews, reflected in behavior coding, and affect the resulting survey estimates.

Belson's respondents gave a range of interpretations to the term "children." "Children" has two basic meanings: young people, regardless of their relation to you, and your offspring, regardless of their age. The exact age cutoff defining "children" in the first sense varies from one situation to the next (there is one age cutoff for haircuts, another for getting into R-rated movies, still another for ordering liquor), and Belson found similar variation across survey respondents. He also found some idiosyncratic definitions (e.g., nervous children, one's own grandchildren). If "children" receives multiple interpretations, then a deliberately vague term like "ill effects" is bound to receive a wide range of readings as well. (It is also not very clear why the item makes an exception for violence in "ordinary Westerns" or what respondents make of this.)

These problems in interpreting survey questions might not result in misleading answers if survey respondents were not so reluctant to ask what specific terms mean or to admit that they simply do not understand the question. Some studies have asked respondents about fictitious issues (such as the "Public Affairs Act") and found that as many as 40% of the respondents are still willing to venture an opinion on the "issue" (Bishop, Oldendick, and Tuchfarber, 1986). In everyday life, when one person asks another person a question, the assumption is that the speaker thinks it is likely or at least reasonable that the hearer will know the answer. As a result, respondents may think that they ought to know about issues like the Public Affairs Act or that they ought to understand the terms used in survey questions. When they run into comprehension problems, they may be embarrassed to ask for clarification and try to muddle through on their own. In addition, survey interviewers may be trained to discourage such questions entirely or to offer only unenlightening responses to them (such as repeating the original question verbatim). Unfortunately, as Belson's results show, even with everyday terms, different respondents often come up with different interpretations; they are even more likely to come up with a wide range of interpretations when the questions include relatively unfamiliar or technical terms.

Tourangeau, Rips, and Rasinski (2000) distinguish seven types of comprehension problems that can crop up on surveys:

1) Grammatical ambiguity.
2) Excessive complexity.
3) Faulty presupposition.
4) Vague concepts.
5) Vague quantifiers.
6) Unfamiliar terms.
7) False inferences.

The first three problems have to do with the grammatical form of the question. "Grammatical ambiguity" means that the question can map onto two or more underlying representations. For example, even a question as simple as "Are you visiting firemen?" can mean two different things: Are you a group of firemen that has come to visit? or Are you going to visit some firemen? In real life, context would help sort things out, but in surveys, grammatical ambiguity can produce differing interpretations across respondents. A more common problem with survey questions is excessive complexity. Here is an example discussed by Fowler (1992):

grammatical ambiguity

> During the past 12 months, since January 1, 1987, how many times have you seen or talked to a doctor or assistant about your health? Do not count any time you might have seen a doctor while you were a patient in a hospital, but count all other times you actually saw or talked to a medical doctor of any kind.

excessive complexity

"Excessive complexity" means that the question has a structure that prevents the respondent from inferring its intended meaning. The main question in the example above lists several possibilities (seeing a doctor, talking to an assistant), and the instructions that follow the question add to the overall complexity. The problem with complicated questions like these is that it may be impossible for respondents to keep all the possibilities and requirements in mind; as a result, part of the meaning may end up being ignored.

faulty presupposition

"Faulty presupposition" means that the question assumes something that is not true. As a result, the question does not make sense or does not apply. For example, suppose respondents are asked whether they agree or disagree with the statement, "Family life often suffers because men concentrate too much on their work." The question presupposes that men concentrate too much on their work; respondents who do not agree with that assumption (for example, people who think that most men are lazy) cannot really provide a sensible answer to the question. All questions presuppose a certain picture of things and ask the respondent to fill in some missing piece of that picture; it is important that the listener does not reject the state of affairs depicted in the question.

vague concepts/ vague quantifiers

The next three problems involve the meaning of words or phrases in the question. As Belson has pointed out, many everyday terms are vague, and different respondents may interpret them differently. As a result, it helps if survey questions are as concrete as possible. For example, an item about children should specify the age range of interest. Note, however, that the attempt to spell out exactly what some vague term covers can lead to considerable complexity. That is the problem with the question above, in which the question tries to spell out the notion of an outpatient doctor visit. Some survey items employ vague relative terms ("Disagree somewhat" or "Very often") in response scales. Unfortunately, respondents may not agree about how often is very often, with the result that different respondents use the scale in different ways. Another source of interpretive

unfamiliar term

difficulty is that respondents may not know what a particular term means. The people who write questionnaires are often experts about a subject, and they may overestimate how familiar the respondents are likely to be with the terminology they themselves use every day. An economist who wants to ask people about their pension plans may be tempted to use terms like "401(k)" or "SRA" without defining them. Unfortunately, such terms are likely to confuse many respondents.

false inference

A number of findings suggest that respondents can also overinterpret survey questions, drawing false inferences about their intent. Consider this question, drawn from the General Social Survey:

> Are there any situations you can imagine in which you would approve of a *policeman* striking an adult male citizen?

It is fairly easy to imagine such circumstances leading to a "yes" answer (just watch any cop show!), but many respondents (roughly 30%) still answer "no." Clearly, many respondents do not interpret the question literally; instead, they respond in terms of its perceived intent—to assess attitudes toward violence by the police. Such inferences about intent are a natural part of the interpretation process, but they can lead respondents astray. For example, several studies suggest that respondents (incorrectly) infer that an item about their overall happiness ("Taken altogether, how would you say things are these days? Would you say that you are very happy, pretty happy, or not too happy?") is supposed to exclude their marriages when that item comes right after a parallel item asking about marital happiness (e.g., Schwarz, Strack, and Mai, 1991). In everyday conversation, each time we speak we are supposed to offer something new; this expectation leads respondents to believe that the general item is about the rest of their lives, which has not been covered yet, apart from their marriage. Unfortunately, sometimes such inferences may not match what the survey designers intended.

This section underscores the fact that language matters. In many circumstances, there is a tension between explicitly defining terms in a question (in an attempt to eliminate ambiguity) and increasing the burden on the respondent to absorb the full intent of the question. More research is needed for determining what level of detail to offer all respondents, how to discern when definitional help is needed by a respondent, and how the interaction between interviewer and respondent affects the interpretation of the verbal content of questions.

7.3.3 Forgetting and Other Memory Problems

Another potential source of error in survey responses is failure to remember relevant information. Sometimes, respondents cannot remember the relevant events at all, and sometimes they remember the events only sketchily or inaccurately. It is useful to distinguish several forms of memory failure, since they can have different effects on the final answers:

1) Mismatches between the terms used in the question and the terms used to encode the events initially
2) Distortions in the representation of the events over time
3) Retrieval failure
4) Reconstruction errors

The first type of memory failure occurs when the terms the respondent uses to encode an event differ so markedly from the terms used in the question that the question does not call to mind the intended memories. For example, a respondent may not think of a glass of wine with dinner as an "alcoholic beverage." As a result, an item that asks about weekly consumption of alcoholic beverages may fail to trigger the relevant memories. Similarly, most of us probably do not think of trips to the hardware store as "shopping"; thus, it is very important that the NCVS item on shopping explicitly mentions hardware stores ("On average, during the last 6 months, that is, since _____, how often have YOU gone shopping? For example, at drug, clothing, grocery, hardware, and convenience stores"). When survey researchers develop questionnaires, they often conduct focus groups to learn how potential survey respondents think and talk about the survey topic. Both comprehension and retrieval improve when the terms used in the questionnaire match those used by the respondents in encoding the relevant events.

A second source of inaccuracy in memory is the addition of details to our representation of an event over time. Most autobiographical memories probably consist of a blend of information we took in initially, while or shortly after we experienced the event, and information we added later on, as we were recounting the event to others who were not there themselves, reminiscing with others who also experienced it, or simply thinking about it later on. For example, when we think back to our high school graduation, our memory includes information we took in at the time, plus information we added later in looking at yearbooks, photographs, or videos of the event. These activities, which memory researchers term "rehearsal," play a pivotal role in maintaining vivid memories (e.g., Pillemer, 1984; Rubin and Kozin, 1984). Unfortunately, it is very difficult for us to identify the source of the information we remember; we cannot always distinguish what we experienced firsthand from what we merely heard about or inferred after the fact. Thus, any distortions or embellishments introduced as we recount our experiences or share reminiscences may be impossible to separate from the information we encoded initially. This sort of "postevent information" is not necessarily inaccurate, but it can be and once it is woven into our representation of an event it is very difficult to get rid of.

Still another source of trouble is retrieval failure—the failure to bring to mind information stored in long-term memory. We already noted one reason for retrieval failure: the question may not trigger recall for an event because it uses terms that differ too much from those used in encoding the event. Another reason for retrieval failure is the tendency to lose one memory among the other memories for similar experiences. Over time, it gets increasingly difficult to remember the details that distinguish one event, say, a specific doctor visit, from other events of the same kind; instead, the events blur together into a "generic memory" for a typical doctor visit, trip to the store, business trip, or whatever (cf. Barsalou, 1988; Linton, 1982). The accumulation of similar events over time means that we have more difficulty remembering specific events as more time elapses. The impact of the passage of time is probably the strongest and most robust finding to emerge from more than 100 years of research on forgetting (Rubin and Wetzel, 1996). Although researchers are still unclear about the exact shape of the function relating forgetting to the passage of time, it is clear that forgetting is more rapid at first and levels off thereafter. The amount forgotten in a given period also depends on the type of event in question; one study shows that people could still remember the names of nearly half of their classmates 50 years later (Bahrick, Bahrick, and Wittlinger, 1975). Figure 7.2 shows the percent of events correctly recalled for different kinds of phenomena. Although classmates are relatively easy to recall, the decay rate over time for recall of grades is very high. The best antidotes to retrieval failure in surveys seem to be providing more retrieval cues and getting the respondents to spend more time trying to remember. Table 7.1 (adapted from Tourangeau, Rips, and Rasinski, 2000) provides a more comprehensive list of the factors affecting recall and their implications for survey design. In short, the events most easily recalled are recent, distinctive, near another easily recalled event, and important in the life of the respondent. Questions that work best have rich, relevant cues and give the respondent time and encouragement to think carefully.

This tendency for recall and reporting to decline as a function of length of recall has yielded an important measurement error model. In the model μ_i is the

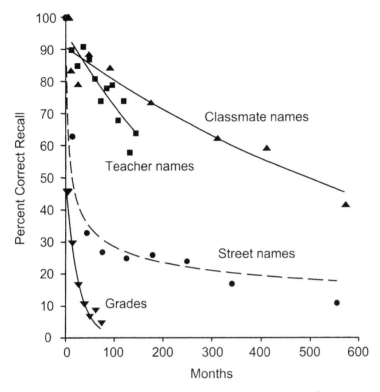

Figure 7.2 Recall accuracy for types of personal information. (Source: Tourangeau, Rips, and Rasinski, 2000.)

number of events experienced by the ith respondent relevant to the survey question. That is, if there were no recall problems, the ith person would report μ_i in response to the question. The model specifies that instead of μ_i, the respondent reports y_i:

$$y_i = \mu_i(ae^{-bt}) + \varepsilon_i$$

where a is the proportion of events that are reported (reflecting concerns about sensitivity or social desirability), b is the rate of decline in reporting as a function of time, and ε_i is a deviation from the model for the ith respondent. The e is just Euler's number, the base of the natural logarithms. Thus, the model specifies that the proportion of events correctly reported exponentially declines (rapid decline in the time segments immediately prior to the interview and diminishing declines far in the past). The literature implies that for events that are distinctive, near an easily recalled temporal boundary, and important in the life of the respondent, a is close to 1.0 and b is close to 0.0. For nonsensitive events that are easily forgotten, a may be close to 1.0, but b is large. As seen in Figure 7.2, the exponential decay model fits some empirical data better than others.

A final form of memory failure results from our efforts to reconstruct, or fill in, the missing pieces of incomplete memories. Such reconstructions are often based on what usually happens or what is happening right now. For instance,

reconstruction

reference period

Table 7.1. Summary of Factors Affecting Recall

Variable	Finding	Implication for Survey Design
Characteristics of Event		
Time of occurrence	Events that happened long ago are harder to recall	Shorten the reference period
Proximity to temporal boundaries	Events near significant temporal boundaries are easier to recall	Use personal landmarks, or life events calendars to promote recall
Distinctiveness	Distinctive events are easier to recall	Tailor the length of the reference period to the target events; use multiple cues to single out individual events
Importance, emotional impact	Important, emotionally involving events are easier to recall	Tailor the length of reference period to the properties of the target events
Question Characteristics		
Recall order	Backward search may promote fuller recall	Not clear whether backward recall is better in surveys
Number and type of cues	Multiple cues are typically better than a single cue; cues about the type of event (what) are better cues about participants or location (who or where), which are better than cues about when they occured	Provide multiple cues; use decomposition
Time on task	Taking more time improves recall	Use longer introductions to questions; slow the pace of the interview

Smith's studies of dietary recall suggested that respondents filled in gaps in their memories for what they actually ate with guesses based on what they usually eat. In a classic study, Bem and McConnell (1974) demonstrated a different strategy. Respondents in that study inferred their past views from what they now thought about the issue. This "retrospective" bias has been replicated several times (e.g., Smith, 1984). Our current state influences our recollection of the past with other types of memory as well, such as our recall of pain, past use of illicit substances, or income in the prior year (see Pearson, Ross, and Dawes, 1992 for further examples). We seem to reconstruct the past by examining the present and projecting it backwards, implicitly assuming that the characteristic or behavior in question is stable. On the other hand, when we remember that there has been a change, we may exaggerate the amount of change.

Many survey items ask respondents to report about events that occurred within a certain time frame or reference period. The first three of our sample items all specify a reference period extending from the moment of the interview to a specific date in the past (such as the date exactly six months before). Such questions assume that respondents are reasonably accurate at placing events in time. Unfortunately, dates are probably the aspect of events that are hardest for people to remember with any precision (e.g., Wagenaar, 1986). Although with some events—birthdays, weddings, and the like—people clearly do encode the date, for most events, the date is not something we are likely to note and remember. Because of the resulting difficulty in dating events, respondents may make "telescoping" errors in which they erroneously report events that actually occurred before the beginning of the reference period. The term "telescoping" suggests that past events seem closer to the present than they are; actually, recent

---

Neter and Waksberg (1964) on Response Errors

In 1964, Neter and Waksberg published a study comparing different designs for reporting past events.

Study design: Two design features were systematically varied: whether the interview was bounded (i.e., the respondents were reminded about their reports from the prior interview) and the length of the recall period (i.e., 1, 3, or 6 months). The context was a survey of the number of residential repair and renovation jobs and the expenditures associated with them, using household reporters.

Findings: With unbounded interviews, there were much higher reports of expenditures than with bounded interviews (a 55% increase). The increase in reports was larger for large jobs. The authors conclude that respondents were including reports of events that occurred before the reference period (this was labeled "forward telescoping"), and that rare events were subject to greater telescoping. Asking people to report events 6 months earlier versus 1 month earlier led to lower reports per month, with smaller jobs being disproportionately dropped from the longer reference periods. For the 6 month reference periods, the number of small jobs was 32% lower per month than for the 1 month reference period. The authors conclude this is a combined effect of failure to report and forward telescoping.

Limitations of the study: There were no independent data on the jobs or expenditures. Hence, the authors based their conclusions on the assumption that the bounded interviews offer the best estimates. Some respondents had been interviewed multiple times and may have exhibited different reporting behaviors.

Impact of the study: This study greatly sensitized designers to concerns about length of reference periods on the quality of reports. It encouraged the use of bounding interviews, for example, as in the National Crime Victimization Survey.

telescoping studies suggest that "backward" telescoping is also common. As more time passes, we make larger errors (in both directions) in dating events (Rubin and Baddeley, 1989).

Despite this, telescoping errors tend, on the whole, to lead to overreporting. For example, in one classic study (see box on page 233), almost 40% of the home repair jobs reported by the respondents were reported in error due to telescoping **bounding** (Neter and Waksberg, 1964). A procedure called "bounding" is sometimes used to reduce telescoping errors in longitudinal surveys. In a bounded interview, the interviewer reviews with the respondent a summary of the events the respondent reported in the previous interview. This is the procedure that was used in the NCVS until 2007 to attempt to eliminate duplicate reports of an incident reported in an earlier interview. The first of seven NCVS interviews asked about victimization incidents in the last six months, but data from this interview were not used in NCVS estimates. Instead, the second interview used first interview incident reports to "bound" the second interview reports. The interviewer asked whether an incident reported in the second interview might be a duplicate report of one in the first interview by checking the first interview reports. Similar procedures are used in later waves, always using the immediately prior interview as a "bound." This procedure sharply reduces the chance that respondents will report the same events in the current interview due to telescoping. Starting in 2007, because of cost pressures on the agency, the first interview data began to be used, with a statistical adjustment, in the annual NCVS estimates.

7.3.4 Estimation Processes for Behavioral Questions

Depending on what they recall, respondents may be forced to make an estimate in answering a behavioral frequency question or a render a judgment in answering an attitude question. Consider two of the examples we gave earlier in the chapter:

> Now turning to business conditions in the country as a whole – do you think that during the next 12 months we'll have good times financially, or bad times, or what? [SOC]

> Now think about the past 12 months, from [DATE] through today. We want to know how many days you've used any prescription tranquilizer that was not prescribed to you or that you took only for the experience or feeling it caused during the past 12 months. [NSDUH]

Some respondents may have a preexisting judgment about the economy that they are ready to report in response to a question like that in the SOC, but most respondents probably have to put together a judgment on the fly. Similarly, only a few respondents are likely to keep a running tally of the times they have abused prescription tranquilizers; the rest must come up with a total through some estimation process. (It is useful to note that NSDUH permits the respondent to answer the question with different metrics: average days per week, average days per month, or total days.) With both attitude and behavioral questions, the need for respondents to put together judgments on the spot can lead to errors in the answers.

Let us first examine behavioral frequency questions like those in the NSDUH. Besides recalling an exact tally, respondents make use of three main estimation strategies to answer such questions:

1) They may remember specific incidents and total them up, perhaps adjusting the answer upward to allow for forgotten incidents ("recall-and-count").
2) They may recall the rate at which incidents typically occur and extrapolate over the reference period ("rate-based estimation").
3) They may start with a vague impression and translate this into a number ("impression-based estimation").

recall-and-count

rate-based estimation

impression-based estimation

For example, one respondent may recall three specific occasions on which he used prescription tranquilizers and report "3" as his answer. Another respondent may recall abusing prescription tranquilizers roughly once a month over the last year and report "12" as her answer. A third respondent may simply recall that he used the drugs a "few times" and report "5" as the answer. The different strategies for coming up with an answer are prone to different reporting problems.

The recall-and-count strategy is prone both to omissions due to forgetting and false reports due to telescoping. Depending on the balance between these two sources of error, respondents may systematically report fewer incidents than they should have or too many. Generally, the more events there are to report, the lower the accuracy of answers based on the recall-and-count strategy; with more events, it is both harder to remember them all and harder to total them up mentally. As a result, respondents tend to switch to other strategies as the reference period gets longer and as they have more incidents to report (Blair and Burton, 1987; Burton and Blair, 1991).

Several studies have asked respondents how they arrived at their answers to behavioral frequency questions like the ones in the NSDUH, and the reported popularity of the recall-and-count strategy falls sharply as the number of events to recall increases. Instead, respondents often turn to rate-based estimation when there are more than seven events or so to report. The literature suggests that rate-based estimation often leads to overestimates of behavioral frequencies. Apparently, people overestimate rates when the rate fluctuates or when there are exceptions to what usually happens.

But the most error-prone strategies for behavioral frequency questions are those based on impressions. When the question uses a closed format, the response options that it lists can affect impression-based estimates. When the answer categories emphasize the low end of the range, the answers tend to be correspondingly lower; when they emphasize the high end of the range, the answers tend to be high. The box on page 236 shows the results of a study that asked respondents how much television they watched in a typical day. Depending on which set of answer categories they got, either 16.2% or

Overreporting and Underreporting

When respondents report things that did not happen at all or report more events than actually occurred, this is called "overreporting." Certain types of things are characteristically overreported in surveys. For example, in any given election, more people say that they voted than actually did. The opposite error is called "underreporting" and involves reporting fewer events than actually took place.

> **Schwarz, Hippler, Deutsch, and Strack (1985) on Response Scale Effects**
>
> In 1985, Schwarz, Hippler, Deutsch, and Strack published the results of several studies measuring the effects of response scales on reporting.
>
> *Study design*: Two different randomized experiments were embedded in larger surveys. A between-subject design administered one form of a question to one-half of the sample, and another, to the other. One experiment used a quota sample of 132 adults; the other, 79 employees of an office recruited into a survey. One-half of the sample received a question about hours spent watching television, with a six-category scale with middle categories 1–1.5 hours and 1.5–2 hours; for the other half-sample, the middle categories were 3–3.5 hours and 3.5–4 hours.
>
> | Low Options | | High Options | |
> |---|---|---|---|
> | Responses | % | Responses | % |
> | < ½ hr | 7.4% | < 2½ hr | 62.5% |
> | ½ to 1 hr | 17.7% | 2½ to 3 hr | 23.4% |
> | 1 to 1½ hr | 26.5% | 3 to 3½ hr | 7.8% |
> | 1½ to 2 hr | 14.7% | 3½ to 4 hr | 4.7% |
> | 2 to 2½ hr | 17.7% | 4 to 4½ hr | 1.6% |
> | >2½ hr | 16.2% | > 4½ hr | 0.0% |
> | Total | 100.0% | | 100.0% |
>
> *Findings*: Respondents receiving the low average scale tended to report watching less television than those receiving the high average scale. For example, 16.2% reported watching more than 2.5 hours per day in the low average scale but 37.5% in the high average scale.
>
> *Limitations of the study*: One of the studies used a quota sample, the other, a group of office workers, limiting the ability to generalize to other survey conditions.
>
> *Impact of the study*: The studies demonstrated that response scales affect reporting of behaviors. Researchers now attempt to choose center categories that are closest to the expected population averages.

37.5% of the respondents said they watched more that two-and-a-half hours per day. Impression-based estimates are also prone to wild values when the question is posed in open-ended format.

7.3.5 Judgment Processes for Attitude Questions

Responding to an attitude question might seem to involve very different cognitive processes from those needed to answer factual items about behavior, but at least some authors (e.g., Tourangeau, Rips, and Rasinski, 2000) argue that there are more similarities than differences between the response processes for the two types of items. The same four types of information from which frequency estimates are derived—exact tallies, impressions, generic information, and specific memories—have their counterparts in attitude questions. For instance, some respondents (economists and people who follow the stock market) may have clearly defined views on which to base an answer to the item about the economy from the Survey of Consumers ("Do you think that during the next 12 months we'll have good times financially, or bad times, or what?"). Others may have a vague impression ("Gee, I read something in the *Wall Street Journal* the other day and it sounded pretty ominous"). Just as we may have only a hazy sense of how often we have done something, we may have an equally vague impression of a person or issue we were asked to evaluate. Or, lacking any ready-made evaluation (even a very hazy one), we may attempt to construct one either from the top down, deriving a position from more general values or predispositions, or from the bottom up, using specific beliefs about the issue to construct an opinion about it. The latter two strategies resemble the use of generic information and the recall-and-count strategy to answer frequency questions.

When respondents do not have an existing evaluation they can draw on, their answers to an attitude question may be strongly affected by the exact wording of the question or by the surrounding context in which the question is placed. Consider these two items, both administered in the early 1950s to gauge public support for the Korean War:

> Do you think the United States made a mistake in deciding to defend Korea or not? [Gallup]

> Do you think the United States was right or wrong in sending American troops to stop the Communist invasion of South Korea? [NORC]

The NORC item consistently showed higher levels of support for the Korean War than the Gallup item did. In a series of experiments, Schuman and Presser (1981) later showed that adding the phrase "to stop a Communist takeover" increased support for U.S. military interventions by about 15 percentage points. Several studies have shown similar wording effects on other topics; there is far more support for increased spending on halting the rising crime rate than on law enforcement, for aid to the poor than for welfare, for dealing with drug addiction than with drug rehabilitation, and so on (Rasinski, 1989; Smith, 1987). The wording of an item can help (and influence) respondents who need to infer their views on the specific issue from more general values; the NORC wording apparently reminded some respondents that the general issue was the spread of Communism and that helped them formulate their judgment about the U.S. role in Korea.

Question context can also shape how respondents evaluate an issue. Most attitude judgments are made on a relative basis. When we evaluate a political figure, say, that evaluation almost inevitably involves comparisons to rival candidates, to other salient political figures, or to our image of the typical politician. The standard of comparison for the judgment is likely to have an impact on which characteristics of the political figure come to mind and, more importantly, on how those characteristics are evaluated. A Democrat may evaluate Bill Clinton's terms in office quite favorably when the standard of comparison is the Reagan administration but less favorably when the standard is that of Franklin Delano Roosevelt.

7.3.6 Formatting the Answer

Once they have generated an estimate or an initial judgment, respondents confront a new problem—translating that judgment into an acceptable format. Survey items can take a variety of formats, and we focus on the three most common:

1) Open-ended questions that call for numerical answers
2) Closed questions with ordered response scales
3) Closed questions with categorical response options

open and closed questions

The examples below, taken from the BRFSS, illustrate each of these formats.

1) Now, thinking about your physical health, which includes physical illness and injury, for how many days during the past 30 was your physical health not good?

2) Would you say that in general your health is:
   1 Excellent
   2 Very good
   3 Good
   4 Fair
   5 Poor

3) Are you:
   1 Married
   2 Divorced
   3 Widowed
   4 Separated
   5 Never married
   6 A member of an unmarried couple

For formats (2) and (3), the interviewer is instructed to "please read" the answer categories (but not the numbers attached to them). Almost all of the items in the BRFSS follow one of these three formats or ask yes–no questions. Yes–no questions are probably closest to closed categorical questions. The BRFSS is not unusual in relying almost exclusively on these response formats; most other surveys do so as well.

Each of the three formats presents their own special challenges to respondents. With numerical open-ended items like those in (1), respondents may have a lot of trouble translating a vague underlying judgment ("I had a pretty bad month") into an exact number ("I felt sick on three days"). There are good reasons why open-ended items are popular with survey researchers. In principle, open-ended items yield more exact information than closed items. Even with finely graded response options, there is inevitably some loss of information when the answer is categorical. Moreover, the answer categories must often be truncated at the high or low end of the range. For example, at one point the BRFSS included an open item asking respondents how many sex partners they have had in the last 12 months. A closed item would have to offer some top category (e.g., "10 or more") that would yield inexact information about the most sexually active part of the population. So relative to open items, closed items lose information because of the grouping and truncation of possible answers.

In practice, however, respondents often seem to act as if open questions were not really seeking exact quantities. Tourangeau and his colleagues, for instance, found that most of the respondents in a survey of sexual behavior who reported 10 or more sexual partners gave answers that were exact multiples of five (Tourangeau, Rasinski, Jobe, Smith, and Pratt, 1997). Survey respondents also report many other quantities as round values, such as how much stress they feel in caring for relatives with disabilities (Schaeffer and Bradburn, 1989) or how long ago they completed an earlier interview (Huttenlocher, Hedges, and Bradburn, 1990). With items asking for percentages, the responses tend to cluster at 0, 50, and 100. Although several things contribute to the use of round numbers (such as fuzziness about the underlying quantity or embarrassment about the value to be reported), the main problem seems to be the sheer difficulty many respondents experience in assigning numbers to their estimates and judgments. Respondents may try to simplify this task by selecting a value from a limited number of ranges; they then report the range they chose as a round value.

Scale ratings, such as item (2) above, also have their characteristic problems. With some types of ratings (e.g., personnel ratings), respondents seem to shy away from the negative end of the scale, producing "positivity bias." With other types of rating scales, respondents tend to avoid the most extreme answer categories. When the scale points have numerical labels, the labels can affect the answers. A study by Schwarz and his collaborators (Schwarz, Knäuper, Hippler, Noelle-Neumann, and Clark, 1991) illustrates this problem. They asked respondents to rate their success in life. One group of respondents used a scale that ranged from –5 to +5; the other group used a scale that ranged from 0 to 10. In both cases, the end points of the scale had the same verbal labels. With both sets of numerical labels, the ratings tend to fall on the positive half of the scale (exhibiting an overall positivity bias), but the heaping on the more positive end of the scale was more marked when the scale labels ran from –5 to +5. Negative numbers, according to Schwarz and his collaborators, convey a different meaning from numbers ranging from 0 upward. A value of 0 implies a lack of success in life, whereas a value of –5 implies abject failure.

positivity bias

At least two other features of rating scales can affect the answers: the labeling and number of options presented (e.g., five points vs. nine points). Krosnick and Berent (1993) conducted a series of studies comparing two types of response scales: ones that labeled only the end categories and ones that provided verbal labels for every category. In addition, they compared typical rating scale items, which present all the response options in a single step, to branching items, which present a preliminary choice ("Are you a Republican, Democrat, or Independent?") followed by questions offering more refined categories ("Are you strongly or weakly Democratic?"). Both labeling and the two-step branching format increased the reliability of the answers. Krosnick and Berent argue that the labels help clarify the meaning of the scale points and that the branching structure makes the reporting task easier by breaking it down into two simpler judgments. Aside from labeling and branching, the sheer number of the response categories can affect the difficulty of the item. With too few categories, the rating scales may fail to discriminate between respondents with different underlying judgments; with too many, respondents may fail to distinguish reliably between adjacent categories. Krosnick and Fabrigar (1997) argue that seven scale points seems to represent the best compromise.

The final common response format offers respondents unordered response categories [like the marital status item given in (3) above]. One problem with this format is that respondents may not wait to hear or read all of the options; instead, they may select the first reasonable answer they consider. For example, with the BRFSS item on marital status a respondent may select the "Never married" option without realizing that the final option ("Member of an unmarried couple") offers a better description of his or her situation. Several studies have compared what happens when the order of the response options is reversed. These studies have found two types of effects: primacy and recency effects. With a "primacy effect," presenting an option first (or at least near the beginning of the list) increases the chances that respondents will choose that option. With a "recency effect," the opposite happens—putting an option at or near the end of the list increases its

primacy effect

recency effect

popularity. Most researchers believe that respondents are likely to consider the response options one at a time and to select the first one that seems to provide an adequate answer. Using the terminology of Krosnick (1999), respondents satisfice rather than optimize; they pick an answer that is good enough, not necessarily the one that is best. The tendency to take such shortcuts would explain why primacy effects are common. The reason that recency effects also occur is that respondents may not consider the answer options in the same order as the questionnaire presents them. When an interviewer reads the questions to the respondent, the last option the interviewer reads may be the first one that respondents think about. (By contrast, when respondents read the questions themselves, they are more likely to read and consider the response categories in order.) Because of this difference, respondents in telephone surveys seem prone to recency effects, whereas respondents in mail surveys are more prone to primacy effects. Response order effects are not just a survey phenomenon; Krosnick has shown that the order in which candidates for political office are listed on the ballot affects the share of the vote each one gets.

7.3.7 Motivated Misreporting

So far, we have focused on respondents' efforts to deal with the cognitive difficulties posed by survey questions, but a glance at many survey questionnaires reveals many difficulties of another sort. Consider these items from the NSDUH:

a) Think specifically about the last 30 days, from _____ up to and including today. During the past 30 days, on how many days did you use cocaine?
c) How long has it been since you last smoked part or all of a cigarette?

sensitive question

It is easy to imagine someone having little difficulty in interpreting (a) or in retrieving and formatting the information it asks for, but still giving the wrong answer. Survey researchers call questions like these "sensitive" or "threatening," and they have become more common on national surveys as researchers attempt to monitor the use of illicit drugs or the behaviors that contribute to the spread of HIV/AIDS. "Sensitive questions" are questions that are likely to be seen as intrusive or embarrassing by some respondents. For example, questions about personal income or about sexual behavior fall in this category. Respondents are more likely to refuse to answer such questions or to give deliberately wrong answers to them. Sensitive questions create a dilemma for respondents; they have agreed to help the researchers out by providing information, but they may be unwilling to provide the information that specific questions are requesting. Respondents often seem to resolve such conflicts by skipping the questions or providing false answers to them. For instance, one study (Moore, Stinson, and Welniak, 1997) has looked at missing data in questions about income. Moore and his colleagues report that more than a quarter of the wage and salary data in the Current Population Survey (CPS) is missing or incomplete. This is roughly ten times the rate of missing data for routine demographic items.

Sometimes, refusing to answer a question may be more awkward than simply underreporting some embarrassing behavior. For example, refusing to answer the NSDUH item about cocaine use [example (a)] is tantamount to admitting one has used cocaine. It may be easier just to deny using cocaine. Among the potentially embarrassing behaviors that seem to be underreported in surveys are the use of illicit drugs, the consumption of alcohol, smoking (especially among teens and pregnant women), and abortion. Respondents may also be reluctant to admit that they have not done something when they feel they should have. As a result, they may overreport certain socially desirable behaviors such as voting or going to church.

Some researchers have attempted to use "forgiving" wording to improve the reporting of sensitive information. Consider, for example, this question about voting:

> In talking to people about elections, we often find that a lot of people were not able to vote because they were not registered, they were sick, or they just didn't have the time. How about you—did you vote in the elections this November? (American National Election Study)

The wording of the voting question encourages respondents to report that they did not vote. Despite this, the findings suggest that such wording does not eliminate overreports of voting. The most important single tactic for improving reports on sensitive topics seems to be removing the interviewer from the question-and-answer process (see Section 5.3.5). This can be accomplished in several ways. First, the sensitive questions can be administered on a paper self-administered questionnaire or presented directly to the respondent by a computer. Both of these forms of self-administration seem to increase reporting of potentially embarrassing information compared to interviewer administration of the questions. Another technique is called the "randomized response technique" (Warner, 1965). With this method, respondents spin a dial or use some other chance device to determine whether they answer the sensitive question or a second, innocuous question ("Were you born in September?"). People seem to be more willing to answer truthfully when the interviewer does not know which question they are answering. Estimates of the prevalence of the sensitive characteristics are based on the known probabilities of the randomizing device assigning the sensitive question and the nonsensitive question. Evaluations of the technique in practice suggest that it reduces some but not all of the bias common in answering sensitive questions.

randomized response technique

7.3.8 Navigational Errors

When the questions are self-administered (for example, in a mail questionnaire), the respondents have to understand both the questions themselves and any instructions the questionnaire includes, about which questions they are supposed to answer, what form their answers should take (for example, "Mark one"), and any other instructions. In fact, in a self-administered questionnaire, an important part of the respondents' job is to figure out where to go next after they have answered a question. To help the respondents find the right path through the questionnaire, questionnaires often include various skip instructions ("If No, go to Question 8"); these verbal instructions are often reinforced by visual and graphical cues, such as boldfacing and arrows. Figure 7.3 is an example adapted from

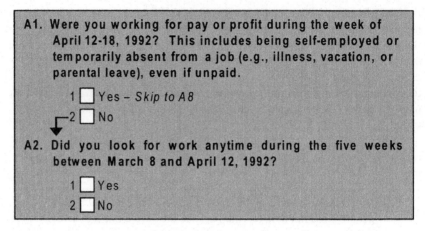

Figure 7.3 Example questions from Jenkins and Dillman (1997).

one discussed by Jenkins and Dillman (1997). This example illustrates several principles that help respondents find their way through questionnaires. Throughout the questionnaire, fonts, boldface, and graphical symbols are used in a consistent way. For example, the questions themselves are set off from other text by the use of bold type; the question numbers (A1 and A2) are made to stand out even more prominently by their position at the extreme left margin. The routing instructions (*Skip to A8*) for the first item (and for subsequent items) are in italics. Spaces that the respondents are supposed to fill in are in white, a color that contrasts with the shaded background. In addition, where possible, arrows rather than verbal instructions are used to convey the intended path (e.g., those who answer "No" to Question A1 are directed to A2 by an arrow).

navigational error

Still, it is easy for respondents to make "navigational errors," to skip items they were supposed to answer or to answer questions they were supposed to skip. Respondents may not notice instructions or, if they do, they may not understand them. As a result, self-administered questionnaires, especially poorly designed ones, often have a higher rate of missing data than interviewer-administered questions do.

With the increase in self-administered questionnaires in ACASI, Web, and e-mail surveys, there is much methodological research needed to discover how respondents react to alternative formats. What properties of formats reduce the burden on respondents? Do low-literacy respondents profit more from different formats than high-literacy respondents? Can formatting act to increase the motivation of respondents to perform their tasks?

7.4 GUIDELINES FOR WRITING GOOD QUESTIONS

There is value in being aware of all the potential pitfalls in questionnaire design; it makes it easier to recognize the problems with a question but it is also useful to have some positive rules for avoiding these problems in the first place. Several textbooks provide guidelines for writing good survey questions, and this section summarizes one of the most comprehensive lists, the one developed by Sudman and Bradburn (1982). (Another source of very good advice about writing survey questions is Converse and Presser, 1986.) Where we believe Sudman and Bradburn's recommendations have not stood the test of time, we drop or amend their original advice. Their recommendations were based empirical findings and, for the most part, they have held up pretty well. Many of them have already been foreshadowed in our discussion of the things that can go wrong in the response process.

Sudman and Bradburn do not offer a single set of guidelines for all survey questions but instead give separate recommendations for several different types of questions:

1) Nonsensitive questions about behavior
2) Sensitive questions about behavior
3) Attitude questions

These distinctions are useful, since the different types of question raise somewhat different issues. For example, sensitive questions are especially prone to deliberate misreporting and may require special steps to elicit accurate answers. Attitude questions are likely to involve response scales, and response scales (as we noted earlier) raise their own special problems. We will deal with each type of question in turn.

7.4.1 Nonsensitive Questions About Behavior

With many questions about behavior, the key problems are that respondents may forget some or all of the relevant information or that their answers may reflect inaccurate estimates. Accordingly, many of the Sudman and Bradburn's guidelines for nonsensitive questions about behavior are attempts to reduce memory problems. Most of their guidelines for nonsensitive questions make equally good sense for sensitive ones as well. Here they are:

1) With closed questions, include all reasonable possibilities as explicit response options.
2) Make the questions as specific as possible.
3) Use words that virtually all respondents will understand.
4) Lengthen the questions by adding memory cues to improve recall.
5) When forgetting is likely, use aided recall.
6) When the events of interest are frequent but not very involving, have respondents keep a diary.
7) When long recall periods must be used, use a life event calendar to improve reporting.
8) To reduce telescoping errors, ask respondents to use household records or use bounded recall (or do both).
9) If cost is a factor, consider whether proxies might be able to provide accurate information.

The first three recommendations all concern the wording of the question. It is essential to include all the possibilities in the response categories because respondents are reluctant to volunteer answers that are not explicitly offered to them. In addition, possibilities that are lumped together in a residual category ("All Others") tend to be underreported. For example, the two items below are likely to produce very different distributions of answers:

1) What is your race?
    White
    Black
    Asian or Pacific Islander
    American Indian or Alaska Native
    Some Other Race

2) What is your race?
    White
    Black
    Asian Indian
    Chinese
    Japanese
    Korean
    Vietnamese
    Filipino
    Other Asian
    Native Hawaiian
    Guamanian
    Samoan
    Some Other Pacific Islander
    American Indian or Alaska Native
    Some Other Race

Unpacking the "Asian and Pacific Islander" option into its components clarifies the meaning of that answer category and also makes it easier for respondents to recognize whether it is the appropriate option for them. As a result, the second item is likely to yield a higher percentage of reported Asian and Pacific Islanders.

Making the question as specific as possible reduces the chances for differences in interpretation across respondents. It is important that the question be clear about who it covers (does "you" mean just the respondent or everyone in the respondent's household?), what time period, which behaviors, and so on. A common error is to be vague about the reference period that the question covers: "In a typical week, how often do you usually have dessert?" Our eating habits can vary markedly over the life course, the year, even over the last few weeks. A better question would specify the reference period: "Over the last month, that is, since [DATE], how often did you have dessert in a typical week?" The interviewer would fill in the exact start date at the time he or she asked the question.

The third recommendation is to use words that everyone understands, advice that is unfortunately far easier to give than to follow. Some more specific guidelines are to avoid technical terms ("Have you ever had a myocardial infarction?") in favor of everyday terms ("Have you ever had a heart attack?"), vague quantifiers ("Often," "Hardly ever") in favor of explicit frequency categories ("Every day," "Once a month"), and vague modifiers ("Usually") in favor of more concrete ones ("Most of the time"). If need be, vague or technical terms can be used but they should be defined, ideally just before the question ("A myocardial infarction is a heart attack; technically, it means that some of the tissue in the heart muscle dies. Have you ever had a myocardial infarction?").

The next five guidelines are all about reducing the impact of forgetting on the accuracy of survey reports. One basic strategy is to provide more retrieval cues to the respondent, either by incorporating them into the question itself or by asking separate questions about subcategories of the overall category of interest. Adding cues may lengthen the question, but, as the fourth guideline notes, it may improve recall as well. Asking separate questions about subcategories is referred to as "aided recall," and this is what the fifth guideline recommends. An example is the NCVS item on shopping, which lists several examples of different places one might shop ("... drug, clothing, grocery, hardware, and convenience stores"); it would also be possible to ask separate questions about each type of retail outlet. Both approaches provide retrieval cues that help jog respondents' memories. It is important that the retrieval cues are actually helpful. Breaking a category down into nonsensical subcategories ("How many times did you go shopping for red things?" "How many times did you go shopping on a rainy Tuesday morning?") can make things worse (see Belli, Schwarz, Singer, and Talarico, 2000). Retrieval cues that do not match how the respondents encode the events can be worse than no cues at all. Another strategy for improving recall is to tailor the length of the reference period to the likely memorability of the events. Things that rarely happen, those that have a major emotional impact, and those that last a long time tend to be easier to remember than frequent, inconsequential, or fleeting events. For example, respondents are more likely to remember a hospital stay for open heart surgery than a 15-minute visit to the doctor for a flu shot. As a result, it makes sense to use a longer reference period to collect information about memorable events, such as hospital stays (where one year might be a reasonable reference period), than about nonmemorable events, such as doctor visits (where two weeks or a month might be reasonable).

aided recall

The tailoring approach has its limits, though. When the survey concerns very routine and uninvolving events (say, small consumer purchases or food intake), the reference period might have to be too short to be practical, say, yesterday. In such cases, it is often better to have respondents keep a diary rather than rely on their ability to recall the events. At the other extreme, a long reference period may be a necessary feature of the survey design. Most panel surveys, for example, cannot afford to visit the respondents very often; generally, they interview panel members every few months or even once a year and so must use long references periods if they intend to cover the whole period between interviews. (For example, the NCVS visits sample dwellings every six months.) When a long reference period has to be used, a life events calendar can sometimes improve recall. A life event calendar collects milestone events about several domains of a person's life, such as marital history, births of children, jobs, and residences. These are recorded on a calendar and help jog respondents' memories about more mundane matters, like how much they were earning at the time, illnesses they experienced, or crime victimizations. The autobiographical milestones recorded on the calendar provide rich chronological and thematic cues for retrieval (Belli, 1998); they can also serve as temporal landmarks, improving our ability to date other events.

These event calendars require the interviewer to engage in less structured interaction with the respondent. There are unanswered research questions about whether such tools increase variability in survey results from interviewer effects (see Section 9.3). This is a ripe area for methodological research.

Another kind of memory error involves misdating events (or "telescoping errors"), and the eighth guideline recommends two methods for reducing these errors. The first is to have respondents consult household records (calendars, checkbooks, bills, insurance forms, or other financial records) to help them recall and date purchases, doctor visits, or other relevant events that may leave some paper trail. The second tactic is called "bounding"; it involves reminding respondents what they already reported in a previous round of a panel survey.

The final recommendation is to use proxy reporters to provide information when data collection costs are an issue. A "proxy" is anyone other than the person about whom the information is being collected. Most surveys ask parents to provide information about young children rather than interviewing the children themselves. Other surveys use a single adult member of the household to report on everyone else who lives there. Allowing proxies to report can reduce costs, since the interviewer can collect the data right away from a proxy rather schedule a return trip to speak to every person in the household. At the same time, proxies differ systematically from self-reporters. They are, for example, more likely than self-reporters to rely on generic information (e.g., information about what usually happens) in answering the questions than on episodic information (e.g., detailed memories for specific incidents). In addition, self-reporters and proxies may differ in what they know. It hardly makes sense to ask parents about whether their teenaged children smoke; the children are likely to conceal this information, especially from their parents. On the whole, though, proxies often seem to provide reliable factual information (e.g., O'Muircheartaigh, 1991).

7.4.2 Sensitive Questions About Behavior

As we noted in Section 7.3.7, some surveys include questions about illegal or potentially embarrassing behaviors, such as cocaine use, drinking, or smoking. Both the NSDUH and the BRFSS are full of such questions. Here are Sudman and Bradburn's guidelines (updated as necessary) for sensitive questions:

1) Use open rather than closed questions for eliciting the frequency of sensitive behaviors.
2) Use long rather than short questions.
3) Use familiar words in describing sensitive behaviors.
4) Deliberately load the question to reduce misreporting.
5) Ask about long periods (such as one's entire lifetime) or periods from the distant past first in asking about sensitive behaviors.
6) Embed the sensitive question among other sensitive items to make it stand out less.
7) Use self-administration or some similar method to improve reporting.
8) Consider collecting the data in a diary.
9) At the end of the questionnaire, include some items to assess how sensitive the key behavioral questions were.
10) Collect validation data.

Open questions about sensitive behaviors have two advantages over closed questions. First, closed questions inevitably lose information (for example, about the very frequent end of the continuum). In addition, closed categories may be taken by the respondents as providing information about the distribution of the behavior in question in the general population and, thus, affect their answers (see the box describing the study by Schwarz and colleagues on page 236).

Sudman and Bradburn recommend longer questions, largely because they promote fuller recall (by giving respondents more time to remember) (Sudman and Bradburn, 1982); they are particularly useful with behaviors that tend to be underreported (such as drinking). For example, to ask about the consumption of wine, they recommend the following wording:

> Wines have become increasingly popular in this country in the last few years; by wines, we mean liqueurs, cordials, sherries, and similar drinks, as well as table wines, sparkling wines, and champagne. Did you ever drink, even once, wine or champagne?

The enumeration of the various types of wine may help clarify the boundaries of the category, but mostly serves to initiate and provide extra time for retrieval. The next recommendation, to use familiar terms for sensitive behaviors (e.g., "having sex" rather than "coitus"), may make respondents more comfortable with the questions but also tends to improve recall, since the terms in the question are more likely to match the terms used to encode the relevant experiences. Interviewers can determine which terms the respondents would prefer to use at the outset of the interview.

"Loading" a question means wording it in a way that invites a particular response, in this case, the socially undesirable answer. Sudman and Bradburn distinguish several strategies for doing this (Sudman and Bradburn, 1982): the "everybody-does-it" approach ("Even the calmest parents get mad at their children sometimes. Did your children do anything in the past week to make you angry?"); the "assume-the-behavior" approach ("How many times during the past week did your children do something that made you angry?"); the "authorities-recommend-it" approach ("Many psychologists believe it is important for parents to express their pent-up frustrations. Did your children do anything in the past week to make you angry?"); and the "reasons-for-doing-it" approach ("Parents become angry because they're tired or distracted or when their children are unusually naughty. Did children do anything in the past week to make you angry?"). The question about voting on page 241 also illustrates this last approach.

loading

The next two recommendations help reduce the apparent sensitivity of the item. In general, it is less embarrassing to admit that one has ever done something or did it a long time ago than to admit one has done it recently. For example, during the 2000 presidential campaign, candidate Bush admitted he had had a drinking problem more than ten years before, and this admission provoked little reaction. It would have been quite a different story for him to admit he had been drinking heavily the day before on the campaign trail. Most survey researchers think that sensitive items should not come at the start of the interview, but only after some less sensitive questions. In addition, embedding one sensitive question (for example, an item on shoplifting) among other more sensitive items (an item on armed robbery) may help make the sensitive item of interest seem less threatening by comparison. Like many judgments, the perception of sensitivity is affected by context.

As we already noted, one of the most effective methods for improving reports about sensitive behaviors is by having the respondent complete a self-administered or a computer-administered questionnaire. Another approach is the randomized response technique in which the interviewer does not know the question the respondent is answering. This method is illustrated below. Respondents pick a red or a blue bead from a box and their selection determines which question they answer:

> (Red) Have you been arrested for drunk driving in the last 12 months?
> (Blue) Is your birthday in the month of June?

The interviewer records either a "yes" or "no" answer, not knowing which question the respondent is answering. Since the researcher knows the probability of a red bead or a blue bead being selected (and the probability the respondent was born in June), an estimate of the proportion of "yes" answers to the drunk driving question can be obtained.

Finally, a third approach is having the respondents keep a diary, which combines the benefits of self-administration with the reduced burden on memory. Diary surveys that require detailed record keeping, however, tend to have lower response rates.

The final two recommendations allow us to assess the level of sensitivity of the questions (by having respondents rate their discomfort in answering them) and to assess the overall accuracy of responses by comparing them to an external benchmark. For example, respondents' survey reports about recent drug use might be compared to the results of a urinalysis.

7.4.3 Attitude Questions

Many surveys ask about respondents' attitudes. Among our six example surveys, only the SOC includes a large number of attitude items. Still, these are a very common class of survey questions, and Sudman and Bradburn present some guidelines specifically for them. Here is our amended version of their list:

1) Specify the attitude object clearly.
2) Avoid double-barreled questions.
3) Measure the strength of the attitude, if necessary using separate items for this purpose.
4) Use bipolar items except when they might miss key information.
5) The alternatives mentioned in the question have a big impact on the answers; carefully consider which alternatives to include.
6) In measuring change over time, ask the same questions each time.
7) When asking general and specific questions about a topic, ask the general question first.
8) When asking questions about multiple items, start with the least popular one.

GUIDELINES FOR WRITING GOOD QUESTIONS

9) Use closed questions for measuring attitudes.
10) Use five- to seven-point response scales and label every scale point.
11) Start with the end of the scale that is the least popular.
12) Use analogue devices (such as thermometers) to collect more detailed scale information.
13) Use ranking only if the respondents can see all the alternatives; otherwise, use paired comparisons.
14) Get ratings for every item of interest; do not use check-all-that-apply items.

The first six of these guidelines all deal with the wording of the questions. The first one says to clearly specify the attitude object of interest. Consider the item below:

> Do you think the government is spending too little, about the right amount, or too much on antiterrorism measures?

It would improve comprehension and make the interpretation of the question more consistent across respondents to spell out what antiterrorism measures the question has in mind (and what level of government). Double-barreled items inadvertently ask about two attitude objects at once. For example, question (a) below ties attitudes about abortion to attitudes about the Supreme Court and (b) ties abortion attitudes to attitudes about women's rights: **double-barreled items**

a) The U.S. Supreme Court has ruled that a woman should be able to end a pregnancy at any time during the first three months. Do you favor or oppose this ruling?
b) Do you favor legalized abortion because it gives women the right to choose?

The answers to double-barreled items are difficult to interpret; do they reflect attitudes to the one or the other issue or both?

The two characteristics of an attitude that are generally of interest are its direction (agree or disagree, pro or con, favorable or unfavorable) and its intensity or strength. The third recommendation is to assess intensity, using a response scale designed to capture this dimension ("Strongly disagree," "Disagree somewhat," etc.), a separate item, or multiple items that can be combined into a scale that yields intensity scores. The fourth recommendation is to use bipolar items except where they might miss some subtle distinction. For example, ask about conflicting policies in a single item rather than asking about each policy alone:

> Should the government see to it that everyone has adequate medical care or should everyone see to his own medical care?

The "bipolar approach" forces respondents to choose between plausible alternatives, thereby discouraging acquiescence. There are times, however, when this approach misses subtleties. For instance, positive and negative emotions are not **bipolar approach** always strongly (negatively) related, and so it may make sense to include separate questions asking whether something makes respondents happy or sad. The fifth recommendation about the wording about attitude items concerns the consequences of including middle (e.g., "neither agree nor disagree") and no-opinion options. In general, these options should be included unless there is some compelling reason not to. (For example, in an election poll, it is important to get those leaning one way or the other to indicate their preferences; under those circumstances, middle and no-opinion options may be omitted.) The next guideline advises that the only way to measure changes in attitudes is to compare apples with apples, that is, to administer the same question at both time points.

The next two recommendations are designed to reduce the impact of question order. If a questionnaire includes both a general question and more specific questions about the same domain, it is probably best to ask the general item first; otherwise the answers to that item are likely to be affected by the number and content of the preceding specific items. (Recall our earlier discussion of how respondents reinterpret a question on overall happiness when it follows one asking about marital happiness.) If the questionnaire includes several parallel questions that vary in popularity (e.g., the GSS includes several similar items asking about support for abortion under different circumstances), the unpopular ones are likely to seem even less appealing when they follow the popular ones. Putting the unpopular items first may yield more revealing answers.

The final six recommendations concern the format of the response scales that are nearly ubiquitous with attitude items. The first of these recommendations (the ninth overall) advises us to use closed attitude items rather than open-ended ones. The latter are simply too difficult to code. The next recommendation gets more specific, suggesting that five to seven verbally labeled scale points be used. Fewer scale points lose information; more tend to produce cognitive overload. The verbal labels help ensure that all respondents interpret the scale in the same way. If the response options vary in popularity, more respondents will consider the less popular ones when they come first than if they come later. If interviewers administer the questions aloud, then respondents probably consider the last option they hear first; in such cases, the least popular option should come at the end.

analogue method

ranking

check-all-that-apply

Other formats are also popular for attitude questions. The last three recommendations concern analogue methods (such as feeling thermometers), rankings, and check-all-that-apply items. When more than seven scale points are needed, an analogue scale may help reduce the cognitive burden. For example, a feeling thermometer asks respondent to assess their warmth toward public figures using a scale that goes from 0 (indicating very cold feelings) to 100 (very warm). Results suggest that respondents are likely to use 13 or so points on the scale. Respondents have difficulty with unanchored numerical judgments (that is one reason they tend to use round numbers) so it helps when the scale has both an upper and a lower limit, as the feelings thermometer does. Respondents can also be asked to rank various objects (e.g., desirable qualities for a child to have). As the thirteenth guideline points out, the ranking task may exceed the cognitive capacity of the respondents unless all of the items to be ranked are displayed on a card that the respondents can look at while they rank the items. When that is impossible (e.g., in a telephone interview), the researchers may have to fall back on comparisons between pairs of objects. The final recommendation discourages researchers from using check-all-that-apply items, since respondents are likely to check only some of the items that actually apply to them (Rasinski, Mingay, and Bradburn, 1994). Asking respondents to say yes or no (agree or disagree, favor or oppose, etc.) to each item on the list reduces this form of satisficing.

7.4.4 Self-Administered Questions

Sudman and Bradburn (and most other questionnaire design texts) focus on questionnaires for face-to-face and telephone interviews. In these settings, a trained interviewer typically mediates between the respondents and the questionnaire. By contrast, with mail questionnaires, it is up to the respondents to figure out which questions to answer, how to record their responses, and how to comply with any other instructions for completing the questionnaire. Jenkins and Dillman (1997; see also Redline and Dillman, 2002) offer several recommendations for improving the chances that respondents will correctly fill out mail and other self-administered questionnaires. Here are their recommendations:

1) Use visual elements in a consistent way to define the desired path through the questionnaire.
2) When the questionnaire must change its conventions part way through, prominent visual guides should alert respondents to the switch.
3) Place directions where they are to be used and where they can be seen.
4) Present information that needs to be used together in the same location.
5) Ask one question at a time.

The "visual elements" mentioned in the first guideline include brightness, color, shape, and position on the page. As we noted in Section 7.3.8, the questionnaire may set question numbers off in the left margin, put the question numbers and question text in boldface, put any instructions in a different typeface from the questions themselves, and use graphical symbols (such as arrows) to help guide the respondents to the right question. When the questionnaire uses the same conventions from beginning to end, it trains respondents in the use of those conventions. Unfortunately, it is sometimes necessary to depart from the conventions set earlier midway through a questionnaire. For example, the questions may switch response formats. According to the second guideline, in such cases, the visual cues should make it very clear what the respondents are supposed to do. Figure 7.4 makes the switch from circling the number of the selected answer to writing in the answer fairly obvious; the contrast between the white space for the response and the shaded background calls attention to what the respondents are now supposed to do and where they are supposed to do it.

The next two guidelines refer to the placement of instructions. The first page of a self-administered form often consists of lengthy instructions for completing the form. Jenkins and Dillman argue that respondents may simply skip over this initial page and start answering the questions; or, if they do read the instructions, the respondents may well have forgotten them by the time they actually need to apply them. Respondents are more likely to notice and follow instructions when

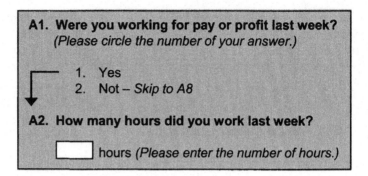

Figure 7.4 Illustration of use of visual contrast to highlight the response box.

they are placed right where they are needed. In Figure 7.4, instructions about how to indicate one's answer (*Please circle the number of your answer*) comes just before the answer options themselves. A related point is to put conceptually related information physically together. For example, a respondent should not have to look at the question and then look at the column headings to figure out what sort of answer is required. All of the information needed to understand the question should be in a single place.

Survey questions often try to cover multiple possibilities in a single item. For example, the question about doctor visits on page 228 is complicated partly because it covers doctors and other medical staff and face-to-face and telephone consultations all in a single question. The temptation to ask multiple questions in a single item can be especially strong in a self-administered questionnaire because adding an item or two can mean adding a page to the form. But asking multiple questions at once can impose a heavy interpretive burden on the respondents, who may be unable to keep the full set of logical requirements in mind. Here is an example discussed by Jenkins and Dillman:

> How many of your employees work full-time and receive health insurance benefits, how many work full-time without health insurance benefits, and what is the average length of time each type of employee has worked for this firm?

This item asks four separate questions: How many full-time employees get health insurance benefits? On average, how long have they worked for the firm? How many full-time employees do not get health insurance benefits? On average, how long have they worked for the firm? Whatever savings in space this achieves is likely to be offset by losses in understanding.

7.5 Summary

Survey respondents engage in a series of processes involving comprehension, memory retrieval, judgment and estimation, and reporting in the course of answering survey questions. Some visually presented self-administered questionnaires also require the respondent to make navigational decisions that can affect the flow of questions. Measurement of behaviors and attitudes seem to raise somewhat different issues at the judgment and estimation step.

The survey methodological literature contains many randomized experiments that demonstrate how measurement errors can arise at each of the steps in the response process. These include encoding problems, where the information sought is not stored in memory in an accessible form; misinterpreting the question because of wording or grammatical problems; forgetting and other memory problems; estimation issues in behavioral frequency questions; and judgment effects arising from question context or the sensitivity of the question. Over time, survey methodology has discovered tools to combat these problems. The tools vary according to whether the question is a nonsensitive behavior question, a sensitive behavior question, or an attitude question.

Alert readers will have noticed that this chapter spends about twice as much time cataloguing the problems that affect questions than in detailing their cures. There are several reasons for this, but the main one is that guidelines have their limitations. Accordingly, we think the principles from which the guidelines are derived are more important than the guidelines themselves.

What are these limitations of guidelines? First, any set of guidelines, no matter how comprehensive, cannot cover every situation. For example, some survey researchers would argue that certain types of questions—questions about causality or questions about our reactions to hypothetical situations—are basically too difficult to yield reliable answers. The guidelines we have presented omit this useful advice. It is impossible to formulate rules that cover every possible contingency. A second problem with guidelines is that every rule has its exceptions. Sudman and Bradburn recommend against check-all-that-apply items, and we generally agree with that advice, but the Census 2000 short form used that approach in collecting data on race. Most survey researchers would agree that this was a better solution than asking people if they are White (Yes or no?), Black (Yes or no?), and so on through the list. The check-all-that-apply format gave people a natural and efficient way to indicate a multiracial background. Still another limitation on guidelines is that they sometimes offer conflicting advice. On the one hand, we are supposed to specify the attitude object clearly but, on the other, we are supposed to avoid double-barreled items. Our guess is that many double-barreled items result from the effort to nail down the attitude object of interest. Similarly, the effort to spell out vague everyday concepts (like doctor visits) can lead to excessive complexity (see our discussion of the question on page 228). The issue of conflicts between guidelines is an important one. Often, such conflicts represent trade-offs between different, equally valid design considerations. For example, including a middle option in an attitude item has the advantage that it lets people who are actually in the middle accurately convey their views; it has the drawback that it provides an easy out for respondents who do not want to work out their position on the issue. It is not always easy to say which of the two considerations should take precedence. Or, to cite another issue, breaking a complicated item into simpler constituents may improve the answers but increase the length of the questionnaire.

Ultimately, guidelines are simply about what will work well in a given situation. These hypotheses should be tested whenever possible. Even the most experienced questionnaire designers like to have data to help them make decisions about questionnaires; after all, the proof of the pudding is in the eating. In the next chapter, we turn to methods for testing questionnaires.

Keywords

acquiescence
aided recall
analogue method
bipolar approach
bounding
check-all-that-apply
closed questions
comprehension
double-barreled items
encoding
estimation
excessive complexity
false inference
faulty presupposition
generic memory
grammatical ambiguity
impression-based estimation
judgment
loading
navigational error
open questions
positivity bias

primacy effect
proxy
questionnaire
randomized response technique
ranking
rate-based estimation
recall-and-count
recency effect
reconstruction
reference period
rehearsal
reporting
retrieval
retrieval failure
retrieval cue
satisficing
sensitive question
social desirability
telescoping
unfamiliar term
vague concepts
vague quantifier

For More In-Depth Reading

Sudman, S., and Bradburn, N. (1982), *Asking Questions: A Practical Guide To Questionnaire Design*, San Francisco: Jossey-Bass.

Sudman, S., Bradburn, N., and Schwarz, N. (1996), *Thinking about Answers: The Application of Cognitive Processes to Survey Methodology*, San Francisco: Jossey-Bass.

Tourangeau, R., Rips, L.J., and Rasinski, K. (2000), *The Psychology of Survey Responses*, Cambridge: Cambridge University Press.

EXERCISES

1) In an effort to gauge public support for energy conservation, the (fictional) Andrews Foundation conducted a recent poll that found that 72% of Americans agreed with the following statement (phrased as an "Agree/ Disagree" question):

   "I would support President Obama's decision to use the U.S. military to help local cities achieve energy independence by installing more energy efficient public lighting."

   a) What are the critical aspects of the attitude measured by this question?
   b) Does it meet the analytic goal of assessing public support for installing more energy efficient public lighting? Why or why not?

2) Using what you know about constructing attitude questions, write a standardized, interviewer administered question that you think will capture the direction and strength of public support for a ground invasion of Iraq involving U.S. troops. You may use more than one question, and any question and response format you desire. Be sure to specify response categories (if any are used), and what information is read to respondents (as opposed to interviewer instructions). Identify any skip patterns with appropriate formatting or notation.

   There are many different ways you could approach this task. To get started, you are free to do a search for questions used in actual surveys on this topic (e.g., you might explore the Gallup or Pew Research Center websites). If you use any items from other surveys; however, make sure that you identify (1) which items these are and where you took them from, (2) how, if at all, they have been revised, and (3) how your revisions meet the analytic goals of your questionnaire. Note: just because a question has been used and is in the public domain does not mean it is a good question. Your questionnaire should be based on the principles and rules of thumb you have learned in this chapter.

3) Write one or two paragraphs discussing how your question or questions are derived from the guidelines provided. Be specific. For example, if you used more than one question, why did you start with the question you did? If you used an 11-point response scale, why did you? How did you handle the trade-offs between the different options available to you, and how might these affect the results of your questionnaire?

4) How does social desirability affect response? Describe two ways you could reduce the effects of social desirability when asking respondents to report their attendance of religious services.

5) Thinking about the cognitive processing models of the response task, describe potential problems you see in the wording and response options for this question:

Many people who own vehicles have regular service work done on them, such as having the oil changed. What kind of service work do you usually have done on your vehicle? Specify one or two. [Respondent reads the following response options to choose from.]

    Oil changes
    Fluid replacement
    Tune-up
    Body repair
    Warranty-related services
    Tire care
    Transmission overhaul
    Air conditioner treatment

6) For each of the items below, diagnose the problem(s) with the questions. Your diagnosis should be either the cognitive model of information processing presented in this chapter. Then, based on that diagnosis, suggest improved wording that solves the problem.

Ex. 1

During the past four weeks, beginning [DATE FOUR WEEKS AGO] and ending today, have you done any housework, including cleaning, cooking, yard work, and household repairs, but not including any activities carried out as part of your job?

Ex. 2

In the past week, how many times did you drink alcoholic beverages?

Ex. 3

Living where you do now and meeting the expenses you consider necessary, what would be the smallest income (before any deductions) you and your family would need to make ends meet each month?

Ex. 4

During the past 12 months, since [DATE], about how many days did illness or injury keep you in bed more than half the day? Include days while you were an overnight patient in a hospital.

7) Briefly give three reasons why it may be wise to avoid questions of the agree–disagree form.

8)   Specify what type of estimation strategy the respondent might use in each of the following cases and its consequence on reported frequency:
   a) Number of times the respondent was hospitalized in the past 2 years

EXERCISES

b) Number of times the respondent ate in a restaurant in the past month
c) Number of time respondent's spouse/partner went on vacation during the past summer

9) Describe briefly two approaches the researcher can use to deal with telescoping.

CHAPTER EIGHT

EVALUATING SURVEY QUESTIONS

8.1 Introduction

Chapter 2 described two sources of error in survey estimates: errors of nonobservation (when the characteristics of the respondents do not match those of the population from which they were drawn) and observation or measurement error (when the answers to the questions are not good measures of the intended constructs). This chapter examines methods for evaluating survey questions and for determining how much measurement error they introduce into survey statistics.

Question evaluation has two components. First, it assesses the issues discussed in Chapter 7, such as how well questions are understood or how difficult they are to answer, which affect the quality of measurement. Survey researchers evaluate question comprehension, difficulty in memory retrieval, and related issues primarily by observing people trying to understand and answer the questions. The assumption is that questions that are easily understood and that produce few other cognitive problems for the respondents introduce less measurement error than questions that are hard to understand or that are difficult to answer for some other reason. Second, question evaluation assesses how well the answers correspond to what we are trying to measure, that is, directly estimating measurement error. To do this, survey methodologists either compare the survey answers to some other measure or repeat the measurement. The comparison to other measures addresses issues of validity or response bias; the repetition of the measure addresses issues of reliability or response variance. We discuss the practical techniques for estimating these errors in Section 8.9.

There are three distinct standards that all survey questions should meet:

1) Content standards (e.g., are the questions asking about the right things?)   **content standards**
2) Cognitive standards (e.g., do respondents understand the questions consistently; do they have the information required to answer them; are they willing and able to formulate answers to the questions?)   **cognitive standards**
3) Usability standards (e.g., can respondents and interviewers, if they are used, complete the questionnaire easily and as they were intended to?)   **usability standards**

Different evaluation methods are relevant to the three standards. After discussing some of the major tools for assessing questions, we will return to the question of how the different tools provide information related to each standard.

Survey Methodology, Second Edition. By Groves, Fowler, Couper, Lepkowski, Singer, and Tourangeau
Copyright © 2009 John Wiley & Sons, Inc.

One of the major advances in survey research in the past 20 years has been increased attention to the systematic evaluation of questions. There are at least five different methods that researchers can use to evaluate draft survey questions:

expert review

1) Expert reviews, in which subject matter experts review the questions to assess whether their content is appropriate for measuring the intended concepts, or in which questionnaire design experts assess whether the questions meet the three standards for questions given above.
2) Focus group discussions, in which the survey designers hold a semi-structured ("focused") discussion with members of the target population to explore what they know about the issues that the questionnaire will cover, how they think about those issues, and what terms they use in talking about them.
3) Cognitive interviews, in which interviewers administer draft questions in individual interviews, probe to learn how the respondents understand the questions, and attempt to learn how they formulate their answers.
4) Field pretests, in which interviewers conduct a small number of interviews (typically, fewer than 100) using sampling and field procedures similar to the full-scale survey and in which (a) debriefings with the interviewers may be held (to gain their insights into the problems they had in asking the questions or those the respondents had in answering them); (b) data may be tabulated and reviewed for signs of trouble (such as items with high rates of missing data); and (c) recordings of interviews may be made and behavior coded (to provide quantitative data about questions that are difficult to read as worded or hard for respondents to answer; see Fowler and Cannell, 1996).
5) Randomized or split-ballot experiments, in which different portions of the pretest sample receive different wordings of questions attempting to measure the same thing (Fowler, 2004; Moore, Pascale, Doyle, Chan, and Griffiths, 2004; Schuman and Presser, 1981; Tourangeau, 2004).

The following sections describe how these evaluation activities are carried out in practice.

8.2 Expert Reviews

As we noted earlier, both subject matter experts and questionnaire design experts may review a draft of the questionnaire. We focus on the role of questionnaire design experts, but emphasize that a review of the questionnaire by substantive experts is also essential to ensure that the questionnaire collects the information needed to meet the analytic objectives of the survey. The experts review the wording of questions, the structure of questions, the response alternatives, the order of questions, instructions to interviewers for administering the questionnaire, and the navigational rules of the questionnaire.

Sometimes, the experts employ checklists of question problems. Over the years, many writers have published lists of principles for good survey questions, starting at least as far back as Payne (1951). Initially, the lists were based primarily on the writers' opinions and judgments. Over time, they have been increasingly informed by cognitive testing, behavior coding of pretest interviews, or psychometric evaluations. We have presented our own list of guidelines (adapted from Sudman and Bradburn, 1982) in Section 7.4.

A difficulty with the items on such lists is that they are subject to interpretation and judgment. Thus, a basic standard for survey questions is that they should be understood in the same way by all respondents, and this understanding should match the one the authors of the question intended. There is no argument about the principle, but even experts looking at the same question can disagree about whether a question is likely to be ambiguous. For that reason, the primary use for a checklist is to guide preliminary review of questions that, in turn, are targeted for some additional form of testing.

There have been several attempts to develop more detailed and explicit systems for assessing potential problems with questions. Lessler and Forsyth (1996) present one of the most detailed, distinguishing more than 25 types of problems with questions, largely derived from a cognitive analysis of the response process similar to the one presented in Section 7.2. Graesser and his colleagues (Graesser, Bommareddy, Swamer, and Golding, 1996) distinguish 12 major problems with questions, most of them involving comprehension issues, and they have developed a computer program that can apply these categories to draft questions, serving as an automated but rough expert appraisal (Graesser, Kennedy, Wiemer-Hastings, and Ottati, 1999). The dozen problems distinguished by Graesser and colleagues include:

1) Complex syntax
2) Working memory overload
3) Vague or ambiguous noun phrase
4) Unfamiliar technical term
5) Vague or imprecise predicate or relative term
6) Misleading or incorrect presupposition
7) Unclear question category
8) Amalgamation of more than one question category
9) Mismatch between the question category and the answer options
10) Difficult to access (that is, recall) information
11) Respondent unlikely to know answer
12) Unclear question purpose

Graesser's model of the question answering process assumes that the listener first decides what type of question has been posed (e.g., a yes–no or a why question); thus, when the type of question is unclear or more than one type is involved, it creates problems for respondents.

8.3 Focus Groups

Before developing a survey instrument, researchers often recruit groups of volunteers to participate in a systematic discussion about the survey topic. A "focus group" discussion is a discussion among a small number (six to ten) of target population members guided by a moderator (see Krueger and Casey, 2000). The group members are encouraged to express their viewpoints and to feel comfortable disagreeing with the perspectives of others. Group members are free to influence one another in their ideas.

focus group

The researcher spends considerable effort structuring the discussion topics in order to target key issues in the subject area. At an early stage of survey development, focus groups might help the researcher learn about how members of the target population understand the concepts in the questionnaire, what terminology they used in discussing them, what common perspectives are taken by the population on key issues, and so on. Groups might discuss reactions to alternative recruitment protocols for the sample members or perceptions of the sponsor of the survey.

The moderator attempts to create an open, relaxed, permissive atmosphere. Good moderators subtly keep the group on target and make seamless transitions across topics. Good moderators encourage all focus group members to participate, drawing out the shy members, and politely closing off the dominant speakers. They listen carefully to each participant's comments, and when new observations are made, seek reactions of other group members to the observations. The moderators are guided by a set of topic areas that are to be discussed. They are not scripted in their questions or probes. Thus, they need to know the research goals quite well in order to answer questions from the group members.

Usually, there is an attempt to choose focus group members who are homogeneous on key dimensions related to the topic. For example, a focus group in preparation for a survey of employment search might separate those persons interested in salaried versus hourly positions. If the survey is to be conducted on diverse subpopulations, then separate focus groups may be mounted for each subpopulation.

Sometimes, focus groups are held in specially designed rooms, with one-way mirrors, to permit the research team to observe the group, and to allow unobtrusive audio/video recording of the group. Sometimes, the product of the group is a set of written notes summarizing key inputs; other times, a full transcript of the group discussion is produced. With videotaped groups, edited videotape segments can summarize the key findings.

Focus groups are common tools in the early stages of questionnaire development, in order to learn what respondents know about the topic of the survey. Focus groups have three main attractions to the questionnaire designer:

1) They are an efficient method for determining what potential respondents know and what they do not know about the survey topics, and how they structure that knowledge. For example, in a survey on health insurance, it is useful to know what types of insurance plans respondents are aware of and what they know (or think they know) about each type of plan. It is also useful to have a sense of which issues or dimensions respondents think are important (and which they think are unimportant). Finally, it is helpful to know how potential respondents think about the issues and how they categorize or group them. For example, in a survey on health insurance, respondents may see health maintenance organizations (HMOs) as very different from other types of health service plans. This sort of information may help the researchers structure the questionnaire to promote the most accurate reporting.

2) Focus groups are also a good method for identifying the terms that respondents use in discussing the topic and exploring how they understand these terms. A key goal in designing survey questions is to use

words that are familiar to the respondents and consistently understood by them. Focus groups provide an excellent opportunity to hear how potential respondents spontaneously describe the issues involved and how they understand candidate phrases or terms.

3) Survey questions ask respondents to describe something they know about. Whether the questions concern subjective states (such as feelings, opinions, or perceptions) or objective circumstances or experiences, the questions are best designed if researchers have a firm grasp of the underlying realities that respondents have to report. Thus, a final function of focus groups is to convey to researchers what respondents have to say on the survey topic. Leading a focus group through the various topics to be covered by a survey and getting a sense of the range of experiences or perceptions that respondents will be drawing upon to form their answers, enables researchers to write questions that fit the circumstances of the respondents.

The strength of the focus group is that it is an efficient way to get feedback from a group of people. There are, however, three main limitations of focus groups:

1) Participants in focus groups are not necessarily representative of the survey population, so one cannot generalize about the distribution of perceptions or experiences from focus groups alone.
2) A focus group is not a good venue for evaluating wording of specific questions or for discovering how respondents arrive at their answers. Focus groups can give a sense of the range and kinds of differences among members of the survey population. However, assessing the wording of specific questions and evaluating the cognitive issues associated with the questions are done more easily with a one-on-one testing protocol.
3) Because the information gathered from the discussion is rarely quantitative, there is the potential for the results and conclusions to be unreliable, hard to replicate, and subject to the judgments of those who are conducting the groups.

Despite these limitations, focus groups are efficient ways of gathering qualitative information about the survey topic from the perspective of the target population prior to imposing the structure of a survey questionnaire.

8.4 Cognitive Interviews

In 1983, the U.S. National Research Council (NRC) held a workshop that brought cognitive psychologists and survey research methodologists together to explore their potential mutual interests. One of the outgrowths of the workshop was that survey researchers began to explore the value of techniques developed by cognitive psychologists to find out how people understand and answer questions (Jabine, Straf, Tanur, and Tourangeau, 1984; Schwarz and Sudman, 1992).

Schuman and Presser (1981) and Belson (1981) had presented evidence indicating considerable misunderstandings of survey questions years before; the papers spawned by the NRC workshop sparked more widespread interest in learning how questions are understood and answered by survey respondents.

One of the methods discussed at the workshop was the use of cognitive interviewing to test survey questions. Cognitive interviewing is based on a technique called "protocol analysis" that was invented by Simon and his colleagues (see, for example, Ericsson and Simon, 1980, 1984). In a "protocol analysis," subjects think aloud as they work on the problems and their verbalizations are recorded. Simon was interested in the processes by which people solve different kinds of problems, such as proving simple mathematical theorems or playing chess. The term "cognitive interviewing" is used somewhat more broadly to cover a range of cognitively inspired procedures. These include:

1) Concurrent think-alouds (in which respondents verbalize their thoughts while they answer a question)
2) Retrospective think-alouds (in which respondents describe how they arrived at their answers either just after they provide them or at the end of the interview)
3) Confidence ratings (in which respondents assess their confidence in their answers)
4) Paraphrasing (in which respondents restate the question in their own words)
5) Definitions (in which respondents provide definitions for key terms in the questions)
6) Probes (in which respondents answer follow-up questions designed to reveal their response strategies)

This list is adapted from a longer list given by Jobe and Mingay (1989); see also Forsyth and Lessler (1992).

As the list suggests, there is no single way that cognitive interviewing is done. Typically, the respondents are paid volunteers and the interview includes some draft questions along with probes or other procedures for discovering how the respondents understood the questions and arrived at their answers. Respondents may also be asked to think aloud as they answer some or all of the questions. The interviewers may be research scientists, cognitive psychologists, experts in survey question methodology, interviewers with special training in question evaluation, or standardized interviewers with no special training.

Different organizations and different interviewers use different techniques or mixes of techniques to gather information in cognitive tests. Some rely on pre-scripted probes; others emphasize think-alouds. Some ask for retrospective protocols immediately after a question is administered; others collect them at the end of the interview. There are also different methods for recording the information from cognitive interviews, which range from the formal (e.g., videotaping the interviews and making and transcribing audiotapes) to the informal (having the interviewer make notes during the interviews).

Although the use of cognitive testing is growing and the technique appears to yield valid insights, there is a critical need for empirical studies of the reliability of findings from cognitive interviews, their value in improving data quality, and the significance of the many variations in the way cognitive interviews are done. DeMaio and Landreth (2004) have conducted the most comprehensive study to date, which showed that different approaches to cognitive testing can produce similar results. Still, their study shows considerable differences among three teams assessing the same questions in which questions they identified as having problems, which problems they thought the questions had, and how they proposed to fix them. In addition, observational studies examining what happens during cognitive interviews indicate considerable variation across interviewers (Beatty, 2004). On the other hand, Fowler (2004) provides examples of questions, revised based on cognitive testing, that result in apparently better data. As yet, though, evidence is limited about the extent to which cognitive testing generally improves survey data (Willis, DeMaio, and Harris-Kojetin, 1999; Forsyth, Rothgeb, and Willis, 2004).

8.5 FIELD PRETESTS AND BEHAVIOR CODING

"Pretests" are small-scale rehearsals of the data collection conducted before the main survey. The purpose of a pretest is to evaluate the survey instrument as well as the data collection and respondent selection procedures. Pretests with small samples (often done by relatively small numbers of interviewers) have been standard practice in survey research for a long time.

Historically, pretests have yielded two main types of information about the survey and the survey questionnaire. First, the views of the interviewers have often been solicited in "interviewer debriefings." These are a bit like focus groups with the pretest interviewers, in which the interviewers present their conclusions about problem questions and other issues that surfaced in the pretest. Often, the interviewers offer suggestions about how to streamline the procedures or improve the questions. The second type of informa-

---

Presser and Blair (1994) on Alternative Pretesting Methods

Presser and Blair compared four pretesting methods.

Study design: Separate pretesting staffs evaluated a common "test" questionnaire of 140 items, in an initial round and a revised round. First, eight telephone interviewers collected 35 first round and 43 second round interviews in a traditional pretest, with a debriefing assessment. Second, the researchers examined behavior coding from the interviews. Third, three cognitive interviewers interviewed a total of about 30 respondents, using follow-up probes and think-aloud techniques. Fourth, two panels of questionnaire experts identified problems in the questionnaire.

Findings: Expert panels identified on average 160 problems compared to about 90 for conventional pretests, cognitive interviews, and behavior coding. Pretests and cognitive interviewing showed great variability over trials. Behavior coding and expert panels were the most reliable in types of problems found. Pretests and behavior coding tended to find interviewer administration problems. Cognitive interviews found comprehension problems and no interviewer problems. The expert panel was the most cost-effective method.

Limitations of the study: Only one questionnaire was used to evaluate the methods. Telephone pretest results may not imply similar findings for face-to-face pretests. There was no distinction made between important and trivial problems. The ability of the researchers to solve the problems was not examined.

Impact of the study: The study led to recommendations to use more expert panels for questionnaire development.

---

pretests

interviewer debriefing

tion to emerge from pretests is quantitative information based on the responses. The data collected during a pretest are often entered and tabulated. The survey designers may look for items that have high rates of missing data, out-of-range values, or inconsistencies with other questions. In addition, items with little variance (that is, items that most respondents answer the same way) may be dropped or rewritten.

behavior coding

Tape recording pretest interviews, then making systematic observations of how questions are read and answered, can also provide useful information about the questions (Oksenberg, Cannell, and Kalton, 1991). "Behavior coding" is the systematic classification and enumeration of interviewer–respondent interaction to describe the observable behaviors of the two persons related to the question-and-answer task. Table 8.1 gives some examples of codes that are used for each question in the questionnaire, for each interview coded.

Using codes like those in Table 8.1, for each interview coded, the behavior coder makes judgments about whether the interviewer reading of the question followed training guidelines and which respondent behaviors were exhibited. The resulting dataset includes the behaviors coded for each question-and-answer sequence for each interview. The question designer then analyzes these data by computing statistics for each question, such as:

1) The percentage of interviews in which it was read exactly as worded.
2) The percentage of interviews in which the respondent asked for clarification of some aspect of the question.

Table 8.1. Examples of Behavior Codes for Interviewer and Respondent Behaviors

Code Category	Description
Interview Questioning Behaviors (choose one)	
1.	Reads question exactly as worded
2.	Reads questions with minor changes
3.	Reads questions so that meaning is altered
Respondent Behaviors (check as many as apply)	
1.	Interrupts question reading
2.	Asks for clarification of question
3.	Gives adequate answer
4.	Gives answer qualified about accuracy
5.	Gives answer inadequate for question
6.	Answers "don't know"
7.	Refuses to answer

3) The percentage of interviews in which the respondent did not initially provide an adequate answer to the question so that the interviewer had to probe or offer an explanation to obtain a codeable answer.

There are various aspects of the question-and-answer process that might be of interest and could be coded. An important area for further research is to identify behaviors that can be reliably coded during interviews and that have implications for the quality of data that are being collected.

Integrating behavior coding into pretesting simply involves tape-recording the interviews, with respondent permission of course, then tabulating the rates at which the behaviors noted above occur. A particular value that behavior coding adds to a standard pretest is that the results are systematic, objective, and replicable. As Fowler and Cannell (1996) report, when two interviewing staffs independently tested the same set of questions, the correlations between the rates of the above behaviors on each question were between 0.75 and 0.90. This indicates that questions consistently and reliably produce high or low rates of these behaviors, regardless of which interviewers are asking the questions.

8.6 Randomized or Split-Ballot Experiments

Sometimes, survey designers conduct studies that experimentally compare different methods of data collection, different field procedures, or different versions of the questions. Such experiments can be done as stand-alone studies or as part of a field pretest. When random portions of the sample get different questionnaires or procedures, as is typically the case, the experiment is called a "randomized" or "split-ballot" experiment. Tourangeau (2004) describes some of the design issues for such studies and cites a number of examples of split-ballot experiments. Of our sample surveys, the NSDUH has been especially active in using split-ballot experiments; it has conducted a number of major split-ballot studies examining how reporting of illicit drug use is affected by different modes of data collection and different wording of the questions (see Turner, Lessler, and Devore, 1992; Turner, Lessler, George, Hubbard, and Witt, 1992; and, for a more recent example, Lessler, Caspar, Penne, and Barker, 2000). Similarly, when the Current Population Survey questionnaire required an overhaul of questions about unemployment, a major experiment was carried out to compare the old version of the questions to the new version (Cohany, Polivka, and Rothgeb, 1994). That way, it was clear how much of the change in the monthly unemployment rate (which is derived from the CPS data) was due to the changeover in the questionnaires.

randomized experiments

split-ballot experiments

Experiments like these offer clear evidence of the impact on responses of methodological features—differences in question wording, question order, the mode of data collection, and so on. Unfortunately, although they can demonstrate that the different versions of the instruments or procedures produce different answers, many split-ballot experiments cannot resolve the question of which version produces better data. An exception occurs when the study also collects some external validation data against which the survey responses can be checked. Results are also interpretable when there are strong theoretical reasons for deciding that one version of the questions is better than another. For example, Turner and his colleagues concluded that self-administration improved reporting of illicit

> **Oksenberg, Cannell, and Kalton (1991) on Probes and Behavior Coding**
>
> In 1991, Oksenberg and coworkers reported a study about evaluating questions using behavior coding.
>
> *Study design*: Six telephone interviewers used a questionnaire of 60 items on health-related topics assembled from questions used on existing surveys. Behavior coding identified some problems with the questions. The researchers revised the questionnaire and took 100 additional interviews.
>
> *Findings*: The three questions below showed the following behavior coding results:
> 1) What was the purpose of that visit (to a health care person or organization)?
> 2) How much did you pay or will you have to pay out of pocket for your most recent visit? Do not include what insurance has paid for or will pay for. If you don't know the exact amount, please give me your best estimate.
> 3) When was the last time you had a general physical examination or checkup?
>
> **Percent of problems per question**
>
> | Question | 1 | 2 | 3 |
> |---|---|---|---|
> | *Interviewer action* | | | |
> | Slight wording change | 2 | 30 | 3 |
> | Major wording change | 3 | 17 | 2 |
> | *Respondent action* | | | |
> | Interruption | 0 | 23 | 0 |
> | Clarification request | 2 | 10 | 3 |
> | Inadequate answer | 5 | 17 | 87 |
> | "Don't know" | 0 | 8 | 12 |
>
> The first question was relatively unproblematic. The second caused both the interviewer and respondent conversational problems. The third had the ambiguous phrase, "general physical examination" and the lack of a clear response format. By changing Question 2 so that it did not appear to be completed prematurely, interruptions were reduced.
>
> *Limitations of the study*: The study did not identify how to fix problems found by behavior coding.
>
> *Impact of the study*: The study showed how some structural problems with questions could be reliably detected from the question-and-answer behaviors.

drug use because reporting of drug use increased. A number of earlier studies had shown that respondents tend to underreport their drug use, so an increase in reporting is likely to represent an improvement. Fowler (2004) describes several other split-ballot experiments in which, despite the absence of validation data, it seems clear which version of the questions produced the better data.

Fowler's examples involve fairly small samples (some based on fewer than 100 cases), because of the time and expense involved. If detecting relatively small differences between experimental groups is important, large samples of respondents may be needed. For many surveys with modest budgets, adding the cost of even a small split-ballot experiment may seem excessive. For these reasons, split-ballot experiments before surveys are not routine. Nonetheless, they do offer the potential to evaluate the impact of proposed wording changes on the resulting data that other testing approaches do not provide.

8.7 Applying Question Standards

The different methods we have discussed vary in the kinds of problems they are best suited to identify. Here we discuss which methods are suitable for evaluating whether the questions meet our three standards for survey questions.

The "content standard" for questions is whether or not they are asking for the right information. This has to be assessed from two perspectives. First, from the point of view of the analysts, the questions must gather the information needed to address the research objectives. The only way to assess this is to ask the experts—the analysts or other subject matter specialists—whether the questions provide the information needed for the analysis. The other key issue is whether the respondents can actually provide this information. Surveys can only provide useful information if respondents can answer the questions with some degree of accuracy. The primary ways to assess how well respondents can answer candidate questions are focus group discussions and cognitive interviews. Through focus groups, we can learn what potential respondents know. From cognitive interviews, we can see whether a particular set of questions can be consistently answered and whether the answers actually provide the information the analysts are seeking. content standard

"Cognitive standards," whether respondents can understand and answer the questions, are most directly assessed by cognitive testing. That is what cognitive interviews are designed to do. However, there are three other question evaluation activities that can make a contribution to identifying cognitive problems with questions: cognitive standard

1) Focus group discussions can identify words that are not consistently understood, concepts that are ambiguous and questions that respondents are unable to answer
2) Expert reviews often can flag ambiguous terms and concepts, as well as response tasks that are difficult to perform, prior to any cognitive testing
3) Behavior coding of pretest interviews can identify questions that are unclear or that respondents have trouble answering

The assessment of "usability," how well a survey instrument can be used in practice, is the primary aim of a field pretest. In addition, prior to a pretest, expert reviews of the questions can identify questions that are likely to pose problems for respondents or interviewers. Usability testing is most valuable in self-administered questionnaires. In controlled laboratory conditions or in typical survey settings, survey staff can observe respondents handling the survey materials, attempting to understand the task, and performing the task. With computer-assisted instruments (Couper, Hansen, and Sadowsky, 1997; Hansen and Couper 2004; Tarnai and Moore, 2004), the computer itself might be used to time keystrokes, to measure the extent of backward movement in the questionnaire, and to compute the rate of illegal entries. Although the laboratory may not turn up all the problems likely to crop up in the field, the problems identified there are likely to be even worse under realistic data collection conditions. usability

8.8 Summary of Question Evaluation Tools

All of the techniques for evaluating survey questions discussed in this chapter have potential contributions to make and all of them have limitations. Here we summarize some of these virtues and limitations.

1) Expert content review provides the important perspective of what the users of the data need in order to meet the analytic objectives of the survey. However, it provides no information about what the best questions are, the ones that respondents can answer most accurately in order to provide the necessary information.
2) Systematic review of the questions by questionnaire design experts is perhaps the least expensive method and the easiest to carry out (Presser

and Blair, 1994; see box on page 265). Still, it is only as good as the experts. Experts may disagree about whether the questions are clear or the response tasks they pose are too difficult for respondents to carry out. As a result, techniques such as cognitive testing with real potential respondents are needed to evaluate the questions empirically. On the other hand, a growing number of characteristics of questions have been identified as consistently posing problems. To the extent that it is possible to expand the list of question characteristics that are likely to cause problems, the value of expert review of the questions can be proportionately increased.

3) Focus group discussions provide an efficient way to get the ideas, perceptions, and contributions of six to ten people at a time about issues of practical relevance to designing the questions. However, because it is a group setting, focus groups are not the best method for investigating how individuals understand the specific questions or how they go about answering them.

4) Cognitive testing is a useful method to find out how individuals understand questions and arrive at their answers to them. However, cognitive testing usually involves a small number of people, who are not necessarily representative of an entire target population. Thus, a major concern about cognitive testing is the generalizability of the results. We cannot know how the distribution of problems or issues found in a group of cognitive interview respondents will generalize to the target population. We also need to be concerned that paid volunteers under laboratory conditions may be able and willing to perform tasks that respondents under realistic survey conditions will not. Finally, there is the danger that different cognitive interviewers may produce different conclusions, perhaps even steering the interviews to produce evidence of problems (Beatty, 2004). A related issue is that cognitive interviews yield unsystematic impressions about the problems with the questions rather than objective data (Conrad and Blair, 1996).

5) Usability testing in laboratory settings has strengths and limitations that parallel those of cognitive testing.

6) Field pretests are the best way to find out how instruments and field procedures work under realistic conditions. Through behavior coding, researchers can gain systematic information about how the question-and-answer process in fact is performed under real conditions. It is generally useful to tabulate the data and to debrief the pretest interviewers. The limitations of field pretests are that, because researchers are trying to replicate realistic survey procedures, there is not great flexibility to probe and understand the nature of the problems that interviewers and respondents face.

One critical question is the extent to which the various techniques produce the same information. Table 8.2 summarizes the results of several studies that compare multiple methods of item evaluation. The first of these studies, done by Presser and Blair (1994) compared the results from conventional pretests, expert reviews, cognitive testing, and behavior coding of a pretest (see box on page 265). They found that there was some overlap in the problems found, but the approaches did not yield the same results. The expert reviews and cognitive inter-

Table 8.2. Studies Comparing Question Evaluation Methods

Presser and Blair (1994)		
Methods Tested	**Criteria**	**Conclusions**
1. Coventional pretest 2. Behavior coding 3. Cognitive interviews 4. Expert panels	• Number of problems found • Type of problem detected (problems were classified into four categories) • Consistency across trials with the same method	1. Conventional pretests and behavior coding found the most interviewer problems. 2. Expert panels and cognitive interviews found the most analysis problems. 3. Expert panels and behavior coding were more consistent across trials and found more types of problems. 4. Behavior coding was most reliable but provided no information about the cause of a problem, did not find analysis problems, or distinguish between respondent-semantic and respondent-task problems. 5. Expert panels were most cost-effective. 6. Most common problems were respondent-semantic.
Willis, Schechter, and Whitaker (1999)		
Methods Tested	**Criteria**	**Conclusions**
1. Cognitive interviewing (done by interviewers at two organizations) 2. Expert review 3. Behavior coding	• Number of problems found • Consistency within and across methods regarding the presence of a problem (measured by the correlation across methods and organizations between the percent of the time items were classified as having a problem) • Type of problems found (based on a five-category coding scheme)	1. Expert review found the most problems. 2. The correlation between behavior coding trials was highest (0.79), followed closely by the correlation between the cognitive interviews done by two organizations (0.68). 3. Across methods of pretesting and organizations, most problems were coded as comprehension/communication; there was a high rate of agreement in the use of subcodes within this category across techniques.
Rothgeb, Willis, and Forsyth (2001)		
Methods Tested	**Criteria**	**Conclusions**
Three research organizations tested three questionnaires, using each of these methods and coding problems according to a classification scheme developed by the authors: 1. Informal expert review 2. Formal cognitive appraisal 3. Cognitive interviewing	• Number of problems found • Agreement across methods based on summary score for each item (summary scores ranged from 0 to 9 based on whether the item was flagged as a problem item by each technique and each organization)	1. Formal cognitive appraisal (QAS) found the most problems but encouraged a low threshold for problem identification. 2. Informal expert review and cognitive interviewing found similar numbers of problems, but found different items to be problematic. 3. Results across organizations were more similar than across techniques: Moderate agreement across organizations in summary scores (r's range from 0.34 to 0.38). 4. Communication and comprehension problems were identified most often by all three techniques.

(continued)

Table 8.2. Studies Comparing Question Evaluation Methods (continued)

Forsyth, Rothgeb, and Willis (2004) (Note: This study is a follow-up to Rothgeb et al., 2001)		
Methods Tested	**Criteria**	**Conclusions**
1. Informal expert review 2. Formal cognitive appraisal (QAS) 3. Cognitive interviewing	• Conducted randomized experiment in a random-digit-dial telephone survey that compared control questionnaire (with the original items pretested in 2001 study) and experimental questionnaire (with revised items designed to fix problems found in the pretest) • Classified items as low, moderate, or high in respondent and interviewer problems, based on behavior coding data and interviewer ratings	1. Items classified as high in interviewer problems during pretesting also had many problems in the field (according to behavior coding and interviewer ratings). 2. Items classified as high in respondent problems during pretesting also had many problems in problems in the field. 3. Items classified as having recall and sensitivity problems during pretesting had higher nonresponse rates in the field. 4. The revised items in the experimental questionnaire produced nonsignificant reductions in item nonresponse and problems found via behavior coding, but a significant reduction in respondent problems (as rated by the interviewers). However, interviewers rated revised items as having more interviewer problems.
DeMaio and Landreth (2004)		
Methods Tested	**Criteria**	**Conclusions**
1. Three cognitive interview methods (three different "packages" of procedures carried out by three teams of researchers at three different organizations) 2. Expert review	• Number of problems identified • Type of problem identified • Technique that identified the problem • Frequency of agreement between organizations/methods	1. The different methods of cognitive interviewing identified different numbers and types of problems. 2. Cognitive interviewing teams found fewer problem questions than did expert reviews, although all three organizations identified as problematic most of the "defective" questions (those for which at least two experts agreed there was a problem of a specific type). 3. The problems identified by the cognitive interviewing teams were also generally found by the experts. 4. Different teams used different types of probes. 5. Upon revising the questionnaires and readministering cognitive interviews, it was found that only one team's questionnaire had fewer problems than the original.

views tended to find more comprehension problems, whereas the pretest found more problems in administering the items (that is, usability issues). The experts flagged the most problems, but not always important problems.

The conclusions of the Presser and Blair study have stood up well over time. For example, a more recent study by Forsyth, Rothgeb, and Willis (2004) took a more critical look at the importance of the problems identified by the various methods and found (as Presser and Blair did) that the techniques find some of the same problems but some unique question problems as well. Experts find the most "problems," but some of those probably do not affect data quality. Weak measures of data quality, a chronic challenge for studies of question design and evaluation, made the results of these evaluation studies inconclusive; we cannot say whether the "problems" really reduce the validity of the answers.

There seems to be little doubt that the different techniques complement each other; each has some obvious strengths and weaknesses compared to the others and they provide information related to different issues. As a result, many surveys use a combination of methods to pretest survey questions. Which specific methods are used often depends on the survey budget and the level of prior experience with the questions. A survey with a new questionnaire but a limited pretesting budget might conduct a few focus groups, an expert review, a round or two of cognitive interviews, and a small field pretest. Expert reviews and cognitive testing can be done cheaply, and both are likely to uncover lots of potential problems (Presser and Blair, 1994; Forsyth, Rothgeb, and Willis, 2004). The focus groups will help align the concepts and terminology used in the questionnaire with those of the respondents. A small field test can detect any operational problems. If an existing questionnaire is being used with minor modifications, the focus groups and field tests might be dropped, and cognitive testing would focus on the new items.

At the other extreme, a large survey is rarely fielded without a correspondingly large field pretest. The risk of a major operational failure is too great to go into the field without substantial pretesting. Census 2000, for example, underwent a series of field tests culminating in a dress rehearsal conducted in 1998 that involved three areas (Sacramento, California; Menominee County, Wisconsin; and an 11 county area including Columbia, South Carolina) and several hundred thousand respondents. The pretesting program that led to the latest NSDUH questionnaire involved several large-scale split-ballot experiments (e.g., Lessler, Caspar, Penne, and Barker, 2000).

However, the major problem with all of these techniques is that they tell us very little, if anything, about how the problems identified affect data quality. There is some evidence that the problems identified by these testing techniques can have major effects on survey estimates. For instance, Fowler (1992) has shown that if key terms in questions are not understood consistently, systematic biases are likely to result. Mangione and his colleagues have also shown that if questions are hard for respondents, so that interviewers have to probe extensively to get adequate answers, it increases the likelihood of interviewer effects and results in with inflated standard errors (Mangione, Fowler and Louis, 1992). Nonetheless, the techniques discussed so far, with the possible exception of split-ballot tests, do not tell us what kind and how much error particular question problems are likely to produce. Moreover, we lack assessments of which testing techniques, either alone or in combination, are best at finding problems that affect survey results. To get estimates on those issues, different kinds of studies and analyses are needed. That is the topic of the next section.

8.9 LINKING CONCEPTS OF MEASUREMENT QUALITY TO STATISTICAL ESTIMATES

Unfortunately, the terminology used within survey methodology for the quality of measurement is not standardized. The two traditions that provide the common terms are psychometrics and sampling statistics. The first focuses on answers to questions by an individual respondent and uses the terms "validity" and "reliability." The second focuses on statistics summarizing all the individual answers and uses the terms "bias" and "variance."

8.9.1 Validity

validity

"Validity" is a term used in somewhat different senses by different disciplines and, even within survey research, it seems to mean different things to different researchers. A common definition of "validity" is the extent to which the survey measure accurately reflects the intended construct; this definition applies in different ways to different items. Unfortunately, this definition does not suggest a specific method of evaluating validity. An early meaning (Lord and Novick, 1968) was based on a simple conceptual model of the measurement process as being one realization of a set of conceptually infinite trials. That is, each survey measurement could (in concept only) be repeated so that each answer from a given respondent to a given question administration is just one trial within that infinite set of trials. Just as we presented in Section 2.3, let

μ_i = the true value of the construct for the ith respondent
Y_{it} = the response to the measure by the ith respondent on the tth trial
ε_{it} = the deviation from the true value of the construct related to the response, Y_i, on the tth trial

Then the conceptual model of the response process is the following. When the question about the construct, μ, is given to the ith respondent in a survey (called the tth trial), instead of providing the answer, μ_i, the respondent provides the answer, Y_{it},

$$Y_{it} = \mu_i + \varepsilon_{it}$$

"Validity" is measured by the correlation (over persons and trials) of two of the terms in the equation above. It is usually defined as the correlation between Y_i and μ_i. This means that a measure is higher in validity when the values of Y_i are on average, closer to those of μ:

$$\text{Validity}(Y) = \frac{\sum_{i,t}(Y_{it} - \bar{Y})(\mu_i - \bar{\mu})}{\sqrt{\sum_{i,t}(Y_{it} - \bar{Y})^2 \sum_i (\mu_i - \bar{\mu})^2}} = \text{Correlation}(Y_i, \mu_i)$$

Given this, validities are expressed by numbers that lie between 0.0 to 1.0; the higher the number, the higher the validity. We will describe two ways of estimating validity in practice: using data external to the survey and using multiple indicators of the same construct in one survey.

Estimating Validity with Data External to the Survey. Consider an item we discussed in the previous chapter:

> During the past 12 months, since _____, how many times have you seen or talked to a doctor or assistant about your health? Do not count any time you might have seen a doctor while you were a patient in a hospital, but count all other times you actually saw or talked to a medical doctor of any kind.

The item is a factual item and, at least in principle, the quality of the answer can be determined with reference to a reasonably well-defined set of facts. The respondent has, in fact, made a certain number of eligible medical visits within the period specified and, although there may be some ambiguity as to whether specific episodes should be included (for instance, should respondents count a consultation with a doctor by telephone?), in principle, these ambiguities could be resolved. For this question, if the respondent had two visits in the past 12 months, μ_i is 2. If Y_{it} is 2, there is no response deviation from the true value for that trial. If, across all respondents, answers similarly agreed with their true number of visits and this occurred in all possible trials, there would be validity of 1.0.

With some survey items, records or other external data might allow us to assess the quality of survey responses. By asserting that whatever value lies in the record for the ith respondent is μ_i, the true value for the respondent can be calculated. If we can further assert that the survey we conduct is one representative of all possible trials, we can compute validity by

$$\text{Estimated validity from one trial} = \frac{\sum_{i=1}^{n}(y_i - \bar{y})(\hat{\mu}_i - \bar{\hat{\mu}})}{\sqrt{\sum_{i=1}^{n}(y_i - \bar{y})^2 \sum_{i=1}^{n}(\hat{\mu}_i - \bar{\hat{\mu}})^2}}$$

where

μ_i = the value of the record variable for the construct, sometimes called the "gold standard" for the study
$\bar{\mu}$ = the mean of the record values across all respondents

This is merely the correlation between the respondent answers and their record values. If the correlation is near 1.0, then the measure is said to have high validity.

Estimating Validity with Multiple Indicators of the Same Construct. A second item is attitudinal, and it is not clear what facts are relevant to deciding whether respondents have answered accurately.

> Now turning to business conditions in the country as a whole, do you think that during the next 12 months we'll have good times financially, or bad times, or what? [Survey of Consumers]

This item is intended to assess economic expectations, and the actual state of the economy twelve months later is not directly relevant to the accuracy of the answer. The set of facts relevant to deciding whether a respondent's answer to this question is a good one is not well defined, and these facts are, in any case, subjective ones. Most respondents do not have some preexisting judgment about the future state of the economy but probably consult various beliefs about the economic trends in answering the question. For instance, respondents may consider their beliefs about the trend in the unemployment rate or about the direction of the stock market. It would be very difficult to say which beliefs are relevant to judging the accuracy of the respondents' answers and we do not have direct access to those beliefs anyway. In the notation above, it is not clear what μ_i is for this question for any respondent. Issues of measuring validity are thus more complex.

Correlating survey answers to some "gold standard" derived from another source is ideal. However, it is unusual in survey research to have accurate external information with which to evaluate answers. Moreover, with surveys that ask about subjective states, such as knowledge, feelings, or opinions, there is no possibility of a "gold standard" independent of the respondent's reports. In the absence of some outside standard, evaluation of the validity of most answers rests on one of three kinds of analyses:

1) Correlation of the answers with answers to other survey questions with which, in theory, they ought to be highly related
2) Comparison between groups whose answers ought to differ if the answers are measuring the intended construct
3) Comparisons of the answers from comparable samples of respondents to alternative question wordings or protocols for data collection (split-ballot studies)

The first approach is probably the most common for assessing validity. For example, if an answer is supposed to measure how healthy a person is, people who rate themselves at the high end of "healthy" should also report that they feel better, that they miss fewer days of work, that they can do more things, and that they have fewer health conditions than those who rate themselves at the low end of the scale. The results of such analyses are called assessments of construct validity (Cronbach and Meehl, 1955). If researchers did not find the predicted relationships, it would cast doubt on the validity of the health measure.

For example, the Mental Health Inventory Five Item Questionnaire (MHI-5) is a widely used series of questions designed to measure current psychological well-being (Stewart, Ware, Sherbourne, and Wells, 1992):

> These questions are about how you feel and how things have been going during the last four weeks. For each question, please

give the one answer that comes closest to the way that you have been feeling. How much of the time during the past four weeks:

a) Have you been a happy person?
b) Have you felt downhearted and blue?
c) Have you been a very nervous person?
d) Have you felt calm and peaceful?
e) Have you felt so down in the dumps that nothing could cheer you up?

The response categories include "all of the time," "most of the time," "a good bit of the time," "some of the time," "a little bit of the time," and "none of the time." All of these questions ask people to describe their psychological well-being. An index is created by assigning a number (for example, from 1 to 6) to each of the six possible answers, reverse scoring the negatively worded items so 6 is always a positive response, and adding the answers across the five questions (producing a score that could range from 5 to 30).

We can evaluate the validity of the MHI-5, to see the extent to which it was measuring what it was intended to measure, by looking at the correlation of the total score with other measures. Other indicators of mental health are used as criteria. This is sometimes called "concurrent validity," because it is based on relationships among attributes measured at the same time. Stewart, Ware, Sherbourne, and Wells (1992) find that the MHI-5 correlates –0.94 with a measure of psychological distress, –0.92 with a measure of depression, –0.86 with a measure of anxiety, +0.88 with a measure of positive affect, +0.69 with a measure of perceived cognitive functioning, and +0.66 with a measure of feelings of belonging. The authors conclude that these patterns of association are in the direction and of the order of magnitude one would expect to find with a summary measure of mental health, and, hence, that there is good evidence for the validity of the measure.

Validity assessment can also be applied using sophisticated modeling approaches that simultaneously evaluate the evidence for validity by looking at the patterns and strength of the correlations across numerous measures (e.g., Andrews, 1984; Saris and Andrews, 1991). Consider the MHI-5 items given earlier. The structural modeling approach depicts answers to each of the five items as reflecting the same underlying construct μ_i at different levels of validity λ_α:

$$Y_{\alpha i} = \lambda_\alpha \mu_i + \varepsilon_{\alpha i}$$

Notice that this model is just another small variant of the base error model of $Y_i = \mu_i + \varepsilon_i$. Here, instead of a response to just one item, the equation describes the responses to many items, each subscripted with a different α. The response to each item is viewed as a function of the underlying construct μ_i, which is described by the coefficient λ_α.

For example, a simple model of the measurement process for the MHI-5 can be presented as a path diagram, as in Figure 8.1. The circle at the top represents the underlying construct μ_i. The arrows emanating out of the circle imply that the construct "causes" the values of the indicators ($Y_{\alpha i}$) as a function of the λ_α, $\alpha = 1$,

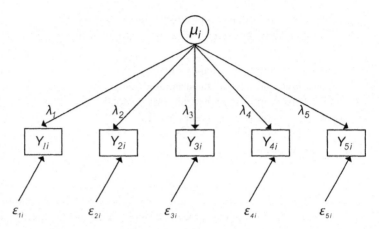

Figure 8.1 Path diagram representing $Y_{ai} = \lambda_a \mu_i + \varepsilon_{ai}$, a measurement model for μ_i.

2, 3, 4, or 5, which are displayed along the arrows. The individual items are presented in squares, one for each of the five items.

Thus, in a more compact form the figure describes five equations:

$$Y_{1i} = \lambda_1 \mu_i + \varepsilon_{1i}$$
$$Y_{2i} = \lambda_2 \mu_i + \varepsilon_{2i}$$
$$Y_{3i} = \lambda_3 \mu_i + \varepsilon_{3i}$$
$$Y_{4i} = \lambda_4 \mu_i + \varepsilon_{4i}$$
$$Y_{5i} = \lambda_5 \mu_i + \varepsilon_{5i}$$

The λ_a are the validity measures for each of the five items. They can be estimated using a variety of techniques (see Andrews, 1984; Saris and Andrews, 1991; Saris and Gallhofer, 2007).

Another approach to assessing validity is to compare the answers of groups of respondents who are expected to differ on the underlying construct of interest. For example, in the United States, we expect Republicans to be more conservative than Democrats. Thus, we would expect registered Republicans to score higher, on average, than registered Democrats on an item designed to measure conservatism. Such evaluations depend critically on our theory about the relationships among certain variables. If, in the above example, registered Republicans were not different from Democrats, it could mean either that the question was a poor measure of conservatism or that Republicans and Democrats do not, in fact, differ on conservatism. In general, it is difficult or impossible to distinguish poor measurement from inaccurate theories in the assessment of construct validity.

These results can only be interpreted in the context of a good theory. If there is no basis for deciding the likely direction of the errors, then we cannot say which results are more accurate or valid when two data collection protocols produce different results.

8.9.2 Response Bias

The most common confusion in terminology about errors associated with questions is how "validity" relates to "bias." As we pointed out earlier, validity is a function of the correlation between the response and the true value. Thus, validity is a property of individual answers to questions. What happens if there is a systematic deviation in responses away from a true value? For example, Sections 5.3.5 and 7.3.7 note consistent underreporting of socially undesirable traits. In some circumstances, systematic underreporting may not lower the correlation between the responses and their true values. For example, if all respondents underreport their weight by five pounds, the correlation between their reported weight and their true weight is 1.0. However, the mean weight of the respondents would be exactly five pounds less than the mean of the true values. The term "bias" is most often used to describe the impact of systematic reporting errors on summary statistics like sample means.

response bias

The split-ballot approach is designed to test for the presence of bias. Sudman and Bradburn (1982) provide an example. They compared two questions about alcohol consumption:

a) On days when you drink alcohol, how many drinks do you usually have—would you say one, two or three, or more?
b) On days when you drink alcohol, how many drinks do you usually have—would you say one or two, three or four, five or six, or seven or more?

The respondents were randomly assigned to answer one of these two questions. Sudman and Bradburn found that those who were asked question (b) were much more likely than those who were asked question (a) to report having more than three drinks on days that they drank alcohol. The researchers were confident that respondents tended to underreport the amount of alcohol they consumed. Given that premise, they concluded that the answers to the second question were more valid than the answers to the first question. Note that in a fashion similar to that above regarding measurement of validity, this measurement of bias forces the researcher to make some assumption about truth. (When some record system is used to compare survey responses, the assumption is that the records are measured without error.)

To illustrate "bias" in survey statistics, we can start with the same measurement model:

$$Y_{it} = \mu_i + \varepsilon_{it}$$

That is, the response to the question departs from the true value by an error. If the error term has a systematic component, its expected value will not be zero. Bias is introduced when the expected value of the survey observations (the Y's) differs from that of the true scores; that is, there is a systematic deviation across all trials and persons between their response and their true value:

$$\text{Bias}(Y_{it}) = \sum_t \left[\sum_i (Y_{it}) - \mu_i \right]$$

This expression is related to the summary statistic, the mean. The average or expected value of an individual response over all persons (the E_i part of the expression above) is merely the mean of the responses.

Thus, this is the same as saying that the mean of the responses is a biased estimate of the mean of the true values:

$$\text{Bias}(\bar{Y}) = \sum_t \left(\frac{\sum_i Y_{it}}{N} \right) - \frac{\sum_i \mu_i}{N}$$

The first term in the expression above is merely the expected value of the response mean over all trials.

The expression of the bias depends on values of the μ_i, but the expression for the validity above just depends on correlation of the μ_i's with the Y_i's. The notion of bias depends on the existence of a true value. Whether "true values" exist for subjective states, such as knowledge, opinions, and feelings, is controversial. Although psychometricians have sometimes tried to define true scores for subjective constructs like attitudes, as we pointed out, it generally is impossible to get a measure of a true score based on an external source. That is, with subjective constructs, we can only compare two or more reports from the respondent. For this reason, the concept of bias technically only applies to measures of objectively verifiable facts or events.

There are two ways, in practice, that response bias is estimated empirically: using data on individual target population elements from sources external to the survey and using population statistics not subject to the survey measurement error.

Using Data on Individual Target Population Elements. When the existence of a true value is plausible, the practical way to estimate response bias of answers to survey questions is to compare them with some external indicator of the true value. For example, some studies of response bias in surveys have used medical records as the indicators of the true values (Cannell, Marquis, and Laurent, 1977; Edwards, Winn, and Collins, 1996). Cannell and his colleagues compared respondents' reports about whether they had been hospitalized in a specific period with hospital records for that same period. Similarly, Edwards and his associates evaluated the quality of reporting of health conditions by comparing survey reports with the conditions recorded in medical records (Edwards et al., 1994). When records exist with which to compare survey answers, researchers are in the best position to evaluate the quality of reporting.

record check study

Let us examine one of these "record check" studies in more detail. In the study by Cannell and colleagues (Cannell, Marquis, and Laurent, 1977), interviews were conducted in households that included someone who had been hospitalized during the year prior to the interview. The households had been selected using the hospital records, and the researchers determined whether the hospitalizations in the records were in fact reported in the health interview. Table 8.3 shows the percentage of known hospitalizations reported by the length of stay (how many days the patient was in the hospital) and by how many weeks prior to the interview the hospitalization occurred. On average, respondents reported only about 85% of their hospitalizations. However, their likelihood of reporting a hospitalization was greatly affected by the length of stay and the recency of the hos-

Table 8.3. Percentage of Known Hospitalizations Not Reported, by Length of Stay and Time Since Discharge

Time Since Discharge	Length of Stay		
	1 day	2–4 days	5 days
1–20 weeks	21%	5%	5%
21–40 weeks	27%	11%	7%
41–52 weeks	32%	34%	22%

Source: Cannell et al., 1977.

pitalization. More recent hospitalizations were reported better than those that occurred earlier; hospitalizations that involved stays of five or more days were reported better than those that were shorter.

This table is a classic demonstration of how memory deteriorates over time and how more major events are easier to remember than those that are less significant (cf. Figure 7.1). Table 8.3 implies a bias in the estimate of a summary statistic—the mean number of hospitalizations in a group of respondents. The bias of that mean would be relatively greater for the mean number of hospitalizations 41–52 weeks before the interview than for the mean of hospitalizations 1–20 weeks before the interview.

Using Population Statistics Not Subject to Survey Response Error. Sometimes, survey data also can be evaluated in the aggregate, even when the accuracy of the individual reports cannot be determined. For example, a 1975 survey studied gambling behavior (Kallick-Kaufman, 1979). The survey included questions about the amount of money respondents had wagered legally at horse racing tracks. The total amount wagered at horse racing tracks is published. The researchers could not evaluate the quality of a respondent's answer directly. They could compare the overall survey estimate to the published total wagers. This allowed them to estimate net bias of the estimate. They found that the estimates from the survey and those from the racetrack records were remarkably similar. As a result, the researchers concluded that respondents were not, on average, under- or overreporting the amount they wagered at racetracks (Kallick-Kaufman, 1979). Similarly, voter behavior is published the day of elections and survey statistics summarizing voter behavior can be compared to the published results. Without the individual results, however, only the bias of summary statistics that are available publicly can be estimated.

8.9.3 Reliability and Simple Response Variance

"Reliability" is a measurement of variability of answers over repeated conceptual trials. Reliability addresses the question of whether respondents are consistent or

reliability

stable in their answers. Hence, it is defined in terms of a variance component, the variability of ε_{it} over all respondents and trials. In notation,

$$\text{Reliability}(Y_{it}) = \frac{E(\mu_i - \bar{\mu})^2}{E(\mu_i - \bar{\mu})^2 + E(\varepsilon_{it} - \bar{\varepsilon})^2} = \frac{\text{Variance of true values}}{\text{Variance of reported values}}$$

If the variance of the response deviations

$$E_{i,t}(\varepsilon_{it} - \bar{\varepsilon})^2$$

is low, this reliability ratio approaches 1.0 and the measure on the population is said to have "high reliability." If the variability in answers over trials is high (thus producing large response deviations variance), then reliability approaches 0.0.

simple response variance

The survey statistics field does not tend to use the term "reliability," but instead uses the phrase, "simple response variance." It is appropriate to think of "simple response variance" as the opposite of "reliability." When an item has high reliability for a population, then it has low simple response variance. (The term "simple response variance" will be contrasted with "correlated response variance," discussed in Section 9.3).

"Reliability" refers to the consistency of measurement either across occasions or across items designed to measure the same construct. Accordingly, there are two main methods that survey researchers use to assess the reliability of reporting: repeated interviews with the same respondent and use of multiple indicators of the same construct.

reinterview

Repeated Interviews with the Same Respondent. These are sometimes called "reinterview studies." These repeated measures can be used to assess simple response variance with the following assumptions:

1) There are no changes in the underlying construct (i.e., μ_i does not change) between the two interviews.
2) All the important aspects of the measurement protocol remain the same (i.e., these are sometimes referred to as "essential survey conditions").
3) There is no impact of the first measurement on the second responses (i.e., there are no memory effects; the second measure is independent of the first).

In practice, there are complications with each of the three assumptions. However, reinterview studies, in which survey respondents answer the same questions in a second interview as they did in a prior interview, are a common approach to obtaining estimates of simple response variance (e.g., Forsman and Schreiner, 1991; O'Muircheartaigh, 1991). When the survey measurement is repeated, the researcher essentially has two responses for each respondent, Y_{i1} and Y_{i2}.

Using reinterview studies, there are several different statistics commonly calculated to measure the consistency of response over trials:

1) Reliability, as defined above
2) The index of inconsistency, which equals (1 − reliability) — **index of inconsistency**
3) The simple response variance, which is merely

$$\frac{1}{2N}\sum_i (Y_{i1} - Y_{i2})^2$$

where Y_{i1} and Y_{i2} are the answers in the interview and reinterview, respectively.

4) The gross difference rate, which for a dichotomous variable is merely twice the simple response variance — **gross difference rate**

None of these measures have distinct advantages over the others, as they are all arithmetic functions of one another. Different survey organizations tend to use different measures of consistency in reporting.

To provide an example of practical uses of such response error statistics, Table 8.4 shows indices of inconsistency for the NCVS (see Graham, 1984). For example, the index of inconsistency for the proportion of the respondents who reported a broken lock or window is 0.146. This corresponds to a reliability coefficient of (1 − 0.146) = 0.854, which is often considered a high level of reliabil-

Table 8.4. Indexes of Inconsistency for Various Victimization Incident Characteristics, NCVS

Question and Category	Index of Inconsistency	
	Point Estimate	95% Confidence Interval
6c. Any evidence offender(s) forced way in building? (multiple response item)		
Broken lock or window	0.146	0.094–0.228
Forced door or window	0.206	0.143–0.299
Slashed screen	0.274	0.164–0.457
Other	0.408	0.287–0.581
13f. What was taken? (multiple-response item)		
Only cash taken	0.276	0.194–0.392
Purse	0.341	0.216–0.537
Wallet	0.189	0.115–0.310
Car	0.200	0.127–0.315
Part of car	0.145	0.110–0.191
Other	0.117	0.089–0.153
7b. Did the person(s) hit you, knock you down, or actually attack you in any way? (single-response item)		
Yes	0.041	0.016–0.108
No	0.041	0.016–0.108

Source: Graham, 1984, Tables 59–60.

ity for survey responses. Note that the indices of inconsistency for the visible evidence and items taken are higher (lower reliability), on average, than the indices for whether the offender attacked the respondent (index of inconsistency = 0.041; reliability = 0.959). This reflects the fact that some types of attributes generated more consistent answers over trials than others.

Using Multiple Indicators of the Same Construct. Another approach to assessing reliability is to ask multiple questions assessing the same underlying construct. This approach is used most often to measure subjective states. This approach makes the following assumptions:

1) All questions are indicators of the same construct (i.e., their expected values are the same).
2) All questions have the same expected response deviations (i.e., their simple response variance or reliability is constant).
3) The measures of the items are independent (i.e., the answer from one item does not influence how the respondent replies to another).

Cronbach's alpha

Cronbach's alpha is a widely used measure of the reliability of such multi-item indices (Cronbach, 1951).

Section 8.9.1 described the five-question mental health index called the MHI-5 and the use of correlations of the index with other mental health indicators as a way of estimating its validity. The reliability of the index generated by combining answers to the items, as measured by Cronbach's alpha (α), depends on the number of items k and their average intercorrelation $\bar{r}$:

$$\alpha = \frac{k\bar{r}}{1+(k-1)\bar{r}}$$

For instance, suppose the correlations of each item with all the other items were as given in Table 8.5. The average of the ten correlations is 0.539, so that α is 0.85:

Table 8.5. Illustrative Intercorrelations among MHI-5 Items

Question	Happy Person	Downhearted, Blue	Nervous	Calm	Down in Dumps
Happy Person	–				
Downhearted, Blue	0.55	–			
Nervous	0.45	0.59	–		
Calm	0.62	0.51	0.54	–	
Down in Dumps	0.49	0.63	0.56	0.45	–

$$\alpha = \frac{k\bar{r}}{1+(k-1)\bar{r}}$$
$$= \frac{5(0.539)}{1+4(0.539)} = 0.854$$

A high value of Cronbach's alpha implies high reliability or low response variance. Unfortunately, it can also indicate that the answers to one item affected the responses to another to induce high positive correlation. A low value can indicate low reliability or can indicate that the items do not really measure the same construct.

The MHI-5 rests on the assumption that answers to each of the five questions reflect at least in part the same underlying construct. The answers reflect both the common construct and some item-specific variance because, for example, while "happy" and "calm" both are good emotional states, they are not identical. The evidence that the items reflect a common construct is the high level of intercorrelations among the five items.

Not all multi-item measures derive their validity in this way. An investigator can define a complex concept and then sample different examples of aspects of it that need not be related in any way. An example of this sort of multi-item scale comes from the Consumer Assessment of Health Plans (CAHPS) surveys, which are designed to measure patient experiences in getting health care. The CAHPS surveys use a number of multi-item composite measures designed to summarize patient reports of their experiences; the results are provided to people who are choosing health plans. One example is a composite called Getting Care Quickly, which consists of four questions asking patients how often they (1) got the help they needed when they called a doctor's office for help or advice, (2) got appointments as soon as they wanted them for routine care, (3) saw a doctor as soon as they wanted when they needed help right away for an illness or injury, and (4) saw a doctor within 15 minutes of their scheduled appointment time. Although all of these questions have a conceptual relationship to getting care right away, there is no particular reason to think that how well an office handles phone requests for information would be strongly related to how long patients wait in the waiting room to see a doctor. These items are put into the same index because the investigators grouped them

---

O'Muircheartaigh (1991) on Reinterviews to Estimate Simple Response Variance

O'Muircheartaigh (1991) examined data from reinterviews of the Current Population Survey (CPS) designed to estimate simple response variance.

Study design: About a week after CPS interviews, a different interviewer reinterviewed a 1/18 sample of respondents. Any eligible respondent reported the data on each occasion. The gross difference rate (GDR) is merely the likelihood of a discrepancy between the first and second report.

Findings: There are higher GDRs when persons report for themselves than about others, for reports about younger persons, and for reports made by nonheads of households.

Limitations of the study: The finding of higher response variance for self-reporters is limited by the fact that there was no random assignment of reporter status. Reporters tend to be those at home more often and thus tend to be unemployed. The study did not attempt to measure response biases, just instability over time. The reinterview is sometimes conducted by more senior interviewers using a different mode than the that of the first interview.

Impact of the study: The study identified systematic influences on stability of answers over repeated trials. It demonstrated the value of reinterview data when studying correlates of response variability in an ongoing survey. The finding about higher instability of self-reporters was plausibly explained by their tendency to undergo more change in their employment. This suggests lower reliability of reports for rapidly changing attributes.

together, not because they are measures of the same underlying process or phenomenon.

When the items are correlated, they do not correlate with one another at a particularly high level. The value of alpha, the measure of internal reliability described above, is only 0.58. However, the composite measure as a whole, formed by summing the responses to the four questions, turns out to be a highly significant predictor of how people rate their overall health care (0.57) and it was found to be highly reliable in providing a measure of this construct for health plans ($r = 0.94$). Thus, despite the fact that the items in the composite are not all measuring the same underlying construct, they do provide a reliable measure, one that is an important predictor of respondents' overall ratings of their health care.

In summary, both reinterviews and use of multiple items to measure reliability or simple response variance require assumptions that can be challenging to uphold. Nonetheless, both techniques are common and useful for survey researchers.

8.10 SUMMARY

Evaluating survey questions has two components. The first is determining whether the right questions are being asked, whether respondents understand them as intended and can answer them without undue difficulty, and whether the questions can be administered easily under field conditions. We described five methods for addressing these issues: expert reviews, focus groups, cognitive testing, field pretests, and split-ballot experiments. Almost all surveys use one or more of these methods in developing questionnaires. The different methods yield somewhat different information. The methods chosen for any particular survey are likely to depend on the specific issues of concern to the survey designers, the budget for the survey, and whether most or all of the questions have been used before. Unfortunately, we have little data on whether the assessments emerging from these techniques really pinpoint the most serious problems affecting the survey responses.

Statistical evaluations of survey questions measure the validity or reliability of the answers and the response variance and bias of summary statistics. There are two principal methods to estimate validity: comparing the survey responses to external data (such as records) or determining whether survey-derived measures follow theoretical expectations.

Reliability and simple response variance are commonly assessed by administering the same question to a respondent twice, once in the main interview and a second time during a reinterview. Another approach is to administer multiple items assessing the same construct in the same interview and to examine the consistency of the answers across the items.

Measuring response bias compares survey responses to some external data, either data for individual respondents or statistics on the target population.

Comparing survey responses to accurate external data generally provides the most useful estimates of the error in the survey data, but such external data are seldom available. (When they are available, it is often for special populations, such as the members of a particular HMO or persons who were hospitalized at a specific hospital, who may not be fully representative of the populations in which the questions will be used.)

The more indirect methods for assessing validity generally provide less-convincing evidence regarding the level of measurement error. Although it is reassuring to be able to say that the answers to the questions have the predicted relationships to other variables, that does not give a quantified estimate of the level of measurement error. Still, such evidence is often the best we can do to evaluate the quality of the answers to survey questions.

KEYWORDS

behavior coding
cognitive interviewing
cognitive standards
content standards
Cronbach's alpha
expert review
focus group
gross difference rate
index of inconsistency
interviewer debriefing
pretest

protocol analysis
randomized experiment
record check study
reinterview
reliability
response bias
simple response variance
split-ballot experiment
usability standards
validity

FOR MORE IN-DEPTH READING

Alwin, D. (2007), *Margins of Error*, New York: Wiley.

Harkness, J., Vijver, F., and Mohler, P. (2002), *Cross-Cultural Survey Methods*, New York: Wiley.

Presser, S., Rothgeb, J., Couper, M., Lessler, J., Martin, E., Martin, J., and Singer, E. (eds.) (2004), *Methods for Testing and Evaluating Survey Questionnaires*, New York: Wiley.

Saris, W. and Gallhofer, I. (2007), *Design, Evaluation, and Analysis of Questionnaires for Survey Research*, New York: Wiley.

Willis, G. (2005), *Cognitive Interviewing: A Tool for Improving Questionnaire Design*, Thousand Oaks, CA: Sage.

EXERCISES

1) Compare and contrast the benefits of three methods used in questionnaire development: cognitive testing, focus groups, and coding interviewer and respondent behavior during field pretests. By "benefits" is meant the acquisition of information about weaknesses in the design of the questionnaire that can be repaired prior to the main survey data collection. Name at least three ways the methods differ in their benefits.

2) Cognitively test the following questions with a friend or two, and think about how well you find they stand up to cognitive standards:

   a) What was your income in the past year?
   b) How often do you exercise—almost every day, several times a week, once a week, once a month, or less often?
   c) Do you favor or oppose universal health insurance in the United States?
   d) In the past year, about how many times did you use an ATM machine for any transaction?
   e) Consider the statement: "I am happier than usual these days." Would you say you strongly agree, generally agree, neither agree nor disagree, somewhat disagree, or strongly disagree with this statement?

3) Name two different ways to find out if a question designed to measure objective facts is producing valid data

4) Describe two different kinds of analyses one could do to assess the validity of answers to questions designed to measure a subjective state, such as happiness or anxiety.

5) For each of the items below, design two probes to be used in a cognitive interview that might help to identify any possible problems with the questions.

   a) During the past four weeks, beginning [date four weeks ago] and ending today, have you done any exercise, including sports, physically active hobbies, and aerobic exercises, but not including any activities carried out as part of your job or in the course of ordinary housework?
   b) How many times each week do you have milk, butter, or other dairy products?
   c) Living where you do now and meeting the expenses you consider necessary, what would be the smallest income (before any deductions) you and your family would need to make ends meet each month?
   d) Some people feel the federal government should take action to reduce the inflation rate even if it means that unemployment would go up a lot. Others feel the government should take action to reduce the rate of unemployment even if it means the inflation rate would go up a lot. Where would you place yourself on this [seven point] scale?

1	2	3	4	5	6	7
Reduce Inflation						Reduce Unemployment

   e) During the past 12 months, since [date], about how many days did illness or injury keep you in bed more than half the day? Include days while you were an overnight patient in a hospital.

EXERCISES

6) For each of the following scenarios, list one pretesting technique that would most directly address the problem at hand, and state why that technique would be useful.

   a) You are beginning to draft questions for a new survey, and you need to know how your target population thinks and talks about the survey topic—what words they use, how they define those terms, and so on. What technique should you employ?

   b) Your primary concern for a survey about to go into the field is that the interviewer–respondent interaction be as standardized as possible. You found during your own very informal testing (administering the questionnaire to a handful of coworkers) that the interaction was somewhat awkward—you were sometimes interrupted with an answer before you had read all of the response categories to the respondent, other times you were asked to repeat questions, and in some cases you were asked what certain words meant (although no standard definition was available). What would be the best way to address this concern in regard to standardization of interviewer–respondent interaction during a pretest?

   c) You are very concerned about potential comprehension and recall problems for a questionnaire you have been asked to finalize for data collection. You will ultimately conduct a large-scale "dress rehearsal" field pretest several months from now, but something must be done before then to improve the questionnaire. Some questions have ambiguous wording, and could easily be interpreted in different ways by different people; other questions ask for information that seems rather difficult to recall for most people. In addition, you suspect that there may be different problems for various subgroups of the population (e.g., those with low levels of education, those from different ethnic groups, etc.). What should you do?

7) You have very limited funds for pretesting, and your client has given you a questionnaire that he/she thinks is "ready to go." Upon looking at it, you see major issues with question wording, flow of topics, and so on. You only have enough money to conduct one field pretest, and you need to convince your client that the questionnaire needs revision prior to the field pretest. You have very limited time in which to address the problems with the questionnaire before starting the field pretest. What should you do?

8) You are developing a questionnaire that asks about complicated topics over long periods of time, and you suspect there may be different problems for various subgroups of the population (e.g., those with low levels of education, those from different ethnic groups, etc.). You've been asked to finalize the questionnaire for data collection. You will ultimately conduct a large-scale "dress rehearsal" field pretest several months from now, but something must be done before then to improve the questionnaire. What should you do?

9) You have two subject matter specialists in your organization who have each written what they consider to be the "best" question on a particular topic.

Only one of these can be included in the final version of the survey. Both people are adamant that their version is better, and you are about to conduct a fairly large-scale field pretest before starting the main survey. How could you address this situation in a way that would appease both of your specialists?

10) Suppose you have used a multi-item series of questions to measure how happy respondents have been in the past week. Each of the four items asks respondents to describe themselves using words with similar meanings, such as joyful, cheerful, and upbeat. The average correlation among the answers to the four questions is 0.60.

   a) What is the value of Cronbach's alpha for this four-item index?
   b) What does Chronbach's alpha measure?

11) Describe briefly two techniques used to measure response bias. What are the pros and cons of each?

12) You conduct a reinterview study in which a new interviewer asks respondents a sample of the questions from a survey two weeks after the original survey. For some questions, the results are almost identical; for others there are some notable differences between the answers that respondents gave to the same questions two weeks apart.

   a) What are the technical terms used for the degree of correspondence between the two answers?
   b) What are four reasons that answers might differ?
   c) What are the implications for the validity of the answers as measures when the answers are not highly consistent?
   d) What are the implications for the validity of the answers as measures when the answers are highly consistent?

CHAPTER NINE

SURVEY INTERVIEWING

When interviewers are used in surveys, they play a central role in the whole process. Consequently, they have great potential to affect survey costs and data quality, for better or for worse. The chapter describes what we know about how interviewers affect data and how design choices that researchers make can reduce or increase the amount of error in survey data attributed to the interviewers.

9.1 THE ROLE OF THE INTERVIEWER

Interviewers play several essential roles in surveys:

1) In household surveys using area probability samples, they build sampling frames by listing addresses.
2) Within selected units, they enumerate eligible persons and implement respondent selection rules.
3) They elicit the cooperation of sample respondents.
4) They help respondents to perform their roles in the interview interaction.
5) They manage the question-and-answer process, asking a set of questions, then asking follow-up questions or probes when respondents fail to answer a question completely.
6) They record the answers that are provided, often by entering them into a computer data file.
7) They edit the answers for correctness and transmit the edited data to the central office of the survey organization.

The roles of the interviewer in sampling and in gaining cooperation are extremely important ones. How well interviewers perform these aspects of their jobs can have important effects on coverage and nonresponse errors. Because enlisting cooperation is one of the most difficult tasks for interviewers to do in general population surveys, it is an important topic for both interviewer training and ongoing supervision. However, for the most part, material on these topics is found in previous chapters. The focus of this chapter is on how interviewers affect data quality once someone has agreed to be interviewed.

Earlier chapters distinguished two types of error in survey data: systematic deviations from a target value or fixed errors, called "biases;" and variances of estimates, reflecting the instability of estimates over conceptual replications. Studies have shown that interviewers can affect both kinds of errors of survey estimates.

9.2 INTERVIEWER BIAS

There are three sets of findings about the sources of systematic interviewer effects on responses:

1) Reduced reporting of socially undesirable traits due to the social presence of the interviewer versus self-administered reporting.
2) Altered reporting when observable traits of the interviewer can be related to the topic of the question.
3) Altered reporting as a function of interviewer experience.

9.2.1 Systematic Interviewer Effects on Reporting of Socially Undesirable Attributes

As we discussed in Chapters 5 and 7, many studies have found that the presence of any interviewer at all can produce biased responses. These studies usually compare data collected by interviewers with data collected by self-administration (e.g., Turner, Forsyth, O'Reilly, Cooley, Smith, Rogers, and Miller, 1998). Questions about sensitive behaviors (e.g., drug use) appear to be most affected by the presence of an interviewer. The theory that best explains these effects is that the "social presence" of the interviewer stimulates the respondents to consider social norms at the judgment phase of their response. The pressure to conform to norms then produces underreporting of socially undesirable attributes. We place these effects on the bias side of the error structure because they appear to be pervasive and systematic.

9.2.2 Systematic Interviewer Effects on Topics Related to Observable Interviewer Traits

There is another body of research that has discovered the influence of observable interviewer characteristics on respondent behavior in a set of restricted situations. These appear to occur when the observable characteristics of the interviewer are perceived as relevant to the question topic. In these situations, it appears that the interviewer characteristics help form the respondents' comprehension of the question as well as their judgment of appropriate answers.

For example, a 1946 study by Robinson and Rhode showed that when questions were asked about anti-Semitic feelings, interviewers with a common Jewish last name or prominent Jewish features decreased the rate at which answers suggesting anti-Semitic feelings or opinions were expressed to interviewers. Similarly, in the United States, African-American interviewers obtain fewer reports from white respondents of hostility or fear toward African Americans (Schuman and Hatchett, 1976). Schuman and Converse (1971) (see box on next page) report that many items dealing with racial relations were unaffected by the race of the interviewer, but those that were most closely related to how the respondents felt about blacks or whites tended to differ by the race of the interviewer. One often-cited study showed that teenagers answer differently depending on the age of the interviewer they are talking with (Ehrlich and Riesman,

1961). Mothers on welfare reported income levels more accurately to middle class interviewers, obviously from outside their neighborhoods, than they did to interviewers who were more demographically similar to them (Weiss, 1968). Both males and females report different gender-related attitudes to female interviewers than to males (Kane and Macaulay, 1993).

Every society tends to have some observable characteristics (e.g., through voice, appearance, or behavioral tendencies) that have social meaning. In the United States, it is observable racial characteristics; in Quebec, it is native language of the interviewer (Fellegi, 1964). When the survey questions asked are relevant to that observable characteristic, the respondents tend to use those characteristics to make a judgment about their answer. Answers tend to be biased toward reports that are judged as compatible with those of the interviewer. When the question topic is not relevant to the observable characteristic, there are no effects.

The above studies dealt with opinions and attitudes. As noted, the concept of biases is complicated, and possibly meaningless, when dealing with subjective phenomena. What can be said is that in some cases answers are systematically different based on the characteristics of an interviewer.

On the other hand, what is perhaps most striking about these studies is how many questions seem to be unaffected by the observable characteristics of the interviewers. It appears that for the most part interviewers are successful in establishing a context that is independent of their observable characteristics. It is only when the questions are directly and personally related to obvious interviewer characteristics that the context is systematically affected in ways that have a discernable influence on the data. Thus, if one is measuring attitudes or behaviors for which race, gender, or

Schuman and Converse (1971) on Race of Interviewer Effects in the United States

Schuman and Converse studied race of interviewer effects in Detroit, Michigan, in 1968, when racial tensions were high.

Study design: Twenty-five black professional interviewers and 17 white graduate student interviewers completed 330 and 165 interviews, respectively, with black heads of households or their spouses. Interviewer's race was randomly assigned to sample segments of about five households each. The interview's 130 questions asked about racial attitudes, work and life experiences, and background variables.

Findings: The majority of the questions (74%) showed no differences by race of interviewer. However, attitudinal questions with racial content consistently showed that more black respondents provided answers consistent with hostility toward whites to a black than to a white interviewer. Effects appeared somewhat larger for blacks of lower socioeconomic status.

Percentage Answering in Given Category by Race of Interviewer

Question	Category	Percent in Category	
		White Int'r*	Black Int'r*
Can trust white people?	Trust most whites	35%	7%
Negro parents work best with Negro teacher?	Yes	14%	29%
Favorite entertainers?	Named only black entertainers	16%	43%

*Int'r = interviewer.

Limitations of the study: Race of interviewer effects could be confounded with age and interviewing experience. The study did not examine variation in effects within interviewer race groups.

Impact of the study: Survey researchers learned to consider race of interviewer as a factor in the perceived intent of questions when they had some racial content.

age is likely to be highly salient, one might want to think about how interviewers will be assigned. In many cases, it is not clear whether interviewers who are similar to respondents will produce better or worse data. In the case of subjective measures, there may not be a standard for what constitutes data quality. In the face of ambiguity, the best approach may be to consider some kind of randomization of assignments of respondents to interviewers. Such a design will not reduce interviewer-related error, but it will make it possible to measure it.

9.2.3 Systematic Interviewer Effects Associated with Interviewer Experience

There is somewhat mixed evidence about the role of experience in how interviewers perform. Although experienced interviewers tend to have better response rates than inexperienced interviewers, it is virtually impossible to say whether that is an effect of experience or simply reflects the fact that interviewers who are not comfortable enlisting cooperation do not stay on to become experienced interviewers. Meanwhile, the evidence suggests that, if anything, experienced interviewers tend to be less careful than new interviewers about reading questions exactly as worded and following protocols (Bradburn, Sudman, and Associates, 1979; Gfroerer, Eyerman, and Chromy, 2002).

For example, in the NSDUH there is evidence that interviewers with experience obtain reports of lower lifetime drug use than those with no NSDUH experience. Table 9.1 presents results of reports using the paper self-administered questionnaires embedded in the earlier NSDUH on reporting any lifetime use. Overall, interviewers who had no prior NSDUH experience produced a 21 per-

Table 9.1. Percentage Reporting Lifetime Use of Any Illicit Substance by Interview Order by Interviewer Experience (1998 NSDUH)

Groups of Interviewers in Order Taken by the Interviewer	Percent Reporting Use of Illicit Substance during Lifetime		Ratio (1)/(2)
	(1) Interviewers with no NSDUH experience	(2) Interviewers with some NSDUH experience	
1–19	40.9%	35.5%	1.15
20–39	38.7%	32.6%	1.19
40–59	38.2%	33.7%	1.13
60–99	39.0%	32.1%	1.21
100+	43.2%	31.7%	1.36
Total	40.1%	33.1%	1.21
Ratio (1–19)(100+)	1.02	1.07	

Source: Hughes, Chromy, Giacoletti, and Odom (2002, Table 8.1).

cent higher prevalence of persons using illicit drugs in their life relative to interviewers who had conducted one or more NSDUH interviews prior to the 1998 study. The rows of the table allow us to see the comparison of the two interviewer groups for the first interviews they conducted in 1998 (the first through the 19th interview) and the very last interviews (the 100th or more interview). In every group defined by the order of the interview in the workload, the more experienced interviewers generate fewer reports of drug use. There also is some tendency for the reports to experienced interviewers to decline as interviews are being completed within an interviewer's workload.

Why would experienced interviewers generate such results? Keep in mind that the mode of data collection in the NSDUH is self-administration, so the interviewers are not asking the questions! Extensive multivariate analysis shows that the contrasts are not the effect of true differences in the characteristics of sample persons assigned. One possibility is that experienced interviewer behavior conveys to the respondent the acceptability of rushing through the questionnaire. If drug use is reported, follow-up questions are asked. Reporting no drug use reduces the time required to complete the instrument. (Later remedial training and use of computer-assisted self-administered techniques appeared to dampen the effect of interviewer experience.)

There is an earlier study that utilized hospital records as a way to measure the bias of survey answers (assuming no error in the records). In the survey, the number of households at which interviewers were expected to complete interviews varied widely from interviewer to interviewer. When the number of interviews completed by each interviewer was tabulated against the percentage of known hospitalizations that respondents reported, it was found there was a large correlation between the two (0.72). The respondents of interviewers who completed the most interviews reported much more poorly than did those whose interviewers had small assignments (Cannell, Marquis, and Laurent, 1977). Another more recent study also found that interviewers tended to get fewer mental health symptoms reported as they conducted more interviews (Matshinger, Bernert, and Angermeyer, 2005).

What are the mechanisms underlying these systematic effects of interviewer experience? More research needs to be done on these findings, but the most common hypothesis involves an unintended consequence of the reward system for interviewers. Ongoing feedback to face-to-face interviewers most often consists of evaluations of response rates and productivity indicators. Less common is feedback about the quality of interviewing affecting measurement errors. Given this, the hypothesis argues, experienced interviewers begin to emphasize respondent cooperation and cost efficiency, while giving less attention to the motivation of respondents to provide high-quality responses. If this hypothesis is true, one would expect fewer effects of this sort in centralized telephone survey data, and future research might focus on mode comparisons to discover the mechanisms underlying the finding.

9.3 INTERVIEWER VARIANCE

Although the systematic effects of interviewers above are predictable and replicated in many studies, surveys also experience effects on responses that vary across interviewers. "Interviewer variance" is the term used to describe that com-

interviewer variance

ponent of overall variability in survey statistics associated with the interviewer. Like notions of sampling variance, interviewer variance incorporates the conceptual model that each survey is but one realization of a large number of possible replications. Different interviewers might be used in different survey realizations. If different interviewers tend to produce different responses, values of survey estimates will vary over replications.

Much of the logic of how interviewers can increase the variability of survey estimates parallels the discussion in Section 4.4 about the effect of clustering on sampling variance. The assumptions for calculating standard errors from simple random samples include that each new observation, new interview, or survey return provides a new independent look at the population as a whole. In the case of clusters, if the people within clusters look more alike, on average, than the population as a whole, then additional observations within the same cluster do not provide as much information; they are not, on average, as much of a new independent look at the population as an observation that is not part of a cluster. In the same way, if interviewers affect the answers they obtain, getting additional interviews taken by that same interviewer does not provide as much new, independent information as would have been the case if another interviewer had done the interview or an interviewer who did not affect the answers had done the interview.

9.3.1 Randomization Requirements for Estimating Interviewer Variance

One practical problem of measuring interviewer variance is that there are two possible reasons that different interviewers would obtain different kinds of average responses in the groups of respondents they interview:

1) Interviewers influence the answers of respondents.
2) The respondents assigned to different interviewers have truly different attributes.

interpenetrated assignment

To estimate purely the effects interviewers have on respondents, it is necessary to remove the effects of real differences among respondents. "Interpenetrated sample assignments" are probability subsamples of the full sample assigned to each interviewer. Interpenetrated workloads permit the measurement of interviewer variance without the contamination of real differences across respondents assigned to different interviewers. Such a design assigns a random subsample of cases to each interviewer. Hence, in expectation, each workload should produce the same estimated survey statistics. Clearly, since each workload is a relatively small sample, there will be large sampling variability across the statistics computed on them. However, if the differences across interviewer workload statistics are larger than those expected from sampling variance alone, there is evidence of interviewer variance when interpenetration is used. The calculation of how much interviewers affect the variance of estimates resembles the approach used to calculate the effects of a clustered sample.

The nature of the interpenetration must parallel the source of interviewer variability that needs to be estimated. For example, in face-to-face surveys in a small community, interviewers are commonly assigned workloads that may take them to all parts of the city. To study interviewer variance in such a design, interviewers might be assigned simple random subsamples of the entire sample. This design measures interviewer variance among all interviewers. In a national face-to-face household survey, interviewers are generally hired from among residents of the primary sampling areas. Thus, each interviewer is commonly assigned cases in only one primary sampling unit (PSU). Interpenetrated designs in such surveys randomize sample segments (clusters of selection housing units) to interviewers working the same PSU, often randomly splitting a segment into two random pieces, each assigned randomly to an interviewer. This design measures interviewer variance among interviewers within the same PSUs. In a centralized telephone interviewing facility, as interviewers generally work different shifts (e.g., Monday, Thursday, Saturday, 3 PM–9 PM), at any one point all active sample numbers are available for assignment to all interviewers working the shift. Interpenetrated assignments in such facilities assign interviewers working the same shift a random set of numbers active during the shift. Such a design measures interviewer variance within shifts in the facility. In short, the interpenetration design should permit measurement of variation across interviewers who could (over replications of the survey) be assigned the same set of sample cases. For example, in the national face-to-face survey, interviewers living in New York would never be assigned sample cases in Los Angeles, so the interpenetrated design should not permit such assignments.

If the sample assignments are not interpenetrated, there is the risk that estimates of interviewer variance might be confounded with true differences among cases assigned to different interviewers. For example, some interviewers may be assigned cases that were disproportionately reluctant to provide the interview. Some interviewers may interview only persons of certain language groups or race groups. Whether these departures affect the nature of interviewer variability is generally not knowable. Only with a design that gives each interviewer equivalent assignment groups through randomization can the design hope to provide interpretable interviewer variance estimates.

9.3.2 Estimation of Interviewer Variance

The estimation of interviewer variance entails the application of a measurement model to the survey data. There are two principal measurement models: one introduced by Hansen, Hurwitz, and Bershad (1961), the other, by Kish (1962). The simpler and most commonly used model is Kish's, and we describe it here. Most simply, it starts with the measurement model we have been discussing in earlier chapters, $y_i = \mu_i + \varepsilon_i$, where μ (mu) is the true value and ε (epsilon) is the response deviation from the true value, inherent in the answer y. It notes that when an interviewer asks a question about an attribute μ, the response departs from the true value by a deviation that is common to all respondents questioned by the jth interviewer and the deviation idiosyncratic to the ith respondent. For the ith respondent, being questioned by the jth interviewer, the answer to the question about μ_i is

$$y_{ij} = \mu_i + b_j + \varepsilon_{ij}$$

or

Reported value = True value + Interviewer deviation + Respondent deviation where b_j is the systematic error due to the jth interviewer, a deviation from the true value μ_i for the ith person. In addition, the model specifies a random error term, uncorrelated with the interviewer effect, labeled ε_{ij}, which is thus an additional deviation that is associated with the respondent when interviewed by the jth interviewer. This same model might be used for interviewer bias as well, which would propose that the expected value of the b_j would be nonzero (i.e., the value of the bias). The most common assumption in the interviewer variance literature is that the expected value of the b_j is zero.

Interviewer variance arises when b_j varies across interviewers. One way of measuring this is to estimate how the total deviations from the true values cluster among the respondents interviewed by the same interviewer. The intraclass correlation (a measure of the correlation of response deviations among respondents of the same interviewer) is used for this purpose.

The strategies for estimating the intraclass correlation, ρ_{int} (rho), and the complexities thereof, have been extensively discussed by Kish (1962), Biemer and Stokes (1991), and Groves (1989). The approach proposed by Kish (1962) has the advantage that it uses estimators from simple analysis of variance. The basic approach to estimating ρ_{int} is

$$\rho_{int} = \frac{\left(\dfrac{V_a - V_b}{m}\right)}{\left(\dfrac{V_a - V_b}{m}\right) + V_b}$$

where

V_a is the between-mean square in a one-way analysis of variance with interviewer as the factor
V_b is the within-mean square in the analysis of variance
m is the total number of interviews conducted by an interviewer

Expected values of ρ_{int} range from $-1/(m-1)$ to 1.0, and estimation with small numbers of interviewers or cases sometimes yields estimated values slightly below 0. A value near 0.0 means that answers to a particular question are unaffected by the interviewers; in our terms, the interviewers obtain the same results (within sampling variability). Higher values of ρ_{int} indicate statistics for which the impact of interviewers on the answers and the amount of interviewer-related error is correspondingly higher. Although the value of ρ_{int} is unrelated to the assignment size of the interviewers, the significance of ρ_{int} for the total variance of a statistic that is interviewer-related is proportionate to the average number of interviews taken per interviewer. The equation for calculating that is

$$\text{interviewer design effect} = \textit{deff}_{int} = 1 + \rho_{int}(m-1)$$

where $\textit{deff}_{int}$ is the extent to which variances are increased from interviewer variability in results and m is the average number of interviews per interviewer. These calculations are only appropriate when assignments to interviewers are randomized in some way so that the effects of the interviewer can be disassociated from the characteristics of the samples assigned to interviewers.

If a simple random sample were drawn, the $deff_{int}$ is merely the inflation of the simple random sample sampling variance of the mean of y that would arise from the interviewer variability. For example, if ρ_{int} is 0.02 and the number of interviews per interviewer is 101, then $deff_{int} = 1 + 0.02(101 - 1)$ or 3.00. This is a 300% increase in the variance of the sample mean or a 73% increase in the standard error of the sample mean. This would imply that an increase in the width of confidence intervals of 73% ($\sqrt{3} = 1.73$).

Groves (1989) calculated an average value of ρ_{int} of about 0.01 across all questions for the various studies he was able to identify for which values had been calculated. With interviewer workloads of 41 interviews each, this average ρ_{int} translates to a $deff_{int}$ of 1.4 or a 40% increase in the variance of a sample mean ($\sqrt{1.4} = 1.18$ or a 18% increase in the standard error). In a survey with larger workloads, say each interviewer completing 100 interviews, the $deff_{int}$ would be 2.0, leading to a 41% increase in standard errors ($\sqrt{2} = 1.41$) The moral is that even small ρ_{int}'s when combined with large workloads can produce major increases in variances of sample statistics.

In any given survey, the extent to which interviewers affect answers will vary from question to question. Questions that ask the interviewers to depart from a specified script appear to be more subject to interviewer effects (e.g., rate of probing, number of words used in open responses). Further, some respondent types appear to be more sensitive than average to interviewer effects (e.g., the elderly). As seen above, under the model, the significance of any particular value of ρ_{int} for the precision of survey estimates depends importantly on the average number of interviews per interviewer.

Kish (1962) on Interviewer Variance

Kish developed a simple estimator for interviewer variance, demonstrated in two face-to-face surveys.

Study design: Twenty male interviewers received one week of training for a survey of auto plant workers. The questionnaire asked about attitudes toward foremen, stewards, the union, management, and various job features. A stratified random sample of 462 workers was assigned at random to the interviewers for interviews in the respondents' homes. A single factor analysis of variance (with interviewer as the factor) was used to compute ρ_{int}.

Findings: ρ_{int} values ranged from 0.00 to 0.09 over 46 variables. With average workloads of 23 interviews, this implies design effects from interviewer variance from 1.0 to 3.0. The table shows some of the variables subject to large interviewer variability.

Design effects reflecting interviewer variance for means on the total sample were higher than those on subclass statistics (e.g., new workers to the plant), following the model $1 + \rho_{int} (m - 1)$. This reflects the fact that m, the number of interviews per interviewer, is smaller for subclasses than for the total sample.

Statistics Showing High Values of ρ_{int}		
Question/Statistic	ρ_{int}	$deff_{int}$
How much should you be earning?	0.092	3.02
>1 reason for not liking company	0.081	2.78
<2 reasons for not liking work he does	0.068	2.50
Total number of criticisms of work	0.063	2.36

Limitations of the study: There were only 20 interviewers in the study, producing rather unstable ρ_{int} values (a 95% confidence interval excluded 0.00 only with ρ_{int} of 0.03 or higher). The survey conditions were atypical of many household surveys.

Impact of the study: The study showed that large increases in the instability of survey estimates were possible from interviewer effects, but that many variables were not sensitive to interviewer influences.

9.4 STRATEGIES FOR REDUCING INTERVIEWER BIAS

Practical tools that act on interviewer bias, if successful for all interviewers in a survey, can also reduce interviewer variance. Hence, the distinction of which tools focus on bias and which on variance is somewhat arbitrary. This section describes some interventions into interviewer behavior that have been demonstrated to reduce the bias of overall survey estimates.

9.4.1 The Role of the Interviewer in Motivating Respondent Behavior

When an interview begins, most respondents have little idea how they are expected to perform. It is up to the interviewer to define what is expected of the respondent. We cited studies in Section 9.2.3 that showed declines in reporting of drug use, known hospitalizations, and psychiatric symptoms associated with interviewer experience and the size of interviewer assignments. One strong hypothesis for why these results occur is that experienced interviewers, and interviewers with more to do, communicate lower expectations for respondent performance.

Another study makes this point even more clearly. This study also had available records about hospitalizations that people in the survey had experienced in the prior year. The measure of data quality was the percentage of known hospitalizations reported by respondents. Interviewers collected data about hospitalizations in two ways. In half of their assigned households, they conducted a complete health interview, including asking questions about hospitalizations. In another set of households, the same interviewers conducted most of the same health interview, but they left a form for respondents to fill out on their own and mail back; that form contained the questions about hospitalizations. When researchers calculated the percentage of known hospitalizations reported by respondents, it was found that the quality of reporting was highly related to who did the interview, regardless of whether respondents reported directly to the interviewer or used the self-administered form. Those interviewers whose respondents reported a high percentage of their hospitalizations to the interviewer directly also had respondents who reported a high percentage of their hospitalizations when they filled out a self-administered form after the interviewer had left (Cannell and Fowler, 1964).

These studies all support the conclusion that interviewers somehow play a motivational role, and that some interviewers are more effective than others at motivating respondents to perform at a high level. The processes by which this occurs emerged from research reported by Fowler and Mangione (1990). Respondents were interviewed again the day after they participated in a health survey interview about the experience. One of the questions that was asked was whether they thought that the interviewer wanted "exact answers" or "only general ideas." In a parallel activity, interviewers filled out questionnaires about their job priorities. Two priorities that were measured were the interviewer's "concern about accuracy" and the interviewer's "concern about efficiency." When the data were analyzed, it was shown that respondents who said that the interviewer wanted "exact answers," rather than "general ideas," were more likely to score high on an index of reporting quality. Most important, respondents' perceptions on this score were significantly related to interviewer priorities. For those interviewers for whom accuracy was the highest priority, respondents were more likely to report that the interviewer wanted "exact answers." In contrast, when the interviewer was someone who reported efficiency was a high priority, their respondents were more likely to say that the interviewer only wanted "general ideas."

These studies taken together show that the interviewers play a crucial role in setting expectations for respondent performance. In turn, what respondents think is expected of them has a significant effect on the quality of the data they provide. Hence, one major source of interviewer differences related to bias in data is the extent to which they communicate high standards to respondents.

9.4.2 Changing Interviewer Behavior

Building on this work, Charles Cannell launched a series of studies to try to reduce the differences between interviewers in the goals they set for respondents. In his work, he tried five different approaches to standardizing the messages that interviewers sent to respondents (Cannell, Marquis, and Laurent, 1977).

Slowing the pace with which the interviewer spoke was one of his first experiments. Preliminary studies indicated that the speed with which interviewers spoke might be related to how respondents thought they were supposed to approach the interview. It is understandable that if interviewers speak very quickly, respondents can infer that going quickly is an interviewer priority. Cannell emphasized the importance of interviewers speaking slowly, and experimented with trying to slow down the pace with which interviewers read questions. This particular intervention was not clearly demonstrated to have a positive effect on data quality.

Experiments with modeling good behavior, however, appeared to be more promising. Henson had interviewers play an audio tape at the beginning of interviews, demonstrating interviewers and respondents speaking slowly and taking great care in answering questions (Henson, Roth, and Cannell, 1977). The respondents who were exposed to the tape seemed to provide better data in some ways than those who were not.

"Systematic reinforcement" showed still more promising results. In these experiments, Cannell programmed interviewers to provide feedback of a positive nature when respondents behaved in ways that were consistent with accurate reporting: asking for clarification, taking their time when answering questions, and providing full and complete answers. In contrast, when respondents answered quickly or after apparently little thought, interviewers provided negative feedback by encouraging respondents to think more carefully. In some experiments, reinforcing appropriate behavior seemed to improve the quality of data.

"Programmed instructions" provided an easier and more direct way to standardize the messages interviewers sent about expectations for performance. In these experiments, interviewers were given a standard introduction to read that emphasized the importance of taking time, thinking about answers, and providing accurate data. The same messages were repeated periodically throughout the interview. When the interviewers read these instructions, there is evidence that the quality of data improved.

"Asking respondents to make a commitment" was perhaps the most innovative of Cannell's experiments. In these studies, after respondents had agreed to the interview and answered a few questions, interviewers stopped the interview process and asked respondents to sign a commitment form. The form essentially committed respondents to provide the most accurate information they could. Respondents were told they could not complete the interview unless they signed the commitment form.

There was concern, when this protocol was first tried, that many respondents would refuse to sign, and the response rates of the study would decline rapidly. However, that proved not to be the case. Almost all respondents agreed to sign the commitment form, and the quality of the resulting data proved to be consistently better for those who signed the commitment form than for those who were not asked to do so.

These techniques no doubt affect respondents in two ways. First, they change their understanding of what is expected. Second, they generate a willingness to perform better by reporting more accurately. Interviewers who set high standards for respondent performance consistently obtain better data; that is, data with less underreporting and, thereby, less systematic error than those who do not communicate such standards. Efforts to standardize interviewers in this respect, so that they are more consistent in communicating high standards, are effective strategies for reducing interviewer-related bias in reporting. Two studies demonstrate these results. Cannell, Groves, Magilavy, Mathiowetz, and Miller (1987) integrated these techniques into a phone survey, compared the results with estimates from the in-person National Health Interview Survey, and found evidence that more events and conditions were reported when these techniques were used. Miller and Cannell (1977) demonstrated that these techniques can also produce less reporting of events or behaviors and that less reporting is probably more accurate. For example, respondents exposed to the experimental procedures reported watching more television but reading fewer books than controls.

Despite a body of evidence that the techniques that Cannell developed are constructive, they are seldom if ever used in routine data collection. On the one hand, they do take some extra time, but not much. Response error is not easy to observe in the average survey. Hence, the value of efforts to improve data quality may not be evident to most investigators or users of data. Nonetheless, the way interviewers orient respondents to the survey task does affect data quality. When thinking about survey interviewing, and how to minimize interviewer-related error, it is a concept that deserves more attention.

9.5 Strategies for Reducing Interviewer-Related Variance

In order to make interviewers as consistent as possible, and to minimize the effect that individual interviewers have on the answers they obtain, there are three general areas to which researchers can attend: the questions they give them to ask, the procedures that they give interviewers to use, and the way they manage interviewers.

One of the most effective ways to reduce interviewer variance is to create questions that do not require the interviewers to vary their behavior over respondents. The variation of importance here concerns clarifying questions and probing inadequate answers.

In addition, there are five areas of interviewer behavior that are generally thought to be important to standardizing the data collection process:

1) Interacting with the respondent in a way that is professional, task oriented, and that minimizes the potential of respondents to adhere to or infer preferences for the kind of answers that are obtained
2) Reading questions exactly as worded
3) Explaining the survey procedures and question-and-answer process to the respondent
4) Probing nondirectively; that is, in a way that does not increase the likelihood of one answer over others
5) Recording answers that respondents give without interpreting, paraphrasing, or inferring what respondents themselves have not said

9.5.1 Minimizing Questions that Require Nonstandard Interviewer Behavior

Questions have a predictable, reliable effect on the interaction between interviewers and respondents. The clearest demonstration of this was a study in which the same instrument was administered in parallel by interviewing staffs at the University of Massachusetts, Boston, and the University of Michigan (Fowler and Cannell, 1996). The interviews were tape recorded, with respondent permission, and the behaviors of respondents and interviewers were coded on a question-by-question basis. The rates at which interviewers read questions as worded were highly correlated between these two staffs across the different questions. Moreover, the rates at which respondents interrupted the questions while they were being read and the rates at which they gave inadequate answers, which required additional probing on the part of the interviewers, were also highly related. These results mean that individual questions have highly predictable effects on the way that interviewers read them and that respondents answer them. Some questions are very likely to be misread, whereas others are not. Some questions are very likely to be answered immediately, with no further interviewer probing, whereas other questions are likely to require clarification and probing to yield an adequate answer.

The importance of this finding was demonstrated in a paper by Mangione, Fowler, and Louis (1992). They studied the relationship between behaviors that were associated with questions and the interviewer-related error for those questions. Their clearest finding was that the more often interviewers had to probe in order to obtain an adequate answer, the greater the interviewer effects on the results. A second finding was that questions answered in narrative form, which are particularly likely to require interviewer probing and also pose recording challenges for interviewers, were also more likely than average to be subject to interviewer effects. These results, taken together, suggest that one of the important ways to minimize interviewer-related error is to design good questions. The standard for a "good question" in this context is one that minimizes the need for interviewer probing in order to get an adequate answer. Characteristics of questions that accomplish that include:

1) Questions that are clearly worded, so that respondents can understand them the first time that they are read.
2) Questions for which the response task is clear. One of the most common reasons that questions require probing is that the question itself does not communicate to respondents how they are supposed to answer.

For example:

Question: "When did you last go to the doctor for your own health care?"

Comment: This question does not specify the terms in which the respondent is supposed to answer.

Alternative Answers: "Last June"
"About 4 months ago"
"Right after my pregnancy"

These are reasonable answers to the above question that technically meet the question's stated requirements. However, answers need to be in the same form across respondents in order to be analyzed. As a result, the interviewer is somehow going to have to intervene to explain to the respondent which of the three kinds of answers that seem acceptable is the one that is acceptable in this survey.

The question also does not specify how exact an answer is desired. A common reason for interviewer probing is that respondents answer more generally than researchers want them to. Specifying a level of precision desired can eliminate that need for interviewer probing. For example:

Question: "What are the best things about living in this neighborhood?"

Comment: In addition to the extra probing that will be required because the question calls for a narrative answer, an ambiguity in the question is how many different points or issues are appropriate for respondents to raise. Groves and Magilavy (1980) found that one of the biggest differences between interviewers that was associated with interviewer effects was the number of points that were made in response to questions of this sort.

At the moment, we do not have a full list of question characteristics that produce a need for nonstandard interviewer behavior. However, because of the strong link between the rates at which questions are probed and the effects of interviewers on results, a clear strategy for improving survey data is to do good pretests of survey instruments, using behavior coding (discussed in Section 8.5). During pretests, questions can be identified that are particularly likely to require interviewer probing, and an effort can be made to improve those questions. Reducing the need for interviewer probing, and other innovative actions, is one of the important ways to minimize the interviewer variance.

9.5.2 Professional, Task-Oriented Interviewer Behavior

From the early days of survey interviewing, researchers have felt that it is important for an interviewer to establish the right kind of relationship with respondents that would allow them to feel free to answer the questions accurately and completely. It has long been acknowledged, however, that rapport was potentially a two-edged sword. On the one hand, a positive relationship with the interviewer could be a key motivating force that would make respondents want to contribute effectively to a research interview. On the other hand, to the extent that the relationship between interviewer and respondent is more personal and less professional, one could conceive of a distortion of the goals of the interview, which in turn could lead to distortion of answers. Van der Zouwen, Dijkstra, and Smit (1991) experimented with training interviewers to be more personal or more formal, and their results were inconclusive.

We have noted that the reporting of personal or sensitive data, such as drug use or health problems, tends to suffer when an interviewer is involved compared with when questions are answered via self-administration. This suggests that the relationship with the interviewer, even the kind of relationship established via the telephone, is a hindrance to the reporting of such information.

There is some evidence from studies reported by Fowler and Mangione (1990) that some level of tension-breaking interaction when the interviewer has a higher educational status than the respondent may be constructive to reporting, but no such evidence exists when respondents are equal or higher in status than interviewers. The issue of rapport becomes particularly interesting in the context of the many interviews that are conducted by telephone. Telephone interviews are shorter, there is less interaction between respondents and interviewers, and respondents obviously have less potential to have information about interviewers and to relate to them personally than when interviews are done in person. Hence, on the telephone the potential for rich rapport to develop is more limited.

Overall, there is a lack of systematic data on the topic of what rapport is, how much is ideal, and how it affects respondent performance and data quality. In an attempt to deal with this, survey organizations tell interviewers to try to be warm and professional at the same time. Operational implications of this goal are:

1) Interviewers should refrain from expressing any views or opinions on the topics covered in the survey instrument.
2) Interviewers should refrain from presenting any personal information that might provide a basis for inferring what their preferences or values might be that are relevant to the content of the interview.
3) Although a little informal chatting about neutral topics, such as the weather or pets, may help to free up communication, for the most part, interviewers should focus on the task.

These guidelines are fairly vague, and the evidence for their importance in producing standardized data is not strong. Nonetheless, at the moment it is still part of the litany of things that most survey organizations are hopeful that interviewers will do to reduce their effects on the data.

9.5.3 Interviewers Reading Questions as They Are Worded

Uniform wording of questions asked is perhaps the most fundamental and universally supported principle of standardized interviewing. There is bounteous evidence that small changes in question wording can affect the answers that respondents give. The easiest way to demonstrate this is to ask comparable samples questions that are similar, with apparently the same objective, but which differ slightly in the way they are worded:

Question A: Should Communists be allowed to speak in public places in the United States?

Question B: Should Communists be forbidden to speak in public places in the United States?

Even though the concepts of "forbidden" and "not allowed" seem nearly identical, when comparable samples were asked these two questions, 50% said Communists should not be allowed to speak in public, whereas only 20% would forbid it. This finding clearly indicates that if interviewers were to be permitted to reword questions from "not allow" to "forbid," it would have a major effect on the survey results (Schuman and Presser, 1981). Rasinski (1989) found that respondents were much more willing to support spending for "aid to the poor" than for "welfare" and for dealing with drug "addiction" than with drug "rehabilitation." Although from one perspective these are equivalent terms, they clearly evoke different connotations among survey respondents.

It is not always the case that small, or even relatively large, changes in question wording will have a big effect on answers. For example, in a similar experiment, Schuman and Presser (1981) found that substituting the term "abortion" for "end a pregnancy" had no effect on answers.

Two other kinds of studies call into question the importance of exact reading of questions. When Groves and Magilavy (1980) studied the relationship between how well interviewers read questions and interviewer-related error, they found no consistent relationship. Fowler and Mangione (1990) reported that questions that are misread at higher than average rates are not subject to higher levels of interviewer-related error.

Having said that, it is still well documented that changing question wording can have potentially big effects on survey results. Giving interviewers freedom to reword questions is generally seen as an invitation to have large amounts of interviewer-related error. If interviewers are asking questions that are worded differently, it is almost certain that sometimes these wording differences will have important effects on the answers and the results.

9.5.4 Interviewers Explaining the Survey Process to the Respondent

Training interviewers to explain the survey process to the respondent is another key skill that is central to the attempt to minimize interviewer variance. This is also one of the less understood and somewhat controversial aspects of what ideal interviewer procedure should be.

When researchers study survey interactions, they often find that there are awkward interactions that result from the standardized interview protocol (Suchman and Jordan, 1990; Houtkoop-Steenstra, 2000; Schaeffer, 1991). Consider the following:

Question: "How would you rate your child's school: excellent, very good, good, fair, poor?"

Answer: "Well, it depends on what you mean. My child is in the second grade, I like her teacher, and it is a very positive environment. I really don't think they are doing very much with math and reading. On the other hand, she is happy, she likes recess and playing."

Comment: The interviewer has a dilemma. The respondent has raised a legitimate question about the question. There is nothing in the question that specifies which aspects of the school she should be rating. Like many people, there are some things about the school she likes, others she does not; what she focuses on will affect what her answer is. This is an occasion when the interviewer needs to train the respondent; by this we mean the interviewer needs to explain to the respondent how the survey process works. In this case, a good answer might look something like the following:

Interviewer: "That is a very legitimate point. The question does not suggest that you focus on any one thing. In this case, you should take into account whatever it is that you think the question implies and give me the answer that is closest to what you think."

Question: "Overall, how would you rate the school your child attends: excellent, very good, good, fair or poor?"

Answer: "I'd say 'not so good.'"

Interviewer: "I'm going to ask you to pick one of the answers I read: excellent, very good, good, fair, poor. I know this may seem artificial, and it is possible that none of those answers exactly captures the way you feel. However, the way a survey works is that we ask people exactly the same questions and have them choose from exactly the same answers. When we do it that way, then we can meaningfully compare the answers that different people give. If we change the questions, or we have people choose their own words, then it gets much harder to actually compare what different people say."

Comment: These explanations are important in creating a standardized interview process. An interview is an unusual form of interaction, one that respondents seldom experience. The fact that interviewers have to ask the same questions, even though they do not seem worded in exactly the right way to fit the respondent's circumstances, can seem artificial or even incompetent. In the same way, when respondents have to choose among answers that do not describe what they have to say very well, it seems as if good communication is being undermined (Schaeffer, 1991; Houtkoop-Steenstra, 2000). Addressing these perceptions directly is a way to make the interview work as it should.

There are three possible approaches to dealing with problems such as these. Writing clear questions is very helpful. There are also those who favor giving interviewers more flexibility to help respondents directly in how to interpret or answer questions. However, one of the best and standardized approaches is to teach interviewers to train respondents.

9.5.5 Interviewers Probing Nondirectively

probe

When a respondent fails to answer a question completely and adequately, interviewers have to do something. The words the interviewer uses to attempt to obtain an adequate answer are called "probes." The standard instruction to interviewers is to probe in a nondirective way. What that means in principle is to probe in a way that does not affect which answer is chosen. Because probing is such a critical part of the interview process, we provide below some examples of the issues that are at stake.

The type of answer required has an effect on the probing challenges confronting an interviewer. If a question is in closed-ended form, which asks respondents to choose from a set of alternatives that are provided, an inadequate answer results anytime the respondent fails to choose one of the answers provided. In such cases, an appropriate probe is to repeat the question entirely or perhaps to repeat the response alternatives.

If a question calls for a numerical answer, then an inadequate answer would be either that the respondent did not answer in numerical terms or provided a range that was wider than seemed appropriate. In that case, the goal of probing is to get respondents to answer in the appropriate terms or to be more exact in their answers.

For open-ended questions, there are three different ways that respondents may answer inadequately:

1) They may not answer the question that was asked or they may answer some other question:

   Question: "What do you think is the biggest problem facing the United States?"

   Answer: "I think there are a lot of problems that are important for the government to address."

   Comment: This clearly does not answer the question. Probably the best probe at this point is to simply repeat the question. That would certainly be a nondirective probe, since it would not change the stimulus in any way.

2) The respondent can answer the question, but the answer can be too vague, so that the result is unclear:

   Question: "What do you think is the biggest problem facing the United States?"

   Answer: "The crime problem."

Comment: Potentially, that is an answer to the question, but the comment is an extremely broad response. It could include white collar crime, crime on the streets, crime in the high places. Most researchers would want to know more about what the respondent had in mind.

Possible probes: "Could you tell me more about that?" or "What do you mean by crime?"

Comment: Both of these probes stimulate the respondent to elaborate and provide more detail. In neither case is there anything about the probe that would increase the likelihood of one kind of an answer over another.

3) Sometimes, questions that are to be answered in narrative form ask respondents to provide several possible answers if they have them. In the example above, the question could have been open ended about the number of problems to be reported:

Question: "What are the most important problems facing the United States today?"

Comment: After the respondent had given the "crime" answer, the interviewer might want to probe for additional possible problems. In that case, once a complete and understandable answer describing a first problem has been given, the best probe would be to use "Anything else?"

4) Contrast the nondirective probes outlined above with some alternative ways of probing that might seem helpful in getting answers but would generally be considered directive.

For an example of a fixed-choice question, consider the following:

Question: "How would you rate your health: excellent, very good, good, fair, or poor?"

Answer: "It hasn't been very good lately."

Directive probe: Well, which answer is closest to your view, fair, or poor?

There are two reasons why most methodologists would consider that to be a poor probe. First, and probably most important, an interviewer has concluded that the answer will be one of the two bottom categories, whereas the respondent has not actually chosen or indicated that is the choice. Although the bottom categories may seem consistent with the answer the respondent gave, the probe essentially limits the response to those two options, which may be inappropriate. Second, there is good evidence that presenting the respondent with all five options affects the respondent's sense of what the options mean. For example, if the response task had only had three categories—good, fair, or poor—the alternative "fair" would almost certainly be more popular because it would be seen as more positive than if it were the fourth response out of five. Given that information, asking the respondent to make a choice when leaving out the top three categories may significantly change the way the respondent sees the alternatives. For both of those reasons, the best approach would be to read the entire question, or at least to read all of the response alternatives in order to get the respondent answer:

> Question: "In the last 12 months, how many times have you been to the doctor's office for medical care for yourself?"
>
> Answer: "Several times."
>
> Directive probe: "Would you say about four times?"
>
> Nondirective probe: "Would you say fewer than four, or four or more?"

In response to the question about the major problem, consider the following:

> Answer: "The crime problem."
>
> Directive probe: "Do you mean like murders and robberies?"
>
> Comment: A directive probe is one that increases the likelihood of some answers over other possible answers. One sure sign of a directive probe is if it can be answered in a "yes" or "no" form. In the last two examples, the directive probe shows the interviewer guessing at the answer intended by the respondent. When the interviewer does that, and asks the question in the yes/no form, it does two things. First, it provides the respondent with a very easy way to provide an answer—to not have to do any work by, for example, trying to remember exactly how many times he or she has been to a doctor's office. Second, interviewers may inadvertently be communicating that they have a preferred answer or that they think that a particular answer is the correct answer. Respondents are always looking for cues to what the interviewer wants. A directive probe can be an open invitation for a respondent to give the interviewer what she seems to want.

Studies have shown that probing well, particularly for questions to be answered in narrative form, is probably the most difficult skill for interviewers to learn (Fowler and Mangione, 1990). It also has been shown that questions that require probing at a high rate are those most likely to be affected by interviewers. As the examples above show, probing is a complex process, and some probing errors are no doubt worse than others. However, the evidence is fairly solid that probing well, and avoiding directive probing, is perhaps the most important interviewer behavior that can be associated with interviewer effects on the variance of estimates (Mangione, Fowler, and Louis, 1992).

9.5.6 Interviewers Recording Answers Exactly as Given

Recording answers exactly is the final aspect of the interviewer's behavior that is part of producing standardized interviews. The interviewer's task in this respect differs across types of questions. When Dielman and Couper (1995) studied recording errors, they found that errors occurred at quite a low rate, and they are lower when answers were entered with a computer than recorded on paper (Lepkowski, Sadosky, and Weiss, 1998).

When questions offer fixed alternatives, interviewers are instructed to get the respondents themselves to choose an answer; interviewers are not supposed to infer which answer respondents probably would want to give:

Question: "How would you rate your health: excellent, very good, good, fair, or poor?"

Answer: "My health is fine."

Comment: At this point, the interviewer is not supposed to infer that that means "good" or "very good"; the interviewer is supposed to repeat the question and the alternatives, and get the respondent to pick an answer.

If respondents are to answer in narrative form, interviewers are supposed to record answers in as near verbatim form as possible. Studies have shown that when interviewers paraphrase or write summaries, they do so differently, in ways that affect the results (Hyman, Cobb, Feldman, and Stember, 1954).

There are survey organizations that ask interviewers to "field code" narrative answers; that is, they ask questions in open-ended form, without providing response alternatives to respondents, then ask the interviewers to classify the narrative answer into a set of categories (see Section 10.2.3). Sometimes, those categories are well structured and this is not a complex task. For example, a question might ask for the number of times the respondent has done something, and the response alternatives might be some groupings of numbers. If the interviewer can get the respondent to give an exact number, or a range that falls completely in one of the categories, there is virtually no discretion in the classification, and it is an easy task to do.

On the other hand, if a question is not well structured, and the categories interviewers are supposed to use are complicated, field coding can be an error-prone activity (see Houtkoop-Steenstra, 2000). Some organizations try to minimize the extent to which interviewers are asked to code narrative answers into categories; they prefer to have interviewers record answers verbatim, then have the answers coded as a separate step by coders who can be supervised and checked. Other organizations are more comfortable with having interviewers make these classification decisions. Interviewers usually receive little training in coding classification; their coding decisions are essentially unsupervised (Collins and Courtenay, 1985). Such questions constitute a very small part of most surveys. However, a good general principle is that minimizing interviewer discretion is likely to reduce interviewer error. Following that, minimizing coding decisions that interviewers make is probably also a good idea.

9.5.7 Summary on Strategies to Reduce Interviewer Variance

Most researchers attempt to make interviewers consistent across all the respondents, and consistent with other interviewers, by giving them a set of procedures designed to minimize their idiosyncratic effects on the data they collect: reading questions as worded, probing nondirectively, giving respondents appropriate training and explanations, managing their interpersonal behavior, and minimizing discretionary recording of answers. However, there is some controversy about whether these procedures are the best for collecting valid data.

9.6 THE CONTROVERSY ABOUT STANDARDIZED INTERVIEWING

The goal of having all people answer exactly the same questions, interpreting them in exactly the same way, under as consistent conditions as possible, seems the right one for maximizing consistent measurement. This is a key principle of "standardized interviewing." Such a goal places high value on replicability of findings. "Replicability" is a property of scientific findings that permits another scientist to obtain the same results, using the same methods, in an independent trial of the study. Replication requires a detailed description of the methods used so that another scientist can repeat them. Being able to say with a high level of confidence what questions were asked and how they were administered is critical to making surveys replicable.

standardized interviewing

replicability

On the other hand, critics have pointed out several concerns about standardization of interviewer behavior:

1) Exposing people to the same words does not necessarily mean they are being exposed to the same meanings. Indeed, having a question that means exactly the same thing to everyone is an unachievable ideal. Possibly, giving interviewers some flexibility to adapt the wording to respondents to make the questions fit individuals better could help to address this concern.
2) A standardized interview is an unnatural interaction; it is not normal conversation. At its worst, it creates cumbersome and redundant interactions. If interviewers had more flexibility to adapt the question asking to the situation, they could create a more natural interaction, one that was likely to be more comfortable to respondents and might produce better data.
3) The straitjacket of standardization seems particularly inappropriate when an interviewer can see clearly that a respondent is misunderstanding a question. Although the prescription for writing better questions is a good idea, and one that should be pursued, inevitably, questions will be imperfect and respondents' understanding will be imperfect. In those cases, the data would be more accurate and more valid if interviewers could improvise interventions to correct misunderstandings that they perceive.

There obviously is legitimacy to all three of these concerns. Consider the following example:

CONTROVERSY ABOUT STANDARDIZED INTERVIEWING 313

> Question: "In the last 12 months, how many times did you go to a doctor's office to get care for yourself?"
>
> Answer: "I have two questions. First, do you want me to include visits to a psychiatrist's office? Second, when I have gone to a doctor's office, but did not actually see the doctor, should that count?"
>
> Comment: These are both reasonable questions. There are right or wrong answers to the questions the respondent has asked. If the results of the survey are to be meaningful at all, the users of the data need to know what decision rules were used when people reported their answers.

It is important to point out that the basic cause of this interaction lies with design of the questions. A good question will spell out what most respondents need to know in order to answer a question accurately.

One could argue that the seeds of the answers to the respondent's questions are in the question: If a psychiatrist is an MD, a psychiatrist's office ought to count as a doctor's office; the question specifies going to a doctor's office, not actually seeing a doctor. So, a careful reading of the question might lead one to a set of concrete answers. Nonetheless, our topic is what an interviewer should do, given this question. Options include:

1) Say "Whatever it means to you."
2) Answer the respondent's question, if the interviewer knows the answer, or give a best guess about how the question should be interpreted.

Although the "Whatever it means to you" response is perfectly appropriate for questions about subjective states (although it may not be very satisfactory if a question contains vague or poorly defined concepts), it seems a fairly lame response when it is obvious that there is a right and a wrong answer. However, those who are concerned about interviewer effects on data worry that when interviewers are given freedom to explain or clarify questions, they are likely to do so in ways that are inconsistent, thereby potentially producing more error than if they allow the respondents to interpret the question themselves. Alternatively, there are those who argue that in instances like that above, most interviewers could help respondents to correctly understand and interpret the question, which in turn would lead to better data. Their argument is that the net benefits of giving interviewers that freedom outweigh any minor costs in interviewer inconsistency.

In fact, there are some examples where interviewers are given more flexibility in administering questions that seem to improve the flow of the interview and improve data quality. For example, when collecting information about who lives in a household, filling in age, gender, marital status, and relationships for each household member, it is essentially impossible to script the interviewer's side of the interaction. In a similar way, establishment surveys often collect similar information from businesses every month. The type of information required is factual, often derived from records. For such surveys, if interviewers are used, the interviewers and respondents often behave as if filling in needed information is a joint project. The interviewer's job is largely to say what is expected, to clarify the terms and decision rules to be used, and to make sure all the needed information is provided. Interviewers often do not read standardized scripts.

event history calendar

Another example of when a more flexible style of interviewing is used and seems appropriate is the event history calendar. An "event history calendar" is a visual device on which the respondents record landmark dates of great salience to them in various life domains (e.g., residential location, educational achievements, job change, anniversaries) and then use the landmarks to aid in recall and dating of events of interest to the survey (Belli, Shay, and Stafford, 2001). When using event history calendars, interviewers are given a good deal of discretion about wording and probing. Interviewers also do much more probing than do interviewers in a more traditional question format. As least one study reported that data quality was improved by this approach and that interviewer-related error was not increased (Belli et al. 2004).

These three examples share two features:

1) The subject matter is factual, so the answers should not be as affected by exact question wording as questions about subjective states might be.
2) It is hard to predict what information people will have to report and in what order it will be the most logical to report it.

There also have been experiments using scripted questions that have evaluated the contribution of letting interviewers help respondents by providing definitions and clarifications to factual questions when they sense that respondents may be misunderstanding or misinterpreting the meaning of the question. The results have been mixed. There is evidence that such interviewer help can improve the accuracy of answers to some questions. On the other hand, it greatly increases the interview length when interviewers do this routinely, the potential to improve the data applies to only a small number of questions, and it has not been demonstrated that in a long, complex survey, with many opportunities for questions to be unclear, interviewers will consistently provide appropriate and correct help (Schober and Conrad, 1997; Conrad and Schober, 2000). Moreover, in the instances in which the most potential for data improvement exists, the problems are likely to be solved by simply improving the design of the questions.

This controversy is not closed. Those who have studied the survey interview as an interaction identify many inherent problems in the notion of standardized interviews, which are summarized in Maynard, Houtkoop-Steenstra, Schaeffer, and van der Zouwen (2002). There may well be certain questions for which giving interviewers more flexibility to help respondents would improve the quality of data, without increasing interviewer-related error. At the same time, there seems to be ample reason for caution in allowing interviewers freedom to change the wording or intervene in ways that fundamentally affect respondent understanding of what questions mean. The bounteous evidence that interviewers behave inconsistently on more clearly defined tasks, such as the kinds of probes they use or even how well they read questions, does not make one sanguine that, on average, interviewers can perform this kind of innovative task appropriately. Writing good questions, then providing training to respondents in how the question-and-answer process in a standardized interview works, are less controversial paths to successful standardized interviews.

9.7 INTERVIEWER MANAGEMENT

In the preceding section, we described how interviewer behaviors can affect interviewer-related error. In this section, we describe other actions that researchers can take to reduce interviewer error. Specifically, we consider:

1) Interviewer selection
2) Interviewer training
3) Interviewer supervision
4) Interviewer workload
5) Interviewers and computer use

9.7.1 Interviewer Selection

Selecting the right interviewers would seem to be an obvious way to maximize interviewer quality. However, there is little evidence that currently known methods of interviewer selection play an important role in this respect.

Interviewing in the United States is most often a part-time job. Pay rates are typically low, often similar to those offered retail establishment salespersons. For that reason, some of the interviewer selection criteria are limited by the characteristics of the labor market that is interested in part-time work.

There certainly are important requirements for the job of interviewer. Reading skills and an ability to articulate clearly are obvious assets. Most productive household interviewing is done in the evenings or on weekends, so availability at those times and willingness to work less than 40 hours a week are necessary characteristics for the majority of interviewing jobs. Further, the job increasingly requires interviewers to be persuasive in gaining cooperation with the survey request and attentive to diverse concerns about survey participation. Because interviewing now typically is computer-assisted, some familiarity with computers and some keyboard skills can be helpful.

Conrad and Schober (2000) on Standardized versus Conversational Interviewing Techniques

Conrad and Schober address how interviewer action to clarify the meaning of questions affects responses.

Study design: Twenty experienced interviewers conducted two different telephone interviews with each of 227 respondents. The first interview with all respondents used standardized interviewing techniques (e.g., reading each question as worded, probing nondirectively); the second interview format, randomly assigned, was standardardized for half the cases and conversational for the other half. Interviewers using conversational techniques read the questions as worded, but then could say whatever they judged would help the respondent understand the intent of the question. Of the 20 interviewers, five conducted the conversational interviews. Change in responses from the first to the second interview and respondent explanations of their answers were examined.

Findings: There were no impacts of interviewing technique on response rates in the second wave. More responses (22%) changed between first and second interview with the conversational interviewing than with the standardized (11%). Respondents followed the survey definitions of purchase categories in 57% of the reports in the first (standardized interview), in 95% of the reports in the second conversational interview, and in 57% of the reports of the second standardized interview. Conversational interviews took 80% longer than standardized interviews.

Limitations of the study: With only five conversational interviewers, the study begs the question of whether the technique is practical for studies with thousands of interviewers. There was no measure of response bias, just a comparison between methods.

Impact of the study: When conducted, the study was one of the best attempts to implement an alternative to strict standardization. It demonstrated some of the potential gains and costs of such approaches and informed future efforts to study more flexible approaches to interviewing.

However, over and above rather practical job requirements, the evidence that interviewer characteristics are associated with better job performance is fairly scant. There is much more research needed on whether altering the selection criteria of interviewers improves the quality of data. Inevitably, such research would also have to consider cost implications of changing the selection criteria.

9.7.2 Interviewer Training

How much training interviewers receive has been shown to be critically important to the way in which interviewers do their jobs. Two experimental studies, using very similar designs, both presented compelling evidence that interviewers who receive little training (less than a day) do not perform at satisfactory levels (Billiet and Loosveldt, 1988; Fowler and Mangione, 1990).

Consider, for example, Table 9.2, taken from the Fowler and Mangione research. The table is derived from the results of a study in which newly recruited interviewers were randomized to one of four training protocols that lasted half a day, two days, five days, or ten days. All training programs included instruction in the survey objectives, question asking, probing on inadequate responses, recording of answers, and general job administrative duties. All the programs lasting two or more days included supervised practice interviewing; the ten day training taught interviewers how to use monitoring forms, how to code, and how to organize their working day efficiently.

After training, interviews taken by the interviewers were tape recorded and coded with respect to how well the questions were read, how well probing and recording were carried out, and the appropriateness of the interpersonal behavior of the interview. Two effects are clear from the table. First, in all respects except recording the answers to fixed-choice questions, interviewers who received one-half-day training were much worse interviewers than the others, and the majority were not satisfactory on the key skills of reading questions and probing. Second,

Table 9.2. Percentage of Interviewers Rated Excellent or Satisfactory for Six Criteria by Length of Interviewer Training

Rating Criterion	Length of Interviewer Training (days)				
	<1	2	5	10	p
Reading questions as worded	30%	83%	72%	84%	<0.01
Probing closed questions	48%	67%	72%	80%	<0.01
Probing open questions	16%	44%	52%	69%	<0.01
Recording closed questions	88%	88%	89%	93%	ns
Recording open questions	55%	80%	67%	83%	<0.01
Nonbiasing interpersonal behavior	65%	95%	85%	90%	<0.01

Souce: Fowler and Mangione (1990), p. 115.

it can also be seen that probing, which is the most difficult task for interviewers, distinctively benefited from the more extensive training.

Analyses that looked at the extent to which interviewers affected data in the Billiet and Loosveldt and Fowler and Mangione studies both suggested that better quality of data was associated with more training.

9.7.3 Interviewer Supervision and Monitoring

Interviewer supervision can also affect data quality. In computer-assisted interview surveys, computer software often examines the data for inconsistencies not already programmed into the instruments, but it is often difficult to discern the role of the interviewer in any inconsistencies found. In contrast, with paper questionnaires, it is common for supervisors to review completed questionnaires, evaluating whether the interviewer followed training guidelines. In computer-assisted modes, supervisors can examine missing data rates across interviewers to see whether there are unusual patterns for some interviewers. In face-to-face surveys, supervisors rarely have ongoing information on how well interviewers are administering the survey protocol. In centralized facilities, there are usually supervisors present in the interviewing room continuously.

The key element of supervision is whether or not the question-and-answer process is systematically monitored. A feature of the two studies above is that some interviewers were randomized to have their interviews tape recorded and reviewed by a supervisor, whereas other interviewers were neither tape recorded nor observed by a supervisor. It makes sense to think that if interviewers are not supervised with respect to the way they carry out the question-and-answer process, they will not perform as well. Both studies produced some evidence that data quality improved when interviewers were tape recorded. Both of those studies were done with in-person interviewers, for whom audio tape recording is the most practical way to monitor the interviewing. A recent development in CAPI software has permitted random subsets of the interview interaction to be digitally recorded using a laptop computer (Biemer, Herget, Morton, and Willis, 2003). If this feature becomes standard practice, then behavior coding of key interview segments could become a routine supervisory tool.

When interviewers work from a centralized telephone facility, of course, it is much easier to have supervisors listen to samples of the interviewers' work. It is standard practice in many centralized telephone facilities for a third party to listen to a sample of interviews, generally evaluating a portion of the interview. Often, the monitor is experienced in interviewer skills, trained in the purposes of the survey, and paid at a higher level than interviewers. Some monitoring procedures identify probability subsamples of interview interactions; others leave the selection of interview segments to monitor discretion. Some monitoring procedures resemble those of behavior coding, producing quantitative data on individual behaviors of interviewers and respondents. Others ask monitors to make overall qualitative evaluations of the interviewer behavior. Some monitoring procedures request that the monitor give feedback to the interviewer, in an attempt to improve their behaviors in future interviews. Others assemble monitoring reports over time and provide feedback on the cumulative results to the interviewer.

How much training and supervision interviewers receive affects survey costs. Training interviewers for three or four days obviously costs four times as much as training them for less than a day. More importantly, monitoring and reviewing samples of interviewers' work on an ongoing basis increases supervisory cost.

There is convincing evidence that at least more than half a day's training is essential to getting interviewers to carry out their roles in the question-and-answer process in a reasonably acceptable way. The two studies cited also present evidence that monitoring interviewer performance on an ongoing basis probably has a positive effect on data quality. At this time, however, we lack good studies in three important areas. First, both of the studies cited dealt with in-person interviews. We lack comparable studies on interviewer performance and data quality for telephone interviews. Second, despite the fact that a study of interviewer monitoring would be comparatively easy in a telephone facility, we lack data on the impact of monitoring interviewer performance on data quality. Third, anecdotal evidence suggests that survey organizations vary widely in the amount and kind of training they provide, and the extent to which they try to supervise the question-and-answer process. The limited evidence we have suggests that training and supervision do have salutatory effects on data, but we very much need more research about how much difference they make and how much variability in training and supervision matters to data quality.

9.7.4 The Size of Interviewer Workloads

The size of interviewer workloads is the final option available to researchers to minimize the effect of interviewers on variance. As seen earlier in this chapter, when interviewers affect the answers they get, the impact on estimates of standard errors is directly related to the average number of interviews per interviewer. This flows from the design effect for interviewer variance, $1 + \rho_{int}(m - 1)$, where m is the average workload size. Consequently, using more interviewers on a particular study, and having them each take fewer interviews, is another way to reduce the effect of interviewers on estimates of standard of errors.

9.7.5 Interviewers and Computer Use

Since the middle of the 1980s, an increasing percentage of interviews has been done with computer assistance. Questions pop up on a screen, interviewers read the questions off the screen, and then they record the answers respondents give directly into a computer. The introduction of the computer into the survey process has introduced several new challenges that deserve additional attention.

First, there is a need for a better understanding of how to maximize the usability of a computer-assisted survey instrument for interviewers. In the beginning, computer-assisted instruments were basically paper instruments transferred onto computer screens. Studies of how interviewers actually use these instruments show that some ways of organizing questions that work well on paper are very inefficient, and create problems, when they are on computers. We need studies of how interviewers actually interact with computer-assisted programs. One finding of studies done so far is that interviewers fail to use many of the aids that are provided to them; there are function keys, designed to be helpful, that interviewers never use (see, e.g., Sperry, Edwards, Dulaney, and Potter, 1998). Making the computer easy to use for interviewers is in the interest of the research process, and we need to develop better principles about how to accomplish that.

Second, we need to learn more about how the presence of a computer affects the respondent–interviewer interaction. We know some straightforward things, such as that interviewers like computer assistance and respondents' ratings of the interview experience are not adversely affected. We also know from observing the respondent–interviewer interactions that the computer becomes a third party to the interaction. Interviewers spend more time looking at the computer than they do at the respondent. What we lack at the moment are studies of whether there are any significant effects on respondent behavior or data quality from introducing this extra actor into the process.

Third, there is no doubt that the birth of computer-assisted interviewing has had an effect on interviewer training. A good portion of general interviewer training now has to be devoted to ensuring that interviewers can use the computer-assisted programs. Portions of training that used to be devoted to lectures, demonstrations, and practice are now spent on tutorials on the computer. On the one hand, computers help interviewers do things, particularly to properly manage complex skip instructions, that were harder to do with paper-and-pencil instruments. On the other hand, if training time is finite, the time that is spent in training interviewers to use computers may be taken out of their training on some of the other critical skills that interviewers need, such as enlisting cooperation, probing, and relating to respondents. Interviewers cannot work unless the computer skills are mastered, whereas they can and do work without being excellent at probing or establishing rapport. This is another subject area in which we need more systematic information.

9.8 Validating the Work of Interviewers

All of the preceding discussion about interviewer management was aimed at reducing interviewer-related error in answers respondents give. This section addresses an issue that is, fortunately, probably not very common but of great concern because of its potential effects on data quality and credibility: interviewer falsification of survey responses or providing false information about how responses were obtained. "Interviewer falsification" means the intentional departure from the designed interviewer guidelines or instructions, unreported by the interviewer, that could result in the contamination of data. "Intentional" means that the interviewer is aware that the action deviates from the guidelines and instructions.

Falsification includes:

1) Fabricating all or part of an interview—the recording of data that are not provided by a designated survey respondent and reporting them as answers of that respondent
2) Deliberately misreporting disposition codes and falsifying process data (e.g., the recording of a refusal case as ineligible for the sample; reporting a fictitious contact attempt)
3) Deliberately miscoding the answer to a question in order to avoid follow-up questions

4) Deliberately interviewing a nonsampled person in order to reduce effort required to complete an interview, or intentionally misrepresenting the data collection process to the survey management

In all large surveys involving many interviewers who are temporary staff, not part of the ongoing research, falsification will probably occur. On the other hand, the rate of falsification is believed to be quite low. There are few published studies of falsification, but Table 9.3 shows the results of a study at the U.S. Census Bureau (Schreiner, Pennie, and Newbrough, 1988). In this work, falsification included taking an interview with an incorrect person in the sample household or taking an interview in a mode not authorized by the study. The table shows that higher rates were obtained in the Housing Vacancy Survey, a one-time data collection, than in the ongoing Current Population Survey and NCVS.

Schreiner, Pennie, and Newbrough (1988) also found that interviewers with less experience were more likely to falsify relative to those with more experience, and that experienced interviewers had more sophisticated patterns of cheating (e.g., falsifying the first wave of a panel survey).

Interviewer falsification appears to be reduced through various survey administrative procedures. The most common method of attempting to reduce falsification uses training of interviewers regarding the importance of the integrity of the data collection protocols. This training is then supplemented with verification procedures during data collection. There are three main types of verification processes:

1) Observational methods
2) Recontact methods
3) Data analysis methods

"Observational methods" mean that a third party (in addition to the interviewer and the respondent) hears and/or sees the interview take place. The most common use of observation is a supervisor listening in to a sample of interviews in centralized telephone facilities. It is common that 5–10% of the interviews are monitored in this fashion. This technique is common practice and generally believed to be an effective deterrent to falsification in such facilities. In face-to-face surveys, some CAPI surveys are now using the built-in microphone of the laptop computer to digitally record randomly chosen pieces of the interview (Biemer, Herget, Morton, and Willis, 2003). These recordings are then reviewed in the cen-

Table 9.3. Percentage of Interviewers Detecting Falsifying for Three Surveys Conducted by the U.S. Bureau of the Census

Survey	Percentage Falsifying
Current Population Survey	0.4%
National Crime Victimization Survey	0.4%
New York City Housing Vacancy Survey	6.5%

tral office in an effort to detect unusual behaviors that might indicate falsification (e.g., laughter and side comments indicating the use of a friend as the respondent).

"Recontact methods" means that another staff member (often a supervisor) attempts to speak with the respondent after the interview is reported, in order to verify that the interview was completed according to the specified protocol. This is the most common tool for dispersed interviewing designs, but the method faces the challenge of achieving high response rates on the verified cases. Face-to-face recontact methods are the most expensive but probably generate the highest response rates, telephone recontact is cheaper but generates lower response rates, and mail self-administered recontact yields the lowest return rate but can be used for a much higher percentage of cases than other methods because of its low unit cost. It is sometimes cost-effective to use a mixed-mode approach for such verification.

"Data analytic methods" of verification examine completed data records resulting from the interviews. Sometimes, process data are examined. For example, if interviews taken by a particular interviewer tend to be very short, if a high percentage of an interviewer's interviews are obtained on the first call, or if the hours per interview are much lower than average, that interviewer can be flagged for special attention and intensive follow-up procedures. Sometimes, patterns of the actual survey responses are examined. Interviews that have unusual patterns of reports (e.g., answers on key branching questions that skip over large portions of the questionnaire, unusual patterns of responses among a set of correlates) can also be flagged for verification.

The American Statistical Association and the American Association for Public Opinion Research (AAPOR) have supported guidelines for good practice in detecting interviewer falsification (seehttp://www.amstat.org/sections/SRMS/). These include some of the following recommendations:

1) A probability sample of interviews should be identified for verification (e.g., a 5–15% sample).
2) Verification questionnaires should include questions about household composition and/or other eligibility requirements; mode of data collection; length of interview; payment of incentive, if any; use of computer during data collection; key topics discussed; and key items, especially those that govern large skips in the interview.
3) If verification fails, all the work of an interviewer should be examined and the interviewer should be removed from data collection activities during that examination.
4) If the preponderance of the evidence points to falsification, personnel actions permissible under the organization's rules should be initiated immediately.
5) All cases found to be falsified need to be repaired, typically by taking a full interview.
6) The technical report of the survey should report the portion of cases in the probability sample that were falsified.

By making reports of falsification a standard part of the documentation of a survey, professional associations are attempting to raise awareness of the problem and its solution.

There are many unanswered questions about interviewer falsification. Because only samples of completed work are verified, the proportion of falsified cases actually found is dependent on the rate of verification sampling. The common belief that cheating is not done consistently by any interviewer but varies within a workload makes the sampling problem quite difficult. What is the optimal rate of sampling? Can adaptive sampling procedures be put in place? What motivates the cheating; are there reward systems that generate higher propensities of cheating (e.g., paying interviewers a fixed amount for each completed interview may encourage falsification more than paying by the hour). These are all questions for which we need better answers. However, at this time, we can say that taking some steps to verify how interviewers carried out their work should be a routine part of interviwer management.

9.9 THE USE OF RECORDED VOICES (AND FACES) IN DATA COLLECTION

This final section of the chapter is about simulated or virtual interviewing. We traditionally speak of data collection as occurring in one of two distinct categories: when an interviewer reads the questions and records the answers, and when the respondent reads the questions and records the answers. However, there are some data collection protocols that may be blurring those lines. Basically, the protocols all involve a prerecorded interviewer (or perhaps a computerized voice) reading the questions aloud to a respondent. Sometimes, the respondent is sitting at a computer keyboard; sometimes, the respondent uses a touchtone telephone to respond; and sometimes, a voice recognition system decodes the answers. However, a common feature of all of these data collection protocols is that the answers are entered by the respondent without the assistance of an interviewer. To further blur the data collection lines, if the respondent is looking at a computer screen, a face can be presented to go with the preprogrammed words. That face can have a gender, a race, and an apparent age—characteristics that might be inferred from a voice alone, but made much more explicit and certain with a picture.

Do these protocols belong in a chapter about interviewing? The existence of this section is testimony to the fact that we think they deserve discussion, albeit briefly.

Computerized interviewing voices, as an adjunct to self-administration, can help respondents whose reading skills are limited understand the questions. They provide the opportunity to have questions administered fluently in multiple languages, something that is often hard to achieve in practice with live interviewers. They have the potential to standardize aspects of administration that are almost impossible to control with live interviewers, such as pronunciation, pace, and intonation, and they make it very easy to minimize spontaneous chatter that may be potentially biasing. If a respondent asks for help, they make it possible for the provided help, such as definitions of terms, to be consistent for all respondents. When a picture is added, they make it possible for the apparent age, gender, and race of the apparent interviewer to be controlled.

These are all potential plusses for some survey tasks, and they are all examples of how computer-enhanced self-administration can address some possible limitations of using interviewers as data collectors. However, for the most part, the evidence is that these strategies are just that: enhanced self-administrative strategies. Respondents report to computerized voices as if they were computers, not as if they were live people. That is a plus, as we have discussed, when respondents are asked to provide potentially embarrassing information. Respondents also do not pay much attention to the characteristics of the images of interviewers when they are provided. Respondents to computerized "females" do not show the same kind of gender-related effects that they show when interviewed by a live interviewer.

However, computerized voices are not good at enlisting cooperation. Respondents are much more likely to break off an interview with a computerized voice on the phone than when they are talking with a live interviewer. While computerized voices can be programmed to reliably follow paths for question sequences, they do not do much individualized problem solving. We do not have good data on this point, but based on the data on cooperation rates and break-offs, one would hypothesize that computers are not as good at motivating respondents as live interviewers.

Research will continue on how best to meld the various ways we have to collect data, and getting computers to simulate some of what interviewers do will be part of the that agenda. However, almost certainly interviewers will be part of the survey process for a long time to come, and how to select, train and manage them to maximize the quality of the resulting data will continue to be an important methodological concern.

9.10 SUMMARY

Interviewers carry the burden of implementing many survey design features affecting coverage, nonresponse, and measurement errors. Some of the measurement errors have systematic causes; some appear to manifest themselves as variability in results from different interviewers. The systematic influences are lower reporting of socially undesirable attributes, greater reporting of answers compatible with expectations based on observable interviewer characteristics (like race, age, and gender), and greater measurement error from interviewers with more experience on the job.

The increased variation in survey results associated with interviewers arises most often when interviewer behavior influencing answers is unguided and thus variable. The significance of interviewer behavior varies from question to question, with some questions showing more interviewer variance than others. Without special measures, these variable effects of interviewers on survey data are not measurable.

Interviewer variance inflates the standard errors of survey statistics, making them less precise. The effects of interviewer error on standard errors are worthy of attention. That is particularly the case when interviewer workloads are large. There are some studies for which only a very small number of interviewers (in the extreme case, only one) do all the interviewing. In those cases, there is a particularly high probability for interviewer effects to be important. In a similar vein, on large national studies, it is not uncommon for interviewers to take 50, 100, or even 200 interviews. In those cases, again, the potential for large interviewer effects on standard errors is great.

Statistical errors induced by interviewers are reduced when interviewers are effective in motivating good respondent behavior, using task-oriented approaches to the interaction, reading questions as worded, explaining the response task to the respondent, probing nondirectively, and recording answers exactly as given. There is some controversy about how standardized the behavior of an interviewer must be to minimize these errors, and that controversy suggests the need for more research into how interviewers affect the response process.

In recent years, there has been renewed interest in question design issues. However, there has been a tendency to neglect the role of the interviewer in collecting high-quality data. It is reasonable to think that some things that interviewers do may become increasingly less important as we rely on telephone interviews, where the interpersonal component is much reduced, and as we rely increasingly on computers to assist in data collection efforts. Yet the respondents' motivation and perceptions of their roles will affect performance no matter what the mode of data collection, and the role of the interviewers in motivating and orienting respondents has been shown to be important.

KEYWORDS

event history calendar
interpenetrated assignment
interviewer variance
probe
replicability
standardized interviewing

FOR MORE IN-DEPTH READING

Conrad, F., and Schober, M. (2008), *Envisioning the Survey Interview of the Future*, New York: Wiley.

Fowler, F. J., and Mangione, T. W. (1990), *Standardized Survey Interviewing: Minimizing Interviewer Related Error*, Newbury Park, CA: Sage.

Maynard, D., Houtkoop, H., Schaeffer, N., and van der Zouwen, J. (2002), *Standardization and Tacit Knowledge: Interaction and Practice in the Survey Interview*, New York: Wiley.

EXERCISES

1) Exactly 2000 interviews are conducted by 40 interviewers. For a particular item, the value of ρ_{int}, the intraclass correlation coefficient associated with the interviewer, is 0.015.

   a) Calculate the value of the interviewer DEFF for that item's mean.
   b) Briefly state in words what this DEFF means.

EXERCISES

2) Name and describe four of the five basic techniques interviewers are taught in order to have their behavior standardized.

3) Which behavior detectable through behavior coding is most associated with the likelihood of significant interviewer effects on answers?

4) For each of the four answers to the question presented, write what, if anything, you think a standardized, nondirective interviewer should say.

   Question: "What is the biggest problem facing America today?"
   Answer: "The Taliban."
   Answer: "I don't know."
   Answer: "Without doubt, crime and terrorism."
   Answer: "Given what is going on abroad, I don't think we can afford to worry about our own problems. We have to devote all of our energies to defeating terrorism."

5) You are the project manager of a large household survey using face-to-face interviewing methods to measure poverty levels in your country. You have carefully constructed an interpenetrated design assigning interviewers to random subsets of sample households within the PSUs in which they work. On average, each interviewer completes 25 interviews. From this design you have computed ρ_{int} (according to the model reviewed in this chapter) that reflects interviewer variance.

   The following table provides a few of the results:

Estimate	ρ_{int}
Percentage of households in poverty	0.09
Pertcentage of households with unemployed adults	0.02

   The project team receives the bad news that the next year's budget may be cut. You are asked to assess the impact on the precision of estimates if the number of interviewers is cut in half but the sample size is kept constant.

   a) If that were the only change made in the design, by what factor would the variance of each of the two estimates above change?
   b) What additional factors would you consider in addressing the proposed change in interviewer numbers?

6) A survey is designed to measure fertility experience and sexual behavior of a household sample of male and female adults. Discuss in one paragraph the issues that would need to be considered in minimizing measurement error related to interviewer gender. That is, what issues would need to be considered in choosing to use all male, all female, or a haphazard mix of male and female interviewers for these topics?

7) Which mode of data collection is less conducive to interviewer cheating: face-to-face or telephone interviewing from a centralized facility? (Give a reason for your answer)

8) You are the manager of a large survey effort, currently engaged in deciding whether to move from face-to-face interviewing to telephone interviewing. You are now engaged in an experimental comparison between the two methods, with half of the sample (about 1000 interviews) devoted to telephone interviews and the other half to face-to-face interviews. The centralized telephone facility uses 25 interviewers and the face-to-face interviewer corps consists of 50 interviewers. The centralized facility uses continuous monitoring of interviewers and a feedback protocol for remedial training when necessary. The face-to-face survey uses supervisory review of completed interviews and observation of one interview for each interviewer. You have introduced an interpenetrated interviewer assignment plan in both modes, in order to measure interviewer variance.

At the end of the survey, you compute interviewer-level intraclass correlations for key survey estimates and find that the telephone half of the sample exhibits average intraclass correlations of 0.025 and the face-to-face half 0.040. Using the criterion of interviewer variance alone, discuss which mode you would prefer and why.

9) You are examining some data from behavior coding of interviews in a study you direct. You are focused on the following question: "In the last 12 months, that is, from April 20, 2008, to the present, how many times have you talked with medical personnel about your health condition? Include both physicians, nurses, and other medical personnel; include telephone conversations and in-person conversations; include both health problems and positive health experiences."

You find that interviewers read the question exactly as worded in 70% of the cases, relatively much less frequently than other questions in the survey. You can

a) Do nothing.
b) Provide remedial training to improve compliance with the standardized delivery.
c) Attempt to alter the question.

Comment on each of the three possible actions.

10) Given what you have read, give arguments pro and con that interviewer training sessions should be much longer and more intensive.

11) What are four things a researcher can do that are likely to minimize the effect of interviewers on survey estimates?

12) Although standardized interviewing is considered a "best practice" throughout the survey industry, there are those who argue against it. Briefly state the main arguments for and against standardized interviewing.

13) Based on your reading:

   a) Define "interviewer variance."
   b) Identify one way in which researchers can assess its impact on measurement error.
   c) Briefly discuss how it can affect the precision of survey estimates.

14) Discuss briefly (2–3 sentences) whether each of the following design features affect the magnitude of interviewer effects:

   a) Form of the question
   b) Interviewer attributes
   c) Use of computer assisted interviewing

15) Give one reason why centralization of interviewing might decrease interviewer's contribution to the total variance (of a statistic) and one reason why it might increase it.

16) For each of the five answers to the question presented, what, if anything, do you think a standardized interviewer should say:

   Question: Overall, how would you rate your health now: excellent, very good, good, fair or poor?
   Answer: Not so good.
   Answer: Well, generally I think my health is pretty good, but today I am still recovering from the flu.
   Answer: I can't run as fast as I used to. Is that what you mean?
   Answer: Well, I do have diabetes, but other than that I would say my health is good.
   Answer: I have aches and pains, like most people my age, but overall, I'd say my health is very good.

17) How would you define a "nondirective" probe?

18) How would you define a "directive" probe?

CHAPTER TEN

POSTCOLLECTION PROCESSING OF SURVEY DATA

10.1 INTRODUCTION

This chapter focuses on survey operations that generally occur after data are collected. These steps have the ultimate goal of facilitating estimates of the target population attributes. In the dawn of surveys, large staffs examined each questionnaire and tallied the answers, but over time the field developed procedures to reduce the burden of these steps.

One way to describe the postsurvey processing steps is to contrast a paper questionnaire survey with one using varieties of computer assistance. Figure 10.1 presents the flow of a survey common to paper questionnaire surveys. Paper instruments record the answers of respondents. Since the products of surveys are numbers, when data do not have a numeric form (e.g., text answers to open questions, a check box next to a response wording), they must be transformed into numbers (an operation called "coding"). After all the data are in numeric form, then they are entered in some way into an electronic file ("data entry" in Figure 10.1).

For much of the data that are already in numeric form, there may be a restricted set of numbers allowed (e.g., 1 = Yes, 2 = No, 8 = Don't Know, 9 = Refused). Some answers should have logical relationships to others (e.g., those reported as "male" should not report that they have had Caesarian section surgery) (an operation called "edit checks" in Figure 10.1). When there is an item with missing data, researchers may choose to place an estimated answer into that item's data field (an operation called "imputation"). Some designs require the use of adjustment weights for statistical estimation of the target population, and these are construct-

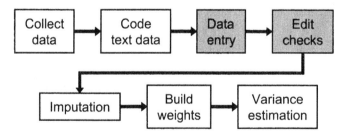

Figure 10.1 Flow of processing steps in paper surveys.

Survey Methodology, Second Edition. By Groves, Fowler, Couper, Lepkowski, Singer, and Tourangeau
Copyright © 2009 John Wiley & Sons, Inc.

ed after the data are collected (an operation called "weight construction"). Initial estimations of precision of survey statistics are often conducted soon after data collection in order to assess the quality of the survey estimates (an operation called "variance estimation"). Although some of these operations can begin as soon as a few questionnaires are returned, they generally are completed only after all questionnaires are received and data collection operations have ended.

The use of computer assistance in data collection (CATI, CAPI, Web) has radically changed these steps of processing, most often by placing the functions within the data collection step itself. Figure 10.2 shows various checks on data that on paper questionnaires are part of activities after the data are collected but, with computer-assisted surveys, are performed simultaneously with the data collection. Note in Figure 10.1 that the data entry and editing steps lie after the "collect data" box. Computer assistance has changed surveys by forcing attention prior to data collection to the traditional post-data-collection activities. (It is useful to note that even in computer-assisted surveys, the researcher may choose to do some editing after data collection; see Section 10.4).

The outcome of the reordering of activities is that computer-assisted surveys must make all the design decisions about data entry protocols and editing during the process of developing the questionnaire. A further implication of the combining of collection, entry, and editing is that surveys collecting only numeric answers (i.e., no text answers) can entirely skip the coding step. Since coders are often a separate unit in most organizations, there appears to be an increasing tendency to avoid entirely the use of open questions and other measures producing text answers. Instead, more and more surveys appear to collect numeric answers exclusively. The exception to this might be the use of the "other, specify" answer for closed questions, which permits the respondent to give an answer not listed in the fixed choices.

This chapter describes six different activities that might occur after data collection:

1) Coding—the process of turning word answers into numeric answers
2) Data entry—the process of entering numeric data into files

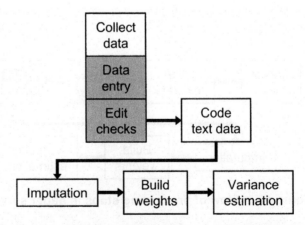

Figure 10.2 Flow of processing steps in computer-assisted surveys.

3) Editing—the examination of recorded answers to detect errors and inconsistencies
4) Imputation—the repair of item-missing data by placing an answer in a data field
5) Weighting—the adjustment of computations of survey statistics to counteract harmful effects of noncoverage, nonresponse, or unequal probabilities of selection into the sample
6) Sampling variance estimation—the computation of estimates of the instability of survey statistics (from any statistical errors that are measurable under the design)

10.2 CODING

"Coding" is the translation of nonnumeric material into numeric data. This section reviews what survey methodology knows, as a field, about how alternative coding approaches affect the quality and cost of survey data. Sometimes, the translation from nonnumeric to numeric is a quite simple one. For example, in a self-administered paper questionnaire:

coding

> Indicate your current status (Check one box):
>
> ☐ Full-time student
> ☐ Part-time student
> ☐ Applicant, acceptance letter received
> ☐ Applicant, acceptance letter not received

Here, the coding step merely involves assigning a distinct number to each of the four possible answers (e.g., 1 = full time, 2 = part time, 3 = applicant with letter, 4 = applicant without letter). These numbers then become the values in a field of the electronic data file eventually produced.

At other times, the translation is a bit more complicated. For example, the National Crime Victimization Survey asks the questions:

> What is the name of the (company/government agency/business/nonprofit organization) for which you worked at the time of the incident?
>
> What kind of business or industry is this?
>
> *Read if necessary: What do they make or do where you worked at the time of the incident?*
>
> What kind of work did you do; that is, what was your occupation at the time of the incident?
>
> *For example: plumber, typist, farmer*
>
> What were your usual activities or duties at this job?

The interviewer types in the words of the respondent in answering these questions. For example, the respondent might answer in response to "What kind of business or industry is this?": "We make plastic inserts for containers of various sizes. The inserts are used to seal liquid inside the containers, so they don't leak. Sometimes the containers have toxic liquids that could harm someone." In self-administered designs, the words would be written or typed by the respondents themselves. The statistics computed based on questions like this might be percentages of persons who fall into certain categories of answers. For example, on the industry answer above, there is a category labeled "plastics material and resin manufacturing" that might be used to characterize the industry. In the North American Industry Classification System (NAICS), this is code 325211. A statistic might be computed on the survey, the percentage of respondents who gave an industry answer that fit into 325211. (More likely, statistics are computed at a higher level of aggregation of codes.)

Coding of text material is crucial for the ability to analyze statistically the results of surveys, but the act of coding itself can produce statistical errors. This can have noticeable effects on survey statistics. For example, Collins (1975) found that in consecutive monthly interviews 32% of the respondents apparently experience a change in occupation codes (using a three-digit coding scheme). Other information about month-to-month job changes suggests that much of this variation might arise from coding variation across months.

It is important to note that not all nonnumeric data collected in surveys are words. They might be visual images (e.g., pictures, video taping), sounds (e.g., audio tapes), or samples of physical materials that need to be described numerically (e.g., soil samples, blood samples).

10.2.1 Practical Issues of Coding

Coding is both an act of translation and an act of summarization. Like all translations, entities in one framework are mapped onto another framework. When the two frameworks are compatible, the mapping can be effortless. When the frameworks are mismatched, the translation task can be complex and subject to error. Coding summarizes data by combining many individual answers into one code category. Like all acts of summarization, someone must decide what level of summarization fits the uses of the coded data.

Our first attention thus must be to the construction of the category framework that is used for classifying the text material. Unfortunately, there is no single accepted terminology for this framework. Some call it a "code structure," some, a "nomenclature," others, a "code list." To be useful, codes must have the following attributes:

code structure

1) A unique number, used later for statistical computing
2) A text label, designed to describe all the answers assigned to the category
3) Total exhaustive treatment of answers (all responses should be able to be assigned to a category)
4) Mutual exclusivity (no single response should be assignable to more than one category)
5) A number of unique categories that fit the purposes of the analyst

For surveys with scientific purposes (when the research is attempting to study causal processes), each of the code categories should link to different parts of key hypotheses. For example, when studying the achievement of supervisory status, the occupational codes should separate supervisors from nonsupervisors. If, on the other hand, the scientific questions involve the educational background required for a job classification, then the coding should permit separation of jobs by educational background.

The total number of code categories to use must be specified. The extreme is obviously when the number of codes is equal to the number of respondents, but such a structure generally yields complicated data because there is no summarization involved. The number of codes must be driven by the intended uses of the variable. In short, the creation of a coding system is a substantively relevant act. Different uses demand different coding structures.

Several coding structures can be used for the same measure. For example, in Britain it is common to produce two codes from answers to the questions about occupation. One of them is a standard occupation classification and the other is a classification of socioeconomic group (a ranking of statuses, which also employs measures from the size of the employer). On questions that elicit multiple answers, separate coded variables may be created for each mention (e.g., "What are the most important problems facing the country today?")

No matter how comprehensive a coding structure might appear to be prior to use, it is typical that at least a few responses cannot be easily assigned to a code. To reduce the occurrence of responses that do not fit prestructured codes, it is common to take a set of responses (perhaps from early returns of completed questionnaires) and use responses from that set to "test" and refine a coding structure. This is generally only partially successful. When a response that does not fit well with existing categories is encountered, the case can sometimes lead to a reconsideration of the existing code structure. When a coding structure is changed during the coding process itself, the researcher is obliged to reexamine coding decisions for prior cases.

Finally, it is important to note that coding structures must be designed to handle all responses, even those judged as uninformative. Hence, it is critical to have a code for those who fail to give an answer that meets question objectives ("not ascertained"). Also, if the questionnaire protocol calls for some respondents to skip a question because it does not apply to them based on their previous answers (for example, questions about the characteristics of a burglary would not be asked of those who said they had not experienced a burglary), it is critical to have a code that identifies those respondents to whom the question does not apply and, hence, there will be no answer in the data file ("inapplicable"). Because these options apply to large numbers of variables, it is common that a consistent set of numeric codes is assigned to them. For example, in a single-digit coding scheme, "9" might be the value routinely assigned to answers that were "not ascertained" and "0" to those of whom the question was not asked because it was inapplicable. In two-digit codes, the corresponding codes might be "99" and "00." In addition, for some variables, investigators may want to single out instances when questions were not answered for a specific reason, such as the respondents "refused" or said they "don't know" the answer. In that case, again, separate specific codes will have to be used to identify those situations.

10.2.2 Theoretical Issues in Coding Activities

The basic act of coding requires a decision about whether two verbal representations are equivalent. For example, are the words "I repair drains, install water pipes, and replace sinks" compatible with occupation label "plumber?" Ideally, the categorizations made by the coder would agree with fully informed categorizations provided by the respondents themselves. Figure 10.3 shows the nature of the problem. Coding accuracy can arise when the choice of the code category by the coder (the bottom-right box) is exactly what was intended by the respondent (the top box). One challenge comes from the fact that the coder does not see what the respondent intends, but only words written down by the interviewer. (In self-administered surveys the respondent's words are examined by the coder.)

To perform the coding task, the coder attempts to discern the intent of the respondent through the words recorded. That is just half of the challenge, because the task of the coder is to map the intended answer of the respondent to one and only one of the code categories. To do this, the coder ideally understands fully the intent of the researcher in defining each code category (the bottom-left box). This understanding is probably a function both of the quality of the descriptions of code categories and the quality of the training about the concepts being described in those words.

10.2.3 "Field Coding"—An Intermediate Design

As we learned in Sections 7.2.4 and 9.5.5, one value of an open question is the freedom it gives respondents to deliver an answer in their own words. This sometimes reveals insights into their comprehension of the question or the structure of their memory that researchers find valuable in understanding the substance of what they are studying. Closed questions present the respondent with a set of responses invented by the researcher, which may or may not fit with the perspective of the respondent.

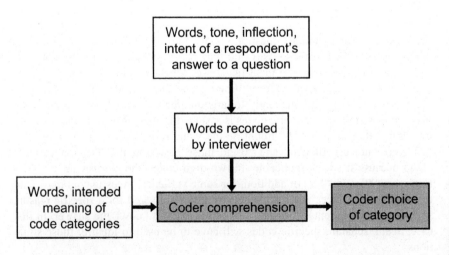

Figure 10.3 Comprehension and judgment task of the coder.

"Field coding" is a process by which the respondent is presented with an open question, for which they form the answer, but the interviewer is asked to code the verbal answer into a numeric category. Field coding thus combines verbal data collection with coding. For example, the National Crime Victimization Survey asks an open question about where a reported victimization occurred and then asks the interviewer to choose the appropriate answer immediately (see Table 10.1).

field coding

Since the respondents are free to answer this question using any descriptor that they find useful, the interviewer is given the burden to interpret the question, review the possible categories of answers that it might fit, probe for any uncertainties they face in that decision, and then choose one category among the 26 different response categories presented. It is not an easy task. On the other hand, relative to a coder back in the office, privy only to the text answer recorded by the interviewer, field coding can be aided by probing and a richer hearing of the answer.

There have been a few studies about how interviewers perform this task relative to the decisions made by coders based on written or tape recorded answers of respondents. These studies are comparisons of two methods of performing the same task and offer estimates of agreement between the two methods. On occupation coding, for example, office coders agree with a gold standard coder at somewhat higher rates than field interviewers (on an occupational code structure of nearly 400 categories), but the differences overall are small (Campanelli, Thomson, Moon, and Staples, 1997). However, there is also evidence that field coding has negative impacts on interviewer behavior (see Maynard, Houtkoop-Steenstra, Schaeffer, and van der Zouwen, 2002; Fowler and Mangione, 1990). All of the lessons of interviewer variance that applied to measurement error (see Section 9.3) also apply to coding. That is, since both interviewers and coders are assigned workloads of multiple cases, their variability of coding performance produces an additional component of instability in survey estimates ("coder variance" is the typical term used for this). For these reasons, the choice of field coding versus office coding is generally a trade-off between cost differences and classification errors.

Collins and Courtenay (1985) on Field versus Office Coding

Collins and Courtenay (1985) compared the results of field coding during data collection to office coding after data collection.

Study design: Seven questions used in prior surveys were randomized within three replicate samples as (1) open questions field coded, (2) open questions office coded, or (3) closed questions. The mix of field and office coding varied over the three samples. Six face-to-face interviewers were randomly assigned to respondents within each of four sample areas. The replicates of size 180–200 persons were compared on response distributions and interviewer variance.

Findings: When office coders were given the same code categories as field coding, there were no systematic differences. Differences that did exist showed larger percentages in a small number of codes for the office coding. The average number of codes used per respondent was similar, but the incidence of "other" answers was higher in the field coding. Interviewer variability accounts on average for about 3% of total variance in responses in field coding (a $deff_{int}$ of 1.5 with workloads of 25) but only 0.6% in office coding.

Limitations of the study: The office coding frame was limited to the same categories as the questionnaire, which probably places unusual limits on the office coders. The office coder variance was not measured, to compare to the $deff_{int}$. There is no discussion of interviewer training used for field coding.

Impact of the study: The study pointed out a limitation of field coding because of the burden of hearing the answer and choosing one from a set of fixed categories in the context of an ongoing interview.

Table 10.1. Illustration of Field Coding in the NCVS for the Question "Where Did This Incident Happen?"

Category Number	Category Definition

In respondent's home or lodging
- 01. In own dwelling, own attached garage, or enclosed porch (Include illegal entry or attempted illegal entry of same)
- 02. In detached building on own property, such as a detached garage, storage shed, etc. (Include illegal entry or attempted illegal entry of same)
- 03. In vacation home/second home (Include illegal entry or attempted illegal entry of same
- 04. In hotel or motel room respondent was staying in (Include illegal entry or attempted illegal entry of same)

Near own home
- 05. Own yard, sidewalk, driveway, carport, unenclosed porch (does not include apartment yards)
- 06. Apartment hall, storage area, laundry room (does not include apartment parking lot/garage)
- 07. On street immediately adjacent to own home

At, in, or near a friend's/relative's/neighbor's home
- 08. At or in home or other building on their property
- 09. Yard, sidewalk, driveway, carport (does not include apartment yards)
- 10. Apartment hall, storage area, laundry room (does not include apartment parking lot/garage)
- 11. On street immediately adjacent to their home

Commercial places
- 12. Inside restaurant, bar, nightclub
- 13. Inside bank
- 14. Inside gas station
- 15. Inside other commercial building such as a store
- 16. Inside office
- 17. Inside factory or warehouse

Parking lots/garages
- 18. Commercial parking lot/garage
- 19. Noncommercial parking lot/garage
- 20. Apartment/townhouse parking lot/garage

School
- 21. Inside school building
- 22. On school property (school parking area, play area, school bus, etc.)

Open areas, on street or public transportation
- 23. In apartment yard, park, field, playground (other than school)
- 24. On street (other than immediately adjacent to own/friend's/relative's/neighbor's home)
- 25. On public transportation or in station (bus, train, plane, airport, depot, etc.)

Other
- 27. Other
- 98. Residue
- 99. Out of universe

10.2.4 Standard Classification Systems

This section describes two commonly used classification systems, one for the measurement of occupation, the other for the measurement of industry. These two variables are ubiquitously measured in social and economic surveys. Their classifications are standardized by international bodies and periodically updated. The standardization is valued because it permits a comparison across surveys of attributes of commonly defined populations.

The Standard Occupational Classification (SOC). In the United States, the standard occupational classification system is managed by the U.S. Department of Labor. It was revised in 1980 and then again between 1994 and 1999 to reflect changes in the economy's organization of occupations. It is used to classify workers into occupational groups for purposes of surveys and administrative record systems. The coding structure contains over 820 occupations, combined to form 23 major groups, 96 minor groups, and 449 broad occupations. Each broad occupation includes detailed occupation(s) requiring similar job duties, skills, education, or experience. Table 10.2 shows the labels for the 23 major groups of the SOC.

The latest revision of the SOC considered four different organizing principles. The first, and the basic concept behind the 1980 SOC, was the type of work performed. The second option was to incorporate the international standard classification of occupations, in recognition of the globalization of work, which differentiated "women's work fields" from others. The third option was to devise a "skill-based system," with detailed differentiation of jobs based on skill utilization in the job. The fourth option identified an "economic-based system," using macroeconomic theories. The new system had the following users in mind when created: education and training planners; jobseekers, students, and others seeking career guidance; various government programs; and private companies wishing to relocate or set salaries. The choice was made to continue the focus on type of work performed, believing that skill-based assessment of jobs was not sufficiently accurate for use as an organizing principle. There was a great effort to update the classifications to reflect new occupational groupings arising since 1980 (e.g., environmental engineers, skin care specialists, concierges, aerobics instructors).

There are various underlying assumptions of the coding scheme that reflect the intent of those constructing the codes (see http://www.bls.gov/soc/socguide.htm):

1) The classification covers all occupations in which work is performed for pay or profit, including work performed in family operated enterprises by family members who are not directly compensated. It excludes occupations unique to volunteers. Each occupation is assigned to only one occupation at the lowest level of the classification.
2) Occupations are classified based upon work performed, skills, education, training, and credentials.
3) Supervisors of professional and technical workers usually have a background similar to the workers they supervise, and are therefore classified with the workers they supervise. Likewise, team leaders, lead workers and supervisors of production, sales, and service workers who spend at least 20% of their time performing work similar to the workers they supervise are classified with the workers they supervise.

Table 10.2. 23 Group Standard Occupational Classification

Category Number	Category Definition
11-0000	Management Occupations
13-0000	Business and Financial Operations Occupations
15-0000	Computer and Mathematical Occupations
17-0000	Architecture and Engineering Occupations
19-0000	Life, Physical, and Social Science Occupations
21-0000	Community and Social Services Occupations
23-0000	Legal Occupations
25-0000	Education, Training, and Library Occupations
27-0000	Arts, Design, Entertainment, Sports, and Media Occupations
29-0000	Healthcare Practitioners and Technical Occupations
31-0000	Healthcare Support Occupations
33-0000	Protective Service Occupations
35-0000	Food Preparation and Serving-Related Occupations
37-0000	Building and Grounds Cleaning and Maintenance Occupation
39-0000	Personal Care and Service Occupations
41-0000	Sales and Related Occupations
43-0000	Office and Administrative Support Occupations
45-0000	Farming, Fishing, and Forestry Occupations
47-0000	Construction and Extraction Occupations
49-0000	Installation, Maintenance, and Repair Occupations
51-0000	Production Occupations
53-0000	Transportation and Material Moving Occupations
55-0000	Military-Specific Occupations

4) First-line managers and supervisors of production, service, and sales workers who spend more than 80% of their time performing supervisory activities are classified separately in the appropriate supervisor category, since their work activities are distinct from those of the workers they supervise. First-line managers are generally found in smaller establishments where they perform both supervisory and management functions, such as accounting, marketing, and personnel work.

5) Apprentices and trainees should be classified with the occupations for which they are being trained, whereas helpers and aides should be classified separately.

6) If an occupation is not included as a distinct detailed occupation in the structure, it is classified in the appropriate residual occupation. Residual occupations contain all occupations within a major, minor, or broad group that are not classified separately.

7) When workers may be classified in more than one occupation, they should be classified in the occupation that requires the highest level of skill. If there is no measurable difference in skill requirements, workers are included in the occupation in which they spend the most time.

8) Data collection and reporting agencies should classify workers at the most detailed level possible. Different agencies may use different levels

of aggregation, depending on their ability to collect data and the requirements of users.

These comments underscore that the purpose of the classification scheme is to separate into different categories those workers who perform similar functions on the job at similar levels of skill. In most uses of the SOC, it is noted that the occupational titles are different from job titles in a work organization. For example, a "production assistant" might perform very different duties in a television studio than in a steel plant. Indeed, a consistent tension in constructing occupational classification is deciding how much of the coding structure should reflect the work setting (e.g., an industry) and how much the nature of skills required and functions performed. This is a great example of how uses of a coding structure should determine its character.

The North American Industry Classification System (NAICS). Like the Standard Occupation Classification system, industrial classifications have traditionally been standardarized by government statistical agencies. As the economies of the world become globalized, the need for consistent definitions of industries is key to comparisons of the macroeconomic structures of different nation states. The most recent revision affecting U.S. statistics is the North American Industry Classification System, unveiled in 1997. This was a collaboration among Canada, Mexico, and the United States, and paralleled a separate effort by the United Nations—the International Standard Industrial Classification System (ISIC, Revision 3).

Like every change in a code structure, the revisions were motivated by a mismatch between the current concept being measured and the then available code structure:

1) The existing classification missed many new business types; it did not reflect the economy of the United States.
2) The adoption of the North American Free Trade Agreement underscored the need for comparable industrial production statistics across Canada, Mexico, and the United States.

NAICS includes 1170 industries of which 565 belong to the service sector. The classification system is designed as a six-digit system that provides comparability among the three countries at the five-digit level. The first two digits represent the highest level of aggregation and are called a "sector." Table 10.4 shows the full nomenclature of the classification system's groupings. The inclusion of a sixth digit allows each country to classify economic activities that are important in the respective countries, but may not be noteworthy in all three countries, and thus increases the flexibility of the system.

How did NAICS change the code structure? Table 10.3 shows that the SIC had only 10 divisions but NAICS has expanded that to 24 divisions. Some of the new sectors represent recognizable parts of the SIC divisions, whereas some SIC divisions have been subdivided. Other NAICS sectors combine lower aggregate parts of SIC divisions. The largest amount of change involved the service sector of the economy, which had grown and become internally differentiated over time.

There were many different ways that the code structure could have been changed. What logic was used to choose this way? All involved accept the goal

Table 10.3. Comparison of SIC Divisions and NAICS Sectors

SIC Divisions	NAICS Code	NAICS Sectors
Agriculture, Forestry, and Fishing	11	Agriculture, Forestry, Fishing and Hunting
Mining	21	Mining
Construction	23	Construction
Manufacturing	31–33	Manufacturing
Transportation, Communications, and Public Utilities	22	Utilities
	48–49	Transportation and Warehousing
Wholesale Trade	42	Wholesale Trade
Retail Trade	44–45	Retail Trade
	72	Accommodation and Food Services
Finance, Insurance, and Real Estate	52	Finance and Insurance
	53	Real Estate, Renting and Leasing
Services	51	Information
	54	Professional, Scientific, and Technical Services
	56	Administrative and Support and Waste Management and Remediation Services
	61	Educational Services
	62	Health Care and Social Assistance

that NAICS should reflect the current structure of the economy. However, the phrase "structure of the economy" can have different definitions or explanations and, thus, an industry classification system could be based on one or more of these concepts. Two main approaches can be derived from economic theory (Economic Classification Policy Committee, 1993): a demand-side and a supply-side classification system. A demand-side, or commodity-oriented, classification system is based on the use of the output of the organizations. A supply-side, or production-oriented, classification would group those establishments together that have the same or a very similar production function. A production-oriented

Table 10.4. NAICS Structure and Nomenclature

2-digit	Sector
3-digit	Subsector
4-digit	Industry Group
5-digit	NAICS Industry
6-digit	National

code structure means that statistical agencies in North America would produce statistics useful for measuring productivity, unit labor costs, and the capital intensity of production; constructing input–output relationships; and estimating employment output relationships and other such statistics that require that inputs and outputs be used together.

The SIC, which was developed in the 1930s, was not based on a consistent conceptual framework. Thus, some industrial codes were demand-based, whereas others were production-based (e.g., sugar products were grouped in three different SICs because of their differences in production processes, whereas other products, such as musical instruments, were combined into one industry although different production processes produced them). The lack of a consistent classification concept made it also difficult to explain why data are grouped in one way rather than another. NAICS is based on a production-oriented economic concept. Economic units that use like processes to produce goods or services are grouped together.

10.2.5 Other Common Coding Systems

There are many other classification frameworks that are important in different substantive areas. Several of the coding schemes are used in the health field, for diagnostic guidance to physicians, for administration of health care finance systems, and for epidemiological statistics. The International Classification of Diseases (ICD) and the International Classification of Diseases, Clinical Modification (ICD-9-CM) are used to code and classify the reasons for death on death certificates and reports written on medical records. The Diagnostic and Statistical Manual for Mental Disorders (DSM) is used in psychiatric epidemiological studies that collect mental health symptoms and code the responses into specific disorders. Some health surveys employ these coding schemes for reports of health conditions.

There are also a variety of coding and classification frameworks for geographic entities. Each decade, the U.S. Office of Management and Budget identifies an official set of metropolitan areas for which special tabulations of demographic and economic data are made. In addition, there are area units, called "tracts," defined for decennial census enumeration purposes, based on city blocks and similar units that have clear boundaries. When defined for each census, these consist of 2500 to 8000 persons and are designed to be homogeneous on population characteristics, economic status, and living conditions. Within tracts are block groups and individual blocks, both of which are internal divisions. In many household surveys, matching census data on block groups and tracts permits the analysis of contextual influences on household or person behavior. This often requires a coding of housing unit addresses to block group and/or tract levels.

Finally, some surveys (e.g., household surveys, agricultural surveys) use Global Positioning System (GPS) technology to mark a sample element, both for ease of location and future data supplementation. The GPS provides a geographical coordinate pair that can then be used to link the sample observation to other data that are spatially identified (e.g., remote sensing imagery, water resource, and land use data).

These last examples share a characteristic: categories of a variable are combined and split in various ways in order to permit linkage of a survey record to some other data source, such that the usefulness of the survey is enhanced.

10.2.6 Quality Indicators in Coding

Processing errors can arise at the coding step when coding structures are poorly conceptualized and/or their implementation is not uniformly executed. We discuss two kinds of quality impacts: (a) weaknesses in the coding structure and (b) coder variance.

Weaknesses in the Coding Structure. When a code category combines two responses that are nonequivalent for the purposes of the analyst, consistent and systematic errors can arise. For example, imagine an analyst who is interested in measuring how the effects of obtaining a college degree on job salaries differ from the effects of obtaining a high school diploma or General Education Development (GED) degree. The GED is a status conferred on a person who successfully passes a test measuring knowledge on a variety of topics taught in high schools. However, because GED recipients may not have performed well in high school (had they enrolled for the full curriculum), they may not have the same set of knowledge that graduates typically have, and may not be treated the same in the job market. If a single code category is used to contain responses of those with a high school diploma and those with a GED degree, it is likely that the difference in salaries compared to college graduates would be greater than with a structure that separated GEDs from high school diplomas.

Coder Variance. Most of the attention in survey methodology to the quality of coding examines the increased variation in survey estimates because of variation in coding decisions made among the human coders. It might be worth reviewing Section 9.3, where there is a discussion of interviewer variance. Much of the same logic that underpins interviewer variance applies also to coder variance. "Coder variance" is a component of the overall variance of a survey statistic arising from different patterns of use of code structures by different coders. The perspective is that individual coders vary in their use of the same coding structure—their tendency to use a given code category, their interpretation of certain response words as cues for a given code category, their tendency to use a residual code (e.g., "not elsewhere specified," "other"). The magnitude of this variance component is often measured by the same type of intraclass correlation used in interviewer vari-

coder variance

CODING

ance studies, based on a design that assigns to coders random subsamples of the full sample.

Table 10.5 shows the result of a coder variance study done in Britain, with ρ values measuring coder effects that average about 0.001. These are smaller than interviewer effects presented in Section 9.3, but we need to keep in mind the effect on the variance of a survey statistic. As with interviewer variance, these small ρ values affect the variance of a statistic through

$$Deff = 1 + \rho_c(m-1)(1-r)$$

where ρ_c is the intraclass correlation for coders, m, is the average number of cases coded by an individual coder, and r is the reliability of a particular code.

In practice, coder workloads are often much larger than interviewer workloads. For example, in the study in Table 10.5 each coder coded 322 cases on average. This implies that the average design effect is $1 + 0.001(322 - 1)(1 - 0.903) = 1.03$ or a 3% inflation of variance on a percentage falling in one of the categories because of coder variation in the use of the code structure.

Few studies have tried to reduce the variation in coder performance, although training is likely to be a key to that reduction. Cantor and Esposito (1992), in a qualitative study of interviewers administering the industry and occupation cod-

Table 10.5. Coder Variance Statistics for Occupation Coding

Code Category	Intraclass Correlation, ρ_c	Reliability Estimate	Design Effect Due to Coders
Managers/administrators	0.0005	0.881	1.02
Professional occupations	-0.0019	0.859	0.91
Associate professionals and technical occupations	0.0018	0.836	1.11
Clerical and secretarial occupations	0.0034	0.935	1.09
Craft and related occupations	0.0001	0.929	1.00
Personal and protective service occupations	0.0025	0.950	1.04
Sales and costumer service occupations	-0.0008	0.888	0.97
Plant and machine operatives	0.0000	0.904	1.00
Other occupations	0.0031	0.943	1.06

Source: Campanelli, Thomson, Moon, and Staples, 1997; design effects assume workload of 322 cases.

ing above, developed various recommendations to improve the quality of the coding:

1) Train interviewers to filter as little of the response as possible.
2) Train interviewers on the importance of obtaining an occupation name.
3) Give interviewers probes to use when listing multiple activities.
4) Train interviewers as coders.
5) Provide interviewers with a reference document to assist in probing.

10.2.7 Summary of Coding

To summarize our treatment of coding, the use of coding of text material to numeric data has declined as an unanticipated consequence of increased use of computer assistance in data collection. Coding structures implicitly or explicitly reflect a conceptual framework for the interpretation of responses to survey questions. To be useful to the analysts, the code structure should separate those responses anticipated to be distinctive for their particular statistical purposes. Hence, sometimes different code structures are used on a single question, producing more than one analytic variable.

Through variation in the application of a given code structure, coders can increase the instability of survey estimates. Well-trained coders appear to generate relatively smaller effects than those that interviewers exhibit. However, since coder workloads are generally much higher than interviewer workloads, large increases in standard errors could sometimes arise from coders.

10.3 Entering Numeric Data into Files

The term "data capture" is often used to describe the process of entering numeric data into electronic files. In practice, the nature of data capture is heavily dependent on mode of data collection. In most computer-assisted surveys, either the interviewer or the respondent conducts this step. On touchtone data entry and voice recognition surveys, the respondent enters the data directly into an electronic file. With paper questionnaires, data entry can be conducted by data entry clerks keying the digits one by one, by mark-character recognition, and by optical character recognition.

Using human data entry operators is a costly design feature and, increasingly, surveys are attempting to use some form of computer assistance to avoid those labor costs. A common feature of human data entry operations is 100% rekeying and verification of entries. When used, there is much evidence that the rate of entry error is quite low. In the 1990 U.S. decennial census, errors were found in 0.6% of the fields (U.S. Bureau of the Census, 1993); in the Survey of Income and Program Participation by the U.S. Census Bureau, errors were found in 0.1% of the fields (Jabine, King, and Petroni, 1990). Hence, although the keying operations can be quite well accomplished, their costs encourage survey researchers to adopt computer-assisted modes of data collection to eliminate human data entry after the field period.

10.4 Editing

Editing is the inspection and alteration of collected data, prior to statistical analysis. This inspection might be the review of a completed paper questionnaire by interviewers, supervisors, clerks, or subject matter specialists in the office, or by computer software. The editing may involve examination of data from a single measure or a set of measures. The goal of editing is to verify that the data have properties intended in the original measurement design.

The term "editing" generally includes the alteration of data recorded by the interviewer or respondent to improve the quality of the data. Some uses of the term editing also include coding and imputation, the placement of a number into a field where data were missing.

editing

Editing is accomplished through different kinds of checks. Among those most frequently conducted are:

1) Range edits (e.g., recorded age should lie between 1 month and 120 years) — *range edit*
2) Ratio edits (e.g., the ratio of gallons of milk produced on the farm and number of cows should lie between x and y) — *ratio edit*
3) Comparisons to historical data (e.g., in the wave 2 interview the number of household members should lie within 2 of the wave 1 count)
4) Balance edits (e.g., the percentages of time spent at home, at work, and at some other location should add to 100.0) — *balance edit*
5) Checks of highest and lowest values in the dataset or other detection of implausible outliers
6) Consistency edits (e.g., if recorded age is less than 12, then marital status should be recorded as "never married") — *consistency edit*

Figure 10.2 shows that computer-assisted interviewing software has permitted most of these edits to be part of the data collection itself. This has the attraction of asking the respondent to clarify and, ideally, resolve any problems. However, as Bethlehem (1998) notes, the interaction required to clarify these errors is sometimes complex (e.g., when the discrepancy involves many variables at once). Further, the length of the interview is increased when edit failures must be resolved, risking premature termination of interviews. Not all edits can be feasibly built into CAI applications (e.g., when edits compare survey responses to large external databases). Finally, if the respondent insists on a pattern of answers that violates the edit checks, then the entire interview, if directed by the researcher's logic, may not reflect the respondent's real condition. For that reason, CAI users have created the distinction of "hard checks" for edits that must be followed, and "soft checks" that alert the interviewer to an improbable pattern of answers but permits their retention after that alert.

The amount of editing that is used on a survey is a function of how much factual data (with logical consistency structures) are collected, whether the sample was drawn using a sampling frame with rich information on it, and whether there exist longitudinal data on the case. For this reason, establishment surveys collecting economic data from companies longitudinally generally utilize the largest amount of editing. As an illustration, around 1990 a study of 95 different Federal government surveys found that the vast majority spent over 20% of their total budget on editing data, and most of these were establishment surveys. Most of the surveys had subject matter specialists review the data after some type of automated or clerical editing.

Editing, if not well organized, can lead to near endless changes in the data, with the threat that data quality can actually be reduced. For example, imagine that a check on recorded age and education yields the finding that a reported 14-year-old has a Ph.D. degree. This appears so unlikely that some change in the data record seems likely. A check between age and occupation shows that the reported 14-year-old is not in the labor force. A check of age and the relationship to the head of the household shows the reported 14-year-old is listed as a "son" of the head. Given that the preponderance of the evidence is that a male son of 14 years was measured, it appears that the education variable might be the culprit. A change is made for education (either setting it to missing data or imputing a value that is appropriate for a 14-year-old male living with his parents). All seems fine until another variable is checked, "Have you ever written anything that was published?" The recorded answer is "yes, a Ph.D. dissertation." It appears that the education variable may have been correct. If so, is it the age variable that is wrong? Is it possible that the record applies to a very rare occurrence: a 14-year-old PhD? Without a system of rules to guide these decisions, it's quite possible to fail to iterate to a solution that could be replicated by anyone, including oneself, a few weeks later.

In an important contribution, Fellegi and Holt (1976) invented a system of editing that integrates editing and imputation and has a fixed set of steps, replicable given a set of editing rules. The technique begins with three assertions:

1) Editing should change the smallest number of data fields necessary to pass all the edit checks applied to the data.
2) Editing changes should seek to maintain the marginal and joint frequencies of data whenever possible.
3) Imputation rules should be deducible from editing rules.

Any edit, whatever its original form, can be broken down into a series of statements of the form, "a specified combination of code values is not permissible." For example, imagine a two-variable data record:

Age (whole years of age, at last birthday)

Marital Status
1) Never married
2) Currently married
3) Divorced
4) Widowed
5) Separated

One plausible edit rule is that if the respondent is less than 12, he/she should report "never married." But this implies that the combination of answers "less than 12" and any other marital status is an edit failure. This might be represented by the expression:

(AGE<12 and MARSTAT=MARRIED) = Failure or
(AGE<12) ∩ (MARSTAT=MARRIED) = F
(AGE<12) ∩ (MARSTAT=DIVORCED) = F
(AGE<12) ∩ (MARSTAT=WIDOWED) = F
(AGE<12) ∩ (MARSTAT=SEPARATED) = F

This notion of edit rules implying other edit rules can be generalized to multiple variable edits. One can derive "implied" edits from explicit edits. "Explicit" edits are rules to which data must conform in each survey record, identified by the researcher at time of editing. "Implied edits" are similar rules, logically deduced given some explicit edit rule that must be followed. For example, imagine a data record with age, whether the respondent is registered to vote, and whether the respondent voted. The following two edits, **explicit edit** **implied edit**

(AGE<18) ∩ (REGISVOTE=YES) = F
(REGISVOTE=NO) ∩ (VOTED=YES) = F

imply another edit check:

(AGE<18) ∩ (VOTED=YES) = F

The Fellegi–Holt approach works by identifying data fields that are involved in the largest number of edit failures for a data record. Those are corrected in a way that produces passed edits with the change made. The procedure is designed so that once those changes are made, the record will pass all explicit and implied edits. It achieves this property by ordering the edit check process by the frequency of edit failures. Software systems for editing have been developed, mostly in government statistical bureaus, to implement the Fellegi–Holt procedures.

Summary of Editing. There is some consensus on desirable properties of editing systems. These include the use of explicit rules linked to concepts being measured; the ability to replicate the results; editing shifted to rule-based, computer-assisted routines to save money; minimal distortion of data recorded on the questionnaire; the desirability of coordinating editing and imputation; and the criterion that when all is finished, all records pass all edit checks. The future of editing will not resemble its past. Editing systems will change as computer assistance moves to earlier points in the survey process and becomes integrated with other steps in the survey. It is likely that editing after data collection will decline. It is likely that software systems will increasingly incorporate the knowledge of subject matter experts.

10.5 WEIGHTING

There is another step in processing after data are collected that has not been affected by the large-scale movement to computer-assisted data collection—the construction of record weights that are used in statistical computations.

Surveys with complex sample designs often also have unequal probabilities of selection, variation in response rates across important subgroups, and departures from distributions on key variables that are known from outside sources for the population. It is common within complex sample surveys to generate weights to compensate for each of these features.

Weights arise in survey sampling in a number of different contexts. The purpose of this discussion is merely to illustrate the kind of adjustments that are often used in complex surveys. There are other weighting procedures than the ones given here. The purpose of this section is illustration and not thorough instruction (see Kalton, 1981; Bethlehem, 2002).

We will illustrate four different types of weighting that are prevalent in complex surveys:

first-stage ratio adjustment

1) Weighting as a first-stage ratio adjustment
2) Weighting for differential selection probabilities
3) Weighting to adjust for unit nonresponse
4) Poststratification weighting for sampling variance reduction (and also undercoverage and unit nonresponse)

10.5.1 Weighting with a First-Stage Ratio Adjustment

In stratified multistage designs, like the NCVS and NSDUH, primary selection units (PSUs) are sampled with probabilities proportionate to some size measure. The measures are available on the sampling frame and are usually estimates of the target population size or some proxy indicator of the target population size.

In epsem (equal probability samples) designs, in each stratum the ultimate number of selected units should be proportional to the target population size in the stratum. For example, imagine in NCVS a stratum of the country that contains 0.5% of the household population. The multistage area probability sample of primary sampling units (counties) and secondary units is conducted using population counts on those units. One county is selected in that stratum. A simple estimate of the population size in that stratum based on the one selected PSU is

$$\text{Estimated Stratum Population Total} = \frac{\text{Population Total in Selected PSU}}{\text{Probability of Selecting PSU}}$$

An estimate that has lower sampling variance is the first-stage ratio adjusted total, which weights all cases in the chosen PSU by

$$W_{i1} = \text{First Stage Ratio Adjustment Weight} = \frac{\text{Stratum Population Total from Frame}}{\text{Population Total for Selected PSU} / \text{Probability of Selecting PSU}}$$

What this accomplishes is a stabilizing of estimates over different selections of PSUs in the stratum, so that the weighted total is consistent over realizations of the sample design (Cochran, 1977).

WEIGHTING

For every respondent case that lies in the selected PSU, there is a new variable created that has the value equal to the first-stage ratio adjustment weight. We could label this as W_{i1}, where the "1" subscript will indicate that this is the first weighting factor in a set of factors; the capital letter is used to remind us that this weight is based on frame population, not sample data. Eventually, when we finish presenting each of the common weighting factors in Section 10.5.4, we will combine them all into one final weight, which is the product of the individual factors.

10.5.2 Weighting for Differential Selection Probabilities

Suppose a sample of 125,000 is to be selected for the NCVS among persons ages 12 years and older, and suppose that among the 285 million persons in the United States, 199.5 million, or 70%, are ages 12 years and older. The overall survey would require a sampling fraction of $f = 125{,}000/199{,}500{,}000 = 1/1596$.

The growing Latino population, and differences in crime victimization between Latino and non-Latino populations, raise the question whether the number of Latino persons in the NCVS, under the overall sample of 125,000, is sufficient. Suppose that one person in eight aged 12 years and older is Latino, or a total population of just about 25 million. If an epsem sample were selected, the sample would have 15,625 Latinos and 109,375 non-Latinos. That is, under proportionate allocation, one person in eight in the sample would be Latino as well. The sample results could be combined across Latino and non-Latino populations with no need for weights to compensate for overrepresentation of one group.

Contrast this design with one deliberately set to greatly increase the number of Latinos in the sample to one-half, or 62,500 Latinos. There would have to be a corresponding drop in the number of non-Latinos in order to maintain the overall sample size of 125,000. This allocation of the sample might be attractive if there were strong desires to estimate the Latino population victimization statistics with much higher precision than would be afforded with a proportionate allocation.

In order to accomplish this sample size distribution, the sampling rate for Latinos would have to be increased dramatically, from 1 in 1596 in the proportionate design to 1 in 399. In addition, the sampling rate for non-Latinos would have to drop to 1 in 2793. The sample would have equal numbers of Latino and non-Latino subjects—62,500 each.

As long as the only work done with the survey contrasted or compared Latino and non-Latino groups, or computed estimates separately for the two groups, the differences in sampling rates between the two groups would not be a problem statistically. There is a problem, though, when the data need to be combined across the groups. The need for combining across groups might arise because of a call for national estimates that ignore ethnicity, or the need for data on a "crossclass" that has sample persons who are Latino as well as non-Latino, such as women.

When combining across these groups to get estimates for the entire population, ignoring ethnicity, something must be done to compensate for the substantial overrepresentation of Latinos in the sample. Weights in a weighted analysis applied to individual values are one way to accomplish this adjustment. Recall that the weighted mean can be computed, when individual level weights are available, as

$$\bar{y}_w = \frac{\sum_{i=1}^{n} w_{i2} y_i}{\sum_{i=1}^{n} w_{i2}}$$

selection weight

As discussed under stratified sampling in Section 4.5, one can use the inverse of the selection probability as a "selection weight" for each case. (Indeed, this is the same expression as that on page 119, but we use w_{i2} instead of just w_i to denote that the selection weight will be viewed as the second weighting component, after first-stage ratio adjustment.) Each Latino in the sample would be assigned a weight of 399, whereas for each non-Latino a weight of 2793 would be used.

In the weighted mean $\bar{y}_w$, the weights (w_{i2}) appear in the numerator and denominator of the expression. From an algebraic point of view, then, the weights may be scaled up or down, and the value of the estimate will not change. That is, what is important for the mean is not the values of the weights, but their relative values. So weights of 399 and 2793 could just as easily be converted to a more easily remembered and checked set of numbers. For example, 2793/399 = 7, so a weight of 1 could be assigned to every Latino case, and 7 to every non-Latino.

This is a large weight differentiation between the two groups. When data are combined across groups, this weighting decreases the contribution of values for a variable for Latinos to 1/7th the contribution of non-Latinos. This adjustment allows the sample cases to contribute to estimates for the total population in a correct proportionate share.

In the survey dataset, each Latino subject would receive a weight of 1 (or 399), whereas each non-Latino would receive a weight of 7 (or 2793). When estimates combine across groups, these weights compensate properly. When separate estimates are computed for each group, these weights can also be used. They will make no difference in comparisons because they are all the same value for everyone in the same group.

10.5.3 Weighting to Adjust for Unit Nonresponse

Surveys like the NCVS are subject to nonresponse. The nonresponse rates will not be the same across all groups of the population. Suppose that in the NCVS, younger persons, aged 12–44 years, respond at an 80% rate, whereas older persons, aged 45 years or older, respond at an 88% rate. The resulting sample will consist of 105,000 of the 125,000 persons in the sample and have the distribution shown in Table 10.6.

The distribution assumes that age, or at least age group, is known for everyone in the sample, respondents and nonrespondents alike. This is an important constraint on the type of nonresponse adjustment described here. In practice, nonresponse adjustment classes must be limited to those that can be formed from variables that are known for every sample person. That is, they cannot use variables that are collected in the survey and available only for respondents.

In the initial sample, there are an equal number of persons aged 12–44 and 45 years or older. However, among the respondents, there is a larger proportion ages 45 years or older. Thus, the nonresponse mechanism has led to an overrepresentation of older persons among the respondents.

Table 10.6. Hypothetical Equal Allocation for Latinos, with Nonresponse Adjustments

	Population Size	Sample Size	Respondents	Response Rate	Nonresponse Adjustment Weight, w_{i3}	Nonresponse Adjusted Weight
Latino	24,937,500	62,500	52,500	0.84		
12–44		31,250	25,000	0.80	1.25	1.25
45+		31,250	27,500	0.88	1.14	1.14
Non-Latino	174,562,500	62,500	52,500	0.84		
12–44		31,250	25,000	0.80	1.25	8.75
45+		31,250	27,500	0.88	1.14	7.95
Total	199,500,000	125,000	105,000			

nonresponse weight

To compensate for this overrepresentation, an assumption is made in survey nonresponse weighting that generates the same kind of weighting adjustments discussed for unequal probabilities of selection. If one is willing to assume that within subgroups (in this case, age groups) the respondents are a random sample of all sample persons, then the response rate in the group represents a sampling rate. This assumption is referred to as the "missing at random" assumption, and is the basis for much nonresponse adjusted weights. The inverse of the response rate can thus be used as a weight to restore the respondent distribution to the original sample distribution. In Table 10.6, the nonresponse adjustment weights are 1.25 and 1.14, respectively, for the younger and older age groups.

These adjustment weights are then used in conjunction with the basic weight that adjusted for unequal probabilities of selection. The base weight for Latinos is 1.0, and for non-Latinos it is 7.0. The nonresponse adjusted weight is obtained as the product of the base weight (which consists of $w_{i1} \times w_{i2}$) and the nonresponse adjustment (w_{i3}), and is shown in the last column of Table 10.6.

If the nonresponse adjusted weights are summed across the 105,000 respondents, the sum of the weights will be equal between the two age groups. In other words, the sample distribution of equal number of persons in each age group has been restored through the weighted distribution.

10.5.4 Poststratification Weighting

A final weighting procedure applied to many surveys is poststratification. Suppose, for example, that after the nonresponse adjustment, the weights are summed separately for males and females, and the sum of the weights for each gender is the same. However, from an outside source, it is known that there are actually slightly more females than males in the population—52% versus 48%. "Poststratification" uses case weights to assure that sample totals equal some external total based on the target population. For statistical efficiency as well as for appearance reasons, it would be helpful if the sums of weights for each gender followed this outside distribution. This can be achieved by a further adjustment to the nonresponse adjusted weights.

poststratification weight

In particular, the male weights need to be deflated, and the female weights inflated. If we deflate the weights for each male respondent by $(0.48/0.50) = 0.96$, and inflate each female respondent's weight by $(0.52/0.50) = 1.04$, the outside population gender distribution will be restored for the weighted sample. In the penultimate column of Table 10.7, the poststratification weight is labeled W_{i4} because it is the fourth weighting factor; the capital letters are used to denote that it is based on full target population information, not sample information.

10.5.5 Putting All the Weights Together

Finally, to obtain a final weight than can incorporate the first-stage ratio adjustment (W_{i1}), the unequal probability of selection adjustment (w_{i2}), nonresponse adjustment (w_{i3}), and poststratification (W_{i4}), a final product of all four weights is assigned to each of eight classes ($W_{i1} \times w_{i2} \times w_{i3} \times W_{i4}$): males ages 12–44 and Latino, males 45 years and older and Latino, and so on. The final weights are shown for each of the eight classes in the last column of Table 10.7. For example, all 12,500 Latino males ages 12–44 receive a final weight of 1.20.

Table 10.7. Weighted Sample Distribution and Poststratification for Hypothetical NCVS Sample by Gender, Age, and Ethnicity

	Respondents	Sum of Nonresponse Adjusted Weights	Weighted Sample Distribution	Population Distribution	Poststratification Weight w_{i4}	Final Weight $w_{i1} \times w_{i2} \times w_{i3} \times w_{i4}$
Male	52,500	250,000	0.50	0.48		
12–44	25,000	125,000				
Latino	12,500	15,625			0.96	1.20
Non-Latino	12,500	109,375			0.96	8.40
45+	27,500	125,000				
Latino	13,750	15,625			0.96	1.09
Non-Latino	13,750	109,375			0.96	7.64
Female	52,500	250,000	0.50	0.52		
12–44	25,000	125,000				
Latino	12,500	15,625			1.04	1.3
Non-Latino	12,500	109,375			1.04	9.1
45+	27,500	125,000				
Latino	13,750	15,625			1.04	1.18
Non-Latino	13,750	109,375			1.04	8.27
Total	105,000		105,000			

> **Ekholm and Laaksonen (1991) on Propensity Model Weighting to Adjust for Unit Nonresponse**
>
> Ekholm and Laaksonen (1991) use the expected value of a response propensity to weight respondent records, compensating for unit nonresponse.
>
> *Study design*: The Finnish Household Budget Survey is a stratified random sample of persons on the population register. The survey estimates total consumption and mean consumption per household of various goods and services. The overall response rate is 70%. The authors estimate a logistic regression predicting the likelihood of being a respondent by using a four-factor model based on household composition, urbanicity, region, and property income, which together form 128 crossclassification cells. After collapsing cells connected with poor fit to the model, they estimate average propensities in 125 separate cells. They compare this to using 35 poststrata for weighting.
>
> *Findings*: Estimated response probabilities ranged from about 0.40 to 0.90, the lowest for older single-person households with no property income living in the capital city area; the highest for households with two young persons with some property income in the middle region. Using the propensity weighting estimator for total persons in the country matches independent data better than the poststratified estimator. Because those households with low propensity are smaller and consume less, the mean consumption per household is lower when weighted with the propensity model.
>
> *Limitations of the study*: There was no external validation of the estimates subjected to the propensity weighting. There was no description of alternative propensity models that may have performed differently.
>
> *Impact of the study*: The study demonstrates that multivariate logistic models, estimating the probability of response for each respondent, can be useful in compensating for unit nonresponse.

This final weight, which combines the four different weights, should appear on each respondent data record as a variable to be used in analysis of the data. Some datasets will present the individual component weights as separate variables, as well as their product.

A question is raised in analysis of survey data about whether the weights must be used in any, or only some, analyses. The question arises because there can be many survey estimates for which the weighted and unweighted values are virtually indistinguishable. If that is the case, why bother with the weights, especially when it is recognized that weights can lead to increase in variances and less precise results?

There are analytic instances in which the weights are not necessary. For example, when comparing two of the final eight weighting classes in the small example shown in Table 10.4, if the weights are the same for all individuals within a class, the weights are not needed.

The problem, as for stratified sampling, arises when results are obtained by combining sample cases across weighting classes. Even though for a particular variable there may be no difference between weighted and unweighted estimates, there is no guarantee that the same lack of difference exists for other estimates. This "no difference" condition is particularly difficult to predict for subclasses, subsets of the sample for which separate estimates are desired.

Analysts could use weights in some analyses and not in others. There may then be inconsistencies between analyses that are difficult to explain or interpret.

Survey statisticians routinely use weights in all analyses of a survey dataset. Estimates for descriptive statistics on the population and its subgroups are thus estimated by taking into account design features such as unequal probabilities of selection and nonresponse. Analytic findings remain consistent with the underlying descriptive estimates as well.

10.6 Imputation for Item-Missing data

The nonresponse adjustments described in the last section are designed to address one form of

IMPUTATION FOR ITEM-MISSING DATA

nonresponse: failure to obtain response from an entire unit chosen for the study. They do not adjust for item nonresponse.

As we discussed in Sections 2.2.8 and 6.6, item nonresponse is a label for failing to obtain data for a particular variable (or item) in an interview or questionnaire when data for other variables in the survey have been obtained. For example, in the NCVS, a respondent who has provided information on crime victimization throughout the interview may refuse to answer the question on family income.

Figure 10.4 (repeated from Chapter 2) represents a summary of these two kinds of nonresponse in a survey. The figure presents a rectangular layout for a survey dataset, with rows representing sample subjects and columns the variables collected for each subject. The cases at the bottom are unit nonrespondents; they have data only on the left, representing just the variables transferred from the sampling frame. Hence, the data records are sorted by unit response status, with rows at the bottom representing survey unit nonrespondents. Even for unit nonresponse, a survey will have some (limited) information about each subject that comes from the sample selection process. In a survey like the NCVS, the sample information will include where the subject's household is located, for example.

Among the respondent records (at the top of the figure), item nonresponse is represented by the absence of a number in a data field. The rate of item-missing data varies over variables. Surveys often find the highest rates of item nonresponse for income and asset kinds of information.

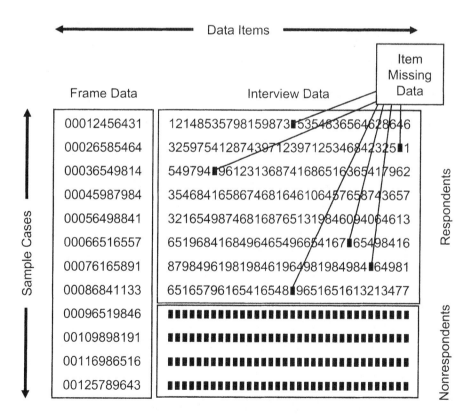

Figure 10.4 Unit and item nonresponse in a survey data file.

There is a significant problem for analysis of datasets with the kind of missing-data structure shown in the figure. It is not clear how best to handle the missing values in the analysis. There are two basic strategies that are used: "ignoring" the problem, or making an adjustment to the data.

One strategy is sometimes called casewise deletion of missing data, or complete case analysis. Item-missing values are ignored by deleting a case (a row) for which there are any missing data for an analysis being conducted. Suppose, for example, that in the NCVS one were regressing the incidence of being a victim of robbery on several predictors, such as age, gender, and family income. Complete case analysis would drop any cases for which any of these four variables in the analysis were missing. That is, if the respondent did not report about whether they had been robbed, did not provide their age or gender, or refused to give a family income, the entire case would be dropped. A missing value on one item means that the other three observed items are ignored.

From an inference viewpoint, in which the survey results are being "projected" to the population, the complete case analysis is actually a form of adjustment. Dropped cases are not exactly being ignored. Instead, there is an implicit imputation or replacement of missing values occurring through the analysis of only complete cases. Effectively, the complete case analysis "imputes" or assigns to each of the missing cases the average or result from all of the complete cases. In other words, missing values are not being ignored but, by inference, the analyst assumes that the result obtained for the respondents applies to the nonrespondents as well.

imputation
The alternative to the complete case analysis approach is to make the imputation explicit. "Imputation" is the placement of one or more estimated answers into a field of a data record that previously had no data or had incorrect or implausible data. Of course, any survey organization that imputes missing values also must indicate which values have been replaced through an "imputation flag" variable for each item so that analysts can decide whether they want to accept or reject the imputed value.

This explicit imputation approach has the advantages that the method for imputation is known, that the analyses will be based on the same number of cases each time, and that all of the data provided by subjects are being used in each analysis. It has disadvantages as well. For many analysts, imputed values are equivalent to using "made up" data, regardless of how well the imputation might be done, and this makes them uncomfortable. Of course, they cannot ignore the fact that an imputation implicitly occurs even if they "ignore" the missing values in complete case analysis. In addition, imputed values are treated in nearly all statistical software systems as real values. Standard errors of estimates are then underestimated, leading to confidence intervals that are too narrow, or to test statistics that are too large.

There are a large number of imputation procedures that have been devised to compensate for missing values. Only a few are discussed here.

Perhaps the simplest method is to replace the missing values for a particular variable by the mean of the item respondent values, say $\bar{y}_r$, for that variable. So, for instance, in the NCVS, missing values of family income are replaced with the average family income (or median or modal value, if income is recorded in categories) for all individuals who reported income. This mean value imputation is roughly equivalent to complete case analysis.

Mean value imputation has several disadvantages. One is that if there are many missing values in the imputed dataset, there is a "spike" in the distribution of the variable at the mean value. The distribution of values in the imputed data is distorted.

mean value imputation

This distortion can be addressed in several ways. One is to add a "stochastic" or random element to the imputed value to reflect the known variability of estimates. The imputed values then have the form $\bar{y}_r + s_i^2$, where s_i^2 may be a value chosen from a normal distribution with mean zero and variance equal to the element variance among nonmissing values for the variable.

The mean value imputation can be done by subgroup as well to provide a more accurate predicted value. For instance, in the NCVS, family income imputation could be done in subgroups formed by race or ethnicity. Mean family incomes would be computed for African American and non-African American subgroups, and every item missing value for an African American is replaced by the mean for African Americans. A stochastic or random value can also be added to this imputed value to avoid distorting the family income distribution for subgroups.

This approach of imputing within subgroups, and adding a randomly generated residual, can be thought of in terms of a regression prediction for the missing value. Family income can be regressed on a dummy variable x_1 for race in the model

$$y_{i(r)} = \beta_0 + \beta_1 x_{1i(r)} + \varepsilon_{i(r)}$$

where only item respondents (designated by the subscript r) are used in the estimation. Then a predicted value is obtained for each missing value as

$$y_{i(r)} = \beta_0' + \beta_1' x_{1i(r)} + \varepsilon_{i(r)}'$$

where β_0' and β_1' are estimated coefficients and the "estimated residual" is a randomly generated variance added to the predicted value.

The regression imputation procedure can also be used with more than one predictor, with a mix of dummy and other predictors on the right-hand side of the equation. That is, a regression imputation can be estimated from respondent data for an item of the form

$$y_{i(r)} = \beta_0 + \beta_1 x_{1i(r)} + \beta_2 x_{2i(r)} + \cdots + \beta_p x_{pi(r)} + \varepsilon_{i(r)}$$

for p predictors x_j, where $j = 1, 2, \ldots, p$.

Regression imputation is sometimes used to impute missing values for a few variables in a survey. It must be adapted to fit the distribution of the imputed variable. If, for instance, family income is reported in several categories, a multinomial logit may be used to obtain predicted probabilities, which can in turn be converted to a category with or without a stochastic contribution.

Regression imputation requires that all values used as predictors are themselves present for the imputation of a given dependent variable. Regression imputation may require that variables be imputed in sequence so that predictors in a model have imputed values themselves.

regression imputation

hot-deck imputation

The regression method has a variety of valuable properties, but large-scale imputation processes often utilize a method developed decades ago—the sequential hot-deck procedure. The hot-deck procedure can be viewed as a form of regression imputation as well, but one in which the predicted residual is "borrowed" from another case in the data set. The procedure is illustrated in Table 10.8.

Suppose that family income is to be imputed for this small dataset, where gender and education are known for each of the 18 individuals in the dataset. Family income is missing for four individuals, and thus reported for 14. The sequential hot deck begins by sorting the data by variables such as gender and education. In Table 10.8, the data are sorted first by gender, and then by education within gender. This places individuals with the same gender or similar education levels next to one another on the list.

The imputation begins by moving through the sorted list in order. Family income for the first case in the data set is examined. If it is missing, it is replaced

Table 10.8. Illustration of Sequential Hot-Deck Imputation for Family Income, Imputed Data, and Imputation Flag Variable

Respondent Number	Gender	Education	Family Income	Hot Value	Imputed Data	Imputation Flag
1	M	9	23	51	23	0
4	M	11		23	23	1
2	M	12		23	23	1
3	M	12	43	23	43	0
7	M	12	35	43	35	0
8	M	12	42	35	42	0
5	M	16	75	42	75	0
6	M	16	88	75	88	0
16	F	10		88	88	1
15	F	12	28	88	28	0
17	F	12	31	28	31	0
18	F	12	35	31	35	0
19	F	12	30	35	30	0
22	F	12		30	30	1
13	F	14	67	30	67	0
14	F	15	56	67	56	0
21	F	15	72	56	72	0
20	F	18	66	72	66	0

by the mean income for the total sample, or the mean for a subgroup that is used in the sorting process. In the illustration, the mean income for responding males (the first subgroup in the sorted data) is $51,000, and it would be imputed to the first case if family income were missing. This initial imputed value is sometimes referred to as the "cold-deck" value.

If the first family income value is not missing, the reported value is stored as the "hot-deck" value, and the next case is examined. If it is missing, it is replaced by the most recently stored "hot value." In the illustration, the second value is missing, and it is replaced by the reported family income from the first case. If, on the other hand, the value is not missing, its reported family income value is stored as the "hot value."

For subsequent cases in the sort sequence, the same process is repeated. If the value is missing, it is replaced by the hot value. If it is reported, it is stored as the hot value. Thus, missing values are imputed by the most recent reported value in the sort sequence. Since the neighboring cases are similar to one another with respect to variables that are related to income, the imputation uses the similarity in sort variables much like predictors in the regression imputation procedure. Further, if a case being imputed is identical in values on the sort variables to the preceding case, which has a reported income value, we can think of this process as imputing the average value for all item respondents with the same values and adding a residual to the average. The residual in this case is "selected" from another subject in that set of cases with identical values on the sort variables.

There are a number of other imputation methods that are used in practice, but it is beyond the scope of this volume to consider them. As noted previously, the imputation process does give data that will tend to underestimate the variances of estimates. This underestimation can be remedied through specialized variance estimation procedures, through a model for the imputation process itself, or through a procedure called multiple imputation. "Multiple imputation" creates multiple imputed datasets, each one based on a different realization of an imputation model for each item imputed (see Rubin, 1987; Little and Rubin, 2002). Variation in estimates across these multiple data sets permits estimation of overall variation, including both sampling and imputation variance.

multiple imputation

10.7 SAMPLING VARIANCE ESTIMATION FOR COMPLEX SAMPLES

Stratification, multistage sample selection, weights, and imputed values are features of survey data that require nonstandard procedures to estimate variances correctly. For instance, as shown in Section 4.5, stratification must be handled in variance estimation by the computation of variances within strata first, and then the combination of estimated variances across strata.

The survey analyst seeking to take into account these kinds of design features must turn to specialized sampling variance estimation procedures and software packages that implement them. Three variance estimation procedures that handle the special features of survey data, and software packages that implement them, are described briefly here.

Survey datasets based on stratified multistage samples with weights designed to compensate for unequal probabilities of selection or nonresponse are sometimes referred to as "complex survey data." Estimation of sampling variances for

complex survey data

statistics from these kinds of data require widely available specialized software systems using one of three procedures: the Taylor series approximation, balanced repeated replication, or jackknife repeated replication.

Taylor series

Taylor Series Estimation. The Taylor series approximation is a commonly used tool in statistics for handling variance estimation for statistics that are not simple additions of sample values. For example, the variance of an odds ratio is difficult to obtain because of the presence of proportions or sample sizes in the denominator of the ratio that are themselves sample estimates.

The Taylor series approximation handles this difficulty by converting a ratio into an approximation that does not involve ratios, but instead is a function of sums of sample values. Taylor series approximations have been worked out analytically for many kinds of statistics, and for stratified multistage sample designs with weights. For example, the variance for a simple ratio mean like

$$\bar{y}_w = \frac{\sum_{i=1}^{n} w_i y_i}{\sum_{i=1}^{n} w_i}$$

using a Taylor series approximation (assuming simple random sampling, for simplicity) is

$$\frac{1}{(\sum w_i)^2}\left[Var(\sum w_i y_i) + \bar{y}_w^2 Var(\sum w_i) - 2\bar{y}_w Cov(\sum w_i y_i, \sum w_i)\right]$$

This looks complicated but is just a combination of the types of calculations that are made for the simple random sample variance estimates (see for example the computations on page 99). Taylor series estimation is perhaps the most common approach to estimating the sampling variance for means and proportions in complex sample designs, mainly because the currently most popular software packages utilize this approach.

balanced repeated replication

jackknife replication

Balanced Repeated Replication and Jackknife Replication. The balanced repeated and jackknife repeated methods take an entirely different approach. Rather than attempting to find an analytic solution to the problem of estimating the sampling variance of a statistic, they rely on repeated subsampling. One can think of repeated replication as being similar to drawing not one but many samples from the same population at the same time. For each sample, an estimate of some statistic, say the mean $\bar{y}_\gamma$, is computed for each sample γ.

An estimate of the mean in the population is computed as an average of the c different sample estimates,

$$\bar{y} = \left(\frac{1}{c}\right)\sum_{\gamma=1}^{c}\bar{y}_\gamma$$

The variance of this estimate is computed as the variability of the separate sample estimates around the mean, or

$$v(\bar{y}) = \left(\frac{1}{c(c-1)}\right) \sum_{\gamma=1}^{c} (\bar{y}_\gamma - \bar{y})^2$$

The strength of this approach to estimating the sampling variance of a statistic is that it can be applied to almost any kind of statistic: means, proportions, regression coefficients, and medians. Using these procedures requires thousands of calculations made feasible only with high-speed computing.

The balanced and jackknife repeated replication procedures are based on slightly different approaches. For both, variation is measured among subsamples that are identified within the full sample. These subsamples are called "replicates" and are drawn from the original sample such that each replicate contains the basic features of the full sample (except the number of sample elements). For example, if a sample were selected using clusters and stratification of those clusters, the subsampling procedure would select sample clusters from each stratum. Many different subsample replicates can be defined; variation in the subsample estimates is the basis of the estimate of the sampling variance for the total sample estimate.

A replicate estimate $\bar{y}_\gamma$ is computed from the first subsample. Then another subsample is drawn from the sample, and the estimate computed again. This subsampling, or replication procedure, is repeated a number of times. The variance estimate is computed following a very similar procedure to that described above. However, rather than using the average of the estimates in the variance estimation, the estimate computed from the total sample is used instead.

The difference between the balanced and jackknife procedures is in how the subsamples are selected. The balanced repeated replication procedure typically identifies a "half sample replicate," selecting half the cases, or half of the clusters, or some other "half sampling" method. The jackknife repeated replication procedure is sometimes called a "drop-out" procedure. It drops a case, or a cluster, from the dataset, to obtain each successive subsample.

The variance estimates obtained from each of these approaches are remarkably similar for a given statistic and dataset. There is little reason to choose one method over the other. The Taylor series approximation is the most commonly used approach, but that does not mean that it provides a more precise or accurate estimate of variance than the other procedures.

These methods are implemented in a variety of special-purpose survey estimation software. Presenting the names and features of this software does not imply an endorsement of one product over another. We describe them here so that the reader can examine the features of the software and make a more informed decision about which to use for their particular purpose.

For complex sample surveys, all of the packages described here require certain information be present on each data record. They require that a stratum, cluster or primary sampling unit, and weight be specified for the estimation. If the selection is stratified random, one can give a case level identification variable instead of a cluster or PSU number. If no weights are needed, a weight variable can be generated that is equal to 1 and used in the analysis without effect on results. And if no stratification has been used, a stratification variable equal to one for all cases can be used. All software does require, though, that there be at least two clusters, or elements, or PSUs, in each stratum in order for variances to be estimated.

CENVAR is a module in a statistical package developed and distributed by the U.S. Bureau of the Census, the Integrated Microcomputer Processing System. CENVAR can estimate variances for means and proportions for the total sample and subgroups using a Taylor series approximation. The system itself is menu driven. It can handle stratified element and stratified multistage sample designs. The integrated package is available for free from the International Programs Center at the Bureau of the Census (see http://www.census.gov for details).

VPLX is a package also developed at the U.S. Bureau of the Census. It handles a wider range of statistics than can CENVAR and incorporates a separate treatment of poststratification if replicate weights are provided. VPLX uses repeated replication procedures, including balanced and jackknife methods. It is associated with another program, CPLX, which has features designed to handle contingency table analysis. VPLX and CPLX are both available for free through the Bureau of the Census website.

EpiInfo is another free package, developed and distributed by the U.S. Centers for Disease Control for epidemiological analysis. Within EpiInfo is a module for computing variance estimates for complex sample survey data. It employs Taylor series approximation methods for means and proportions for the total sample and subclasses. EpiInfo uses a screen oriented Windows-based system. The program can be downloaded from the Centers for Disease Control website at http://www.cdc.gov/epiinfo/.

All of the remaining software systems described here require payment of a licensing fee, and thus cannot be downloaded for free.

SAS version 9, a widely used statistical software system, has added several procedures (PROCs) that estimate variances for complex sample designs. One procedure handles basic means and proportions, whereas the other handles linear regression. The variance estimation is through a Taylor series approximation. More information about the SAS system is available at http://www.sas.com.

STATA is a programmable statistical analysis system (see http://www.stata.com). It has a wide range of statistical procedures available, and a number of those are also available for complex sample survey data. The latter procedures are known in STATA as the "svy" or survey versions of various commands. The available statistical procedures include descriptive statistics such as means and proportions, as well as linear, logistic, multinomial logit, probit, Cox proportional hazards, and Poisson regression.

SUDAAN is a system for complex sample survey data variance estimation developed at the Research Triangle Institute (see http://www.rti.org for details). Users have a choice of Taylor series or repeated replication procedures, and a wide range of analytic methods, from descriptive statistics and ratio estimators to linear, logistic, multinomial logit, probit, and Cox proportional hazards regression. The system is keyword-driven like SAS, and uses a syntax that is very similar to SAS. Further, there are versions of SUDAAN that are compatible with SAS, and can actually be loaded into the SAS system as a separate set of commands.

Finally, WesVar is a package developed by Westat, Inc. for repeated replication variance estimation of complex sample survey data (see http://www.westat.com for details). The program handles a variety of estimation problems, from totals, means, and other descriptive statistics to linear regression. It also has a feature to estimate the effect of poststratification. There is an older version of the program available for free, but the most up-to-date version does require payment of a licensing fee.

10.8 SURVEY DATA DOCUMENTATION AND METADATA

Rarely is a survey dataset constructed for a single analyst. Rarely is analysis on a survey dataset done once, immediately after the data collection, and then never again. Instead, survey data are often analyzed and reanalyzed by many people over many years. Indeed, large survey data archives, like that of the Interuniversity Consortium for Political and Social Research (http://www.icpsr.org) or the Roper Center (http://www.ropercenter.uconn.edu/) contain thousands of survey datasets available for analysis, some of them dating to the first part of the 20th century.

Given this, survey designers have to plan for the use of the data by many people, both within a project and across the world. This requires documentation of all the basic properties of a dataset that a user would want to understand before analyzing the data. The term used most often for this information about the data (or "data about data") is "metadata." "Metadata" is a term that describes all the information that is required by any potential user to effectively gain knowledge from the survey data. If one places no bounds on who the user of the data might be, the ability to define what metadata are required is equally boundless. In the survey context, Colledge and Boyko (2000) list the common types of metadata that are required:

metadata

1) Definitional—describing target population, sampling frames, question wording, and coding terminology
2) Procedural—describing protocols for training interviewers, developing the frame, selecting respondents, and collecting the data
3) Operational—containing evaluations of the procedures like item-missing data rates, edit failure rates, average length of interviews, and number of cases completed per interviewer
4) Systems—related to dataset format, file location, retrieval protocol, definitions of data record fields, and type of sample element asked about the data item

This statement of metadata types is a radical departure from the early days of surveys when an electronic data file, described by a printed codebook, and a memorandum describing the sample and field design were the standard documentation products of a survey. However, the basic codebook is still a part of routine documentation.

codebook

For example, Figure 10.5 reproduces a part of the NCVS codebook for the 2001 data. Note that it has different entries for each variable in the data file. In this section of the data record, the fields relate to individual questions asked of the household respondent. The figure shows two fields—one for VAR V2081, a screener question about vandalism, and one for VAR V2082, a report on what was damaged. Each field has a header that supplies some definitional information (name of variable and question wording) and some systems information [data record location (e.g., COL 140 WID 1, for starting column and width of field),

VAR V2081   VANDALISM AGAINST HOUSEHOLD   NUMERIC
COL 140 WID 1   MISSING 9   HOUSEHOLD DATA

Source code 557

Q.46a Now I'd like to ask about vandalism that may have been committed during the last 6 months against your household. Vandalism is the deliberate, intentional damage to or destruction of property. Examples are breaking windows, slashing tires, and painting graffiti on walls. Since _____ __, 19__, has anyone intentionally damaged or destroyed property owned by you or someone else in your household? (Exclude any damage done in conjunction with incidents already mentioned.)

1. Yes
2. No
3. Refused
8. Residue
9. Out of universe

VAR V2082   LI VANDALISM OBJECT DAMAGED   NUMERIC
COL 141 WID 1   MISSING 9   HOUSEHOLD DATA

Source code 558

Q.46b What kind of property was damaged or destroyed in this/these act(s) of vandalism? Anything else?
MARK (X) ALL PROPERTY THAT WAS DAMAGED OR DESTROYED BY VANDALISM DURING REFERENCE PERIOD.

Lead-in code

(Summary of single response entries for multiple response question. Detailed responses are given in VARS V2083–V2092)

1. At least one good entry in one or more of the answer category codes 1–9
8. No good entry (out of range) in any of the answer category codes 1–9
9. Out of universe

Note: For a "Yes–NA" entry, the lead-in code is equal to 1, the category codes are equal to 0 and the residue code is equal to 8.

Figure 10.5 Illustration of printable codebook section for the National Crime Victimization Survey.

missing data code values, level of the data set (i.e., HOUSEHOLD versus person-level data), and nonmissing data code values and definitions]. In capital letters for V2082 there is some procedural metadata "MARK (X) ALL PROPERTY THAT WAS DAMAGED OR DESTROYED BY VANDALISM DURING REFERENCE PERIOD," which is an instruction to the NCVS interviewer on how to document the respondent's answer. Finally, for V2082 there is some further procedural information that concerns an editing and recoding step that took place after data were collected. (Summary of single response entries for multiple response question. Detailed responses are given in VARS V2083–V2092). This reminds the user that this data field was a postsurvey processing step, combining individual responses provided into a single summary measure for analysis. Codebooks are key tools for the data analyst to make the bridge between the data collection protocol and the individual data items themselves.

With the ability to move from paper codebooks to electronic survey documentation, the world of metadata has become greatly elaborated in the last few years. With the hyperlinking capacity of Web presentation, several pieces of metadata can be linked together to make available simultaneously information from various levels of aggregation. The vision of research and development in this area is that a user could query a Web-based site of study documentation in several ways. If the users were novices to the study, they could inquire about what measures were taken in a particular substantive domain (e.g., socioeconomic status) and retrieve a list of variables. The list would be hyperlinked to specific metadata on the individual variables: the question wording, past study uses of the question, documented measures of simple response variance on the item, analyses in the literature that have used the item previously, as well as the traditional metadata on codes used, missing data, and data record location. Such a structure of metadata would anticipate the needs of large classes of users and then design links among the information that they would need to facilitate faster answers from survey data.

This movement toward elaborated metadata raises previously unasked questions for the survey methodologist. Burdens and opportunities abound. The survey methodologist has the opportunity to vastly increase the audience for measures of survey quality. Reinterview studies generating simple response variance estimates can be easily available to users. The multiple versions of pretest questions, with associated evaluations from behavior coding, cognitive interviews, or other questionnaire developments can be hyperlinked to the question wording actually used. When a user is considering analysis involving that variable, all of the information regarding its measurement qualities can be presented. The burden of this new world is that part of the survey design must now include the design of metadata. At the time the basic conceptual, target population, measurement, and sampling issues are being discussed, the designer must consider what information will make the resulting datasets most powerful in the hands of diverse users.

10.9 SUMMARY

Survey data are generally not ready for immediate analysis after they are collected. However, the steps required to make respondent answers ready for statistical analysis depend on the mode of data collection. Computer assistance has injected some steps that were originally done following data collection into the data collection step itself.

Editing is a process of cleaning data and removing obvious mistakes. Increasingly, with computer-assisted data collection, editing is built into the data collection process. To avoid introducing more errors into the data through editing, structured systems of editing have been constructed. These have permitted a level of documentation and replicability of the editing step that researchers value.

Coding, the transformation of text data into numeric data requires the researcher to invent a code structure that reflects the intended analytic use of the data. Every different use may require a different coding structure. Statistical errors can be introduced at the coding step, both systematic distortions of statistics and loss of precision due to coder variation in the application of the codes. Relative to the measurement of interviewer variance, it is relatively easy to induce designs that provide quantitative estimates of coder error.

Differential weighting of data records in statistical analysis attempts to reflect unequal probabilities of selection into the sample, adjust for omissions in the sampling frame, adjust for omissions from nonresponse of sample persons, and improve the precision of estimates by using information about the target population. Weights based on sample data generally act to increase the values of standard errors of the survey statistics, but are used in the expectation that the bias of the statistics from noncoverage and nonresponse become lower. Weights using frame population attributes generally act to reduce standard errors.

Imputation is a process of replacing missing data with values later used in statistical estimation. It is most often a procedure for item-level missing data, not unit nonresponse. Imputation is introduced to reduce the bias of item nonresponse. With rich auxiliary data that is informative about the nature of the missing observations, this can be achieved. Imputation reuses information from the respondent cases, at the cost of higher standard errors for estimates based on imputed data, relative to a complete dataset.

Sampling variance estimation from survey datasets allows the researcher to reflect all the properties of the design that yield measurable sources of variable error. There are many alternative statistical computational approaches to sampling variance estimation. Standard statistical software packages are slowly adding to their set of programs statistical routines that reflect the stratification, clustering, and differential weighting commonly used in surveys.

Survey documentation and metadata are vital products of a survey and need careful consideration at the point of survey design so that the data will be useful. The nature of survey documentation is rapidly changing because of the increased power of the Web and the hyperlinking capabilities to link related documentation together.

KEYWORDS

balance edit
balanced repeated replication
consistency edit
code structure
codebook

coding
coder variance
complex survey data
editing
explicit edits

explicit edits	multiple imputation
field coding	nonresponse weight
first-stage ratio adjustment	poststratification weight
hot-deck imputation	range edit
implied edits	ratio edit
imputation	regression imputation
mean value imputation	selection weight
metadata	Taylor series

FOR MORE IN-DEPTH READING

Biemer, P., and Lyberg, L. (2003), *Introduction to Survey Quality*, New York: Wiley.

Lyberg L, Biemer, P., Collins, M., de Leeuw, E., Dippo, C., Schwarz, N., and Trewin, D. (eds.) (1997), *Survey Measurement and Process Quality*, New York: Wiley.

Särndal, C.E., and Lundström, S. (2005), *Estimation in Surveys with Nonresponse*, New York: Wiley.

EXERCISES

1. Identify one strength and one weakness of field coding over its alternative of documenting the verbal response and coding it after data collection.

2. You are examining the data entered on an item measuring the number of miles driven per year in a travel survey about automobile trips. You encounter three cases from the total survey of 20,000 cases:

   a) One case reported 17,500 miles but 1750 was entered
   b) Another case reported 17,599 miles but 17,588 was entered
   c) A third case reported 17,599 miles but 15,799 was entered

   In a separate analysis you determine that the full sample mean number of miles driven is 15,004. The following quality assurance techniques are used to identify such keying errors:

   a) Examining the distribution of keyed data and identifying outliers
   b) Drawing a 10% sample of cases and verifying them, to estimate and correct the error
   c) Doing a 100% rekeying of the responses and verifying each, correcting errors

   For each of the three techniques, comment on the cost and effectiveness of eliminating errors in the keyed responses.

3. Using what you have read, describe the general steps in constructing weights to be used in estimation in surveys.

   a) Define what "selection weight" means.
   b) Identify each adjustment step to the base weight.

   For each adjustment step, indicate its effect on the magnitude of the weight (i.e., does it tend to increase or decrease the value of the weight resulting from the previous step?)

4. Identify two problems with the use of the respondent mean as an imputed value for item-missing data in surveys.

5. Consider the following sample of $n = 20$ subjects from a probability sample, with base weights to adjust for unequal probabilities of selection:

Base Weight	Gender	Age	No. of Chronic Conditions
4.4	M	24	0
4.2	M	37	1
2.4	M	66	2
3.0	M	57	3
2.0	M	23	2
2.4	M	26	0
2.6	F	28	1
3.0	F	32	1
3.0	F	39	0
1.1	F	40	1
1.3	F	41	0
1.2	F	47	2
1.4	F	43	1
1.5	F	48	1
1.1	F	53	3
1.4	F	38	3
1.8	F	63	4
1.1	F	68	1
1.1	F	73	2
1.0	F	78	3

   a) Compute an unweighted mean number of chronic conditions and a weighted mean number of chronic conditions that accounts for unequal probabilities of selection. Compare the two means and comment on the reason for similarity or difference.
   b) From a recent Census projection for the population of inference, the following values of w_{i4} are obtained: male under 40, 0.22; male 40 years or

EXERCISES

older, 0.24; female under 40, 0.22; and female 40 years or older, 0.32. Compute poststratification adjustment factors to the base weights for each of these four groups that obtain a weighted distribution, which matches the Census distribution.

c) What type(s) of survey errors could be adjusted for in the process of post-stratification?

6. You have completed a survey measuring the following variables, with their associated response outcomes, in a state permitting driver's licensing at age 16 (here "driver's license status" refers to licensing in that state):
Construct all edit checks possible for these five variables. For each edit check, state the logic that underlies the check.

Gender	Driver's License Status	Age	Empoyment Status	Drive Car as Part of Job
Female	Licensed	< 16	Working	Yes
Male	Not Licensed	16 or older	Sutden	No
			Not Working	Inapplicable (Student or Not Working)

7) Explain conceptually how the balanced repeated replication and jackknife approaches to sampling variance estimation differ from the Taylor series approach.

8) Briefly explain how the "hot-deck" approach to imputation works and compare its strengths and weaknesses relative to imputing the respondent mean.

9) What is multiple imputation? What is the advantage of multiple imputation versus single imputation?

10) Review the websites for the NSDUH and the BRFSS.

a) Describe the poststratification procedures of each survey.
b) What kinds of estimates in each of the surveys are likely to be most affected by the poststratification factors?

CHAPTER ELEVEN

PRINCIPLES AND PRACTICES RELATED TO ETHICAL RESEARCH

11.1 Introduction

Unlike the previous chapters, this chapter addresses issues that are not directly related to the statistical error properties of survey results. Instead, it describes a context within which most societies have sanctioned research, especially research involving humans. It describes norms and rules that guide various choices that survey methodologists must make. Finally, it reviews methodological research about survey design features implied by the norms.

We can distinguish two broad aspects of ethical practice especially relevant for survey research. The first has to do with standards for the conduct of research, that is, practices recommended by the survey profession. The second pertains to the researcher's obligations toward others, especially respondents but also clients and the larger public.

This chapter touches briefly on standards for the conduct of research, and on the researcher's obligations toward clients and the larger public. Its focus, however, is on the relationship between the researcher and the respondent, and on the ethical issues that arise for the researcher in that relationship. Because we rely on the goodwill of the target population in carrying out our work, protecting the interests of respondents is really a matter of self-interest as well.

11.2 Standards for the Conduct of Research

Like researchers in general, survey researchers are held to general standards of scientific conduct. As we noted in Chapter 1, survey methodology is a field of scientific inquiry, but also part of a larger profession of survey research. A number of international survey professional organizations exist, among them WAPOR, the World Association for Public Opinion Research, and ESOMAR, the World Organization for Market Research. In the United States, the organization that comes closest to representing all survey researchers, regardless of their place of employment, is the American Association for Public Opinion Research (AAPOR), which counts academic, commercial, and government researchers among its members. Membership in AAPOR obligates members to adhere to its

Code of Ethics (http://www.aapor.org), which is one of the few such instruments to provide a mechanism for punishing violations, and we use it here as an example. It is couched in general terms, simply asserting that members will "exercise due care" in carrying out research and will "take all reasonable steps" to assure the validity and reliability of results. However, it was augmented in May of 1996 with standards that embody current "best practices" in the field of survey research, and another a set of practices that AAPOR condemns. Members are urged to adhere to the first and refrain from the second.

Among the "best practices" are scientific methods of sampling, which ensure that each member of the population has a measurable chance of selection; follow-up of those selected to achieve an adequate response rate; careful development and pretesting of the questionnaire, with attention to known effects of question wording and order on response tendencies; and adequate training and supervision of interviewers. The preceding chapters of this book offer detailed guidance with respect to such "best practices," and they will not be further discussed here except to note that they are also counted among the *ethical* obligations of survey researchers. Unless the research can yield valid conclusions, it is deemed unethical to burden respondents with a request for participation.

Among practices defined as "unacceptable" by AAPOR (though, like the roster of best practices, not officially part of the AAPOR Code of Ethics) are fund-raising, selling, or canvassing under the guise of research; representing the results of a poll of self-selected or volunteer respondents (for example, readers of a magazine who happen to return a questionnaire, callers to an 800- or 900-number poll, or haphazard respondents to a Web-based survey) as if they were the outcome of a legitimate survey; and "push polls," which feed false or misleading information to respondents under the pretense of taking a poll. For example, a push poll might ask if the respondent would object to learning that candidate X has been accused of child abuse, is unkind to pet dogs, and so on. The motivation for asking these questions is not to discover public opinion about these issues but, rather, to plant seeds of distrust and disapproval in voters' minds. Revealing the identity of research participants without their permission is also listed as an unacceptable practice.

Like other scientists, survey researchers are also held to general standards of scientific conduct. In the United States, the federal executive branch department that funds most research on human subjects (much of it biomedical) is the Department of Health and Human Services. Within that department, the Office of Research Integrity (ORI) oversees scientific misconduct, which consists of **plagiarism**, **falsification**, or **fabrication** in proposing, performing, reviewing research, or in reporting research results. These terms have been defined by the office as shown in Table 11.1.

Little reliable information exists about the extent of research misconduct either in academic or in nonacademic settings, and it is difficult to know whether the occasional case that makes newspaper headlines is the tip of the iceberg or the rare exception (Hansen and Hansen, 1995). Like research on deviant behavior more generally, research on research misconduct is difficult to do and is subject to both nonresponse and measurement error. Some recent research in this area, however, suggests that research misconduct may be more widespread than scientists like to believe (Swazey, Anderson, and Lewis, 1993; Martinson, Anderson, and De Vries, 2005; Titus, Wells, and Rhoades, 2008).

There is, in addition, a small but consistent stream of attention given to **interviewer falsification**. "Interviewer falsification" means the *intentional* departure

Table 11.1. Key Terminology in Research Misconduct

Term	Definition
Fabrication	Making up data or results and recording or reporting them.
Falsification	Manipulating research materials, equipment, or processes, or changing or omitting results such that the research is not accurately represented in the research record.
Plagiarism	Both the theft or misappropriation of intellectual property and the substantial unattributed copying of another's work. It includes the unauthorized use of a privileged communication, but it does not include authorship or credit disputes.

from the designed interviewer guidelines or instructions, unreported by the interviewer, that could result in the contamination of data. "Intentional" means that the interviewer is aware that the action deviates from the guidelines and instructions.

Falsification includes:

1) Fabricating all or part of an interview—the recording of data that are not provided by a designated survey respondent and reporting them as answers of that respondent
2) Deliberately misreporting disposition codes and falsifying process data (e.g., the recording of a refusal case as ineligible for the sample; reporting a fictitious contact attempt)
3) Deliberately miscoding the answer to a question in order to avoid follow-up questions
4) Deliberately interviewing a nonsampled person in order to reduce effort required to complete an interview, or intentionally misrepresenting the data collection process to the survey management

In all large surveys involving many interviewers who are temporary staff, not part of the ongoing research, falsification will probably occur. On the other hand, the rate of falsification is believed to be quite low. There are few published studies of falsification, but Table 11.2 shows the results of a study at the U.S. Census Bureau (Schreiner, Pennie, and Newbrough, 1988). In this work, falsification included taking an interview with an incorrect person in the sample household or taking an interview in a mode not authorized by the study. The table shows that higher rates were obtained in the Housing Vacancy Survey, a one-time data collection, than in the ongoing Current Population Survey and National CrimeVictimization Survey.

Schreiner, Pennie, and Newbrough (1988) also found that interviewers with less experience were more likely to falsify relative to those with more experience and that experienced interviewers had more sophisticated patterns of cheating (e.g., falsifying the first wave of a panel survey). More recent studies are summa-

Table 11.2. Percentage of Interviewers Detected Falsifying for Three Surveys Conducted by the U.S. Bureau of the Census

Survey	Percentage Falsifying
Current Population Survey	0.4%
National Crime Victimization Survey	0.4%
New York City Housing Vacancy Survey	6.5%

Source: Schreiner, Pennie, and Newbrough, 1988.

rized in Schaefer, Schraepler, Mueller, and Wagner (2005), and methods of preventing data fabrication in telephone research are discussed in Smith et al. (2004).

11.3 Standards for Dealing with Clients

According to the AAPOR Code of Ethics, ethical behavior toward clients requires, in the first place, undertaking only those research assignments that can reasonably be accomplished, given the limitations of survey research techniques themselves or limited resources on the part of the researcher or client. Second, it means holding information about the research or the client confidential, except as expressly authorized by the client or as required by other provisions of the Code. Chief among the latter is the requirement that researchers who become aware of serious distortions of their findings publicly disclose whatever may be needed to correct these distortions, including a statement to the group to which the distorted findings were presented. In practice, this provision of the Code may obligate researchers to issue a corrective statement to a legislative body, a regulatory agency, the media, or some other appropriate group if they become aware of the release of such distorted results. Understandably, this provision of the Code may at times create conflicts of interest for researchers if a client releases distorted findings from a survey or poll they have conducted.

A well-known example of such a correction occurred when Bud Roper, then head of the Roper Organization, learned of the finding of an extraordinarily high percentage of respondents questioning whether the Nazi extermination of the Jews had really happened. The result of the survey, conducted by his organization for the American Jewish Committee, was in all likelihood due to the way the question had been worded—"Does it seem possible or does it seem impossible to you that the Nazi extermination of the Jews never happened?"—to which 22% gave answers implying that they did not think the extermination happened. At his own expense, Roper replicated the survey with alternative wording and publicly corrected the error. In the words of one account, Roper said, "We finally did a survey at our expense, a full repeat, with just that one question changed, and you know what we got?—the percentage of people who think the Holocaust never happened?—1%" (see Krazit, 2001). Such an action is potentially costly in terms of one's reputation and financial resources, and, indeed, goes beyond the requirement under the Code, which would have simply demanded a statement from Roper concerning the likelihood of error.

11.4 STANDARDS FOR DEALING WITH THE PUBLIC

The AAPOR Code of Ethics attempts to satisfy the profession's obligations to the larger public by mandating the disclosure of minimal information about a poll or survey whose findings are publicly released.

Table 11.3 lists the eight key elements of the AAPOR disclosure requirements. They include the responsibility of revealing the sponsor and location of the survey. They then include some key descriptions of design features that are related to different error sources in the survey results:

1) Coverage error—the researcher must reveal the target population and the sampling frame
2) Sampling error—the researcher must reveal the sample design and size
3) Nonresponse error—the researcher must reveal the completion rate (a component of the response rate)
4) Measurement error—the researcher must reveal the method of data collection and the wording of questions and instructions

Thus, at a minimum level, these items of information convey some idea of the error likely to be associated with estimates from a given survey, or, conversely, the degree of confidence warranted by the results. The Code requires that

Table 11.3. Elements of Minimal Disclosure (AAPOR Code)

1.	Who sponsored the survey, and who conducted it
2.	The exact wording of questions asked, including any preceding instruction or explanation that might reasonably be expected to affect the response
3.	A definition of the population under study and a description of the sampling frame
4.	A description of the sample selection procedure
5.	Size of sample and, if applicable, completion rates and information on eligibility criteria and screening procedures
6.	The precision of the findings, including, if appropriate, estimates of sampling error and a description of any weighting or estimating procedure used
7.	Which results, if any, are based on parts of the sample rather than the entire sample
8.	Method, location, and dates of data collection

Source: http://www.aapor.org.

this information be included in any report of research results, or, alternatively, that it be made available at the time the report is released. In other words, although researchers are not required by the Code of Ethics to follow the best professional practices, as discussed at length in this book, they are required to disclose the practices they use. Thus, the Code relies on the "marketplace of ideas" to drive out bad practices over time.

As an example of the implementation of the disclosure practice, on the SOC website for the overall study (http://www.sca.isr.umich.edu/), a reader can find an overall description of the survey, a detailed description of the sample design, an image of the questionnaire, and a description of the computations underlying the key statistics of the survey.

11.5 STANDARDS FOR DEALING WITH RESPONDENTS

11.5.1 Legal Obligations to Survey Respondents

In the United States, the *legal* foundation for the protection of human subjects of research, including survey respondents, is the Research Act of 1974 (P.L. 93-348, July 12, 1974). Canada and Australia have guidelines for research on human subjects that are similar to those in the United States. In Canada, all research involving human subjects must be reviewed by Research Ethics Boards (http://www.nserc.ca/institution/mou_sch2_e.htm); in Australia, such review boards are known as Human Research Ethics Committees (http://www.health.gov.au/). In Europe there does not seem to be any regulatory protection for human subjects of research outside biomedical areas. The European Union does, however, have regulations designed to safeguard the confidentiality of personal data (http://www.coe.int).

In the United States, the National Research Act of 1974 led to the development of Regulations for the Protection of Human Subjects of Research [Code of Federal Regulations, 45 CFR 46 May 30, 1974, 46.3 (c)], most recently revised in 1991. These regulations require colleges, universities, and other institutions receiving federal funds to establish Institutional Review Boards (IRBs) to safeguard the rights of research volunteers, including respondents to surveys. "Institutional Review Boards" are committees of researchers and local community representatives that review proposed research on human subjects and pass judgment on whether the rights of subjects have been adequately protected.

Institutional Review Board

According to the Federal Regulations for the Protection of Human Subjects, surveys are exempt from IRB review unless information is recorded in such a way that human subjects can be identified, either directly or indirectly, and disclosure of responses could reasonably be damaging to respondents' reputation, employability, or financial standing, or put them at risk of civil or criminal liability (45 CFR 46.101). But in recent years, partly in response to the deaths of several volunteers in clinical trials at well-known universities, and the ensuing public outrage and government shutdown of research at the universities involved, IRBs intensified their review of research protocols and appeared to be making less use of the opportunities provided in the Regulations for expediting review of research or exempting studies from review altogether (Citro, Ilgen, and Marrett, 2003). A national movement toward accreditation by the Association for the Accreditation of Human

Research Protection Programs (AAHRP) may help to counteract this trend.

At present, only surveys conducted at U.S. institutions that receive federal funding for research are subject to the Regulations for the Protection of Human Subjects. Thus, most commercial surveys are exempt from their provisions (including, for example, the requirement for obtaining informed consent from respondents). But this situation may change. Legislation introduced in 2002 in both houses of Congress would bring all research in the United States, regardless of funding, under the purview of the Office of Human Research Protection. The implications of such a change for the conduct of commercially funded survey research would be far reaching.

11.5.2 Ethical Obligations to Respondents

The ethical, as distinct from the legal, principles for protecting the rights of respondents are rooted in the Belmont Report (National Commission for the Protection of Human Subjects of Biomedical Research, 1979). That Report, issued in 1979, was the work of the National Commission for the Protection of Human Subjects of Biomedical and Behavioral Research, created under the National Research Act of 1974.

The movement for the protection of human subjects of research in the United States can be traced to gross violations of subjects' rights by biomedical researchers, especially by German scientists during the Nazi era but also by American scientists in the Tuskegee syphilis study (Faden and Beauchamp, 1986; Katz, 1972; Tuskegee Syphilis Study Ad Hoc Advisory Panel, 1973) and several other experiments involving medical patients.

> **The Tuskegee Study of Syphilis**
>
> Under sponsorship of the U.S. Public Health Service, the Tuskegee Study recruited poor black southern men with syphilis for a longitudinal study of the course of the disease. Begun in 1932, a time when no effective treatment for syphilis was available, it was continued by government scientists even after the discovery of penicillin, without informing subjects of the existence of a new and effective treatment. In fact, subjects were deceived from the beginning, and led to believe that they were being treated when, in fact, they were not.
>
> About 600 men were chosen, all of them poor African-Americans. Of these, 399 had syphilis, but they were simply told they had "bad blood," which in the local jargon could have meant anemia. They were treated with placebos. Many men went blind; some became insane.
>
> The study was finally ended in 1972. The United States Government did not formally acknowledge responsibility for scientific misconduct in this study or apologize to the victims until 1993, and the Tuskegee study has been a major reason for distrust of the government and the medical establishment on the part of African-Americans.

Although these early violations of human rights occurred in biomedical studies, some social science research (e.g., research by Laud Humphreys observing homosexual acts in public toilets; Humphreys, 1970), as well as other social psychological research involving deception (e.g., Zimbardo's studies of simulated prison settings; Haney, Banks, and Zimbardo, 1973; see also http://www.prison-exp.org/) also aroused public concern about potential harm to subjects. Milgram (1963), in a very important study of obedience to authority prompted in part by concern about the atrocities committed by German civilians under Hitler, enlisted subjects as "collaborators" of "scientists" who were ostensibly teaching unseen volunteers by using electric shock. The subjects were directed by the "scientists" (in reality, Milgram's research assistants) to increase the shock they delivered in response to inaccurate answers on a memory test, in spite of audible cries of pain from the recipients, who were also Milgram's assistants. In reality, no shocks were delivered or received by anyone, but the real subjects of the experiment—those who believed they were delivering electric shock to others—showed varying willingness to inflict pain at the direction of the supposed authorities, and they also experienced varying amounts of distress when they were debriefed by Milgram and his associates at the end of the experiment.

These diverse concerns led to the passage of the National Research Act in 1974 and to the codification and adoption of the Federal Regulations in the same year. In 1991, the various rules of seventeen Federal agencies were reconciled and integrated as Subpart A of 45 CFR 46, otherwise known as the Common Rule.

The Belmont Report advanced three principles for the conduct of all research involving human subjects: beneficence, justice, and respect for persons. The principle of "beneficence" requires researchers to minimize possible harms and maximize possible benefits for the subject, and to decide when it is justifiable to seek certain benefits in spite of the risks involved or when the benefits should be foregone because of the risks. The extensive attention to risks and harms in the Code of Federal Regulations reflects this principle of beneficence.

The principle of "justice" aims to achieve some fair balance between those who bear the burdens of research and those who benefit from it. In the 19th and early 20th centuries, for example, indigent patients largely bore the burdens of medical research, whereas the benefits of improved medical care went largely to affluent private patients. The excessive reliance on psychology students as subjects in psychological experiments might also be seen as violating the principle of justice, quite aside from questions that might be raised about the generalizability of the results (cf. Rosenthal and Rosnow, 1975).

The third principle, "respect for persons," gives rise to the ethical requirement for informed consent, which may be defined as the "knowing consent of an individual or his legally authorized representative . . . without undue inducement or any element of force, fraud, deceit, duress, or any other form of constraint or coercion" (U.S. Department of Health, Education, and Welfare, 1974, p. 18917).

As M. B. Smith pointed out in an article titled "Some Perspectives on Ethical/Political Issues in Social Science Research" (Smith, 1979), the twin pillars of ethical and legal obligations to respondents—beneficence and respect for persons—derive from very different philosophical systems and fit only "awkwardly and uncomfortably" with each other. Smith notes that one of these frameworks is

> . . . the libertarian, voluntaristic, "humanistic" frame captured in the tag phrase "informed consent." . . . It is good if people make their significant decisions for themselves, and they ought even to have the right to make self-sacrificial choices that result in harm to themselves. It is bad for them to be coerced, conned, or manipulated, even for their own good. The other frame, that of the participant's welfare or harm, fits in the "utilitarian" tradition. . . . It does not depend upon the assumption of free will. It sounds more objective, and lends itself to expression in the currently fashionable terms of cost/benefit analysis. (p. 11)

Smith goes on to point to problems that arise for the conduct of social science research under both of these frames, problems we consider in the following sections of this chapter when we discuss the ethical principles of respect for persons and beneficence in more detail.

11.5.3 Informed Consent: Respect for Persons

Many people think the purpose of informed consent is to protect human subjects of research, including respondents, from harm. As a result, the argument is often advanced that if there is no risk of harm, there should be no need for obtaining informed consent. But as noted above, the real purpose of obtaining consent is to give respondents and other subjects meaningful control over information about themselves, even if the question of harm does not arise. For example, informing potential respondents of the voluntary nature of their participation in an interview is essential to ethical survey practice, even if this leads some of them to refuse to participate. Similarly, obtaining respondents' permission to record an interview should be standard practice, even if no harm is likely to come to them as a result. Currently, many states permit such recording with only the interviewer's consent.

Table 11.4 presents the essential elements of informed consent required under the Common Rule. These include descriptions of the purposes of the research, benefits and potential harm from participation, terms of confidentiality of the data, and explicitly noting that participation is voluntary.

Under specified circumstances, IRBs may waive some or all of these elements, or even waive the requirement to obtain informed consent entirely.

Table 11.4. Essential Elements of Informed Consent

1.	A statement that the study involves research, and explanation of the purposes of the research and the expected duration of the subject's participation, a description of the procedures, and identification of any procedures that are experimental
2.	A description of any foreseeable risks or discomfort
3.	A description of any benefits to the subject or others that may reasonably be expected
4.	A disclosure of appropriate alternative procedures or courses of treatment
5.	A statement describing the extent, if any, to which confidentiality of records identifying the subject will be maintained
6.	For research involving more than minimal risk, an explanation of whether and what kind of compensation or treatment is available if injury occurs
7.	An explanation of whom to contact with further questions about the research, subjects' rights, and research-related injury
8.	A statement that participation is voluntary and the subject may discontinue participation at any time without penalty or loss of benefits.

> **Project Metropolitan**
>
> In 1966, a group of Swedish researchers began Project Metropolit by identifying a sample of school children, all born in the year 1953. Data included birth certificates, interviews with parents, recurrent questionnaire-based surveys among the participants, and a set of population register data. One purpose of the study was to understand the relationship between social conditions and subsequent health conditions.
>
> The study continued assembling data through the mid-1980s. However, many of the research subjects were unaware of their participation in the research because initial interviews were done when they were children and subsequent measurements were embedded in surveys whose stated purpose was unrelated to Project Metropolit. Further, they were unaware that data stored in their population register records were routinely added to update their study records. On February 10, 1986, the Stockholm newspaper, *Dagens Nyheter*, ran a story that described the study and emphasized the lack of informed consent.
>
> A public debate began, with many, including members of Parliament, expressing outrage at the lack of consideration of the rights of the subjects to know about how data about them were being assembled and used. Given the turmoil, a government board ordered anonymization of the data, inhibiting future linkages. (During this period of public debate, response rates to other surveys in Sweden, completely unrelated to Project Metropolit, declined.)

The medium of communication and documentation of informed consent in the biomedical field is the written consent form, signed by the respondent. An IRB may waive the requirement for a signed consent form if (1) the only record linking the subject to the research is the consent document and the principal risk is a breach of confidentiality, or (2) the research presents no more than minimal risk of harm and involves no procedures for which written consent is normally required outside the research context.

Many, perhaps most, surveys are characterized by the second of these two features, but some clearly are not. For example, surveys that ask questions about illegal or stigmatizing behavior (like NSDUH) may put participants at risk of harm if the information is disclosed, either inadvertently or deliberately. Respondents have a right to be informed about these risks before they participate, and because of potential conflicts of interest, researchers should not be the only ones deciding how much information should be disclosed to respondents, and in what form. In these circumstances, it is appropriate for IRBs to require that a written document describing the risks, as well as protections against them and recourse in case of injury, be made available to respondents. It is also appropriate for IRBs to require documentation that such a form has actually been provided to respondents before their participation. Whether this documentation must take the form of a consent form *signed by the respondent* is, however, a separate question.

Although codes of ethics of some major professional associations (for example, the American Statistical Association and the American Sociological Association) make mention of the requirement for informed consent, AAPOR's code does not, noting only that "We shall strive to avoid the use of practices or methods that may harm, humiliate, or seriously mislead survey respondents." It is not difficult to understand this omission. Unlike much biomedical research, the quality of surveys depends on the response rates they achieve. As a result, the need to gain the respondent's cooperation puts a premium on brief, engaging survey introductions that make it difficult to convey all the required elements of informed consent, and there is evidence that requiring a signature to document consent significantly reduces response rates (Singer, 1978, 2003).

Further, Presser (1994) has argued that the uses to which a survey will be put are often impossible to predict and even harder to communicate to respondents with any accuracy. And although full disclosure of sponsorship and purpose, for example, may be highly desirable from an ethical point of view, it may also bias responses to a survey (Groves, Presser, and Dipko, 2004). It may do so not only because some people will choose not to participate in surveys sponsored by certain organizations or with a particular purpose, but also because those who do participate may distort their responses in order to thwart or advance the purpose as they understand it. From still another standpoint, the question is how much detail should be communicated in order to inform respondents adequately. Should they, for example, be told in advance that they are going to be asked about their income near the end of the interview, or is it enough to tell them that they need not answer a question if they choose not to?

11.5.4 Beneficence: Protecting Respondents from Harm

Probably the most important risk of harm to survey respondents stems from disclosure of their responses—that is, from deliberate or inadvertent breaches of confidentiality. The more sensitive the information requested, for example, HIV status or whether the respondent has engaged in some illegal activity, the greater the harm that a breach of confidentiality might entail. Not unexpectedly, respondents require strong assurances of confidentiality when they are asked to provide such data about themselves (Singer, Von Thurn, and Miller, 1995).

A variety of threats to the confidentiality of survey data exist. Most common is simple carelessness: not removing identifiers from questionnaires or electronic data files, leaving cabinets unlocked, or not encrypting files containing identifiers. Although there is no evidence that respondents have been harmed as a result of such negligence, it is important for data collection agencies to raise the consciousness of their employees with respect to these issues, provide guidelines for appropriate behavior, and ensure that these are observed.

Less common, but potentially more serious, threats to confidentiality are legal demands for identified data, either in the form of a subpoena or as a result of a Freedom of Information Act (FOIA) request. Surveys that collect data about illegal behaviors such as drug use, for example, are potentially subject to subpoena by law enforcement agencies. To protect against this, researchers and programs like the NSDUH studying mental health, alcohol and drug use, and other sensitive topics, whether federally funded or not, may apply for Certificates of Confidentiality from the Department of Health and Human Services. Such Certificates, which remain in effect for the duration of the study, protect researchers from being compelled in most circumstances to disclose names or other identifying characteristics of survey respondents in federal, state, or local proceedings (42 CFR 2a.7, "Effect of Confidentiality Certificate"). The Certificate protects the identity of respondents; it does not protect the data.

certificate of confidentiality

Additional protection for the confidentiality of statistical data is provided in legislation signed into law in 2002. In 1971, the President's Commission on Federal Statistics recommended that "Use of the term 'confidential' should always mean that disclosure of data in a manner that would allow public identification of the respondent or would in any way be harmful to him is prohibited" and that "data are immune from legal process." The Commission further recommended that "legislation should be enacted authorizing agencies collecting data for statistical purposes to promise confidentiality as defined above." Since that

confidentiality

time, efforts have been made to shore up legal protection for the confidentiality of statistical information, as well as to permit some limited sharing of data for statistical purposes among agencies. In December 2002, such legislation, the Confidential Information Protection and Statistical Efficiency Act of 2002, was finally signed into law. However, concerns about the effects of the Homeland Security Act, signed into law in 2001, on promises of confidentiality have yet to be resolved, as do the effects of the so-called "Shelby Amendment" (PL 105-277, signed into law October 21, 1998), designed to facilitate public access to research data that are subsequently used in policy formation or regulation.

When surveys are done expressly for litigation (rather than statistical use), demands for data from one party or the other are not uncommon, and protection from a "Certificate of Confidentiality" or the newly enacted law may not be available. Presser (1994) reviews several such cases. In such a situation, the researcher may have little recourse except to turn over the data, thus violating the pledge of confidentiality, or go to jail. Presser (1994) discusses several actions that might be taken by professional survey organizations to enhance protections of confidentiality in such cases, for example, mandating the destruction of identifiers in surveys designed for adversarial proceedings, as some survey firms engaged in litigation work already do. If this became standard practice in the survey industry, individual researchers would be better able to defend their own use of such a procedure.

A final threat to data confidentiality comes from what is referred to as "statistical disclosure." "Statistical disclosure" means the identification of an individual (or of an attribute) through the matching of survey data with information available outside the survey. So far, the discussion in this chapter has been in terms of explicit respondent identifiers: name, address, and Social Security number, for example. But there is increasing awareness that even without explicit identifiers, reidentification of respondents may be possible given the existence of high-speed computers, external data files containing names and addresses and information about a variety of individual characteristics to which the survey data can be matched, and sophisticated matching software. Longitudinal studies may be especially vulnerable to such reidentification.

Technologies that make it easier to identify respondents, or more difficult to conceal their identity, are increasingly being used in conjunction with survey data. Such technologies, DNA samples, biological measurements, and geospatial coordinates, for example, all complicate the problem of making data files anonymous and heighten the dilemma of researchers who want to increase access to the data they collect while protecting their confidentiality. At present, no generally accepted guidelines exist for how to protect the confidentiality of data files containing such variables except to release them as restricted files subject to special licensing, bonding, or access arrangements or to create synthetic, or imputed, data files. We discuss these issues in Section 11.8.2.

In view of the diverse threats to confidentiality reviewed above, there is a growing consensus among researchers that absolute assurances of confidentiality should seldom, if ever, be given to respondents (National Research Council, 1993). The "Ethical Guidelines for Statistical Practice" of the American Statistical Association (ASA), for example, caution statisticians to "anticipate secondary and indirect uses of the data when obtaining approvals from research subjects." At a minimum, the Guidelines urge statisticians to provide for later independent replication of an analysis by an outside party. They also warn researchers not to imply protection of confidentiality from legal processes of discovery unless they have been explicitly authorized to do so. Those collecting sensitive data should make every effort to obtain a Certificate of Confidentiality, which would permit them to offer stronger assurances to respondents. Even such Certificates, however, cannot guard against the risk of statistical disclosure.

11.5.5 Efforts at Persuasion

An issue that merits brief discussion here is the dilemma created by the need for high response rates from populations that spend less time at home, have less discretionary time, and are less inclined to cooperate in surveys. This dilemma leads survey organizations to attempt more refusal avoidance and conversion, making greater use of incentives both as an initial inducement to respond and as a way of persuading respondents who refuse. Participants see both follow-up calls that attempt to persuade them without the offer of payment and the offer of money to those who initially refused as undesirable (Groves, Singer, Corning, and Bowers, 1999). They also regard the paying of refusal conversion payments as unfair, but awareness of such payments does not appear to reduce either the expressed willingness to respond to a future survey or actual participation in a new survey (Singer, Groves, and Corning, 1999).

From an ethical perspective, however, the practice raises some questions. Consider the following examples:

> *Example 1*: A contractor gets a multimillion dollar award for a survey of older people. He promises a response rate of 80%. When the field period is almost over, the response rate is still slightly below 80%, and the researcher authorizes interviewers to offer $100 to potential respondents in order to convert their refusals.

> *Example 2*: Researchers are interviewing doctors and genetic counselors in a study of knowledge about, and attitudes toward, genetic testing. In order to increase response rates, they decide to offer physicians $25 for a half-hour interview. But since genetic counselors responded at a higher rate to the pilot test, no incentives are offered to them.

One issue raised by both of these examples is that of equity: Respondents who are less cooperative are rewarded for their lack of cooperation, whereas those who do cooperate with the survey request receive no reward. (Economists look at this issue from a different perspective, however, arguing that the survey has greater utility for those who respond even without an incentive and, therefore, there is no need to offer an additional incentive to them.)

But there may be methodological issues, as well. For example, there is evidence that those brought into a survey by means of an exceptionally large refusal conversion payment, as in the first example above, are wealthier and have a larger number of assets than those who cooperate without the refusal conversion payment, so that the sample yields more accurate data (Juster and Suzman, 1995). Other research indicates that monetary incentives may compensate for respondents' lack of interest in a topic, and that they are therefore important in reducing nonresponse bias (Groves, Couper, Presser, Singer, Tourangeau, Piani Acosta, and Nelson, 2006).

A somewhat different issue is raised when the size or kind of incentive and the situation of the potential respondent make it difficult for the person to refuse. This may be true, for example, if prisoners are offered a reduction in their sentence in return for their cooperation in a study, or if drug users are offered a monetary incentive, or if people who are ill with a particular disease but cannot afford medical care are offered such care in return for their participation in a trial of an experimental drug. Here, one must ask whether the offer of a particular incentive challenges the presumption of voluntary consent, in effect rendering the incentive coercive (see Singer and Bossarte, 2006; Singer and Couper, 2008). Institutional review boards would be very unlikely to approve the protocol of a study proposing to use these kinds of incentives.

11.6 Emerging Ethical Issues

Surveys conducted on the Internet have shown explosive growth in the last several years, and such surveys bring an entirely new set of ethical problems to the fore, or add new dimensions to old problems. For example, maintenance of privacy and confidentiality in such surveys requires technological innovations (for example, "secure" websites and encryption of responses) beyond those required by older modes of data collection. Other ethical issues simply do not arise with older modes of conducting surveys, for example, the possibility that a respondent might submit multiple answers, thus deliberately biasing the results. This becomes an ethical issue for the researchers because a method subject to misleading findings must be avoided under the AAPOR Code. For still other ethical issues, such as obtaining informed consent, surveys on the Web both pose new challenges (for example, how can one be sure that no minors are participating in a survey? How does one best obtain consent for recording covert behavior such as keystrokes?) and new possibilities for solutions (for example, an interactive consent process in which the respondent must demonstrate comprehension of a module before going on to the next one). For a comprehensive discussion of these and related ethical issues, see American Psychological Association (2003) and Singer and Couper (forthcoming).

11.7 Research About Ethical Issues In Surveys

As we have noted, probably the two most important ethical issues facing survey researchers are obtaining respondents' informed consent for participation and maintaining the confidentiality of their replies, since a breach of confidentiality is the most likely way in which respondents might be harmed. In this section, we discuss empirical research on how these issues affect survey participation.

11.7.1 Research on Informed Consent Protocols

Section 11.5.3 has described the key ingredients of informed consent for surveys. Implementing the ingredients in practical survey protocols is not simple, however, as various methodological research studies have shown. There are three areas of methodological research related to informed consent that can inform practice:

1) Research on respondents' reactions to the content of informed consent protocols
2) Research on informed consent in methodological surveys, especially those involving some type of deception
3) Research on written versus oral informed consent, especially that studying the effect of asking for a signature from respondents

Research on Respondents' Reactions to Informed Consent Protocols. Singer (1978) gave random half-samples two descriptions of the content of an interview (see box page 386). Half of them were given a brief, vague description of the survey, a study of leisure time and the way people feel about it. The other half were given a fuller description of the interview, including a statement that it included some questions about the use of alcoholic drinks and drugs and about sex. The study found that differences in the information respondents were given about the survey content had no effect on their response rate (or refusal rate), or on the quality of their responses. But respondents who were given more information about the content of the survey ahead of time expressed less embarrassment when asked, on a self-administered debriefing questionnaire, how upset or embarrassed they had been about answering different kinds of questions on the survey. They were also more likely to report that they expected the kinds of questions they actually got.

A later study using essentially the same questionnaire (Singer and Frankel, 1982) found no effect on response rate or quality of including a vague sentence about the study's methodological purpose in the introduction. Following the main part of that study, respondents were asked whether participants in a study should be told about its aims, even if this information might change the nature of their replies. One-half of the sample was presented with that dilemma in terms of consequences for knowledge only; the other half, in terms of policy and budgetary implications as well. Overwhelmingly, both groups said that researchers should tell people what the study was about beforehand, regardless of the consequences for research outcomes. When only knowledge was involved, 68% chose this option, and 25% said researchers should tell them afterwards; when money was also involved, 65% chose this option, whereas 29% said, "Tell after." Only 5% in each group said researchers should not tell potential participants about the purpose at all, and between 1 and 2% in each group said they should not do the study. These preferences are clearly at odds with the practices of most survey researchers, which usually provide rather cursory descriptions of the purposes of a study in order to avoid biasing responses and to keep the introduction short.

Despite the personal nature of some of the questions asked in both the 1978 and 1982 studies, only about a quarter of the sample said that there were some questions in the interview that the research organization "had no business asking." Here, the experimental conditions made a difference in the responses. Those given a detailed statement of what the content of the interview would be were significantly *less* likely to say that the researchers had no business asking some of the questions than those given only a brief general description, but those who were told about the methodological purpose were significantly *more* likely to say

> **Singer (1978) on Comprehension of Informed Consent**
>
> Singer (1978) studied how consent protocols affected survey error.
>
> *Study design*: A 2 x 3 x 3 factorial design varied the (1) length and detail of the study description (detailed and long versus vague and short); (2) level of confidentiality pledge (none, qualified, or absolute); and (3) asking for a signature on a consent form (none, signature after interview, or signature before interview). About 2000 adults were assigned to one of the 18 treatment groups and approached by interviewers who were assigned cases in all groups. The interview questions concerned leisure activities, mental health, drinking, marijuana use, and sexual behavior.
>
> *Findings:* Among the three factors, only asking for signatures affected unit nonresponse (71% response rate for the no signature group versus 65–66% for the other two groups). About 7% of those asked for a signature refused, although they were willing to be interviewed. Only the confidentiality manipulation affected item nonresponse, with less item-missing data on sensitive items under the highest confidentiality treatment. However, the results do suggest that asking for a signature before the interview leads to fewer reports of sensitive behaviors.
>
> *Limitations of the study:* Since each interviewer administered all treatments, there could have been errors of assignment of treatment to the sample case. With no external validation data, analysis of response quality relies on untestable assumptions.
>
> *Impact of the study:* The importance of the confidentiality pledge for missing data on sensitive items suggested that the content of the pledge may be more important for surveys measuring such items. The fact that some persons are willing to be interviewed but not willing to sign a consent form documented a link between the consent protocol and nonresponse rates.

so than those who were not. A look at the response distributions for the four experimental conditions indicates that it is the combination of a lack of description about sensitive content plus a statement of methodological purpose that is responsible for this negative reaction, suggesting that respondents found this particular purpose insufficient to justify the sensitive questions asked when they were not warned about them ahead of time. When those respondents who said the researchers "had no business asking" certain questions were asked which questions had offended them, sex and income were mentioned most frequently (see Table 11.5).

Although the design of this study has a number of flaws—the questions were posed by interviewers rather than being self-administered, for example—the results give an indication of how sensitive respondents regard various survey topics, and what the effects of informing them in advance about those topics are likely to be.

Although, as pointed out above, survey experiments have found little, if any, evidence that providing more and less information about content or purpose to respondents affects their rates of participation, laboratory experiments have found much stronger effects for both of these variables. Berscheid, Baron, Dermer, and Libman (1973), for example, gave subjects progressively more information about actual psychological experiments that had been carried out, for example, those by Asch (1956) and Milgram (1963). They found that as subjects were given more information about the experiment, they were less likely to volunteer. Gardner (1978) told subjects that they could discontinue a task if the noise level bothered them too much. This information wiped out the deleterious effects of noise on performance seen without this information. King and Confer (reported in Horn, 1978) gave subjects different information about the true purpose of an experiment, which was to see whether use of the pronouns "I" and "we" could be increased by verbal reward. For

Table 11.5. Self-Reports on Type of Questions Considered Offensive to the Respondent Among Respondents Saying Researchers "Had No Business Asking" Sensitive Questions

Question	Percentage Finding It Offensive
Questions about sex	19.2%
Questions about income	9.4%
Questions about drinking	5.1%
Questions about marijuana	4.2%
Questions about mood, physical health	1.2%

Source: Singer, 1984.

those given accurate and complete information, the hypothesized experimental effect was eliminated.

Thus, these laboratory experiments suggest that giving respondents more information about a study they are about to take part in can affect their willingness to participate as well as the study's results. In at least some cases, therefore, the dilemma for survey researchers is a real one: Whether to inform potential respondents fully and risk reducing their participation (and potentially affecting the results), or whether to violate the ethical obligation to obtain informed consent in order to try to protect response rates and avoid contaminating the results. (Note that such a strategy does not always succeed, either. Aronson and Carlsmith, 1969, for example, point out that if subjects are not provided with a plausible explanation for a study, they may form their own hypotheses about what the experimenter is trying to get at, and may then behave in such a way as to support or refute these hypotheses.)

Couper, Singer, Conrad, and Groves (2008) have experimented with informing respondents about the likely risk (probability) that their answers to survey questions would be disclosed to others, along with their names and addresses—in other words, that a breach of confidentiality would occur. Using vignettes to describe different levels of risk and different types of sensitive and nonsensitive surveys, they found that statements of the probability of disclosure had no effect on respondents' reported willingness to take part in the survey described. However, when they were told about some of the possible harms that might result from disclosure, they reported significantly less willingness to participate in surveys with a higher likelihood of disclosure.

Questions about how much information to provide respondents in connection with the survey are not easy to answer. Smith (1979, p. 15) argues that the investigator's vested interest in the advancement of knowledge is likely to produce bias in reaching a decision on these issues, and that there is, therefore, "good reason for jury-like reconsideration by review groups." We would argue that participants' preferences should be taken into account by IRBs, though not in the form of a "community representative," as is so often the case. Rather, we recommend continuing systematic research to determine the norms and preferences of participants in different kinds of survey research. These preferences need not be decisive in whether or not research is approved, but they should be given serious consideration. Past research is, on the whole, reassuring on this point.

To summarize the literature on the effects of information about the interview on cooperation, there is little research evidence that the amount of information about content of the interview affects either cooperation or data quality. However, laboratory experiments suggest that the effects of being forewarned about the sensitive content of questions may be more substantial. There are modest advantages of more elaborate assurances of confidentiality on cooperation or data quality when surveys focus on sensitive topics. More and better research on the public's preferences and tolerance in this area would be highly desirable. This is especially true because in the absence of such information, members of IRBs must substitute their subjective judgments about what respondents would find objectionable for respondents' own preferences in this regard.

Research on Informed Consent Complications in Methodological Studies. Smith (1979, p. 13) captures the inherent conflict between ethical obligations and commonly accepted research procedures by noting that many practices that seem "essential to well-designed inquiry capable of producing dependable knowledge," for example, deception in psychological experiments and unobtrusive observation, are incompatible with any strict interpretation of informed consent. There are many survey methodological studies in which the key scientific question concerns respondent or interviewer reactions to different stimuli. This is true, for example, in studies of question wording involving different levels of cueing (see Section 7.2.2), research on incentive effects on nonresponse (see Section 6.6), and research on interviewer supervisory practices (see Section 9.7). What are the appropriate practices with respect to informed consent in these studies? The respondents *could* be fully informed about the desire to study how they would change their behaviors given different experimental stimuli. However, typically, researchers fear that fully informing respondents before measurement would itself distort the study findings.

> *Example*: In order to understand better the low mail return rate to the 1990 census, the U.S. Census Bureau commissioned the Survey of Census Participation. The survey asked a variety of demographic and attitude questions and also asked whether or not the respondent's household had mailed back a census form. Afterwards, the Census Bureau linked the survey data with the decennial census file, matching the responses of survey participants with their census questionnaire (Couper, Singer, and Kulka, 1998).

In this case, we have no information about what the consequences of failing to obtain consent to linkage might have been. It is possible, for example, that the records of some percentage of the Survey Census of Participation respondents could not have been matched to their census forms if permission had been requested (and denied). If there had been enough of these, and if they had differed substantially from those who gave permission, the findings of the study might have been distorted. At the same time, the responses of those who agreed to the match might have differed somewhat from their actual responses if they had known that their survey answers were going to be compared with census records, and this, too, might have distorted the study's findings. Even if respondents had been asked for permission after the survey had been completed, those who knew their answers did not correspond with the Census Bureau records might have been more likely to refuse, thus leading the researchers to conclude there was more consistency between the survey and the records than was in fact the case. Here, too, more and better research is needed on the effect of variations in informed consent statements on nonresponse and measurement error. If such effects are substantial, ways of satisfying ethical requirements while maximizing scientific values will have to be devised. Currently, informed consent may be waived if the IRB finds that the research poses minimal risk to respondents, that respondents will not be harmed by the waiver, that the research could not practically be carried out without it, and that respondents will be fully debriefed at the conclusion of the study.

Research on Written versus Oral Informed Consent. So far, we have been talking about the ethical obligation to obtain informed consent for research participation from respondents. Does this obligation extend to the requirement to obtain a signed informed consent form?

Evidence from several studies documents the harmful consequences of requiring signed consent forms for survey participation. The earliest such study was done by Singer (1978), in which the request for a signature to document consent reduced the response rate to a national face-to-face survey by some 7 percentage points. But most respondents who refused to sign the form were, in fact, willing to participate in the survey; it was the request for a signature that deterred them from participation. There was virtually no difference in response rate by whether the request for a signature came before or after the survey. Similar findings are reported by Trice (1987), who found that subjects asked to sign a consent form responded at a lower rate than when no signature was required. More recently, an experiment by Singer (2003) found that some 13% of respondents who said they would be willing to participate in a survey were not willing to sign the consent form that would have been required for their participation.

Given these deleterious effects on response rate of requiring a respondent's signature to document consent, we would argue that such a signature should almost never be required. Such a signed consent form protects the institution rather than the respondent. Instead, a functional alternative should be used. For example, the interviewer can be required to sign a form indicating that an informed consent statement has been read to the respondent and a copy given to him or her to keep. Survey firms should be required to audit the truthfulness of these affidavits, just as they are required to audit whether or not an interview has actually been conducted. Only when a signed consent form is clearly required by the Common Rule should a signature be required. Relaxing the requirement for a signature does not imply relaxing the requirement for adequately informing respondents of potential risks and of the voluntary nature of their participation. On the contrary, when surveys carry more than minimal risk, much more scrutiny is warranted of whether enough information has been provided to respondents, whether they understand what they are being told, and whether they are given

minimal risk

enough time to make an informed judgment about whether or not to participate. The situation is quite otherwise when risk is minimal or nonexistent. Under those circumstances, answering the survey questions is probably sufficient evidence that the respondent's "consent" has, in fact, been both voluntary and informed.

The following is an example of an introduction used in a minimal risk survey, the Survey of Consumers, done at the University of Michigan. "I am calling from the University of Michigan in Ann Arbor. Here at the University we are currently working on a nationwide research project. We are interested in what people think and feel about how the economy is doing. We would like to interview you for our study and I was hoping that now would be a good time to speak to you." Before the interview begins, the interviewer is required to read, "This interview is confidential and completely voluntary. If we should come to any question that you do not want to answer, just let me know and we will go on to the next question." If the respondent acknowledges this information and agrees to proceed with the interview, informed consent has been obtained.

Summary of Research on Informed Consent in Surveys. Survey researchers most often use introductory statements orally delivered by an interviewer or written into a self-administered form to fulfill the provisions of the regulations regarding informed consent. Methodological research to date suggests that the content of these statements has only modest effects on cooperation and data quality measures. At the same time, respondent embarrassment and feelings of being upset by question content can be ameliorated by mentioning the sensitive nature of questions in the introduction. As noted above, however, more and better research is needed on the effects of informed consent statements.

If they are required to sign a written consent statement, there is replicated evidence that some persons willing to respond to a survey may refuse because of the signature request. IRBs should be encouraged to waive the requirement for a signature to document consent whenever permitted by the Common Rule. For those surveys judged to carry more than minimal risks, research is needed on ways of reducing the harmful effects of the signature request on cooperation.

11.7.2 Research on Confidentiality Assurances and Survey Participation

As already noted (see Section 11.5.4), a breach of confidentiality is the most likely way through which harm might come to survey respondents. Many surveys ask about sensitive topics (e.g., income or alcoholic beverage consumption), as well as about stigmatizing and even illegal behavior. The disclosure of such information might subject a respondent to loss of reputation, employment, or civil or criminal penalties. But there are at least two other reasons for taking a pledge of confidentiality very seriously. First, the breach of a confidentiality pledge would violate the principle of respect for those consenting to participate in research, even if the disclosure involved innocuous information that would not result in any social, economic, legal, or other harm (see Citro, Ilgen, and Marrett, 2003, Chapter 5). Second, a breach of confidentiality threatens survey research itself, since concerns about privacy and confidentiality are among the reasons most often given by potential respondents for refusing to participate in surveys. Section 11.7.2 reviews the research evidence for this assertion. Although the number of experiments is not large, and many have been sponsored by the U.S. Census Bureau, limiting their generalizability, all of them have shown statistically significant, if small, negative effects on survey cooperation.

The earliest study (National Research Council, 1979) was designed to investigate how changing the length of time for which the Census Bureau assured the confidentiality of census returns might affect the census mail return rate. The survey introduction varied the assurance of confidentiality respondents were given. A random fifth of the sample received an assurance of confidentiality "in perpetuity"; another, an assurance for 75 years; a third, for 25 years; a fourth received no mention of confidentiality; and a fifth group was told that their answers might be given to other agencies and to the public. Among those persons who were reached and to whom the introductory statement, including the experimental treatment, was read, refusals varied monotonically from 1.8 percent for those who received the strongest assurance to 2.8 percent for those told their answers might be shared with other agencies and the public; this increase, though small, was statistically significant.

Two studies have examined the relationship between census return rates and concerns about privacy and confidentiality expressed in an attitude survey. The first matched responses to the Survey of Census Participation, a face-to-face attitude survey carried out in the summer of 1990, a few months after the decennial census, with mail returns to the census from the same households. Both an index of privacy concerns and an index of confidentiality concerns significantly predicted census mail returns, together explaining 1.3 percent of the variance net of demographic characteristics (Singer, Mathiowetz, and Couper, 1993). The study was repeated for the 2000 census, using respondents to a Gallup random digit dial survey in the summer of 2000 and matching their responses to census returns from the same households (Singer, Van Hoewyk, and Neugebauer, 2003). Attitudes about privacy and confidentiality were estimated to account for 1.2 percent of the variance in census mail returns, almost identical to the 1990 result. The relationships were stronger in one-person households, where the respondent to the survey and the person who returned the census form had a higher probability of being identical, than in the sample as a whole. The belief that the census may be misused for law enforcement purposes, as measured by an index of responses to three questions, was a significant negative predictor of census returns.

Findings from two other studies support the conclusion that privacy concerns reduce participation in the census. Martin (2006, Table 4) shows that respondents who received a long form or were concerned about privacy were more likely to return an incomplete census form or fail to return it at all. And an experiment by Junn (2001) reported that respondents who were asked questions designed to raise privacy concerns about the census were less likely to respond to long-form questions administered experimentally than respondents who were given reasons for asking intrusive questions, or those in a control group, who received neither type of experimental treatment.

Research by Bates, Dahlhamer, and Singer (2008), which analyzes interviewers' coding of "doorstep concerns" expressed by households selected for the National Health Interview Study, shows that privacy and confidentiality concerns significantly predict interim refusals but not final refusals, suggesting that if interviewers can successfully address such concerns, they need not lead to nonparticipation. Because these data come from face-to-face surveys, they also suggest that the relation of privacy concerns to final refusals may differ between modes, in part because interviewers may be less able to counter them effectively on the telephone. [A Response Analysis Survey (RAS) of a small number of respondents and nonrespondents to the American Time Use Survey (ATUS), which is conducted by phone, found greater confidentiality concerns among nonrespondents—32%, compared with 24% among respondents. Because the RAS response rate among nonrespondents to the ATUS was very low (32%, compared with 93% for respondents), the actual difference between respondents and norespondents is probably larger (O'Neill and Sincavage, 2004).]

A final set of experiments that bears on the relationship between privacy/confidentiality concerns and survey participation involved the impact of asking for a Social Security number (SSN). A 1992 Census Bureau experiment found that requesting SSNs led to a 3.4 percentage point decline in response rates to a mail questionnaire, and an additional 17 or so percentage point increase in item nonresponse (Dillman, Sinclair, and Clark, 1993). In experiments conducted in connection with the 2000 census, when SSNs were requested for all members of a household, the mail return rate declined by 2.1 percentage points in High Coverage Areas, which make up about 81 percent of all addresses, and 2.7 percent in Low Coverage Areas, which contain a large proportion of the country's black and Hispanic populations as well as renter-occupied housing units. Some 15.5 percent of SSNs were missing when a request for the SSN was made for Person 1 only, with increasing percentages missing for Persons 2–6 when SSNs were requested for all members of the household (Guarino, Hill, and Woltman, 2001, Table 5).

The experiments discussed so far have been embedded in ongoing surveys or have involved matching survey and census responses; they thus have a great deal of external validity but limit the kind of additional information that can be obtained. From a series of other studies, many of them laboratory experiments, we know that when questions are sensitive, involving sexual behavior, drug use, other stigmatizing behavior, financial information, and the like, stronger assurances of confidentiality elicit higher response rates or better response quality (Berman, McCombs, and Boruch, 1977; Singer, Von Thurn, and Miller, 1995). On the other hand, when the topic of the research is innocuous, stronger assurances of confidentiality appear to have the opposite effect, leading to less reported willingness to participate, less actual participation, and greater expressions of suspicion and concern about what will happen to the information requested (Frey, 1986; Singer, Hippler, and Schwarz, 1992; Singer, Von Thurn, and Miller, 1995).

11.8 Administrative and Technical Procedures for Safeguarding Confidentiality

Breaches of confidentiality can occur in a variety of ways (National Research Council, 2005; Ch. 4). Here, we consider two over which survey organizations can exert some control: simple carelessness and statistical disclosure limitation.

11.8.1 Administrative Procedures

To prevent confidentiality breaches resulting from carelessness, many survey organizations have developed formal written pledges that employees are required to sign as a condition of their employment. For example, the form used by the

> **Institute for Social Research**
> **University of Michigan**
>
> **PLEDGE TO SAFEGUARD RESPONDENT PRIVACY**
>
> I have read the Institute for Social Research Policy on Safeguarding Respondent Privacy and pledge that I will strictly comply with that policy. Specifically:
>
> > I will not reveal the name, address, telephone number, or other identifying information of any respondent (or family member of a respondent or other informant) to any person other than a member of the research staff directly connected to the study in which the respondent is participating.
> >
> > I will not reveal the contents or substance of the responses of any identifiable respondent or informant to any person other than a member of the staff directly connected to the study in which the respondent is participating, expect as authorized by the project director or authorized designate.
> >
> > I will not contact any respondent (or family member, employer, other person connected to a respondent or informant) except as authorized by a member of the staff directly connected to the project in which the respondent is participating.
> >
> > I will not release a dataset (including for unrestricted public use or for other, restricted, uses) except in accordance with policies and procedures established by ISR and the Center with which I am affiliated.
>
> I agree that compliance with this pledge and the underlying policy is: (1) a condition of my employment (if I am an employee of ISR), and/or (2) a condition of continuing collaboration and association with ISR (if I am not an employee of ISR, such as a student, visiting scholar, or outside project director or coprincipal investigator, etc.).
>
> If I supervise non-ISR employees who have access to ISR respondent data (other than unrestricted public release datasets), I will ensure that those employees adhere to the same standards of protection of ISR respondent privacy, anonymity, and confidentiality, as required by this pledge and the associated policy.
>
> Signature: _____
>
> Typed or printed name: _____ Date: _____

Figure 11.1 Pledge made by research team members about respondent privacy.

Institute for Social Research appears in Figure 11.1. It pledges protection of respondents' privacy and confidentiality and states that fulfillment of the pledge is a requirement of continued employment. Each employee must renew the pledge of commitment to the policy once a year.

In U.S. government agencies, enforcement of the pledge involves additional legal penalties. For example, staff working on surveys in U.S. Federal statistical agencies face uniform penalties for violation of the pledge of confidentiality: prison sentences of up to 5 years and fines of up to $250,000.

Another necessary ingredient to fulfill the pledge of confidentiality to respondents is a set of workplace rules that take that pledge very seriously. For example, the Institute for Social Research at the University of Michigan has created an Institute-wide Committee on Confidentiality and Data Security. In April 1999, the Committee issued a document, "Protection of Sensitive Data: Principles and Practices for Research Staff," which discusses 14 principles applying to both paper and electronic files (see Table 11.6). These principles acknowledge the fact

Table 11.6. Principles and Practices for Protection of Sensitive Data

1. Evaluate risks. Materials containing direct identifiers require a great deal of care; files containing edited and aggregated data may or may not need to be treated as confidential.
2. Assess the sensitivity of all data under your control.
3. Apply appropriate security measures. Remove direct identifiers such as name, address, and Social Security number. Questionnaires or tapes containing personally sensitive information, for example about drug use or medical conditions, should be stored in locked cabinets, as should questionnaires containing responses to open-ended questions that may reveal the identity of the respondent or others.
4. Do not include identifying personal information on self-administered questionnaires. Provide a separate return envelope for such information.
5. Store cover sheets with personal information about respondents in locked cabinets.
6. Physically secure electronic files just as you do their paper copies.
7. Take special care to secure hard disks containing sensitive material.
8. Segregate sensitive from nonsensitive material on your hard disk.
9. Consider encryption of sensitive material.
10. Consider the costs and benefits of security measures.
11. Know the physical locations of all your electronic files.
12. Know the backup status of all storage systems you use.
13. Be aware that e-mail can be observed in transit.
14. Take care when you erase files; most such files can be recovered unless special precautions are taken.

Source: ISR Survey Research Center, *Center Survey,* April 1999, pp. 1,3.

that the greatest threat to respondent confidentialitiy probably arises from carelessness rather than deliberate misconduct.

11.8.2 Technical Procedures

Although it is easy to give a pledge of confidentiality to survey respondents concerning the information they provide, it is more difficult to fulfill the pledge. The researcher gives a pledge to the respondent about data protection and stores some version of their reports in files for analysis. These files are often used by analysts other than those who directly gave the pledge to the respondents. Indeed, there are now large data archives that contain thousands of data files from surveys conducted over the past few decades. Increasingly, as the availability of administrative and other large databases grows, survey researchers need to guard against the possibility that a "data intruder" might attempt to match those administrative record systems with a survey dataset in order to identify a respondent and gain knowledge of the associated personal information.

Although for many survey datasets it may stretch the imagination to describe a motive for attempting to identify a survey respondent, for others, it is simple. Take the example of the National Survey of Drug Use and Health, which sometimes takes interviews with adolescents within a home. Parents must give their consent for the interview, and thus know that their children have participated. They also know the topic of the survey. It is quite easy to imagine that many parents might want to know whether their child reported drug use to the interviewer. If their desire to know this is intense and they have the requisite skills to manipulate computer files, they might detect the record of their child. This would be the case of an intruder with very detailed personal knowledge of the respondent.

"Disclosure limitation" is the term used to describe procedures that evaluate the risk of identifying an individual respondent and attempt to reduce that risk. There are essentially two ways to reduce the risk of disclosure:

1) Restricting access to the data only to those subscribing to a pledge of confidentiality
2) Restricting the contents of the survey data that may be released

Restricting Access to the Data. Most research projects address several questions simultaneously and can be used by researchers taking different perspectives to make different discoveries. This is especially true for survey data, which often measure hundreds of respondent attributes.

If the dataset contains rich information about each respondent, then it may be possible to discover the identity of persons even if the dataset does not contain names and addresses. For example, a person with a unique (or even very rare) set of attributes, known to be in the sample, may be easily identified by a multivariate cross-tabulation of survey data (e.g., a 12-year-old college student, who has physical handicaps, who has a dual major in physics and art history, within a dataset of college students). Those variables might be important to maintain for all cases in the file but releasing them publicly without restriction may threaten the pledge of confidentiality.

In such cases, some survey organizations extend coverage of the pledge of confidentiality to a wider set of analysts through licensing or sworn agent arrangements. In these arrangements, new analysts swear to abide by the same pledge as the initial investigator. These arrangements may involve legal commitments between research institutions, they may require periodic inspections of data security conditions at the remote site, and they may specify monetary fines or other penalties if the remote user violates the pledge. Another model is illustrated by the U.S. Census Bureau, which has established a set of research data centers where researchers who have received special approved status under the Bureau's confidentiality legislation can remotely access data stored in Census Bureau files. All output from the analysis is then scrutinized by an onsite Census Bureau employee to minimize disclosure risk.

Restricting the Contents of the Survey Data That May Be Released. Common practical steps can be taken by survey researchers to limit the likelihood of inadvertent disclosure:

1) As soon as practical, separate the names, addresses, phone numbers, or other directly identifying information from the respondent data.
2) Restrict the level of geographic detail coded into the file (this may be an analytic variable, such as a city name, some contextual data, or a primary sampling unit code) so that respondent data records cannot be identified.
3) Examine quantitative (both univariate and multivariate) data for outliers that may lead to identification (e.g., reported values of income that are very high and thus might lead, when supplemented with other information, to a disclosure).

When examination of the data suggests properties that threaten inadvertent disclosure, a variety of statistical procedures can reduce the risk of disclosure while limiting the impact on survey estimates.

Some basic concepts occur repeatedly in the ongoing research on disclosure limitation. First, a "population unique" is the term for an element in the target population that, in a cross-tabulation of some set of categorical variables, is the only element in a cell (i.e., it has a unique combination of values on the given variables). A "sample unique" is the equivalent case within a given sample of a target population (see Fienberg and Makov, 1998). These are relevant concepts because population uniques are more susceptible to detection in a dataset than elements with a combination of attributes that are very common in the target population. If the sample size is small relative to the population size, there may be many *sample* uniques that are not *population* uniques.

Second, distinctions are made between the "risk of disclosure" and "harm from disclosure" (Lambert, 1993). The "risk of disclosure" is a function of the probability of identifying the person connected with a released data record. The "harm from disclosure" is a function of what information is revealed by the disclosure and the consequences for the person. Lambert notes that if intruders believe that an identification has been made, they behave in the same way whether or not they are correct. Unfortunately, the targeted persons can be harmed by an intruder revealing false information about them, as well as by true information.

When the unaltered data records are subject to intolerable risks of reidentification (however the researcher defines "intolerable"), a variety of methods can be used to alter them prior to release of a public use data file. These include:

1) Geographic thresholds
2) Data swapping
3) Recoding
4) Additive noise (perturbation)
5) Imputation

"Geographic thresholds" determine the size of the smallest geographic area for which data are released. Currently, the U.S. Census Bureau does not identify geographic areas with a population smaller than 100,000 (Hawala, 2000; Zayatz, 2007). Since precise knowledge of the location of a sample unit greatly facilitates reidentification, limiting geographic detail is an important consideration for survey researchers.

"Data swapping" is the exchange of reported data values across data records (Fienberg, Steele, and Makov, 1996). Data swapping allows the survey organization to report honestly that an intruder has no assurance that the information found will actually pertain to the targeted individual. The real uncertainty is a function of the percentage of data items and percentage of cases involved in the swapping. However, the perceived uncertainty can be much higher when the swapping rates are not revealed by the researcher. Clearly, the challenge in swapping is whether the values of statistics computed from the dataset *after* swapping are close to those obtained *before* the swapping.

data swapping

"Recoding" changes the values of cases that are outliers. These cases have a greater likelihood of being sample uniques because of their extreme values. Recoding is performed to place them into a category shared by other cases. For example, "top coding" is a technique for taking the largest values on a variable and giving them the same code value in the data record (e.g., place all persons with incomes of $250,000 per year or higher in one category). This is a direct intervention to reduce the number of cases that are likely to be uniques. The loss of information involves certain classes of statistics whose estimates are affected by the tails of distributions. Other statistics (e.g., the percentage of persons with incomes less than $100,000) are unaffected by the aggregation. Other recoding methods include rounding the actual value of some variables, such as age or income, to some predetermined common value. For example, the U.S. Census Bureau rounds dollar amounts from $8 to $999 to the nearest $10.

recoding

"Perturbation methods" use statistical models to alter individual data values. For example, the value of a randomly generated variable might be added to each data record's value on some items. If this procedure is publicly disclosed, the data intruder cannot know how close the value in the public use file is to the real value associated with a targeted individual. If the average value of the variable is 0.0, sample means are maintained, albeit at the cost of higher total variance of the means (because of the added "noise" of the random variable). If covariances between variables are to be unaffected by the perturbation, then joint perturbations need to be performed, preserving the covariance. As the goal involves more variables, the complications increase. There are a near infinite number of ways to perturb the data. Generally, the greater the amount of noise added to the variables, the greater the protection but the higher the loss of information.

perturbation methods

"Imputation methods" replace the value of a variable reported by the respondent with another value, based on an imputation process. Rubin (1993) was the first to suggest the extreme form of imputation for disclosure avoidance—the creation of a totally synthetic dataset. Similar ideas were suggested by Fienberg

imputation methods

(1994). This totally imputed dataset would not contain any of the actual reports of any respondent. In some sense, this offers total protection. The imputation models thus take the full burden of building a dataset that produces statistics with all the key properties of the original dataset. Rubin proposes that multiple imputed datasets be created simultaneously, so that the variance implications of the imputation can be assessed empirically. A variation on total imputation imputes only the sensitive values or the identifiers of sensitive cases in the dataset to be protected (Little, 1993).

These ideas have begun to receive critical empirical tests. Abowd and Woodcock (2001), for example, have shown that synthetic data generated by sequential regression imputation have many of the same statistical properties as the original data. The U.S. Census Bureau currently uses full synthesis (generating synthetic values for all records) as well as partial synthesis (generating synthetic values only for a subset of records), and has the capability to synthesize demographic as well as establishment data. A synthetic data file linking data from the Survey of Income and Program Participation (SIPP) with benefit histories from the Social Security Administration and IRS-supplied longitudinal employee–employer earnings reports is available for testing by researchers (Abowd, Stinson, and Benedetto, 2006).

When the distribution of variables is highly skewed, disclosure limitation methods extend beyond just microdata to summary statistics as well. For example, table cells from establishment surveys contain the sum of a value of interest (such as the value of shipments) for all establishments contributing to a cell. But some of these cells may contain a small number of cases, revealing attributes of known enterprises. As Felsö, Theeuwes, and Wagner (2001) note, there are two common rules for identifying a cell as potentially sensitive:

1) (n, k) rule—if a small number (n) of respondents contribute a large percentage (k) to the total cell value, then the cell is judged sensitive.
2) p-percent rule—if any contributing establishment's value (to that cell) can be estimated within p-percent, that cell is judged sensitive.

suppression

The most common practice the authors find is the (n, k) rule. After a cell is identified as sensitive, various alterations can be made. In one, "suppression" rules may be used to omit some statistics from the table. For example, if a few firms (say, 3) represented a large proportion of total sales presented in the cell of a table (say, 70% or more), the value in the cell would be suppressed and not reported. If the cell total were published, it might be possible to identify a firm (and thereby its revenue) from the published statistics. As an alternative to suppression, categories of one of the variables may be recoded or combined to prevent disclosure. Other techniques, such as adding noise to cell values or to the underlying microdata, can also be used to protect establishment tabular data.

In the future, it is quite likely that new methods will be used to protect survey data from abuses, while extending the benefits of survey data to broad populations of users. As these are developed, the resources available to data intruders will also increase. Much research is needed in this area, including research on respondents' perceptions of the risks of harm involved and of how best to communicate such risks to them (Couper, Singer, Conrad, and Groves, 2008).

11.9 Summary and Conclusions

Societal norms and rules about the conduct of scientific activities apply to survey methodologists: sanctions against fabrication of data, falsification of research reports, and plagiarism of others' work. These sanctions extend to staff supporting the research, including survey interviewers. Interviewer falsification risks are small but persistent in data collection, especially face-to-face interviewing. However, the falsification rates appear to be reduced by training in scientific ethics, observation of interviewers' work, recontact of respondents, and inspection of collected data (see Section 9.8).

Survey professional organizations, such as AAPOR, create codes of ethics to guide the behavior of survey methodologists, including dealing with clients, the public, and survey respondents. The most fully elaborated system of rules applies to treatment of survey respondents. Three principles guide such treatment: beneficence, justice, and respect for persons. The principle of beneficence motivates efforts to avoid harming any survey respondent. The principle of justice involves equitable spreading of the burden of participation in surveys, which is often inherent in probability sampling approaches. The principle of respect for persons gives rise to concerns about giving sample persons the right to refuse participation after being informed of the content of the survey and its potential risks and benefits. For much academic and government research in the United States, IRBs oversee the implementation of these principles in all studies involving human subjects.

Under the Common Rule, there are specific requirements for informed consent protocols (see Table 11.4). Informed consent procedures in surveys most often involve a verbal description of the nature of the survey content and the procedures for keeping data confidential. The amount of information given about the survey content appears to have only modest effects on cooperation and data quality, though laboratory experiments have shown larger effects. Requiring survey respondents to sign a written consent form appears to generate refusals among people willing to be interviewed but leery of the signature requirement. Because requesting a written signature can harm the quality of survey results, written consent should not be required in surveys that pose only a minimal risk of harm, as defined under the Common Rule.

Protecting survey respondents from harm, the beneficence principle, has led to procedures that support the pledge of confidentiality to the respondent. These procedures include training and formal commitments made by staff, and in Federal agencies, strong legal sanctions against breaches of the confidentiality pledge. They include a variety of methods for extending the pledge of confidentiality to users of the data who are not part of the original research team. They include efforts to strip from analytic datasets all explicit identifiers of respondents. Finally, they include a growing number of statistical methods to limit the risk of reidentification of a survey respondent's data record in publicly released survey data files. The risk of disclosure is much higher for data records that have unique or near unique combinations of values across multiple variables. The techniques used to reduce the risk of disclosure include data swapping, recoding methods, additive noise methods, and imputation methods. When there is concern that cross-tabulations from a survey might risk reidentification of a sample unit (most common in business surveys), cell suppression and methods to combine cells are common.

Survey researchers must be proactive to fulfill the ethical guidelines described above. These are not procedures that come to mind naturally to the untrained. It is common for new members of survey teams to receive orientation to the issues surrounding the treatment of human subjects and to some of the basic principles. The U.S. National Institutes of Health has a Web-based training module, admittedly more oriented to biomedical research, which is used to train researchers (see http://www.nihtraining.com/ohsrsite/). Ultimately, the protection of respondents' rights depends on the survey research staff involved. For that reason, successful building of a working environment with strong norms about the importance of informed consent, avoidance of harm to respondents, and fulfillment of confidentiality pledges is the most efficient way of achieving the ethical principles outlined above. Efforts at supervision and enforcement can only supplement such internalized norms.

The practices involving ethical treatment of human subjects in survey research are rapidly evolving, not only because of societal concerns about abuses, but also because developments by survey methodologists are giving researchers new tools to reduce research misconduct, fulfill confidentiality pledges, and improve informed consent procedures in surveys.

KEYWORDS

beneficence
certificate of confidentiality
confidentiality
data swapping
fabrication
falsification
harm from disclosure
imputation methods
informed consent
interviewer falsification
institutional review board

justice
minimal risk
perturbation methods
plagiarism
population unique
recoding
respect for persons
risk of disclosure
sample unique
statistical disclosure
suppression

FOR MORE IN-DEPTH READING

Citro, C., Ilgen, D., and Marrett, C. (2003), *Protecting Participants and Facilitating Social and Behavioral Sciences Research,* Washington, DC: National Academy Press.

Doyle, P., Lane, J., Theeuwes, J., and Zayatz, L. (2001), *Confidentiality, Disclosure, and Data Access: Theory and Practical Applications for Statistical Agencies,* Amsterdam: North-Holland.

EXERCISES

1) Many U.S. university campuses offer Web-based training for researchers in the ethical treatment of human subjects. If your campus has such a training

module, complete it. If your campus does not, complete the training on the website of the National Institutes of Health, http://www.nihtraining.com/ohsrsite/.

2) What issues of informed consent are raised by each of the situations below? If you were a member of an IRB, would you approve the research activity described, or would you require some change or addition? Justify your answer for each of the cases described.

   a) As part of a face-to-face survey, the interviewer is asked to observe how many books are visible in the respondent's living room.
   b) As part of a methodological study in a telephone survey, the first few moments of the interaction between the interviewer and the potential respondent are recorded for later analysis.
   c) Before asking for a survey interview as part of a study of children's interactions with parents, the interviewer makes structured observations of how the parent and child interact in a public park. The observations are used to select the sample and become part of the data collected for each respondent.
   d) As part of a study of whether vote intentions are affected by the sponsor of the survey, some respondents are told that the sponsor is an organization different from the one actually conducting the survey.

3) A researcher wants to determine how accurately women report having had an abortion. From medical records, she obtains the names and addresses of women who have had an abortion at a particular clinic, and then sends them a letter inviting them to participate. Among other things, the letter states:

"The aim of the investigation is the collection of data on the health of our women and the factors influencing it, as well as how satisfied women are with the organization of medical care and their opinion about ways it might be reorganized. The substantive part of the research is a survey in which data are collected directly from people. Respondents, including yourself, were selected from the address register by a method of random selection. . . ."

   a) What, if any, ethical principles do you think this experiment violates?
   b) What consequences do you think this procedure will have for the findings the researcher came up with?
   c) Could the research have been done differently? How?
   d) Suppose you were a respondent and discovered how your name had come into the sample. What would your reaction be?
   e) Suppose this was a study of how accurately people report their voting behavior, and a sample of names had been drawn from voter registration lists but respondents were not told how their names had fallen into the sample. Do you think any ethical principles were violated in this study? How would you, as a respondent, feel if you found out how the sample was really drawn?

4) As part of their course assignment, undergraduate students in introductory social psychology classes are sent to various areas of New York City to ask

1520 passersby one of a variety of simple requests. They are told to ask for different kinds of help and to ask for it in different ways. The varying responses to their requests provide a first approximation to answering questions about the prevalence of altruistic compliance and the factors influencing it.

    a) What, if any, ethical principles are violated by this kind of research?
    b) What are the advantages of this kind of research?
    c) Could the research be done in some other way? How?
    d) If you were on a human subjects committee, would you approve it? Why, or why not?

5) As part of a face-to-face survey of homeless, displaced persons, a $100 incentive to respond to the survey is offered. The IRB disapproves the research protocol on the grounds that the incentive is coercive. What response would you make to the IRB?

6) "Push polls" involve a political campaign telephoning many persons and asking survey questions that have false premises designed to leave an unfavorable impression of the opposing candidate, for example, "If I told you that (Candidate A) had been convicted of drug smuggling and murder, would that change your views of whether he should be elected?" What ethical research principles does this practice violate?

7) You have completed a survey of children 9–12 years old, preceded by obtaining written informed consent from the parents and agreement of the sampled children to conduct a one-on-one interview of the children. (That is, no parent is allowed to observe the interview.) After the interview, one parent asks to see their child's answers. What ethical issues does this request raise? How could they be avoided?

8) On the next page is a template of an informed consent process for a Web survey. Identify each of the components of informed consent that appear in this template. Do you see any deficiencies in how this template fulfills the ethical requirements of informed consent?

Welcome to the CAREGIVING: STRESSES AND SUPPORT Survey (HUM00001234)

Dr. John Jones and Dr. Sarah Smith of the University of Michigan, Department of Psychology, invite you to be a part of a research study that looks at the stresses that people experience when they are providing care to a seriously ill family member. The purpose of the study is to design better support programs for caregivers. We are asking you to participate because you recently attended a meeting of a UM Caregivers Support Group.

If you agree to be part of the research study, you will be asked to complete an online survey about your experiences as a caregiver. We expect this survey to take 30 to 45 minutes to complete.

Although you may not receive any direct benefit for participating, we hope that this study will contribute to the improvement of social support systems for those who provide care to others.

Your responses to this survey are anonymous, meaning that the researchers will not be able to link your survey responses to you. The survey software does not collect identifying information about you or your computer. We plan to publish the results of this study, but will not include any information that would identify you.

Participating in this study is completely voluntary. Even if you decide to participate now, you may change your mind and stop at any time. You may choose to not answer a question or skip any part of the study. Simply click "Next" at the bottom of the survey page to move to the next question.

If you have questions about this research study, you can contact John Jones, Ph.D., University of Michigan, Department of Psychology, 123 East Hall, Ann Arbor, MI 48104, (734) 123-4567, jjones@umich.edu.

The University of Michigan Behavioral Sciences Institutional Review Board has determined that this study is exempt from IRB oversight.

By clicking on the link below, you are consenting to participate in this research survey.

www.caregivingsurvey.net

If you do not wish to participate, click the "x" in the top corner of your browser to exit.

CHAPTER TWELVE

FAQS ABOUT SURVEY METHODOLOGY

12.1 INTRODUCTION

To this point, the book has presented many basic principles and practices of survey methodology. The text in all the chapters described how these principles act together to define a framework for survey quality. The keywords define the technical nomenclature of the field. The boxes describing some of the classic articles in the field provide a sense of the key types of scientific inquiry in the field. Finally, the exercises at the end of the chapters simulate the kinds of questions survey methodologists address in their work.

However, you may still have unanswered questions that prevent you from integrating the material in your mind. This happened to others when they read drafts of the book before publication. Their comments helped us revise the chapters, but some of their questions were more global in nature, not fitting well in any one chapter. In addition, there are some questions that survey methodologists hear routinely that also are fundamental to the field.

We have written this chapter in a completely different style than that of the other chapters. It is a list of frequently asked questions and answers (FAQs). The questions are in **boldface** type. You can glance over the pages of the chapter looking for questions of interest. When relevant, the answers may refer to material we presented in other sections of the book.

12.2 THE QUESTIONS AND THEIR ANSWERS

With all the errors described in this book, how can anything work? Can anyone really believe the results of any surveys?

Survey Methodology is a text about how to minimize statistical error and maximize the credibility of survey results. As such, the book focuses on the scientific questions of why error arises and what can be done to reduce it. Each chapter essentially describes a number of potential threats to the quality of survey statistics and what survey methodology teaches us about how to address those threats.

In that context, it might be easy to overestimate the magnitude of error in surveys. Certainly, numerous surveys are done poorly, by the standards articulated in this book, and they produce data of dubious quality. However, many surveys have incorporated excellent procedures, and their results are often very accurate. Probability sample surveys are routinely done that collect data from respondents who collectively look very much like the population from which they were drawn. Survey questions are carefully tested to identify those that are difficult to understand and answer. Questions are improved so that, for the most part, they can be answered consistently and accurately by respondents. Interviewers receive extensive training, are carefully supervised, and largely carry out the administration of surveys in the way that is prescribed.

Survey methodological research demonstrates that properly conducted surveys can achieve very high quality results. For example, since 1998 the jobs count from the CES lies within 0.2% of the actual jobs count based on administrative data. The assessments in the NAEP, when studied by experts in the areas of verbal and quantitative competence, get high marks. We showed in Figure 1.3 how effectively the SOC unemployment expectation predicts actual unemployment that occurs later. The BRFSS successfully detected the obesity epidemic that hit the United States (see Figure 1.4). Surveys that use the lessons of this text can provide very useful information.

On the other hand, when the lessons of this text are ignored in survey design and execution, surveys can yield very misleading results. See the discussion of the Web survey of the National Geographic Society on page 97.

You didn't tell us about other ways of collecting information, like ethnography, in-depth interviewing, or laboratory experiments. Do you think the survey is the best way to study humans?

No. Survey methodology is our field of expertise and the focus of the book. There are many other approaches to the study of human thought and behavior that we would employ, depending on the question at hand. The choice of method needs to be derived from the research objectives, not vice versa. Surveys produce quantitative estimates, but some knowledge cannot easily be quantified. Thus, we do not recommend only quantitative or only qualitative approaches to gathering information or attaining understanding. Indeed, in Sections 7.3.2 and 8.3–8.4 specifically (and throughout the text) we describe the results of methodological studies that use nonsurvey methods to study the properties of survey statistics. Survey methodologists use diverse methods to understand how surveys work. In fact, we believe surveys work best when supplemented with other methods.

Surveys are rather blunt instruments of information gathering. They are powerful in producing statistical generalizations to large populations. They are weak in generating rich understanding of the intricate mechanisms that affect human thought and behavior. Other techniques are preferred for that purpose. Surveys also work best within populations that share some basic language and cultural knowledge. In more diverse populations, other techniques are preferable.

How much does a survey cost?

This question is a little like "How much does a house cost?" There are mansions that cost many millions of dollars, and there are small cottages that cost tens of thousands. Either might be what is needed by a particular purchaser.

You now know that many aspects of quality in survey statistics require complicated efforts. If sampling frames are not readily available (as in area probability samples), the researcher must develop them, at considerable expense. If high precision (low sampling variance) is required for the statistics, large samples are generally required, increasing sampling and data collection costs. If complicated contingencies exist among the questions asked, CAI applications or complicated postcollection editing may be required. If the target population is not motivated to respond or incapable of using a self-administered form, then an interviewer corps must be recruited, trained, and supervised. If the target population is a small subset of the sampling frame, considerable costs in screening to identify eligible respondents may be needed. If the target population is difficult to contact, repeated follow-ups that cost money may be required. If the target population is reluctant, incentives or refusal conversion efforts might be mounted. When none of these challenges exists, a survey can be relatively inexpensive. When they are all present, a survey attempting to reduce the various errors of observation and nonobservation will be costly. Unless one prespecifies all those design features, surveys are clearly not something you want constructed using "the lowest bidder," because the easiest way to achieve low cost is to ignore one or more of the key threats to good survey estimates.

How much confidence can I have in poll results? How much confidence can I have in the results of Internet surveys?

Confidence in survey results should be a function of their purpose, their design, and their implementation. Thus, your confidence in the results must query both design and implementation. If you have no information about the design and implementation, then it is impossible to make a judgment about the statistics. (That is why the AAPOR disclosure requirements reviewed in Section 11.4 are important.)

If the target population, sampling frame, sample design, question evaluation, method of data collection, unit and item nonresponse properties, and editing procedures are documented, then you can begin to form a judgment. The judgment would be formed using all the lessons of the chapters of this text.

The lack of documentation of a survey is perhaps the most frequent impediment to judging how credible a survey statistic is. It is also important to note that all the lessons of the earlier chapters demonstrate that different statistics from the same survey may have very different coverage, nonresponse, and measurement error properties. Hence, there are no "good surveys" or "bad surveys"; there are only good survey *statistics* and bad survey *statistics.*

What is the most important source of error in surveys? What are the most important features of an accurate survey? How do you quickly determine how good a survey is?

There are several possible meanings to these questions:

1) What is the error source that has the greatest potential to harm the quality of survey estimates?

There is no one answer to this question. A survey statistic can be ruined by coverage error, sampling error, nonresponse error, or measurement error.

2) What error source is commonly most harmful to the quality of survey estimates?

Common mistakes include self-selected samples, poorly worded questions, and inappropriate modes of data collection. Self-selected samples, such as those persons who respond to requests for answers in a magazine or call a 900 number, have repeatedly been shown to be remarkably different from the population as a whole (again, the NGS survey is an example). Somewhat more subtly, mail surveys are often done that produce a low rate of return. It typically is found that those who responded, when most people will not respond, are quite different than the target population as a whole in ways relevant to the survey (e.g. Fowler, Gallagher, Stringfellow, Zaslavsky, Thompson, and Cleary, 2002).

Questions that are not understood as the investigators intended also produce major distortions in survey estimates. In attitude measurement, it is not uncommon to find questions that set up contexts or use structures that bias answers.

The importance of using appropriate methods when asking people to answer questions with potentially sensitive content is illustrated well in the mode research described in Section 5.3.5, showing higher rates of reported drug use and stigmatizing sexual behavior in the ACASI relative to interviewer administration.

A key theme of this book is that all error sources must be jointly considered. Which error is of greatest concern will depend on the target population, the materials and staff resources available, and the purposes of the survey.

How do you draw a sample of a target population without a frame?

We used one illustration of sampling a target population of persons in the United States that has no sampling frame of persons: use of area probability sampling.

There are many target populations that do not have up-to-date lists of elements. Like area sampling techniques, the first solution is to find a well-enumerated frame to which the target population can be linked. Researchers who want to sample students, but lack a comprehensive list of students, first sample schools, then go to the school and make lists of students from which to sample. Customers of a retail store could be sampled by enumerating the various entrances to a particular store, sampling possible times when people could shop in the store, and then developing a scheme for systematically sampling the individuals who enter the different doors at different times. In the same way, one could sample beach users by drawing samples of land areas on the beach, developing a scheme for sampling times, then surveying those individuals who were located on the selected beach areas during the particular sampled times.

How big should my sample be? What is the minimum sample size to use?

The sample design should permit conclusions to be made with a level of uncertainty tolerable for key survey statistics within the cost constraints of the survey. That sentence may not seem to answer the question, but let us parse it. First, the sample size is just one aspect of the sample design (others being stratification, clustering, and assignment of probabilities of selection). All four affect standard errors of statistics and, thus, all four must be jointly considered (see Chapter 4).

Second, the standard error should be viewed relative to the decisions to be made based on the survey. If the actual value of a statistic is "X," would the same conclusion be made by the investigator if the survey estimated "0.5X," "0.8X," and "0.9X"? Answers to those questions yield some notion of what standard error is needed (i.e., one that can distinguish 0.5X from X or 0.9X from X). Third, the sample design must be chosen with key analytic goals in mind. Most surveys are designed to make numerous estimates, not just one, and the requirements for precision are likely to differ for each statistic. In addition, estimates often are required for small subgroups of the population, not just the population as a whole. If the different analytic goals yield different ideal sample designs, then some compromise among the goals must be made. Fourth, it is possible that the sample design required by the analytic goals cannot be conducted within the survey budget. Then other compromises must be made.

Having given an answer about what *to* do, here is what *not* to do. Do not use the sample size that another survey used, as if that validates a procedure (e.g., just because the Gallup Poll uses 1000 interviews does not mean that makes sense for your survey). Do not use a sample size that is the same fraction of your population as used in some other population (i.e., remember the lesson of Section 4.3 that sample size requirements are almost independent of frame population size).

What evidence is there that nonprobability samples are really bad?

If you review Section 4.2, we note that the two benefits of probability sampling are:

1) Important types of sample statistics are unbiased.
2) We can estimate the sampling variance (standard error) from one realization of the sample design.

The unbiasedness property applies to *sampling* bias, not to biases from coverage, nonresponse, and measurement error. One of the strengths of probability sampling is that samples are selected by some statistical rule rather than being based on volunteers or those who are most available. Some, but not all, nonprobability sampling relies on one or more of those features, which increases the chances of a biased sample. However, because in practice all surveys are subject to those biases (generally to some unknown level), the first property of probability sampling is not of paramount value. Hence, the greater value is the ability to estimate standard errors with some assurance that they reflect the potential variability of results over replications of the sample.

This does not imply that each individual sample realization of a probability sample is better than each individual sample realization of a similar nonprobability sample. When the sample designer has full knowledge of what attributes of the frame population are correlated with key survey statistics *and* successfully balances the sample on those attributes, a nonprobability design can produce sample statistics that closely resemble those of the population. Such samples, using "quotas" for different subgroups (e.g., age, gender, or race groups) are common. When a statistic is measured that has different correlates, unrelated to those of the quota, considerable error can be introduced. Hence, nonprobability samples, to be superior to probability samples, require more knowledge of the joint distributions of variables in the target populations. Usually, we do not have that knowledge. In other words, a nonprobability sample *may* yield results that are similar to probability samples, but one is never sure when a particular nonprobability approach will work and when it will not.

I hear claims that volunteer Internet panels give the same results as surveys with probability samples? Is that true?

Volunteer Internet panels (or access panels) consist of email IDs sent by persons who at one point reported a willingness to participate in some surveys administered over the Internet. The panel sponsor sends to selected members e-mail requests with imbedded URL links to a website containing a survey questionnaire. There are three issues in examining estimates from a volunteer internet panel: (a) how were the panel members recruited, (b) how are the panel members rewarded for their participation, and (c) what are the cumulative survey experiences of the panelists?

The power of the set of panelists to reflect the full adult population is a function of whether persons with different values on survey variables are members of the panel. In probability samples, all persons in the target population have a known chance of being a panelist. In some Internet panels, the members have sought out their participation (and thus are likely to be very interested in survey participation); in others, they have responded to a notification of rewards if they respond. If the sole motivation of response is the reward, there is some concern that their answers to survey questions may not be thoughtful ones.

Further, there is a concern that rewards for participation have created a set of persons who choose to be members of many panels simultaneously. It is common to have respondents from volunteer Internet panels in the United States report an average of between 5–8 panel memberships. Some studies have shown that very small proportions of Internet users represent large proportions of panel respondents. It appears that for some estimates, these characteristics of recruitment, reward, and answering behavior produce very misleading results. For example, panelists with long tenure in the panel appear to give different answers on some items than short-term panel members. For other purposes, the panels seem to generate results similar to other techniques. At this writing, methodological research has not identified the mechanisms that lead to biased results. Without a deep understanding of the motivation to participate in volunteer Internet panels, the researcher must evaluate their results through empirical comparisons to more theoretically sound methods.

Pollsters report that their results have a margin of error of "plus or minus x percentage points." What does that mean and how helpful is that in knowing how accurate the results really are?

The "margin of error" for a survey reported in popular media is usually a 95% confidence interval under simple random sampling for an estimated percentage equaling 50% (e.g., if 50% of the respondents reported that they planned to vote for Obama in the 2008 election). For example, if the number of completed interviews is 1000 then the margin of error is

$$2\sqrt{(0.50)(0.50)/1000} = 0.0316$$

or 3 percentage points. The percentage, 50%, is typically chosen because it maximizes the "$p(1 - p)$" in the numerator of the sampling variance expression. Hence, all percentages would have 95% confidence intervals less than or equal to that stated number.

Such margins of error need to be scrutinized when the design is not simple random sampling (e.g., unequal probabilities are involved, or clustering is introduced). Each of those departures from simple random sampling tends to increase standard errors, thus making the margin of error an underestimate. In addition, percentages computed on a subsample of the full survey will tend to have larger margins of error. The computation is relevant only for statistics that are percentages; it gives no information for statistics that are means (other than percentages), differences of means, or totals. Finally, variable errors of coverage, nonresponse, or measurement are typically not included in the margin of error. No biases of coverage, sampling, nonresponse, or measurement error are included in the margin of error. Therefore, it is important for readers to understand that the reported "margin of error" reflects only one component of the total error that potentially can affect the estimates.

None of my statistics texts mention weighting or sampling variance computations that reflect the clustering of the design. Why is that?

We know that the statistical formulations in this text are different from those in most introductory applied or mathematical statistics texts. Many of the analytic statistics that are described in those texts assume that the process that generates the sample is one of sampling with replacement or, equivalently, that the observations being summarized with the statistics are a set generated by an ongoing stable process. The set of observations, therefore, all have identical properties and are independent of one another. This text, however, treats more complicated processes of generating the set of observations—ones involving stratification, clustering, and unequal probabilities of selection. The statistics described in this text have average values over replications of the sample design that are functions of this stratification and clustering and the probabilities of selection.

Both what you learned in your statistics classes and what you learned from this text are correct, given the assumptions about how the sample was generated. Most surveys use sample designs involving some stratification, clustering, or unequal probabilities of selection. Hence, the statistics in this book reflect those features.

What is the best mode of data collection for a survey?

They can all be the best mode, in a particular time and place, for a particular purpose. Choosing a mode for data collection is one of those decisions, like that of sample size, that requires thoughtful analysis of the range of issues that can affect the estimates and survey costs. As was discussed in great detail in Chapter 5, there are certain situations in which some modes would be unwise. For example, if a significant portion of the target population lacked Internet access or skills, the Internet would probably be a poor choice of mode, at least as a sole mode of data collection. If the survey was aimed at getting some estimates that potentially were very sensitive or personal, the involvement of some kind of self-administration, with or without an interviewer, might seem like a good idea. The population characteristics, the content, the nature of contact information that is available, and the availability of staff talent and resources will all affect what the best mode may be for any particular survey. There are excellent surveys done by all of the modes discussed in Chapter 5, plus combinations thereof. The task of the survey methodologist is to do a thoughtful analysis of the various factors that affect data quality and costs. Then the methodologist must decide, from a total survey error perspective, which mode for collecting data will best serve the survey's purposes.

Are surveys increasingly moving toward self-administered modes? Will interviewer-assisted surveys die?

There have been many claims about the imminent demise of interviewer-administered surveys. We will not guess about the long-run future of human measurement. The benefit of self-administered modes is that there are increasing numbers of methods to use (e.g., e-mail and Web). The value of self-administration for measurement of sensitive attributes is now well demonstrated (see Section 5.3.5).

Although self-administered modes have strengths, there also will continue to be situations in which interviewers will be critical to executing a successful survey. For example, when there is no sampling frame that includes contact information (i.e., name, address, or e-mail address), interviewers may be needed to locate the sample. When sample persons need external motivation to agree to the survey requests, interviewers are valuable. Finally, there are certain kinds of questions that benefit from interviewers, especially open questions calling for extensive memory searches. Thus, although self-administered modes reduce measurement error for sensitive topics and can be cost-effective, for some purposes interviewers play critical rules in executing surveys.

Are the various models of measurement error really used in surveys?

Sometimes, but probably not often enough. You have learned in Chapters 7, 8, and 9 that various statistical models describe the process of response formation and that their parameters could be estimated given appropriate designs. The design features that are needed to estimate parameters in the models (e.g., interpenetrated interviewer assignments, reinterviews, and multiple indicators) require additional work on the part of the researcher. When these features are implemented, they increase survey costs. For both those reasons, they are not routinely implemented. Survey methodologists look for opportunities to gather auxiliary data about the measurement process in order to produce estimates of the various kinds of measurement error. The use of computer-assisted methods offers great promise to introduce such features at nearly zero marginal cost to the overall survey.

What is the minimum response rate needed to produce an acceptable survey result?

Unfortunately, this has an answer that could vary for different statistics in the same survey. Chapter 6 discussed how the nonresponse rate is only an indicator of the *risk* of nonresponse error. The difference in the survey statistic between the respondents and the nonrespondents is another linkage to nonresponse error. This characteristic of a survey statistic is not measurable in general. When the causes of survey participation are related to the survey statistic, nonrespondents tend to have different values than respondents on that statistic. When that is the case, very high response rates are required to achieve low nonresponse error. When the causes of survey participation are unrelated to the survey statistic, nonrespondents and respondents, by definition, will have similar values on the statistic. In those cases, survey estimates will be similar regardless of the response rate. Most surveys do not (because they cannot) inform us about which case applies.

Despite the ambiguity of their meaning, response rates are routinely reported and are commonly used as a quality indicator. Hence, the credibility of survey statistics is often linked to response rates. In the absence of information about the relationship of nonresponse to survey estimates, it is probably fair to say that the higher the response rate, the lower the average *risk* of significant error due to nonresponse for all the estimates based on a survey.

Will the declining response rates to surveys mean the death of surveys?

First, as Section 6.2 notes, response rates for household surveys do appear to be declining more rapidly than those of establishment surveys. Second, government surveys like the NCVS and NAEP continue to achieve quite high response rates. Third, other countries like the United Kingdom and the Netherlands have historically experienced much lower response rates for household surveys, yet the use of surveys as a basic information gathering tool is thriving there. Fourth, survey protocols for data collection have always had to change as the population has changed and society has changed. When the number of single-person households grew, households with children declined, and more women joined the work force, it became harder to find people at home. Survey organizations had to greatly increase the number of calls they made to get interviews. For phone interviews, sometimes 20 or 30 calls are made to find those difficult-to-reach individuals.

Rather than declining in use, the trend in the last 10 or 15 years has been for surveys to be used increasingly to gather information about the populations in the United States and throughout the world. Although there is evidence that some of the data collection protocols that worked in the past work less well to elicit cooperation, it seems almost certain that new data collection protocols and procedures will evolve.

Is writing survey questions as complex as survey methodologists argue it is? After all, we use questions every day as part of our normal lives.

The key difference between questions in everyday conversation and survey questions is that survey questions are designed to produce answers to the same questions that can be tabulated to provide statistical summaries. For example, you may want to compute a percentage of those answering these questions who have specific characteristics: they are democrats, they are left handed, they favor abortion, or they have visited a doctor in the last year. To do this:

1) Everyone has to answer the same questions; the words in the questions have to mean the same thing to everyone who answers them (this leads to standardized wording).
2) There must be answers that can be summarized and tabulated in some consistent way; people must provide answers in comparable terms (this leads to fixed response questions and probing rules to obtain adequate responses in open questions).

The discovery from survey methodology that apparently minor changes in question wording produce large differences in answers is no longer controversial. Not all the lessons of that research are known by the lay public, but many probably apply to everyday speech. However, the stakes of misunderstanding in everyday conversation are typically lower than in survey interviews. The meandering focus of everyday conversation permits both speakers to learn from later exchanges whether they did indeed misunderstand each other and to correct misunderstandings. The common ground they share (and often have built up over many prior exchanges) gives them great freedom to seek clarification from each other. Survey interviews are much more focused interactions; they need initial question wording with minimal ambiguity and maximal comprehension the first time the question is heard, so that understanding does not require interviewers to clarify the questions very often.

Self-administered modes put even more burden on question wording. In these modes, there is no real chance to clarify any misunderstandings. The question wording and presentation have to communicate to all the intent of the researcher.

In short, although people routinely ask questions, they do not usually use words and phrasing that will be universally understood in the same way and that would be answered in ways that could be tabulated.

Are multi-item measures better than single-item measures?

Multi-item measures are attractive when single item measures have known inadequacies. In psychological measurements, as we note in Chapter 7, sometmes many items can be constructed to reflect the same underlying concept (e.g., just as mathematical ability can be measured with many different arithmetic problems). Here the researcher believes that the construct in focus is measured only with some uncertainty with a single item. Using multiple items reduces this uncertainty under this perspective. Other multi-item batteries take a different perspective. They view the construct as consisting of separate components (e.g., salary, tips, bonuses, interest on savings, or dividends on stock owned as components of income). Here, the multiple items generally add to the aggregate report of a quantity.

Why should we cognitively test questions that have been used in previous surveys?

Cognitive evaluation of questions is comparatively new in survey research. As a result, the vast majority of questions that were developed and used in surveys prior to 1990 were not evaluated from a cognitive perspective. Moreover, even questions developed into the 1990s were not universally cognitively tested. As a result, there are many questions with a long history of use in various fields that do not meet current cognitive standards.

Researchers can argue for the value of using questions that have been used before. Sometimes, replicating a previously used item is valuable because it permits measuring change over time. Alternatively, a previously used question may have been demonstrated to have some degree of validity based on correlation analyses that were done. A third, but somewhat less scientifically justifiable basis for replicating questions is the sense that having been used before and "working" by some unspecified standards gives a question some credibility.

Regardless of which of these bases is used to justify the use of an item, if an item has not been cognitively tested, there is good reason to test it before using it again in a survey.

If a previously used item is cognitively tested, there are three possible outcomes. First, it may be tested and found to be a good question from a cognitive perspective. In that case, everyone is reassured. Second, the question may be found to have significant problems from a cognitive point of view. Once the problems are identified, the researcher may decide that the reasons for using the item, such as measuring trends or its past psychometric performance, do not justify using a question that is seriously flawed. After all, the next survey will be the baseline against which future trends will be measured. There is a strong case for using the best question possible each time a survey is done. Third, in the event that the researcher decides that the value of using the item outweighs the flaws detected in cognitive testing, at least the researcher and other users of the data will be aware of the question's limitations. Thereby, conclusions based on the use of the item can be tempered appropriately.

What is a reasonable protocol for testing of questions prior to a survey?

It is common to use a multiphase testing protocol. If the survey covers areas that are relatively new to the researcher or is covering new ground for the particular target population that is being studied, conducting a few focus groups with members of the target population is almost always a cost-effective way to begin a survey process. An invaluable first step is learning the vocabulary that the target population uses in discussing topics in the survey.

Once a draft set of questions has been designed, expert review from both a content and methodological perspective is also very cost-effective. Content experts can make sure that the questions will yield the information they need for analysis, and the methodological experts can flag potential problem questions. In some cases, principles of good question design will permit improving the questions immediately. The expert review can also flag possible issues in the questions that can be evaluated empirically.

Cognitive testing of questions has become a standard part of the development of a new survey instrument. Even a few cognitive interviews, fewer than ten, can identify important problems with comprehension of questions or with the response formation task. Corrections based on the testing can make a significant contribution to the final value of the survey data.

In addition, field pretests continue to have a role. Two additions to the standard pretest can significantly increase the value of the results. First, for interview surveys, tape recording the interviews and coding the behavior of interviewers and respondents during the pretest interviews is a fairly low-cost way to get systematic data on how well questions are asked and answered. It also provides a quantitative indicator of questions that are posing problems and would likely benefit from revision. More elaborately, the inclusion of small randomized experiments in question wording in the pretest can provide empirical evidence of the effects of question wording on the data. Such tests are particularly valuable when revisions of previously used questions are being considered or when there are two versions of a question that are under consideration. In order to provide meaningful data, the number of cases included in the pretest would necessarily have to be larger than is usually the case, perhaps in the 100 to 200 range. However, the added information that is provided often would make such an investment a good one.

Which is more important: cognitive testing or postsurvey measurement of validity or response bias associated with the question?

On the one hand, the ultimate test of whether or not we have done a good job of measuring something is an agreement of the answers with some external criterion. Thus, when choosing between two questions, if one agrees much more highly with a good indicator of the construct being measured (see Section 8.9), science would dictate choosing the more valid measure, regardless of how the cognitive testing came out. If indicators of response deviations at the respondent level show negligible error, there may be sufficient evidence of good measurement quality.

On the other hand, it is more often the case that researchers lack a "gold standard" measure of the construct they are trying to measure. In the absence of a good validating standard, assessing how well respondents understand the questions and are able to formulate answers should, in fact, be a rather direct way to reduce measurement errors. Having stated this strong theory, we must go on to say that there is not a great deal of evidence that improving questions from a cognitive point of view produces more valid measurements.

Thus, clearly, the answer to this question is that both cognitive standards and psychometric standards are important when evaluating questions. In theory, they should be closely related. We very much need more and better studies to help document and understand those relationships.

What does it mean to say that a question has been validated?

Many questions that arise out of psychological measurement use construct validity as their quality standard (see Section 8.9.1). Sometimes, the initial inventors of the questions conduct validity studies (using external measures, or multiple indicators), improve the measures, run the validity studies again, and then report the measures as having been "validated."

First, the studies often assess validity by looking at the correlation between the answers to a question and some external measures of the construct the question is designed to measure. True "gold standards" are extraordinarily rare. Moreover, correlations of 1.0 with a gold standard virtually never happen. Thus, validity is a matter of degree. Researchers collect evidence that leads them to think that at least some of the variance of the answers truly measures the target construct.

Second, the validation study results (and the very notion of validity) are dependent on the population measured. Validation studies are often conducted on populations dissimilar to those with which the questions are later used.

Third, survey methodology has discovered that question performance sometimes depends on the context of the question and method of data collection.

So, when researchers say they are using a validated "measure," it means that for some population at some point they collected some evidence that some portion of the variation in answers reflected the target construct. It does not inoculate the next user against any of the measurement errors we have described.

Are interviewers really an important source of survey error? If so, why do researchers so seldom report that error? Do interviewers really have much effect on the quality of survey estimates?

It is obvious that the answer to a survey question should not vary by who asks the question. However, when studies assess the extent to which answers are related to interviewers, the average intraclass correlation is around 0.01 (Groves, 1989; Fowler and Mangione, 1990). That may seem to be a small number. However, the effect on standard errors of estimates is:

$$\sqrt{1 + \rho_{int}(m-1)}$$

where m is the number of interviews taken on average by an interviewer (see Section 9.3) What that means is that if interviewers average 50 interviews each for a given survey, the standard errors of sample means for the average question, if the average ρ_{int} is 0.01, should be inflated by a factor of 1.22. If the average interviewer on a survey takes 100 interviews, as happens often for large national surveys, standard errors for that same average question need to be inflated by a factor of 1.41.

Research has shown that interviewer-related error can be reduced by proper training and supervision, as well as by designing questions that do not require a lot of interviewer intervention in order to get an answer. However, virtually all surveys that use interviewers include questions that are significantly affected by interviewers. Unfortunately, very few studies use interpenetrated sample assignments to interviewers, so researchers usually are unable to estimate the effects of interviewers on their statistics.

Some of my econometrician friends say that case weights in a regression model don't make any sense at all. Is that true?

As we noted in a question above, survey statistics employ statistical estimation that reflects the survey design. Econometrics generally uses a different inferential framework, one appropriate to describing infinite processes that produce values on the dependent variable. For example, consider the simple regression equation:

$$y_i = \beta_0 + \beta_1 x_i + \varepsilon_i$$

used by an econometrician to test the theory that x causes y. Traditional OLS estimation can produce unbiased estimates of β_i if the model is correctly specified. The survey methodologist might suggest using selection weights, say in the case of a survey in which older persons were sampled at higher rates than younger persons. The econometrician might argue that if age affects the estimates of the coefficients, then the model must be misspecified and age should be added as one of the causes.

The survey methodologist is using the regression model to summarize a relationship in the frame population, not generally making causal inferences. Thus, two different purposes produce two different approaches. One practical approach is that of Dumouchel and Duncan (1983), who argue that differences between weighted and unweighted estimates of the coefficients might be a prompt to reconsider the model specification. The debate between reflecting the sample design through weights and standard errors reflecting clustering while estimating model coefficients is detailed in Brewer and Mellor (1973) and Groves (1989, pp. 279–290).

When is it and when is it not important to weight general population survey data to adjust for differences in probabilities of selection?

Selection weights are recommended in Sections 4.5.2 and 10.5 for all designs in which sample members are selected with different probabilities. The value of coverage and nonresponse adjustments depends on the assumptions of the procedures, typically whether respondents and nonrespondents (and those covered and not covered) are equivalent on the survey statistic within weighting classes. This assumption is generally not testable with the survey data themselves. Hence, there is no way to evaluate the assumption. There *is* a practical step that many take: running key statistics weighted and unweighted and computing appropriate standard errors. If the conclusions reached are the same with both runs, they use and present the analysis with the smaller estimated standard error.

Finally, it is important to note that statistics sensitive to the sum of the weights (e.g., chi-square statistics and standard error estimates) may be misleading when using some statistical software systems without some alteration (see Chapters 4 and 10).

Almost all of the examples and research cited in the book is based in the United States. Do all the same problems apply to all countries of the world?

No. Countries vary in the resources available to survey researchers and in the nature of the cultural reactions to survey measurement. In countries with population registers that are available as sampling frames, much of the complex sample design that we described is not necessary. In countries with telephone or Internet infrastructure that covers only small parts of the country, methods of data collection using such infrastructure are threatened by larger errors of nonobservation. In countries with low literacy rates and/or unreliable postal systems, self-administered questionnaires may be ineffective. In countries with cultures that do not permit someone to engage in an honest discussion with a stranger on topics of interest to the stranger, surveys may be threatened.

It is our hope and belief that many of the design *principles* described in this text will be applicable in other environments. The *application* of the principles, however, are likely to need careful adaptation to the constraints and opportunities each situation provides.

I just want to do a survey. Do I need to be trained as a survey methodologist in order to do a good one?

As we noted in Chapter 1, survey methodology has identified principles underlying the performance of alternative survey designs. These principles are often derived from combinations of the conceptual frameworks within statistics, psychology, sociology, and computer science. It is difficult to acquire knowledge of that mix without broad and deep reading of the research literatures in survey methodology. Unfortunately, these principles do not suggest that one should choose the same design features for each survey. The desirable mixes of the principles require a shaping to the key purposes of the survey.

Knowledge of the field is necessary for assurance that the mix chosen is optimal, given the current state of knowledge. Thus, survey methodologists are valuable to the extent to which they possess such understanding. Like all fields, novices who aspire to immediate action are well served in seeking consultation from survey professionals, especially at the design stage of a survey, when key decisions could benefit from past methodological research findings.

REFERENCES

Abowd, J., and Woodcock, S. (2001), "Disclosure Limitation in Longitudinal Linked Data." In Doyle, P., Lane, J., Theeuwes, J., and Zayatz, L., eds., *Confidentiality, Disclosure, and Data Access: Theory and Practical Applications for Statistical Agencies,* Amsterdam, North-Holland, pp. 215–277.

Abowd, J. M., Stinson, M., and Benedetto, G., "Final Report to the Social Security Administration on the SIPP/SSA/IRS Public Use File Project," November 2006, http://www.census.gov/sipp/synth_data.html.

Alreck, P., and Settle, R. (1995), *The Survey Research Handbook,* New York: McGraw-Hill.

Alwin, D. (2007), *Margins of Error,* New York: Wiley.

American Association for Public Opinion Research (2000), *Standard Definitions: Final Dispositions of Case Codes and Outcome Rates for Surveys,* Ann Arbor, Michigan: AAPOR.

American Psychological Association (2003), "Psychological Research Online: Opportunities and Challenges," Working Paper Version 3/31/03, Washington, DC: American Psychological Association.

Anderson, M. (1990), *The American Census: A Social History,* New Haven: Yale University Press.

Andrews, F. (1984), "Construct Validity and Error Components of Survey Measures: A Structural Modeling Approach," *Public Opinion Quarterly,* 48, pp. 409–422.

Aneshensel, C., Frerichs, R., Clark, V., and Yokopenic, P. (1982), "Measuring Depression in the Community: A Comparison of Telephone and Personal Interviews," *Public Opinion Quarterly,* 46, pp. 110–121.

Aquilino, W. (1992), "Telephone Versus Face-to-Face Interviewing for Household Drug Use Surveys," *International Journal of the Addictions,* 27, pp. 71–91.

Aronson, E., and Carlsmith, J. (1969), "Experimentation in Social Psychology," in Lindzey, G., and Aronson, E. (eds.), *Handbook of Social Psychology,* 2nd ed., vol. 2, pp. 1–79, Reading, MA: Addison-Wesley.

Asch, S. (1956), "Studies of Independence and Conformity," *Psychological Monographs,* 70, No. 416.

Atrostic, B., and Burt, G. (1999), "What Have We Learned and a Framework for the Future," in *Seminar on Interagency Coordination and Cooperation,* Statistical Policy Working Paper 28, Washington, DC: Federal Committee on Statistical Methodology.

Babbie, E. (1990), *Survey Research Methods* (2nd edition), Belmont, CA: Wadsworth.

Babbie, E. (2004), *The Practice of Social Research*, Belmont, CA: Wadsworth.

Bahrick, H., Bahrick, P., and Wittlinger, R. (1975), "Fifty Years of Memory for Names and Faces: A Cross Sectional Approach," *Journal of Experimental Psychology: General*, 104, pp. 54–75.

Barsalou, L. (1988), "The Content and Organization of Autobiographical Memories," in Neisser, U., and Winograd, E. (eds.), *Remembering Reconsidered: Ecological and Traditional Approaches to the Study of Memory*, pp. 193–243, Cambridge, U.K.: Cambridge University Press.

Bates, N., Dahlhammer, J., and Singer, E. (2008), "Privacy Concerns, Too Busy, or Just Not Interested: Using Doorstep Concerns to Predict Nonresponse," *Journal of Official Statistics,* 24, pp. 591–612.

Battaglia, M., Link, M., Frankel, M., Osborn, L., and Mokdad, A. (2008), "An Evaluation of Respondent Selection Methods for Household Mail Surveys," *Public Opinion Quarterly*, 72, pp. 459–469.

Beatty, P. (2004), "The Dynamics of Cognitive Interviewing," in Presser, S. et al. (eds.), *Questionnaire Development Evaluation and Testing Methods*, New York: Wiley.

Beatty, P., and Herrmann, D. (2002), "To Answer or Not to Answer: Decision Processes Related to Survey Item Nonresponse," in Groves, R., Dillman, D., Eltinge J., and Little, R. (eds.), *Survey Nonresponse*, pp. 71–85, New York: Wiley.

Beebe, T., Harrison, P., McRae, J., Anderson, R., and Fulkerson, J. (1998), "An Evaluation of Computer-Assisted Self-Interviews in a School Setting," *Public Opinion Quarterly*, 62, pp. 623–632.

Béland, Y., and St-Pierre, M. (2008), "Mode Effects in the Canadian Community Health Survey: A Comparison of CATI and CAPI," in Lepkowski, J., Tucker, C., Brick, J., de Leeuw, E., Japec, L., Lavrakas, P., Link, M., and Sangster, R. (eds.), *Advances in Telephone Survey Methodology*, New York: Wiley, pp. 297–314.

Belli, R. (1998), "The Structure of Autobiographical Memory and the Event History Calendar: Potential Improvements in the Quality of Retrospective Reports in Surveys," *Memory*, 6, pp. 383–406.

Belli, R., Shay, W., and Stafford, F. (2001), "Event History Calendars and Question Lists," *Public Opinion Quarterly*, 65, pp. 45–74.

Belli, R., Schwarz, N., Singer, E., and Talarico, J. (2000), "Decomposition Can

Harm the Accuracy of Behavioral Frequency Reports," *Applied Cognitive Psychology*, 14, pp. 295–308.

Belson, W. (1981), *The Design and Understanding of Survey Questions*, Aldershot: Gower Publishing.

Belson, W. (1986), *Validity in Survey Research*, Aldershot: Gower Publishing.

Bem, D., and McConnell, H. (1974), "Testing the Self-Perception Explanation of Dissonance Phenomena: On the Salience of Premanipulation Attitudes," *Journal of Personality and Social Psychology*, 14, pp. 23–31.

Berk, R. (1983), "An Introduction to Sample Selection Bias in Sociological Data," *American Sociological Review*, 48, pp. 386–398.

Berlin, M., Mohadjer, L., Waksberg, J., Kolstad, A., Kirsch, I., Rock, D., and Yamamoto, K. (1992), "An Experiment in Monetary Incentives," in *Proceedings of the Survey Research Methods Section of the American Statistical Association*, pp. 393–398, Washington, DC: American Statistical Association.

Berman, J., McCombs, H., and Boruch, R. (1977), "Notes on the Contamination Method: Two Small Experiments in Assuring Confidentiality of Response," *Sociological Methods and Research*, 6, pp. 45–63.

Bernard, C. (1989), *Survey Data Collection Using Laptop Computers*, Paris: Institut National de la Statistique et des Études Economiques (INSEE), Report No. 01/C520.

Berscheid, E., Baron, R., Dermer, M., and Libman, M. (1973), "Anticipating Informed Consent: An Empirical Approach," *American Psychologist*, 28, pp. 913–925.

Bethlehem, J. (1998), "The Future of Data Editing," in Couper, M., Baker, R., Bethlehem, J., Clark, C., Martin, J., Nicholls II, W., and O'Reilly, J. (eds.), *Computer Assisted Survey Information Collection*, pp. 201–222, New York: Wiley.

Bethlehem, J. (2002), "Weighting Nonresponse Adjustments Based on Auxiliary Information," in Groves, R., Dillman, D., Eltinge, J., and Little, R. (eds.), *Survey Nonresponse*, pp. 275–288, New York: Wiley.

Biemer, P. and Lyberg, L. (2003), *Introduction to Survey Quality*, New York: Wiley.

Biemer, P. and Stokes, L. (1991), "Approaches to the Modeling of Measurement Errors," in Biemer, P., Groves, R., Lyberg, L., Mathiowetz, N., and Sudman, S. (eds.), *Measurement Errors in Surveys*, pp. 487–516, New York: Wiley.

Biemer, P., Herget, D., Morton, J., and Willis, G. (2003), "The Feasibility of Monitoring Field Interview Performance Using Computer Audio Recorded Interviewing (CARI)," in *Proceedings of the Survey Research Methods Section of the American Statistical Association*, pp. 1068–1073, Washington, DC: American Statistical Association.

Billiet, J., and Loosveldt, G. (1988), "Interviewer Training and Quality of Responses." *Public Opinion Quarterly*, 52, pp. 190–211.

Bishop, G., Hippler, H., Schwarz, N., and Strack, F. (1988), "A Comparison of Response Effects in Self-Administered and Telephone Surveys," in Groves, R., Biemer, P., Lyberg, L., Massey, J., Nicholls II, W., and Waksberg, J. (eds.), *Telephone Survey Methodology*, pp. 321–340, New York: Wiley.

Bishop, G., Oldendick, R., and Tuchfarber, A. (1986), "Opinions on Fictitious Issues: The Pressure to Answer Survey Questions," *Public Opinion Quarterly*, 50, pp. 240–250.

Blair, E., and Burton, S. (1987), "Cognitive Processes Used by Survey Respondents to Answer Behavioral Frequency Questions," *Journal of Consumer Research*, 14, pp. 280–288.

Blumberg, S., Luke, J., Cynamon, M., and Frankel, M. (2008), "Recent Trends in Household Telephone Coverage in the United States." In Lepkowski, J., Tucker, C., Brick, J., de Leeuw, E., Japec, L., Lavrakas, P., Link, M., and Sangster, R. (eds.), *Advances in Telephone Survey Methodology*, pp. 56–86, New York: Wiley.

Blumberg, S., and Luke, J. (2008), *Wireless Substitution: Early Release of Estimates From the National Health Interview Survey, July–December 2007*, www.cdc.gov/nchs.

Bogen, K. (1996), "The Effect of Questionnaire Length on Response Rates: A Review of the Literature," in *Proceedings of the Survey Research Methods Section of the American Statistical Association*, Alexandria, VA: American Statistical Association, pp. 1020–1025.

Booth, C. (1902–1903), *Life and Labour of the People of London*, London and New York: MacMillan Co.

Bosnjak, M., and Tuten, T. (2001), "Classifying Response Behaviors in Web-Based Surveys," *Journal of Computer-Mediated Communication*, 6(3) (http://jcmc.indiana.edu).

Botman, S., and Thornberry, O. (1992), "Survey Design Features Correlates of Nonresponse," in *Proceedings of the Survey Research Methods Section of the American Statistical Association*, pp. 309–314, Alexandria, VA: American Statistical Association.

Boyle, J., Kilpatrick, D., Acinerno, R., Ruggiero, K., Resnick, H., Galea, S., Koenan, K., and Galernter, J. (2007), "Biological Specimen Collection in an RDD Telephone Survey: 2004 Florida Hurricanes Gene and Environment Study," in *Proceedings of the 9th Conference on Health Survey Research Methods*, Hyattsville, MD: National Center for Health Statistics.

Bradburn, N., Sudman, S., and Associates (1979), *Improving Interview Method and Questionnaire Design*, San Francisco: Jossey-Bass.

Brehm, J. (1993), *The Phantom Respondents: Opinion Surveys and Political Representation*, Ann Arbor: University of Michigan Press.

REFERENCES

Brewer, K., and Mellor, R. (1973), "The Effect of Sample Structure on Analytical Surveys," *Australian Journal of Statistics*, 15, pp. 145–152.

Brick, M., Montaquila, J., and Scheuren, F. (2002), "Estimating Residency Rates," *Public Opinion Quarterly*, 66, pp. 18–39.

Brøgger, J., Bakke, P., Eide, G., and Guldvik, A. (2002), "Comparison of Telephone and Postal Survey Modes on Respiratory Symptoms and Risk Factors." *American Journal of Epidemiology*, 155, pp. 572–576.

Brunner, G., and Carroll, S. (1969), "The Effect of Prior Notification on the Refusal Rate in Fixed Address Surveys," *Journal of Marketing Research*, 9, pp. 42–44.

Bureau of the Census (2002), *Voting and Registration in the Election of November 2000, Current Population Reports*, P20-542, Washington, DC: U.S. Bureau of the Census.

Burton, S., and Blair, E. (1991), "Task Conditions, Response Formulation Processes, and Response Accuracy for Behavioral Frequency Questions in Surveys," *Public Opinion Quarterly*, 55, pp. 50–79.

Campanelli, P., Thomson, K., Moon, K., and Staples, T. (1997), "The Quality of Occupational Coding in the UK," in Lyberg L., Biemer, P., Collins, M., de Leeuw, E., Dippo, C., Schwarz, N., and Trewin, D. (eds.), *Survey Measurement and Process Quality*, pp. 437–457, New York: Wiley.

Cannell, C., and Fowler, F. (1964), "A Note on Interviewer Effect in Self-Enumerative Procedures," *American Sociological Review*, 29, p. 276.

Cannell, C., Groves, R., Magilavy, L., Mathiowetz, N., and Miller, P. (1987), An Experimental Comparison of Telephone and Personal Health Interview Surveys," *Vital and Health Statistics*, Series 2, 106, Washington, DC: Government Printing Office.

Cannell, C., Marquis, K., and Laurent, A. (1977), "A Summary of Studies," *Vital and Health Statistics*, Series 2, 69, Washington, DC: Government Printing Office.

Cannell, C., Miller, P., and Oksenberg, L. (1981), "Research on Interviewing Techniques," in Leinhardt, S. (ed.), *Sociological Methodology 1981*, pp. 389–437, San Francisco: Jossey-Bass.

Cantor, D., and Esposito, J. (1992), "Evaluating Interviewer Style for Collecting Industry and Occupation Information," *Proceedings of the Section on Survey Research Methods, American Statistical Association*, pp. 661–666.

Catlin, O., and Ingram, S. (1988), "The Effects of CATI on Costs and Data Quality: A Comparison of CATI and Paper Methods in Centralized Interviewing," in Groves, R., Biemer, P., Lyberg, L., Massey, J., Nicholls II, W., and Waksberg, J. (eds.), *Telephone Survey Methodology*, pp. 437–450, New York: Wiley.

Centers for Disease Control (2005), *BRFSS User's Guide*, http://www.cdc.gov/brfss/pdf/userguide.pdf.

Centers for Disease Control (2008), *BRFSS Questionnaires,* http://www.cdc.gov/brfss/questionnaires/pdf-ques/2008brfss.pdf.

Citro, C., Ilgen, D., and Marrett, C. (2003), *Protecting Participants and Facilitating Social and Behavioral Sciences Research,* Washington, DC: National Academy Press.

Cochran, W. (1961), "Comparison of Methods for Determining Stratum Boundaries," *Bulletin of the International Statistical Institute,* 38, pp. 345–358.

Cochran, W. (1977), *Sampling Techniques,* New York: Wiley.

Cohany, S., Polivka, A., and Rothgeb, J. (1994), "Revisions in the Current Population Survey Effective January 1994," *Employment and Earnings,* February, pp. 13–37.

Colledge, M., and Boyko, E. (2000), *UN/ECE Work Session on Statistical Metadata (METIS),* Washington, November 28–30.

Collins, C. (1975), "Comparison of Month-to-Month Changes in Industry and Occupation Codes with Respondent's Report of Change: CPS Mobility Study." Response Research Staff Report 75-6, May 15, 1975, U.S. Bureau of the Census.

Collins, M., and Courtenay, G. (1985), "A Comparison Study of Field and Office Coding," *Journal of Official Statistics,* 1, pp. 221–227.

Conrad, F., and Blair, J. (1996), "From Impressions to Data: Increasing the Objectivity of Cognitive Interviews," in *Proceedings of the Section on Survey Research Methods, Annual Meetings of the American Statistical Association,* Alexandria, VA: American Statistical Association, pp. 1–10.

Conrad, F., and Schober, M. (2000), "Clarifying Question Meaning in a Household Telephone Survey," *Public Opinion Quarterly,* 64, pp. 1–28.

Conrad, F., and Schober, M. (2008), *Envisioning the Survey Interview of the Future,* New York: Wiley.

Conrad, F., Brown, N., and Cashman, E. (1998), "Strategies for Estimating Behavioral Frequency in Survey Interviews," *Memory,* 6, pp. 339–366.

Converse, J. (1987), *Survey Research in the United States,* Berkeley: University of California Press.

Converse, J., and Presser, S. (1986), *Survey Questions: Handcrafting the Standardized Questionnaire,* Thousand Oaks, CA: Sage.

Conway, M. (1996), "Autobiographical Knowledge and Autobiographical Memories," in Rubin, D. (ed.), *Remembering Our Past,* pp. 67–93, Cambridge, U.K.: Cambridge University Press.

Cook, C., Heath, F., and Thompson, R. (2000), "A Meta-Analysis of Response Rates in Web- or Internet-Based Surveys," *Educational and Psychological Measurement,* 60, pp. 821–836.

Couper, M. (1996), "Changes in Interview Setting under CAPI," *Journal of Official Statistics,* 12, pp. 301–316.

REFERENCES

Couper, M. (2000), "Web Surveys: A Review of Issues and Approaches," *Public Opinion Quarterly,* 64, pp. 464–494.

Couper, M. (2001), "The Promises and Perils of Web Surveys," in Westlake, A. et al. (eds.), *The Challenge of the Internet,* pp. 35–56, London: Association for Survey Computing.

Couper, M. (2008a), "Technology and the Survey Interview/Questionnaire," in Schober, M. F., and Conrad, F. G. (eds.), *Envisioning the Survey Interview of the Future,* New York: Wiley, pp. 58–76.

Couper, M. (2008b), *Designing Effective Web Surveys.* New York: Cambridge University Press.

Couper, M., Baker, R., Bethlehem, J., Clark, C., Martin, J., Nicholls, W., and O'Reilly, J. (1998), *Computer Assisted Survey Information Collection,* New York: Wiley.

Couper, M., Blair, J., and Triplett, T. (1999), "A Comparison of Mail and E-Mail For a Survey of Employees in Federal Statistical Agencies," *Journal of Official Statistics,* 15, pp. 39–56.

Couper, M., and Groves, R. (2002), "Introductory Interactions in Telephone Surveys and Nonresponse," in Maynard, D., Houtkoop-Steenstra, H., Schaeffer, N., and van der Zouwen, J. (eds.), *Standardization and Tacit Knowledge: Interaction and Practice in the Survey Interview,* pp. 161–177, New York: Wiley.

Couper, M., Hansen, S., and Sadosky, S. (1997), "Evaluating Interviewer Use of CAPI Technology, in Lyberg, L., Biemer, P., Collins, M., Dippo, C., and Schwarz, N. (eds.), *Survey Measurement and Process Quality,* pp. 267–286, New York: Wiley.

Couper, M., Kapteyn, A., Schonlau, M., and Winter, J. (2007), "Noncoverage and Nonresponse in an Internet Survey," *Social Science Research,* 36, pp. 131–148.

Couper, M., and Nicholls II, W. (1998), "The History and Development of Computer Assisted Survey Information Collection Methods," in Couper, M., Baker, R., Bethlehem, J., Clark, C., Martin, J., Nicholls II, W., and O'Reilly, J. (eds.), *Computer Assisted Survey Information Collection,* pp. 1–22, New York: Wiley.

Couper, M., and Rowe, B. (1996), "Computer-Assisted Self-Interviews," *Public Opinion Quarterly,* 60, pp. 89–105.

Couper, M., Singer, E., Conrad, F.G., and Groves, R. (2008), "Risk of Disclosure, Perceptions of Risk, and Concerns about Privacy and Confidentiality as Factors in Survey Participation." *Journal of Official Statistics,* 24, pp. 255–75.

Couper, M., Singer, E., and Kulka, R. (1998), "Participation in the 1998 Decennial Census: Politics, Privacy, Pressures," *American Politics Quarterly,* 26, pp. 59–80.

Cronbach, L. (1951). "Coefficient Alpha and the Internal Structure of Tests," *Psychiatrika*, 16, pp. 297–334.

Cronbach, L., and Meehl, P. (1955), "Construct Validity in Psychological Tests," *Psychological Bulletin*, 52, pp. 281–302.

Csikszentmihalyi, M., and Csikszentmihalyi, I. (eds.) (1988), *Optimal Experience: Psychological Studies in Flow of Consciousness*, New York: Cambridge University Press.

Curtin, R. (2003), "Unemployment Expectations: The Impact of Private Information on Income Uncertainty," *Review of Income and Wealth*, 49, pp. 539–554.

Curtin, R. (2003), *Surveys of Consumers: Sample Design*, http://www.sca.isr.umich.edu/.

Curtin, Richard T. (2003), *Surveys of Consumers: Survey Description*, http://www.sca.isr.umich.edu/.

de la Puente, M. (1993), "A Multivariate Analysis of the Census Omission of Hispanics and Non-Hispanic Whites, Blacks, Asians and American Indians: Evidence from Small Area Ethnographic Studies," in *Proceedings of the Survey Research Methods Section, American Statistical Association*, pp. 641–646, Alexandria, VA: American Statistical Association.

de Leeuw, E. (1992), *Data Quality in Mail, Telephone and Face-to-Face Surveys*, Amsterdam: TT-Publikaties.

de Leeuw, E. (2005), "To Mix or Not to Mix Data Collection Modes in Surveys," *Journal of Official Statistics*, 21, pp. 233–255.

de Leeuw, E., Callegaro, M., Hox, J., Korendijk, E., and Lensvelt-Mulders, G. (2007), "The Influence of Advance Letters on Response in Telephone Surveys: A Meta-Analysis," *Public Opinion Quarterly*, 71, pp. 413–443.

de Leeuw, E., and de Heer, W. (2002), "Trends in Household Survey Nonresponse: A Longitudinal and International Comparison," Chapter 3 in Groves, R., Dillman, D., Eltinge, J., and Little, R. (eds.), *Survey Nonresponse*, pp. 41–54, New York: Wiley.

de Leeuw, E., and van der Zouwen, J. (1988), "Data Quality in Telephone and Face-to-Face Surveys: A Comparative Meta-Analysis," in Groves, R., Biemer, P., Lyberg, L., Massey, J., Nicholls II, W., and Waksberg, J. (eds.), *Telephone Survey Methodology*, pp. 283–299, New York: Wiley.

DeMaio, T., and Landreth, A. (2004), "Do Different Cognitive Interview Methods Produce Different Results?" in Presser, S. et al. (eds.), *Questionnaire Development Evaluation and Testing Methods*, New York: Wiley.

Deming, W. (1950), *Some Theory of Sampling*, New York: Dover.

Denscombe, M. (2008), "The Length of Responses to Open-Ended Questions; A Comparison of Online and Paper Questionnaires in Terms of a Mode Effect," *Social Science Computer Review*, 26, pp. 359–368.

Deutskens, E., de Ruyter, K., and Wetzels, M. (2006), "An Assessment of Equivalence between Online and Mail Surveys in Service Research," *Journal of Service Research*, 8, pp. 346–355.

Dielman, L., and Couper, M. (1995), "Data Quality in CAPI Surveys: Keying Errors," *Journal of Official Statistics*, 11, pp. 141–146.

Dillman, D. (1978), *Mail and Telephone Surveys: The Total Design Method*, New York: Wiley.

Dillman, D., Eltinge, J., Groves, R., and Little, R. (2002), "Survey Nonresponse in Design, Data Collection and Analysis," in Groves, R., Dillman, D., Eltinge, J., and Little, R. (eds.), *Survey Nonresponse*, pp. 3–26, New York: Wiley.

Dillman, D., Sinclair, M., and Clark, J. (1993), "Effects of Questionnaire Length, Respondent Friendly Design, and a Difficult Question on Response Rates for Occupant-Addressed Census Mail Surveys," *Public Opinion Quarterly*, 57, pp. 289–304.

Dillman, D., Smyth, J., and Christian, L. (2009), *Internet, Mail, and Mixed-Mode Surveys: The Tailored Design Method*. New York: Wiley.

Dillman, D., and Tarnai, J. (1988), "Administrative Issues in Mixed-Mode Surveys," in Groves, R., Biemer, P., Lyberg, L., Massey, J., Nicholls II, W., and Waksberg, J. (eds.), *Telephone Survey Methodology*, pp. 509–528, New York: Wiley.

Doyle, P., Lane, J., Theeuwes, J., and Zayatz, L. (2001), *Confidentiality, Disclosure, and Data Access: Theory and Practical Applications for Statistical Agencies*, Amsterdam: North-Holland.

DuMouchel, W., and Duncan, G. (1983), "Using Sample Survey Weights in Multiple Regression Analyses of Stratified Samples," *Journal of the American Statistical Association*, 78, pp. 535–543.

Economic Classification Policy Committee (1993), *Issues Paper No. 1: Conceptual Issues*, Washington, DC: Bureau of Economic Analysis.

Edwards, P., Roberts, I., Clarke, M., DiGuiseppi, C., Pratap, S., Wentz, R., and Kwan, I. (2002), "Increasing Response Rates to Postal Questionnaires: Systematic Review," *British Medical Journal*, 324, pp. 1183–1192.

Edwards, W., Winn, D., Kurlantzick, V., Sheridan, S., Berk, M., Retchin, S., and Collins, J. (1994), "Evaluation of National Health Interview Survey Diagnostic Reporting," *Vital and Health Statistics*, Series 2, No. 120, Hyattsville, MD: National Center for Health Statistics.

Edwards, W., and Cantor, D. (1991), "Toward a Response Model in Establishment Surveys," in Biemer, P., Groves, R., Lyberg, L., Mathiowetz, N., and Sudman, S. (eds.), *Measurement Errors in Surveys*, pp. 221–236, New York: Wiley.

Edwards, W., Winn, D., and Collins, J. (1996), "Evaluation of 2-week Doctor Visit Reporting in the National Health Interview Survey," in *Vital and Health*

Statistics, Series 2, No. 122, Hyattsville, MD: National Center for Health Statistics.

Ekholm, A., and Laaksonen, S. (1991), "Weighting via Response Modeling in the Finnish Household Budget Survey," *Journal of Official Statistics,* 7, pp. 325–377.

Ericsson, K., and Simon, H. (1980), "Verbal Reports as Data," *Psychological Review,* 87, pp. 215–251.

Ericsson, K., and Simon, H. (1984), *Protocol Analysis: Verbal Reports as Data,* Cambridge, MA: MIT Press.

Erlich, J., and Riesman, D. (1961), "Age and Authority in the Interview," *Public Opinion Quarterly,* 24, pp. 99–114.

Etter, J.-F., Perneger, T., and Ronchi, A. (1998), "Collecting Saliva Samples by Mail," *American Journal of Epidemiology,* 147, pp. 141–146.

Faden, R., and Beauchamp, T. (1986), *A History and Theory of Informed Consent,* New York: Oxford University Press.

Federal Committee on Statistical Methodology (1994), *Working Paper 22: Report on Statistical Disclosure Limitation Methods,* Statistical Policy Office, Office of Information and Regulatory Affairs, Office of Management and Budget, Washington, DC.

Federal Register (1991), "Federal Policy for the Protection of Human Subjects," June 18, pp. 280002–280031.

Fellegi, I., and Holt, T. (1976), "A Systematic Approach to Automatic Edit and Imputation," *Journal of the American Statistical Association,* 71, pp. 17–35.

Fellegi, I. (1964), "Response Variance and Its Estimation," *Journal of the American Statistical. Association,* 59, pp. 1016–1041.

Felsö, F., Theeuwes, J., and Wagner, G. (2001), "Disclosure Limitations Methods in Use: Results of a Survey," in Doyle, P., Lane, J., Theeuwes, J., and Zayatz, L. (eds.), *Confidentiality, Disclosure, and Data Access: Theory and Practical Applications for Statistical Agencies,* Amsterdam: North-Holland/Elsevier.

Fienberg, S. (1994), "A Radical Proposal for the Provision of Micro-Data Samples and the Preservation of Confidentiality," Carnegie Mellon University, Department of Statistics, Technical Report 611, Pittsburgh, PA: Carnegie Mellon University.

Fienberg, S., and Makov, U. (1998), "Confidentiality, Uniqueness, and Disclosure Limitation for Categorical Data," *Journal of Official Statistics,* 14, pp. 385–397.

Fienberg, S., Steele, R., and Makov, U. (1996), "Statistical Notions of Data Disclosure Avoidance and Their Relationship to Traditional Statistical Methodology: Data Swapping and Log-Linear Models," in *Proceedings of the Bureau of the Census 1996 Annual Research Conference,* pp. 87–105, Washington, DC: U.S. Bureau of the Census.

REFERENCES

Forsman, G., and Schreiner, I. (1991), "The Design and Analysis of Reinterview: An Overview," in Biemer, P., Groves, R., Lyberg, L., Mathiowetz, N., and Sudman, S. (eds.), *Measurement Errors in Surveys*, pp. 279–301, New York: Wiley.

Forsyth, B., and Lessler, J. (1992), "Cognitive Laboratory Methods: A Taxonomy," in Biemer, P., Groves, R., Lyberg, L., Mathiowetz, N., and Sudman, S. (eds.), *Measurement Errors in Surveys*, pp. 393–418, New York: Wiley.

Forsyth, B., Rothgeb, J., and Willis, G. (2004), "Does Pretesting Make a Difference?" in Presser, S. et al. (eds.), *Questionnaire Development Evaluation and Testing Methods*, New York: Wiley.

Fowler, F. (1992), "How Unclear Terms Affect Survey Data," *Public Opinion Quarterly*, 56, pp. 218–231.

Fowler, F. (1995), *Improving Survey Questions*, Thousand Oaks, CA: Sage Publications.

Fowler, F. (2001), *Survey Research Methods*, (3rd Edition), Thousand Oaks, CA: Sage Publications.

Fowler, F. (2004), "Getting Beyond Pretesting and Cognitive Interviews: The Case for More Experimental Pilot Studies," in Presser, S. et al. (eds.), *Questionnaire Development Evaluation and Testing Methods*, New York: Wiley.

Fowler, F., and Cannell, C. (1996), "Using Behavioral Coding to Identify Cognitive Problems with Survey Questions," in Schwarz, N., and Sudman, S. (eds.), *Answering Questions*, pp. 15–36, San Francisco: Jossey-Bass.

Fowler, F., Gallagher, P., Stringfellow, V., Zaslavsky, A., Thompson, J., and Cleary, P. (2002), "Using Telephone Interviews to Reduce Nonresponse Bias to Mail Surveys of Health Plan Members," *Medical Care*, 40, pp. 190–200.

Fowler, F., and Mangione, T. (1990), *Standardized Survey Interviewing: Minimizing Interviewer-Related Error*, Beverly Hills, CA: Sage Publications.

Frankel, L. (1983), "The Report of the CASRO Task Force on Response Rates," in Wiseman, F. (ed.), *Improving Data Quality in a Sample Survey*, pp. 1–11, Cambridge, MA: Marketing Science Institute.

Frey, J. (1986), "An Experiment with a Confidentiality Reminder in a Telephone Survey," *Public Opinion Quarterly*, 50, pp. 267–269.

Fuchs, M., Couper, M., and Hansen, S. (2000), "Technology Effects: Do CAPI Interviews Take Longer?" *Journal of Official Statistics*, 16, pp. 273–286.

Gardner, G. (1978), "Effects of Federal Human Subjects Regulations on Data Obtained in Environmental Stress Research," *Journal of Personality and Social Psychology*, 36, pp. 628–634.

Gaziano, C. (2005), "Comparative Analysis of Within-Household Respondent Selection Techniques," *Public Opinion Quarterly*, 69, pp. 124–157.

Gfroerer, J., Eyerman, J., and Chromy. J. (eds.) (2002), *Redesigning an Ongoing National Household Survey: Methodological Issues*, DHHS Pub. No. SMA 03-3768, Rockville, MD: SAMHSA.

Gottfredson, M., and Hindelang, M. (1977), "A Consideration of Telescoping and Memory Decay Biases in Victimization Surveys," *Journal of Criminal Justice*, 5, pp. 205–216.

Goyder, J. (1985), "Face-to-Face Interviews and Mail Questionnaires: The Net Difference in Response Rate," *Public Opinion Quarterly*, 49, pp. 234–252.

Goyder, J. (1987), *The Silent Minority: Nonrespondents on Sample Surveys*, Boulder, CO: Westview Press.

Graesser, A., Bommareddy, S., Swamer, S., and Golding, J. (1996), "Integrating Questionnaire Design with a Cognitive Computational Model of Human Question Answering," in Schwarz, N., and Sudman, S. (eds.), *Answering Questions*, pp. 143–174, San Francisco: Jossey-Bass.

Graesser, A., Kennedy, T., Wiemer-Hastings, P., and Ottati, V. (1999), "The Use of Computational Cognitive Models to Improve Questions on Surveys and Questionnaires," in Sirken, M., et al. (eds.), *Cognition in Survey Research*, pp. 199–216, New York: Wiley.

Graham, D. (1984), "Response Errors in the National Crime Survey: July 1974–June 1976", in Lehnen, R., and Skogan, W. (eds.), *The National Crime Survey: Working Papers*, pp. 58–64, Washington, DC: Bureau of Justice Statistics.

Griffin, D., Fischer, D., and Morgan, M. (2001), "Testing an Internet Response Option for the American Community Survey," Paper presented at the annual meeting of the American Association for Public Opinion Research, Montreal, Quebec, May.

Groves, R. (1979), "Actors and Questions in Telephone and Personal Interview Surveys," *Public Opinion Quarterly*, 43, pp. 190–205.

Groves, R. (1989), *Survey Errors and Survey Costs*, New York: Wiley.

Groves, R. (2006), "Nonresponse Rates and Nonresponse Bias in Household Surveys," *Public Opinion Quarterly*, 70, pp. 646–675.

Groves, R., and Couper, M. (1998), *Nonresponse in Household Interview Surveys*, New York: Wiley.

Groves, R., and Kahn, R. (1979), *Surveys by Telephone: A National Comparison with Personal Interviews*, New York: Academic Press.

Groves, R., and Lyberg, L. (1988), "An Overview of Non-Response Issues in Telephone Surveys," in Groves, R., Biemer, P., Lyberg, L., Massey, J., Nicholls II, W., and Waksberg, J. (eds.), *Telephone Survey Methodology*, pp. 191–212, New York: Wiley.

Groves, R., and Magilavy, L. (1980), "Estimates of Interviewer Variance in Telephone Surveys," in *Proceedings of the Survey Research Methods Section of the American Statistical Association*, pp. 622–627, Alexandria, VA: American Statistical Association.

Groves, R., and Peytcheva, E. (2008) "The Impact of Nonresponse Rates on Nonresponse Bias: A Meta-Analysis," *Public Opinion Quarterly*, 72, pp. 167-189.

Groves, R., Presser, S., and Dipko, S. (2004), "The Role of Topic Salience in Survey Participation Decisions," *Public Opinion Quarterly*, 68, pp. 2–31.

Groves, R., Singer, E., and Corning, A. (2000), "Leverage-Salience Theory of Survey Participation: Description and an Illustration," *Public Opinion Quarterly*, 64, pp. 299–308.

Groves, R., Singer, E., Corning, A., and Bowers, A. (1999), "A Laboratory Approach to Measuring the Joint Effects of Interview Length, Incentives, Differential Incentives, and Refusal Conversion on Survey Participation," *Journal of Official Statistics*, 15, pp. 251–268.

Groves, R., Dillman, D., Eltinge, J., and Little, R. (eds.) (2002), *Survey Nonresponse*, New York: Wiley.

Groves, R., Wissoker, D., Greene, L., McNeeley, M., and Montemarano, D. (2000), "Common Influences on Noncontact Nonresponse Across Household Surveys: Theory and Data," paper presented at the annual meetings of the American Association for Public Opinion Research.

Groves, R., Couper, M., Presser, S., Singer, E., Tourangeau, R., Piani Acosta, G., and Nelson, L. (2006), "Experiments in Producing Nonresponse Bias," *Public Opinion Quarterly*, 70, pp. 720–736.

Guarino, J., Hill, J., and Woltman, H. (2001), *Analysis of the Social Security Number-Notification Component of the Social Security Numbers, Privacy Attitudes, and Notification Experiment*, Washington, DC: U.S Census Bureau.

Gwartney, P. (2007), *The Telephone Interviewer's Handbook: How to Conduct Standardized Conversations*, New York: Wiley.

Haney, C., Banks, C., and Zimbardo, P. (1973), "Interpersonal Dynamics in a Simulated Prison," *International Journal of Criminology and Penology*, 1, pp. 69–97.

Hansen, B., and Hansen, K. (1995), "Academic and Scientific Misconduct: Issues for Nurse Educators," *Journal of Professional Nursing*, 11, pp. 31–39.

Hansen, M., Hurwitz, W., and Bershad, M. (1961), "Measurement Errors in Censuses and Surveys," *Bulletin of the International Statistical Institute*, 38, pp. 359–374.

Hansen, M., Hurwitz, W., and Madow, W. (1953), *Sample Surveys Methods and Theory*, Vols. I and II, New York: Wiley.

Hansen, S., and Couper, M. (2004), "Usability Testing as a Means of Evaluating Computer-Assisted Survey Instruments," in Presser, S. et al. (eds.),

Questionnaire Development Evaluation and Testing Methods, New York: Wiley.

Harkness, J., Vijver, F., and Mohler, P. (2002), *Cross-Cultural Survey Methods*, New York: Wiley.

Hartley, H. (1962), "Multiple Frame Surveys," in *Proceedings of the Social Statistics Section, American Statistical Association*, pp. 203–206, Alexandria, VA: American Statistical Association.

Hawala, S. (2000), "On the Variation of the Percent of Uniques in a Microdata Sample and the Sample Size. Statistical Research Division, United States Census Bureau (unpublished memo).

Heberlein, T., and Baumgartner, R. (1978), "Factors Affecting Response Rates to Mailed Questionnaires: A Quantitative Analysis of the Published Literature," *American Sociological Review*, 43, pp. 447–462.

Heckman, J. (1979), "Sample Selection Bias as a Specification Error," *Econometrica*, 47, pp. 153–161.

Henson, R., Roth, A. and Cannell, C. (1977), "Personal Versus Telephone Interviews: The Effects of Telephone Re-Interviews on Reporting of Psychiatric Symptomatology," in Cannell, C., Oksenberg, L., and Converse, J. (eds.), *Experiments in Interviewing Techniques: Field Experiments in Health Reporting*, 1971–1977, pp. 205–219, Hyattsville, MD: U.S. Department of Health, Education and Welfare, National Center for Health Services Research.

Hill, C., Donelan, K., and Frankel, M. (1999), "Within-Household Respondent Selection in an RDD Telephone Survey: A Comparison of Two Methods," paper presented at the 1999 meeting of the American Association for Public Opinion Research, St. Petersburg, FL.

Hippler, H., Schwarz, N., and Sudman, S. (1987), *Social Information Processing and Survey Methodology*, New York: Springer-Verlag.

Hochstim, J. (1967), "A Critical Comparison of Three Strategies of Collecting Data from Households," *Journal of the American Statistical Association*, 62, pp. 976–989.

Holmberg, A., Lorenc, B., and Werner, P. (2008), "Optimal Contact Strategy in a Mail-and-Web Mixed Mode Survey," Paper presented at the General Online Research Conference (GOR'08), Hamburg, March.

Horn, J. (1978), "Is Informed Deceit the Answer to Informed Consent?" *Psychology Today*, May, pp. 36–37.

Horvitz, D., Weeks, M., Visscher, W., Folsom, R., Massey, R., and Ezzati, T. (1990), "A Report of the Findings of the National Household Seroprevalence Survey Feasibility Study," *Proceedings of the American Statistical Association, Survey Research Methods Section*, pp. 150–159.

Houtkoop-Steenstra, H. (2000), *Interaction and the Standardized Survey Interview: The Living Questionnaire*, Cambridge, U.K.: Cambridge University Press.

Hox, J., and de Leeuw, E. (2002), "The Influence of Interviewers' Attitude and Behavior on Household Survey Nonresponse: An International Comparison," in Groves, R., Dillman, D., Eltinge, J., and Little, R. (eds.), *Survey Nonresponse*, pp. 103–120, New York: Wiley.

Hox, J., and de Leeuw, E. (1994), "A Comparison of Nonresponse in Mail, Telephone, and Face-to-Face Surveys: Applying Multilevel Modeling to Meta-Analysis," *Quality and Quantity*, 28, pp. 329–344.

Hughes, A., Chromy, J., Giacoletti, K., and Odom, D. (2002), "Impact of Interviewer Experience on Respondent Reports of Substance Use," in Gfroerer, J., Eyerman, J., and Chromy, J. (eds.), *Redesigning an Ongoing National Household Survey*, pp. 161–184, Washington, DC: Substance Abuse and Mental Health Services Administration.

Humphreys. L. (1970), *Tearoom Trade: Impersonal Sex in Public Places*, Chicago: Aldine.

Huttenlocher, J., Hedges, L., and Bradburn, N. (1990), "Reports of Elapsed Time: Bounding and Rounding Processes in Estimation," *Journal of Experimental Psychology: Learning, Memory, and Cognition*, 16, pp. 196–213.

Hyman, H., Cobb, J., Feldman, J., and Stember, C. (1954), *Interviewing in Social Research*, Chicago: University of Chicago Press.

Iannacchione, V, Staab, J., and Redden, D. (2003), "Evaluating the Use of Residential Mailing Addresses in a Metropolitan Household Survey," *Public Opinion Quarterly*, 67, pp. 202–210.

ISR Survey Research Center (1999), *Center Survey*, April, pp. 1, 3.

Inter-university Consortium for Political and Social Research (2001), *National Crime Victimization Survey, 1992–1999*, ICPSR 6406, Ann Arbor: ICPSR.

Jabine, T., King, K., and Petroni, R. (1990), *SIPP Quality Profile*, Washington, DC: U.S. Bureau of the Census.

Jabine, T., Straf, M., Tanur, J., and Tourangeau, R. (eds.) (1984), *Cognitive Aspects of Survey Methodology: Building a Bridge between Disciplines*, Washington, DC: National Academy Press.

Jenkins, C., and Dillman, D. (1997), "Towards a Theory of Self-Administered Questionnaire Design," in Lyberg, L., Biemer, P., Collins, M., de Leeuw, E., Dippo, C., Schwarz, N., and Trewin, D. (eds.), *Survey Measurement and Process Quality*, pp. 165–196, New York: Wiley.

Jobe, J., and Mingay, D. (1989), "Cognitive Research Improves Questionnaires," *American Journal of Public Health*, 79, pp. 1053–1055.

Johnson, T., Hougland, J., and Clayton, R. (1989), "Obtaining Reports of Sensitive Behaviors: A Comparison of Substance Use Reports from Telephone and Face-to-Face Interviews," *Social Science Quarterly*, 70, pp. 174–183.

Jordan, L., Marcus, A., and Reeder, L. (1980), "Response Styles in Telephone and Household Interviewing: A Field Experiment," *Public Opinion Quarterly*, 44, pp. 210–222.

Junn, J. (2001), "The Influence of Negative Political Rhetoric: An Experimental Manipulation of Census 2000 Participation," paper presented at the Midwest Political Science Association, Chicago.

Juster, F., and Smith, J. (1997), "Improving the Quality of Economic Data: Lessons from the HRS and AHEAD," *Journal of the American Statistical Association*, 92, pp. 1268–1278.

Juster, F., and Suzman, R. (1995), "An Overview of the Health and Retirement Study," *Journal of Human Resources*, 30, pp. S7–S56.

Kahn, R., and Cannell, C. (1958), *Dynamics of Interviewing*, New York: Wiley.

Kallick-Kaufman, M. (1979), "The Micro and Macro Dimensions of Gambling in the United States," *The Journal of Social Issues*, 35, pp. 7–26.

Kalton, G. (1981), *Compensating for Missing Survey Data*, Ann Arbor, MI: Institute for Social Research.

Kane, E., and Macaulay, L. (1993), "Interviewer Gender and Gender Attitudes," *Public Opinion Quarterly*, 57, pp. 1–28.

Katz, J. (1972), *Experimenting with Human Beings*, New York: Russell Sage Foundation.

Kish, L. (1949), "A Procedure for Objective Respondent Selection Within the Household," *Journal of the American Statistical Association*, 44, pp. 380–387.

Kish, L. (1962), "Studies of Interviewer Variance for Attitudinal Variables." *Journal of the American Statistical Association*, 57, pp. 92–115.

Kish, L. (1965), *Survey Sampling*, New York: Wiley.

Kish, L. (1988), "Multipurpose Sample Designs," *Survey Methodology*, 14, pp. 19–32.

Kish, L. and Frankel, M. (1974), "Inference from Complex Samples" (with discussion), *Journal of the Royal Statistical Society*, Set. B, 36, pp. 1–37.

Kish, L., and Hess, I. (1959), "Some Sampling Techniques for Continuing Survey Operations," *Proceedings of the Social Statistics Section, American Statistical Association*, pp. 139–143.

Kish, L., Groves, R., and Krotki, K. (1976), *Sampling Errors for Fertility Surveys*, World Fertility Survey Occasional Paper 17, The Hague, Voorburg: International Statistical Institute.

Kormendi, E., and Noordhoek, J. (1989), *Data Quality and Telephone Surveys*, Copenhagen: Danmark's Statistik.

Krazit, T. (2001), "Like Father, Like Son," *Public Perspective*, 13, pp. 13–16.

Krosnick, J. (1991), "Response Strategies for Coping with the Cognitive Demands of Attitude Measures in Surveys," *Applied Cognitive Psychology*, 5, pp. 213–236.

Krosnick, J. (1999), "Survey Research," *Annual Review of Psychology*, 50, pp. 537–567.

Krosnick, J. (2002), "The Causes of No-Opinion Responses to Attitude Measures in Surveys: They Are Rarely What They Appear to Be," in Groves, R., Dillman, D., Eltinge, J., and Little, R. (eds.), *Survey Nonresponse*, pp. 87–100, New York: Wiley.

Krosnick, J., and Alwin, D. (1987), "An Evaluation of a Cognitive Theory of Response-Order Effects in Survey Measurement," *Public Opinion Quarterly*, 51, pp. 201–219.

Krosnick, J., and Berent, M. (1993), "Comparisons of Party Identification and Policy Preferences: The Impact of Survey Question Format," *American Journal of Political Science*, 37, pp. 941–964.

Krosnick, J., and Fabrigar, L. (1997), "Designing Rating Scales for Effective Measurement in Surveys," in Lyberg, L., Biemer, P., Collins, M., de Leeuw, E., Dippo, C., Schwarz, N., and Trewin, D. (eds.), *Survey Measurement and Process Quality*, pp. 141–164, New York: Wiley.

Krueger, R., and Casey, M. (2000), *Focus Groups: A Practical Guide for Applied Research*, Beverly Hills, CA: Sage Publications.

Kuusela, V., Callegaro, M., and Vehovar, V. (2008), "The Influence of Mobile Telephones on Telephone Surveys," Chapter 4 in Lepkowski, J., Tucker, C., Brick, J., de Leeuw, E., Japec, L., Lavrakas, P., Link, M., and Sangster, R. (eds.), *Advances in Telephone Survey Methodology*, pp. 87–112, New York: Wiley.

Lambert, D. (1993), "Measures of Disclosure Risk and Harm," *Journal of Official Statistics*, 9, pp. 313–331.

Larson, R., and Richards, M. (1994), *Divergent Realities: The Emotional Lives of Mothers, Fathers, and Adolescents*, New York: Basic Books.

Lepkowski, J., Sadosky, S., and Weiss, P. (1998), "Mode, Behavior, and Data Recording Error," in Couper, M., Baker, R., Bethlehem, J., Clark, C., Martin, J., Nicholls II, W., and O'Reilly, J. (eds.), *Computer Assisted Survey Information Collection*, pp. 367–388, New York: Wiley.

Lepkowski, J., and Groves, R. (1986), "A Mean Squared Error Model for Dual Frame, Mixed Mode Survey Design," *Journal of the American Statistical Association*, 81, pp. 930–937.

Lepkowski, J., Tucker, C., Brick, J.M., de Leeuw, E., Japec, L., Lavrakas, P., Link, M., and Sangster, R. (2008), *Advances in Telephone Survey Methodology*, New York: Wiley.

Lessler, J., Caspar, R., Penne, M., and Barker, P. (2000), "Developing Computer-Assisted Interviewing (CAI) for the National Household Survey on Drug Abuse," *Journal of Drug Issues*, 30, pp. 19–34.

Lessler, J., and Forsyth, B. (1996), "A Coding System for Appraising Questionnaires," in Schwarz, N., and Sudman, S. (eds.), *Answering Questions*, pp. 259–292, San Francisco: Jossey-Bass.

Lessler, J. and Kalsbeek, W. (1992), *Nonsampling Error in Surveys*, New York: Wiley.

Levy, P., and Lemeshow, S., (2008), *Sampling of Populations: Methods and Applications*, 4th Edition, New York: Wiley.

Lievesley, D. (1988), "Unit Non-Response in Interview Surveys," London: Social and Community Planning Research, unpublished working paper.

Likert, R. (1932), "A Technique for Measurement of Attitudes," *Archives of Psychology*, 140, pp. 5–53.

Link, M., and Mokdad, A. (2005), "Alternative Modes for Health Surveillance Surveys: An Experiment with Web, Mail, and Telephone," *Epidemiology*, 16, pp. 701–704.

Link, M., and Mokdad, A.H. (2006), "Can Web and Mail Survey Modes Improve Participation in an RDD-Based National Health Surveillance?" *Journal of Official Statistics*, 22, pp. 293–312.

Linton, M. (1982), "Transformations of Memory in Everyday Life," in Neisser, U. (ed.), *Memory Observed*, pp. 77–91, San Francisco: Freeman.

Little, R., and Rubin, D. (2002), *Statistical Analysis with Missing Data*, 2nd Edition, New York: Wiley.

Little, R. J. (1993). "Statistical Analysis of Masked Data," *Journal of Official Statistics,* 9, pp. 407–426.

Lord, F., and Novick, M. (1968), *Statistical Theories of Mental Test Scores*, Reading, MA: Addison-Wesley.

Lozar-Manfreda, K., Bosnjak, M., Haas, I., and Vehovar, V. (2008), "Web Surveys Versus Other Survey Modes: A Meta-Analysis Comparing Response Rates," *International Journal of Market Research*, 50, pp. 79–104.

Lyberg L, Biemer, P., Collins, M., de Leeuw, E., Dippo, C., Schwarz, N., and Trewin, D. (eds.) (1997), *Survey Measurement and Process Quality*, New York: Wiley.

Mangione, T., Fowler, F., and Louis, T. (1992), "Question Characteristics and Interviewer Effects," *Journal of Official Statistics*, 8, pp. 293–307.

Mangione, T., Hingson, R., and Barrett, J. (1982), "Collecting Sensitive Data: A Comparison of Three Survey Strategies," *Sociological Methods and Research*, 10, pp. 337–346.

Martin, E. (1999), "Who Knows Who Lives Here? Within-Household Disagreeements as a Source of Survey Coverage Error," *Public Opinion Quarterly*, 63, pp. 220–236.

Martin, E. (2006), "Privacy Concerns and the Census Long From: Some Evidence from Census 2000," #2006-10 in Research Report Series, U.S. Census Bureau (http://www.census.gov/srd/papers/pdf/rsm2006-10.pdf).

Martin, J., O'Muircheartaigh, C., and Curtice, J. (1993), "The Use of CAPI for

Attitude Surveys: An Experimental Comparison with Traditional Methods," *Journal of Official Statistics*, 9, pp. 641–662.

Martinson, B., Anderson, M., and de Vries, R. (2005), "Scientisits Behaving Badly." *Nature*, 435, June 9, pp. 737–738.

Matschinger, H., Bernert, S., and Angermeyer, M. (2005), "An Analysis of Interviewer Effects on Screening Questions in a Computer Assisted Personal Mental Health Interview," *Journal of Official Statistics* 21, pp. 657–674.

Maynard, D., Houtkoop-Steenstra, H., Schaeffer, N., and van der Zouwen, H. (eds.) (2002), *Standardization and Tacit Knowledge: Interaction and Practice in the Survey Interview*, New York: Wiley.

McCabe, S., Boyd, C., Couper, M., Crawford, S., and d'Arcy, H. (2002), "Mode Effects for Collecting Alcohol and Other Drug Use Data: Web and US Mail," *Journal of Studies on Alcohol*, 63, pp. 755–761.

McHorney, C., Kosinski, M., and Ware, J. (1994), "Comparison of the Costs and Quality of Norms for the SF-36 Health Survey Collected by Mail Versus Telephone Interview: Results from a National Survey," *Medical Care*, 32, pp. 551–567.

Merkle, D. and Edelman, M. (2002), "Nonresponse in Exit Polls: A Comprehensive Analysis," in Groves, R., Dillman, D., Eltinge J., and Little R. (eds.), *Survey Nonresponse*, pp. 243–258, New York: Wiley.

Merkle, D., Edelman, M., Dykeman, K., and Brogan, C. (1998), "An Experimental Study of Ways to Increase Exit Poll Response Rates and Reduce Survey Error," paper presented at the 1998 AAPOR conference.

Milgram, S. (1963), "Behavioral Study of Obedience," *Journal of Abnormal and Social Psychology*, 67, pp. 371–378.

Miller, P., and Cannell, C. (1977), "Communicating Measurement Objectives in the Interview," in Hirsch et al. (eds.), *Strategies for Communication Research*, pp. 127–152, Beverly Hills: Sage Publications.

Mokdad, A., Ford, E., Bowman, B., Dietz, W., Vinicor, F., Bales, V., and Marks, J. (2003), "Prevalence of Obesity, Diabetes, and Obesity-Related Health Risk Factors, 2001," *Journal of the American Medical Association*, 289, pp. 76–79.

Mokdad, A., Serdula, M., Dietz, W., Bowman, B., Marks, J., and Koplan, J. (1999), "The Spread of the Obesity Epidemic in the United States, 1991–1998," *Journal of the American Medical Association*, 282, pp. 1519–1522.

Moore, J., Pascale, J., Doyle, P., Chan, A., and Griffiths, J. (2004), "Using Field Experiments to Improve Instrument Design," in Presser, S. et al. (eds.) *Questionnaire Development Evaluation and Testing Methods*, pp. 189–207, New York: Wiley.

Moore, J., Stinson, L., and Welniak, E. (1997), "Income Measurement Error in Surveys: A Review," in Sirken, M., Herrmann, D., Schechter, S., Schwarz,

N., Tanur, J., and Tourangeau, R. (eds.), *Cognition and Survey Research*, pp. 155–174, New York: Wiley.

Morton-Williams, J. (1993), *Interviewer Approaches*, Aldershot, U.K.: Dartmouth.

Mulry, M. (2007), "Summary of Accuracy and Coverage Evaluation for the U.S. Census 2000," *Journal of Official Statistics*, 23, pp. 345–370.

National Bioethics Advisory Commission (2001), *Ethical and Policy Issues in Research Involving Human Participants*, Vol. 1: Report and Recommendations, Bethesda, MD: National Bioethics Advisory Commission.

National Commission for the Protection of Human Subjects of Biomedical and Behavioral Research (1979), *Belmont Report: Ethical Principles and Guidelines for the Protection of Human Subjects of Research*, Washington, DC: U.S. Government Printing Office.

National Endowment for the Arts (1998), *1997 Survey of Public Participation in the Arts*, Research Division Report 39, Washington, DC: National Endowment for the Arts.

National Research Council (1979), *Privacy and Confidentiality as Factors in Survey Response*, Washington, DC: National Academy Press.

National Research Council (1993), *Private Lives and Public Policies: Confidentiality and Accessibility of Government Statistics*, Washington, DC: National Academy Press.

National Research Council (2003). *Protecting Participants and Facilitating Social and Behavioral Sciences Research*, Washington, DC: National Academy Press.

National Research Council (2006), *Expanding Access to Research Data: Reconciling Risks and Opportunities*, Washington, DC: National Academy Press.

Nealon, J. (1983), "The Effects of Male vs. Female Telephone Interviewers," *Proceedings of the Survey Research Methods Section of the American Statistical Association*, pp. 139–141.

Neter, J., and Waksberg, J. (1964), "A Study of Response Errors in Expenditures Data from Household Interviews," *Journal of the American Statistical Association*, 59, pp. 17–55.

Neyman, J. (1934), "On the Two Different Aspects of the Representative Method: The Method of Stratified Sampling and the Method of Purposive Selection," *Journal of the Royal Statistical Society*, 97, pp. 558–625.

Nicholls II, W., Baker, R., and Martin, J. (1997), "The Effect of New Data Collection Technologies on Survey Data Quality," in Lyberg, L., Biemer, P., Collins, M., de Leeuw, E., Dippo, C., Schwarz, N., and Trewin, D. (eds.), *Survey Measurement and Process Quality*, pp. 221–248, New York: Wiley.

O'Muircheartaigh, C. (1991), "Simple Response Variance: Estimation and Determinants," in Biemer, P., Groves, R., Lyberg, L., Mathiowetz, N., and Sudman, S. (eds.), *Measurement Errors in Surveys*, pp. 551–574, New York: Wiley.

O'Neill, G., and Sincavage, J. (2004), "Response Analysis Survey: A Qualitative Look at Response and Nonresponse in the American Time Use Survey," retrieved December 14, 2006, from www.bls.gov/ore/pdf/st040140.pdf, Washington, DC: Bureau of Labor Statistics.

O'Toole, B., Battistutta, D., Long, A., and Crouch, K. (1986), "A Comparison of Costs and Data Quality of Three Health Survey Methods: Mail, Telephone and Personal Home Interview," *American Journal of Epidemiology*, 124, pp. 317–328.

Oksenberg, L., Cannell, C., and Kalton, G. (1991), "New Strategies of Pretesting Survey Questions," *Journal of Official Statistics*, 7, pp. 349–366.

Oksenberg, L., Coleman, L., and Cannell, C. (1986), "Interviewers' Voices and Refusal Rates in Telephone Surveys," *Public Opinion Quarterly*, 50, pp. 97–111.

Olson, K. (2006), "Survey Participation, Nonresponse Bias, Measurement Error Bias, and Total Bias, *Public Opinion Quarterly*, 70, pp. 737–758.

Pastore, A. and Maguire, K. (eds.) (2008), *Sourcebook of Criminal Justice Statistics* [Online]. Available at http://www.albany.edu/sourcebook/.

Payne, S. (1951), *The Art of Asking Questions*, Princeton, NJ: Princeton University Press.

Pearson, R., Ross, M., and Dawes, R. (1992), "Personal Recall and the Limits of Retrospective Questions in Surveys," in Tanur, J. (ed.), *Questions About Questions: Inquiries into the Cognitive Basis of Surveys*, pp. 65–94, New York: Russel Sage.

Pillemer, D. (1984), "Flashbulb Memories of the Assassination Attempt on President Reagan," *Cognition*, 16, pp. 63–80.

Presser, S. (1990), "Measurement Issues in the Study of Social Change," *Social Forces*, 68, pp. 856–868.

Presser, S. (1994), "Informed Consent and Confidentiality in Survey Research," *Public Opinion Quarterly*, 58, pp. 446–459.

Presser, S. and Blair, J. (1994), "Survey Pretesting: Do Different Methods Produce Different Results?" in Marsden, P. (ed.), *Sociology Methodology*, 24, pp. 73–104, Washington DC: American Sociological Association.

Presser, S., Rothgeb, J., Couper, M., Lessler, J., Martin, E., Martin, J., and Singer, E. (eds.) (2004), *Methods for Testing and Evaluating Survey Questionnaires*, New York: Wiley.

Rand, M. and Rennison, C. (2002), "True Crime Stories? Accounting for Differences in our National Crime Indicators," *Chance*, 15, pp. 47–51.

Rasinski, K. (1989), "The Effect of Question Wording on Support for Government Spending," *Public Opinion Quarterly*, 53, pp. 388–394.

Rasinski, K., Mingay, D., and Bradburn, N. (1994), "Do Respondents Really 'Mark All That Apply' on Self-Administered Questions?" *Public Opinion Quarterly*, 58, pp. 400–408.

Redline, C., and Dillman, D. (2002), "The Influence of Alternative Visual Designs on Respondents' Performance with Branching Instructions in Self-Administered Questionnaires," in Groves, R., Dillman, D., Eltinge, J., and Little, R. (eds.), *Survey Nonresponse*, pp. 179–195, New York: Wiley.

Richman, W., Kiesler, S., Weisband, S., and Drasgow, F. (1999), "A Meta-Analytic Study of Social Desirability Distortion in Computer-Administered Questionnaires, Traditional Questionnaires, and Interviews," *Journal of Applied Psychology*, 84, pp. 754–775.

Rizzo, L., Brick, J., and Park, I. (2004), "A Minimally Intrusive Method for Sampling Persons in Random-Digit Dial Surveys," *Public Opinion Quarterly*, 68, pp. 267–274.

Robinson, D., and Rohde, S. (1946), "Two Experiments in an Anti-Semitism Poll," *Journal of Abnormal and Social Psychology*, 41, pp. 136–144.

Robinson, J., Ahmed, B., das Gupta, P., and Woodrow, K. (1993), "Estimation of Population Coverage in the 1990 United States Census Based on Demographic Analysis," *Journal of the American Statistical Association*, 88, pp. 1061–1071.

Robinson, J., Neustadtl, A., and Kestnbaum, M. (2002), "Why Public Opinion Polls Are Inherently Biased: Public Opinion Differences Among Internet Users and Non-Users," paper presented at the Annual Meeting of the American Association for Public Opinion Research, St. Petersburg, FL, May.

Rosenthal, R., and Rosnow, R. (1975), *The Volunteer Subject*, New York: Wiley.

Rothgeb, J., Willis, G., and Forsyth, B. (2001), "Questionnaire Pretesting Methods: Do Different Techniques and Different Organizations Produce Similar Results?" in *Proceedings of the Survey Research Methods Section*, http://www.amstat.org/sections/SRMS/Proceedings/.

Rubin, D. (1987), *Multiple Imputation for Nonresponse in Surveys*, New York: Wiley.

Rubin, D. (1993), "Discussion of Statistical Disclosure Limitation," *Journal of Official Statistics*, 9, pp. 461–468.

Rubin, D., and Baddeley, A. (1989), "Telescoping is Not Time Compression: A Model of the Dating of Autobiographical Events," *Memory and Cognition*, 17, pp. 653–661.

Rubin, D., and Kozin, M. (1984), "Vivid Memories," *Cognition*, 16, pp. 81–95.

Rubin, D., and Wetzel, A. (1996), "One Hundred Years of Forgetting: A Quantitative Description of Retention," *Psychological Review*, 103, pp. 734–760.

Saris, W., and Andrews, F. (1991), "Evaluation of Measurement Instruments Using a Structural Modeling Approach," in Biemer, P., Groves, R., Lyberg, L., Mathiowetz, N., and Sudman, S. (eds.), *Measurement Errors in Surveys*, pp. 575–597, New York: Wiley.

Saris, W., and Gallhofer, I. (2007), *Design, Evaluation, and Analysis of Questionnaires for Survey Research.*, New York: Wiley.

Särndal, C., and Lundström, S. (2005), *Estimation in Surveys with Nonresponse*, New York: Wiley.

Schaefer, C., Schraepler, J-P., Mueller, K., and Wagner, G. (2005), "Automatic Identification of Faked and Fraudulent Interviews in Surveys by Two Different Methods," in *Proceedings of the Survey Research Methods Section, American Statistical Association*, pp. 4318–4325, Alexandria, VA, American Statistical Association.

Schaefer, D., and Dillman, D. (1998), "Development of a Standard E-Mail Methodology: Results of an Experiment," *Public Opinion Quarterly*, 62, pp. 378–397.

Schaeffer, N. (1991), "Conversation with a Purpose or Conversation? Interaction in the Standardized Interview," in Biemer, P., Groves, R., Lyberg, L., Mathiowetz, N., and Sudman, S. (eds.), *Measurement Errors in Surveys*, pp. 367–393, New York: Wiley.

Schaeffer, N., and Bradburn, N. (1989), "Respondent Behavior in Magnitude Estimation," *Journal of the American Statistical Association*, 84, pp. 402–413.

Schober, M., and Conrad, F. (1997), "Does Conversational Interviewing Reduce Survey Measurement Error," *Public Opinion Quarterly*, 61, pp. 576–602.

Schober, S., Caces, M., Pergamit, M., and Branden, L. (1992), "Effects of Mode of Administration on Reporting of Drug Use in the National Longitudinal Survey," in Turner, C., Lessler, J., and Gfroerer, J. (eds.), *Survey Measurement of Drug Use: Methodological Studies*, pp. 267–276, Rockville, MD: National Institute on Drug Abuse.

Schreiner, I., Pennie, K., and Newbrough, J. (1988), "Interviewer Falsification in Census Bureau Surveys," in *Proceedings of the Survey Research Methods Section of the American Statistical Association*, pp. 491–496, Washington, DC: American Statistical Association.

Schuman, H. (1997), "Polls, Surveys, and the English Language," *The Public Perspective*, April/May, pp. 6–7.

Schuman, H., and Converse, J. (1971), "The Effects of Black and White Interviewers on Black Responses in 1968," *Public Opinion Quarterly*, 35, pp. 44–68.

Schuman, H., and Hatchett, S. (1976), "White Respondents and Race-of-Interviewer Effects," *Public Opinion Quarterly*, 39, pp. 523–528.

Schuman, H., and Presser, S. (1981), *Questions and Answers in Attitude Surveys:*

Experiments in Question Form, Wording, and Context, New York: Academic Press.

Schwarz, N., and Sudman, S. (eds.) (1992), *Context Effects in Social and Psychological Research*, New York: Springer.

Schwarz, N., Hippler, H.-J., Deutsch, B., and Strack, F. (1985), "Response Categories: Effects on Behavioral Reports and Comparative Judgments," *Public Opinion Quarterly*, 49, pp. 388–395.

Schwarz, N., Knauper, B., Hippler, H.-J., Noelle-Neumann, E., and Clark, F. (1991), "Rating Scales: Numeric Values May Change the Meaning of Scale Labels," *Public Opinion Quarterly*, 55, pp. 618–630.

Schwarz, N., Strack, F., Hippler, H., and Bishop, G. (1991), "The Impact of Administration Mode on Response Effects in Survey Measurement," *Applied Cognitive Psychology*, 5, pp. 193–212.

Schwarz, N., Strack, F., and Mai, H. (1991), "Assimilation and Contrast Effects in Part–Whole Question Sequences: A Conversational Logic Analysis," *Public Opinion Quarterly*, 55, pp. 3–23.

Sheppard, J. (2001), *2001 Respondent Cooperation and Industry Image Study: Privacy and Survey Research*, Council of Marketing and Opinion Research, Cincinnati, OH: CMOR.

Shih, T.-H., and Fan, X. (2008), "Comparing Response Rates from Web and Mail Surveys: A Meta-Analysis," *Field Methods*, 20, pp. 249–271.

Short, J., Williams, E., and Christie, B. (1976), *The Social Psychology of Telecommunications*, New York: Wiley.

Singer, E. (1978), "Informed Consent: Consequences for Response Rate and Response Quality in Social Surveys," *American Sociological Review*, 43, pp. 144–162.

Singer, E. (1984), "Public Reactions to Some Ethical Issues of Social Research: Attitudes and Behavior," *Journal of Consumer Research*, 11, pp. 501–509.

Singer, E. (2002), "The Use of Incentives to Reduce Nonresponse in Household Surveys," in Groves, R., Dillman, D., Eltinge, J., and Little, R. (eds.), *Survey Nonresponse*, pp. 163–177, New York: Wiley.

Singer, E. (2003), "Exploring the Meaning of Consent: Participation in Research and Beliefs about Risks and Benefits," *Journal of Official Statistics*, 19, pp. 273–285.

Singer, E., and Bossarte, R. (2006), "Incentives for Survey Participation: When Are They Coercive?" *American Journal of Preventive Medicine*, 31, pp. 411–418.

Singer, E., and Couper, M. (2008), "Do Incentives Exert Undue Influence on Survey Participation? Experimental Evidence," *Journal of Empirical Research on Human Research Ethics*, 3, pp. 49–56.

Singer E., and Couper, M., "Ethical Considerations in Internet Surveys," forthcoming in L. Kaczmirek, M. Das, and P. Ester, eds., *Social Research and the Internet*.

Singer, E., and Frankel, M. (1982), "Informed Consent in Telephone Interviews," *American Sociological Review*, 47, pp. 116–126.

Singer, E., Mathiowetz, N., and Couper, M. (1993), "The Role of Privacy and Confidentiality as Factors in Response to the 1990 Census," *Public Opinion Quarterly*, 57, pp. 465–482.

Singer, E., and Presser, S. (2007), "Privacy, Confidentiality, and Respondent Burden as Factors in Telephone Survey Nonresponse," in Lepkowski, J., Tucker, C., Brick, J., de Leeuw, E., Japec, L., Lavrakas, P., Link, M., and Sangster, R. (eds.), *Advances in Telephone Survey Methodology*, New York: Wiley.

Singer, E., Groves, R., and Corning, A. (1999), "Differential Incentives: Beliefs About Practices, Perceptions of Equity, and Effects on Survey Participation," *Public Opinion Quarterly*, 63, pp. 251–260.

Singer, E., Hippler, H-J., and Schwarz, N. (1992), "Confidentiality Assurances in Surveys: Reassurance or Threat," *International Journal of Public Opinion Research*, 4, pp. 256–68.

Singer, E., Van Hoewyk, J., and Maher, M. (2000), "Experiments with Incentives in Telephone Surveys," *Public Opinion Quarterly*, 64, pp. 171–188.

Singer, E., Van Hoewyk, J., and Neugebauer, R. (2003), "Attitudes and Behavior: The Impact of Privacy and Confidentiality Concerns on Participation in the 2000 Census," *Public Opinion Quarterly*, 65, pp. 368–384.

Singer, E., Von Thurn, D., and Miller, R. (1995), "Confidentiality Assurances and Survey Response: A Review of the Experimental Literature," *Public Opinion Quarterly*, 59, pp. 266–277.

Singleton, R., and Straits, B. (2005). *Approaches to Social Research*, 4th edition, New York: Oxford University Press.

Sirken, M. (1970), "Household Surveys with Multiplicity," *Journal of the American Statistical Association*, 65, pp. 257–266.

Smith, A. (1991), "Cognitive Processes in Long-Term Dietary Recall," *Vital and Health Statistics*, Series 6, No. 4 (DHHS Publication No. PHS 92-1079), Washington, DC: U.S. Government Printing Office.

Smith, M. (1979), "Some Perspectives on Ethical/Political Issues in Social Science Research," in Wax, M., and Cassell, J. (eds.), *Federal Regulations: Ethical Issues and Social Research*, pp. 11–22, Boulder, CO: Westview Press.

Smith, P., MacQuarrie, C., Herbert, R., Cairns, D., and Begley, L. (2004), "Preventing Data Fabrication in Telephone Survey Research," *Journal of Research Administration*, 35, pp. 13–21.

Smith, T. (1983), "The Hidden 25 Percent: An Analysis of Nonresponse on the 1980 General Social Survey," *Public Opinion Quarterly*, 47, pp. 386–404.

Smith, T. (1984), "Recalling Attitudes: An Analysis of Retrospective Questions on the 1982 General Social Survey," *Public Opinion Quarterly*, 48, pp. 639–649.

Smith, T. (1987), "That Which We Call Welfare By Any Other Name Would Smell Sweeter: An Analysis of the Impact of Question Wording on Response Patterns," *Public Opinion Quarterly*, 51, pp. 75–83.

Sperry, S., Edwards, B., Dulaney, R., and Potter, D. (1998), "Evaluating Interviewer Use of CAPI Navigation Features," in Couper, M., Baker, R., Bethlehem, J., Clark, C., Martin, J., Nicholls II, W., and O'Reilly, J. (eds.), *Computer-Assisted Survey Information Collection*, pp. 351–366, New York: Wiley.

Steiger, D., and Conroy, B. (2008), "IVR: Interactive Voice Response," in de Leeuw, E. D., Hox, J. J., and Dillman, D. A. (eds.), *International Handbook of Survey Methodology*. New York: Lawrence Erlbaum, pp. 285–298.

Stewart, A., Ware, J., Sherbourne, C., and Wells, K. (1992), "Psychological Distress/Well-Being and Cognitive Functioning Measures," in Stewart, A., and Ware, J. (eds.), *Measuring Functioning and Well-Being: The Medical Outcomes Study Approach*, pp. 102–142, Durham, NC: Duke University Press.

Suchman, L., and Jordan, B. (1990), "Interactional Troubles in Face-to-Face Survey Interviews," *Journal of the American Statistical Association*, 85, pp. 232–241.

Sudman, S., and Bradburn, N. (1982), *Asking Questions: A Practical Guide to Questionnaire Design*, San Francisco: Jossey-Bass.

Sudman, S., Bradburn, N., and Schwarz, N. (1996), *Thinking About Answers: The Application of Cognitive Processes to Survey Methodology*, San Francisco: Jossey-Bass.

Swazey, J., Anderson, M., and Lewis, K. (1993), "Ethical Problems in Academic Research," *American Scientist*, 81, pp. 542–553.

Sykes, W., and Collins, M. (1988), "Effects of Mode of Interview: Experiments in the UK," in Groves, R., Biemer, P., Lyberg, L., Massey, J., Nicholls II, W., and Waksberg, J. (eds.), *Telephone Survey Methodology*, pp. 301–320, New York: Wiley.

Taylor, B., and Rand, M. (1995), *"The National Crime Victimization Survey Redesign: New Understandings of Victimization Dynamics and Measurement,"* paper prepared for presentation at the 1995 American Statistical Association Annual Meeting, August 13–17, 1995 in Orlando, Florida (http://www.ojp.usdoj.gov/bjs/ncvsrd96.txt).

Tarnai, J. and Dillman, D. (1992), "Questionnaire Context as a Source of Response Differences in Mail vs. Telephone Surveys," in Schwarz, N., and Sudman, S. (eds.), *Context Effects in Social and Psychological Research*, pp. 115–129, New York: Springer-Verlag.

Tarnai, J., and Moore, D. (2004), "Methods for Testing and Evaluating CAI

Questionnaires," in Presser, S. et al. (eds.), *Questionnaire Development Evaluation and Testing Methods*, New York: Wiley.

Thornberry, O., and Massey, J. (1988), "Trends in United States Telephone Coverage Across Time and Subgroups," in Groves, R., Biemer, P., Lyberg, L., Massey, J., Nicholls II, W., and Waksberg, J. (eds.), *Telephone Survey Methodology*, pp. 25–50, New York: Wiley.

Thurstone, L., and Chave, E. (1929), *The Measurement of Attitude*, Chicago: University of Chicago.

Titus, S., Wells, J., and Rhoades, L. (2008), "Repairing Research Integrity," *Nature*, 453, June 19, pp. 980–982.

Tourangeau, R. (1984), "Cognitive Science and Survey Methods," in Jabine, T., Straf, M., Tanur, J., and Tourangeau, R. (eds.), *Cognitive Aspects of Survey Design: Building a Bridge Between Disciplines*, pp. 73–100, Washington, DC: National Academy Press.

Tourangeau, R. (2004), "Design Considerations for Questionnaire Testing and Evaluation," in Presser, S. et al. (eds.), *Questionnaire Development Evaluation and Testing Methods*, New York: Wiley.

Tourangeau, R., Rasinski, K., Jobe, J., Smith, T., and Pratt, W. (1997), "Sources of Error in a Survey of Sexual Behavior," *Journal of Official Statistics,* 13, pp. 341–365.

Tourangeau, R., Rips, L., and Rasinski, K. (2000), *The Psychology of Survey Response*, Cambridge: Cambridge University Press.

Tourangeau, R., Shapiro, G., Kearney, A., and Ernst, L. (1997), "Who Lives Here? Survey Undercoverage and Household Roster Questions," *Journal of Official Statistics*, 13, pp. 1–18.

Tourangeau, R., and Smith, T. (1996), "Asking Sensitive Questions: The Impact of Data Collection, Question Format, and Question Context," *Public Opinion Quarterly*, 60, pp. 275–304.

Tourangeau, R., Steiger, D., and Wilson, D. (2002), "Self-Administered Questions by Telephone: Evaluating Interactive Voice Response," *Public Opinion Quarterly*, 66, pp. 265–278.

Traugott, M., Groves, R., and Lepkowski, J. (1987), "Using Dual Frame Designs to Reduce Nonresponse in Telephone Surveys," *Public Opinion Quarterly*, 51, pp. 522–539.

Trice, A. (1987), "Informed Consent: Biasing of Sensitive Self-Report Data by Both Consent and Information," *Journal of Social Behavior and Personality*, 2, pp. 369–374.

Tucker, C., Lepkowski, J., and Piekarski, L. (2002), "List-Assisted Telephone Sampling Design Efficiency," *Public Opinion Quarterly*, 66, pp. 321–338.

Turner, C., Forsyth, B., O'Reilly, J., Cooley, P., Smith, T., Rogers, S., and Miller, H. (1998), "Automated Self-interviewing and the Survey Measurement of Sensitive Behaviors," in Couper, M., Baker, R., Bethlehem, J., Clark, C.,

Martin, J., Nicholls II, W., and O'Reilly, J. (eds.), *Computer Assisted Survey Information Collection*, pp. 455–473, New York: Wiley.

Turner, C., Lessler, J., and Devore, J. (1992), "Effects of Mode of Administration and Wording on Reporting of Drug Use," in Turner, C., Lessler, J., and Gfroerer, J. (eds.), *Survey Measurement of Drug Use: Methodological Studies*, pp. 177–220, Rockville, MD: National Institute on Drug Abuse.

Turner, C., Lessler, J., George, B., Hubbard, M., and Witt, M. (1992), "Effects of Mode of Administration and Wording on Data Quality," in Turner, C., Lessler, J., and Gfroerer, J. (eds.), *Survey Measurement of Drug Use: Methodological Studies*, pp. 221–244, Rockville, MD: National Institute on Drug Abuse.

Turner, C., Lessler, J., and Gfroerer, J. (1992), *Survey Measurement of Drug Use: Methodological Studies*, Washington, DC: National Institute on Drug Abuse.

Tuskegee Syphilis Study Ad Hoc Advisory Panel, (1973), *Final Report*, Washington, DC: U.S. Department of Health, Education, and Welfare.

U.S. Bureau of Justice Statistics (1994), *National Crime Victimization Survey (NCVS) Redesign: Questions and Answers*, NCJ 151171, Washington, DC: Bureau of Justice Statistics.

U.S. Bureau of the Census (1993), "Memorandum for Thomas C. Walsh from John H. Thompson, Subject: 1990 Decennial Census—Long Form (Sample Write-in) Keying Assurance Evaluations," Washington, DC: U.S. Bureau of the Census.

U.S. Bureau of Labor Statistics (2003), *BLS Handbook of Methods*, http://www.bls.gov/opub/hom/home.htm.

U.S. Bureau of Labor Statistics, *Monthly Labor Review*, http://www.bls.gov/opub/mlr/mlrhome.htm.

U.S. Census Bureau (2003), "Noninterview Rates for Selected Major Demographic Household Surveys," memorandum from C. Bowie, August 25, Xerox.

U.S. Census Bureau (2006) *Voting and Registration in the Election of 2004*, Current Population Reports, P20-556, Washington, DC: U.S. Census Bureau.

U.S. Department of Education. National Center for Education Statistics (2003), *NCES Handbook of Survey Methods*, NCES 2003–603, by Lori Thurgood, Elizabeth Walter, George Carter, Susan Henn, Gary Huang, Daniel Nooter, Wray Smith, R. William Cash, and Sameena Salvucci. Project Officers, Marilyn Seastrom, Tai Phan, and Michael Cohen. Washington, DC.

U.S. Dept. of Health, Education, and Welfare (1974), "Protection of Human Subjects," *Federal Register*, 39(105), May 30, Pt. II, pp. 18914–18920.

van Campen, C., Sixma, H., Kerssens, J., and Peters L. (1998), "Comparisons of the Costs and Quality of Patient Data Collection by Mail Versus Telephone Versus In-Person Interviews," *European Journal of Public Health*, 8, pp. 66–70.

van der Zouwen, J., Dijkstra, W., and Smit, J. (1991), "Studying Respondent-Interviewer Interaction: The Relationship between Interviewing Style, Interviewer Behavior, and Response Behavior," in Biemer, P., Groves, R., Lyberg, L., Mathiowetz, N., and Sudman, S. (eds.), *Measurement Errors in Surveys*, pp. 419–438, New York: Wiley.

van Leeuwen, R., and de Leeuw, E. (1999), "I Am Not Selling Anything: Experiments in Telephone Introductions," paper presented at the International Conference on Survey Nonresponse, Portland, OR.

Vehovar, V., Batagelj, Z., Lozar Manfreda, K., and Zaletel, M. (2002), "Nonresponse in Web Surveys," in Groves, R., Dillman, D., Eltinge, J., and Little, R. (eds.), *Survey Nonresponse*, pp. 229–242, New York: Wiley.

Vinovskis, M. (1998), *Overseeing the Nation's Report Card: The Creation and Evolution of the National Assessment Governing Board* (NAGB), Washington, DC: U.S. Government Printing Office.

Wagenaar, W. (1986), "My Memory: A Study of Autobiographical Memory Over Six Years," *Cognitive Psychology*, 18, pp. 225–252.

Walker, A., and Restuccia, J. (1984), "Obtaining Information on Patient Satisfaction with Hospital Care: Mail Versus Telephone," *Health Services Research*, 19, pp. 291–306.

Warner, J., Berman, J., Weyant, J., and Ciarlo, J. (1983), "Assessing Mental Health Program Effectiveness: A Comparison of Three Client Follow-up Methods," *Evaluation Review*, 7, pp. 635–658.

Warner, S. (1965), "Randomized Response: A Survey Technique for Eliminating Evasive Answer Bias," *Journal of the American Statistical Association*, 60, pp. 63–69.

Weeks, M., Kulka, R., Lessler, J., and Whitmore, R. (1983), "Personal Versus Telephone Surveys for Collecting Household Health Data at the Local Level," *American Journal of Public Health*, 73, pp. 1389–1394.

Weisberg, H., (2005), *The Total Survey Error Approach: A Guide to the New Science of Survey Research*, Chicago: The University of Chicago Press.

Weiss, C. (1968), "Validity of Welfare Mothers' Interview Responses," *Public Opinion Quarterly*, 32, pp. 622–633.

Wells, G. (1993), "What Do We Know About Eyewitness Identification?" *American Psychologist*, 48, pp. 553–571.

Willis, G., DeMaio, T., and Harris-Kojetin, B. (1999), "Is the Bandwagon Headed to the Methodological Promised Land? Evaluating the Validity of Cognitive Interviewing Techniques," in Sirken, M., et al. (eds.), *Cognition in Survey Research*, pp. 133–154, New York: Wiley.

Willis, G., Schechter, S., and Whitaker, K. (2000), "A Comparison of Cognitive Interviewing, Expert Review, and Behavior Coding: What Do They Tell Us?" in *Proceedings of the Section on Survey Research Methods, American Statistical Association,* pp. 28–37. Alexandria, VA: American Statistical Association.

Willis, G. (2005), *Cognitive Interviewing: A Tool for Improving Questionnaire Design,* Thousand Oaks, CA: Sage.

Wilson, T., and Hodges, S. (1992), "Attitudes as Temporary Constructions," in Martin, L., and Tesser, A. (eds.), *The Construction of Social Judgments*, pp. 37–66, New York: Springer-Verlag.

Witte, J., Amoroso, L., and Howard, P. (2000), "Method and Representation in Internet-Based Survey Tools: Mobility, Community, and Cultural Identity in Survey2000." *Social Science Computer Review*, 18, pp. 179–195.

Yu, J., and Cooper, H. (1983), "A Quantitative Review of Research Design Effects on Response Rates to Questionnaires," *Journal of Marketing Research*, 20, pp. 36–44.

Zayatz, L. (2007), "Disclosure Avoidance Practices and Research at the U.S. Census Bureau: An Update," presented at Workshop on Ensuring Access and Confidentiality Protection for Highly Sensitive Data, Institute for Social Research, University of Michigan, October 3.

INDEX

A bold page number indicates the defintion of a term.

AAPOR 31, 83, 184, 212, 321, 371–376, 380, 399, 407
   Code of Ethics 372–376, 380, 384
ACASI 15, **16, 151**, 152, 153, 156, 158, 169, 172, 174, 176, 224, 242
Access impediments 165, 167, 177, **193–196,** 204
Access panel (see opt-in Internet panel)
Acquiescence **172, 224**, 249–250
Administrative records 8–9, 20, 33, 75, 154–155, 195
Aided recall 243, **245**
American Association for Public Opinion Research 31, 83, 184, 212, 321, 371–376, 380, 399, 407
   Code of Ethics 372–376, 380, 384
American Statistical Association 31, 321, 380, 382
Analogue method 249, **250**
Analytic statistic **2**, 61–62, 111, 191, 411
Area
   frame **73**, 81, 88, 91–93
   probability sample **6**, 11–12, 15, 21, **73**, 76, 110–111, 129–131, 164, 348–354, 408
   sampling 71, 76, **128**, 408
   sampling frame 76, 81
ASA 31, 321, 380, 382
Audio computer-assisted self-interviewing 15, **16, 151**, 152, 153, 156, 158, 169, 172, 174, 176, 224, 242

Balance edit **345**
Balanced repeated replication 359–**360**, 361
Behavior coding 226, 260, 265, **266**–272, 304, 317, 365
Behavioral Risk Factor Surveillance System 24–27, 29–30, 72, 82, 105, 132, 159, 174, 188, 217, 237–239, 246, 406
Belmont Report 377–378
Beneficence **378**, 381–382, 399
Bias **52**–54, 60–62, 78, 118, 274, 366, 387
   acquiesence **172, 224**, 249–250
   coding 54
   coverage **55**–56, 79, 87–88

interviewer 292–295, 298, 300–302,
nonresponse **59**, 165, 167–168, 183, 189–192, 196, 200–210, 384
positivity **239**
retrospective 233
response **52**–53, 259, **279**–281, 315, 408, 436
sampling **56**–58, 78, **98**, 136, 409
social desirability 157–158, 162–163, **168**–170, 172, 177, 209, **224**
Bipolar approach 248–**249**
Bounding 233, **234**, 246
BRFSS 24–27, 29–30, 72, 82, 105, 132, 159, 174, 188, 217, 237–239, 246, 406

CAPI 15, **16**, 151–153, 156–158, 160, 166, 168–170, 224, 317, 320, 330
CASI 152–**153**, 156–158, 166, 169
CATI 25, 28, **151**, 152, 157, 160–162, 166, 168, 170, 175, 330
Cell phones 82–83, 164
Census 3–4, 7, 55, 73–75, 130–131, 135, 205, 341–342, 389, 391
Certificate of confidentiality **382**–383
CES 27–30, 40–42, 46, 52, 70, 75, 85, 113, 159, 165, 174–177, 218, 222
Channels of communication 153, **156**–157, 161, 177
Check-all-that-apply 249, **250**, 253
Closed question 169, 209, 223, 237, 243, 246, 247, 316, 330, 334, 335
Cluster sample **58**, 102, **106**, 108–113, 134
Clustering 12, **72**, 77–83, 86–90, 105, 107–113, 163, 296, 366, 408, 418
Code of Ethics 372–376, 380, 384
Code structure **332**–333, 335, 339–341, 342–343, 366
Codebook **363**–365
Coder variance 335, **342**–343
Coding 7, **54**, 311, 329–330, **331**–344, 345, 363, 366
   coder variance 335, **342**–343
   field coding 311, 334, **335**, 338
Cognitive
   interviewing 260, 263, **264**–265, 269–273, 365, 416

451

Cognitive *(continued)*
  standards **259**, **269**, 415–416
  testing 260, 264–265, 269–270, 273, 415, 416,
Commitment 301–302,
Common Rule 378–379, 389–390, 398–399
Complex survey data 111, **359**
Comprehension 94, 156, 208, 218–219, **220**, 224, 227, 229, 249, 261, 265, 271, 273, 292, 334, 386, 414
Computer assistance 11, 28, 330, 365
Computer-assisted
  interviewing 18, 157, 170, 172–173, 319, 345
  personal interviewing 15, **16**, 151–153, 156–158, 160, 166, 168–170 ,224, 317, 320, 330
  self-interviewing 152–**153**, 156–158, 166, 169
  telephone interviewing 25, 28, **151**, 152, 157, 160–162, 166, 168, 170, 175, 330
Computers and interviewers 318–319
Concurrent validity 277
Confidence
  interval 101, 104–105, 119–120, 356, 410
  limits **101**–102, 104, 109, 121
Confidentiality 155, 197, 207, 376, 379–380, **381**, 384, 392–397
  assurances 381, 386, 388, 390–392
  certificate of, **382**–383
Consistency edits 345
Construct **41**–53, 59, 62–63, 274–286, 414, 416–417
  validity **50**, 51, 63, 276, 278, 416
Contactability 192–193, 202
Content standards **259**
Conversational interviewing 312–315
Correlated response variance 282
Cost 12, 17, 30, 33, 71, 76, 92, 106, 109–112, 128–129, 150, 153, 158–159, 164, 173–177, 205, 246, 271, 317, 321, 406–407
Coverage **70**, 90
  bias 55–56, 79, 87–88
  error 4, **55**–56, 62, 69–94, 349, 375, 407
  survey population **68**
  target population **44**, **67**, 69, 162, 349, 379–380
CPS 27, 163, 186, 224, 250
Cronbach's alpha **264**
Cross-section survey 16
CSAQ **139**

Current Employment Statistics 27–30, 40–42, 46, 52, 70, 75, 85, 113, 159, 165, 174–177, 218, 222
Current Population Survey 27, 163, 186, 224, 250

Data
  analytic methods **321**
  swapping **397**, 399
de facto residence rule **74**, 81
de jure residence rule **74**, 75, 81
deff 298–299, 335, 343
Descriptive statistic **2**, 61, 191, 210, 354, 362
Design effect **109**–112, 116–121, 298–299, 335, 343
  coder 343
  deff 298–299, 335, 343
  interviewer 297–300
  roh **110**–112, 298
Directive probing 310
Disclosure
  harm from, 380–384, 387, **396**
  limitation 392, **395**–396, 398
  risk 382–383, 387, 390, 392, 395, **396**–399
Disk by mail **152**
Disproportionate allocation 25, 120–122
Double-barreled items 248, **249**, 253
Duplication **72**, 77, 79–88

Edit
  balance 345
  consistency 345
  explicit 347
  implied 347
  range 345
Editing 44, 49–50, 53–54, 320, 331, 345–347,365–366
  Fellegi-Holt 347
  outlier 44, 53, 345, 396–397
  outlier detection **44**
Effective sample size **112**
Effects of clustering 109, 112
Element sample **2**
Elements 2, 40, 51, 54–59, **69**–72, 76–93, 98, 99–103, 106–129, 184–185, 361–363, 396
Encoding **218**–219, 222, 225, 230, 253
epsem **103**, 114, 127, 348–349
Equal probability selection method **103**, 114, 127, 348–349
Equity 383
Error **3**, 30, 33, 39–69
  coverage 4, **55**–56, 62, 69–94, 349, 375, 407

INDEX 453

interviewer variance **295**–304, 312, 318, 323, 335
interviewer-related 292–314, 319–323
measurement **40**, 48, **52**, 62, 63, 121–122, 135, 157, 162, 172, 175–176, 208–210, 225, 230, 253, 259, 280, 295, 372, 375, 407, 412
nonresponse 48, 58, **59**–60, 153, 183–211, 375, 407, 413
processing 48, **53**, 342
response 53, 209, 233, 283, 302
sampling **10**, 35, 48, **56**, 58, 63–65, 95, 97, 179, 208, 213, 375, 407
statistical **3**, 51, 62, 324, 366
validity 48, **50**–51, 63, 259–273, **274**–286, 416–417
Errors of
coverage 4, **55**–56, 62, 69–94, 349, 375, 407
nonobservation **40**, 54–60, 72, 152, 162, 175, 418–419
nonresponse 48, 58, **59**–60, 153, 183–211, 375, 407, 413
observation **40**, 50–54, 162
postsurvey adjustment 42, 46–**47**, 48, 59–60, 149, 202, **208**
measurement **40**, 48, **52**, 62, 63, 121–122, 135, 157, 162, 172, 175–176, 208–210, 225, 230, 253, 259, 280, 295, 372, 375, 407, 412
response 53, 209, 233, 283, 302
sampling **10**, 35, 48, **56**, 58, 63–65, 95, 97, 179, 208, 213, 375, 407
validity 48, **50**–51, 63, 259–273, **274**–286, 416–417
Estimation 48, 53, 99, 103–137, 218, 222, 225, 235, 297, 329–331, 359–366, 417
Event history calendar **314**
Excessive complexity 227, **228**, 253
Expert review **260**, 269–273, 415
Explicit edits **347**
Extremeness **172**

Fabrication **372**–374, 399
Face-to-face 7, 11, 12, 15, 43, 106, 150–178, 187–207, 265, 295–299, 321, 389
False inferences 227, **228**
Falsification 319–322, **372**–373
Faulty presupposition 227–**228**
Fellegi-Holt editing **347**
Field coding 311, 334, **335**, 338
Finite population **54**, 69, 103–104

correction **103**–104
First-stage ratio adjustment **348**–353
Fixed costs **173**–178
Focus group 203, 229, 260, **261**–265, 269–273, 286, 415
Foreign element 54, **93**
Fractional interval 123–**124**
Frame 11, 15, 18, 21, 25, 28, 30, 37, 42, **45**–48, 54–60, 63–64, 66, 69–94, 113, 121, 132, 134, 137, 149, **160**–162, 164, 177, 184, 190, 210, 291, 345, 348, 355, 363, 366, 375, 407–408, 412, 418
area sampling **73**, 81, 88, 91–93
coverage 4, **55**–56, 62, 69–94, 349, 375, 407
elements 2, 40, 51, 54–59, **69**–72, 76–93, 98, 99–103, 106–129, 184–185, 361–363, 396
half-open interval **88**–90
household 11, 15, 18, 25, **70**, 74–76, 81–84, 93–94, 106–108, 129–130, 132–137, 164–169
ineligible unit **54**–55, **72**, 76, 167
overcoverage **54**, 74–76
rare population 84–86, **87**
telephone coverage 81–83, 164–165
undercoverage **54**–55, **72**–96, 164–165, 352

Generic memory **230**
Grammatical ambiguity **227**
Gross difference rate **283**–285

Half-open interval **88**–90
Harm from disclosure 380–384, 387, **396**
Hot-deck imputation **358**
Household 11, 15, 18, 25, **70**, 74–76, 81–84, 93–94, 106–108, 129–130, 132–137, 164–169
Housing unit **70**–95

ICPSR 363
ICR **151**–152
Ignorable nonresponse 191, 210
Implied edits **347**
Impression-based estimation **235**
Imputation 47, 209–210, 328–331, 346, 354–359, 397–399
  hot-deck **358**
  mean value 356–**357**
  multiple **359**
  regression **357**–358, 398
Inability to participate 192
Incentives 166, 198, 202, 205, 206, 383–384, 407

Index of consistency **283**–284
Ineligible unit **54**–55, **72**, 76, 167
Inference 30–31, 39, **40**–64, 69, 99–102, 356, 418
Informed consent 377, **378**–400
Institutional Review Board **376**, 384
Instructions 241–242, 251–253, 302
Intelligent character recognition **151**–152
Interactive voice response **151**–156, 175–176
Internet 83–85, 152, 157–166, 365–366, 384, 410
    surveys 28, 83–84, 93, 97, **151**–178, 195, 242, 330, 384, 406, 410
Interpenetrated design **296**–297, 412, 417
Interviewer
    debriefing 260, **265**,
    design effect 297–300
    effects 153–154, 185, 273, 291–324
    experience 292, 294–295
    falsification 319–322, **372**–373
    probing 169, 302–319
    rapport 304–305
    reading questions 294, 303, 305–306, 312–324
    selection 315–316
    supervision 315, 317–318
    training 315, 316–319
    variance **295**–304, 312, 318, 323, 335
        estimation 297–299
        interpenetrated design 297
        roh 295–304
Interviewer-administered 15, 25, 153–178, 242, 412
Interviewer-related error 292–324
Interviewer-respondent interaction 153–157
Interviewing
    cognitive 260, 263, **264**–265, 269–273, 365, 416
    computer-assisted 18, 157, 170, 172–173, 319, 345
    conversational 312–315
    face-to-face 7, 11, 12, 15, 43, 106, 150–178, 187–207, 265, 295–299, 321, 389
    pace of 169, 232, 301, 322
    standardized **312**–314
Intracluster homogeneity **109**–113
IRB **376**, 384
Item missing data **45**–46, 161–162, 169, 208–210, 354–359
    navigational error 241–242
    item nonresponse **45**–46, 161–162, 169, 208–210, 354–359
IVR **151**–159, 175–176

Jackknife replication **360**–363
Judgment 156, 218–**219**, **222**–223, 236–237
Justice **378**

Landline telephone 72, 81–83, 132–134, 164
Leverage-salience theory 199–211
Loading **247**
Longitudinal survey 28, 150, 175, 234

Mail 4, 14, 28, 81, 93, 150–154, 156–157, 159–181, 188, 192–193, 195, 201, 204, 207, 240–242, 251, 300, 321, 388, 391–392, 408
Maintaining interaction **203**
Margin of error 410
Mean value imputation **357**
Measurement 4, 6, 9, 16, 24, 31, 33, 41–42, **43**, 44–59, 61–66, 69, 83, 91–93, 109, 121–122, 135, 149, 153–155, 158–159, 168, 170, 172
    construct validity **50**, 51, 63–64, 276, 278, 416
    error **40**, 48, **52**, 62, 63, 121–122, 135, 157, 162, 172, 175–176, 208–210, 225, 230, 253, 259, 280, 295, 372, 375, 407, 412
Model 279, 297
Meta-analysis 157, 162, 166–167, 170, 190
Metadata **363**–367
Method of data collection 149–150, 158, 171, 174, 201, 375, 407, 417
    ACASI 15, **16, 151**, 152, 153, 156, 158, 169, 172, 174, 176, 224, 242
    CAPI 15, **16**, 151–153, 156–158, 160, 166, 168–170 ,224, 317, 320, 330
    CASI 152–**153**, 156–158, 166, 169, 181
    CATI 25, 28, **151**, 152, 157, 160–162, 166, 168, 170, 175, 330
    disk by mail **152**
    face-to-face 7, 12, 43, 94, 106, 150–178, 187–188, 193, 195, 197, 199, 202, 204, 206–207, 224, 251–252, 265, 295–297, 299, 317, 320–321, 335, 389, 391, 399
    mail 4, 14, 28, 81, 93, 150–154, 156–157, 159–181, 188, 192–193, 195, 201, 204, 207, 240–242, 251, 300, 321, 388, 391–392, 408
    SAQ **153**, 157, 172

INDEX

T-ACASI **151**, 153
TDE **152**
telephone 1, 4, 7, 12, 17–18, 24,
   28–30, 43, 45, 54, 56, 71–72, 74,
   76–83, 91–93, 97–99, 132–134,
   150–178, 184–185, 188, 193,
   195–207, 240, 250–252, 265,
   268, 275, 295, 297, 305, 315,
   317–318, 320–324, 374, 392, 418
text-CASI **153**
video-CASI **153**
Milgram 377–378, 386
Minimal risk **389**, 390, 399
Minimizing error 33, 39
Missing data 45–46, 53, 161–162,
   168–170, 177, 209–210, 240,
   242, 260, 266, 317, 329
Mixed-mode 134, 151, 158, 174–177,
   181, 321
Mobile phone 82–83, 164
Mode 30, **138**, 147–148, 170
   comparisons 158, 160–161, 170, 176,
      295
   data collection, 8, 30, **47**, 155, 162,
      163, 168, 170–171, 174, 176,
      180, 192, 207, 267, 295, 321,
      324, 344, 365 411–412
      ACASI 15, **16, 151**, 152, 153, 156,
         158, 169, 172, 174, 176, 224, 242
      CAPI 15, **16**, 151–153, 156–158,
         160, 166, 168–170 ,224, 317,
         320, 330CASI 152–**153**,
         156–158, 166, 169, 181
      CATI 25, 28, **151**, 152, 157,
         160–162, 166, 168, 170, 175, 330
   disk by mail **152**
   face-to-face 7, 12, 43, 94, 106,
      150–178, 187–188, 193, 195,
      197, 199, 202, 204, 206–207,
      224, 251–252, 265, 295–297,
      299, 317, 320–321, 335, 389,
      391, 399
   mail 4, 14, 28, 81, 93, 150–154,
      156–157, 159–181, 188,
      192–193, 195, 201, 204, 207,
      240–242, 251, 300, 321, 388,
      391–392, 408
   SAQ **153**, 157, 172
   T-ACASI **151**, 153
   TDE **152**
   telephone 1, 4, 7, 12, 17–18, 24,
      28–30, 43, 45, 54, 56, 71–72, 74,
      76–83, 91–93, 97–99, 132–134,
      150–178, 184–185, 188, 193,
      195–207, 240, 250–252, 265,
      268, 275, 295, 297, 305, 315,
      317–318, 320–324, 374, 392, 418
   text-CASI **153**video-CASI **153**
Motivating respondents 154, 300, 323
Multi-mode surveys 134, 151, 158,
   174–177, 181, 321
Multiple
   frame sampling **91**
   imputation **359**
   indicators 275, 282, 284, 412, 416
Mappings **77**
Multiplicity sampling **90**–91
Multistage sampling 128

NAEP 20, 22–24, 29–30, 50, 59, 69, 71,
   106–107, 116, 129, 154–155,
   159, 174, 185, 217–218, 406, 413
National Assessment of Educational
   Progress 20, 22–24, 29–30, 50,
   59, 69, 71, 106–107, 116, 129,
   154–155, 159, 174, 185,
   217–218, 406, 413
National Crime Victimization Survey 8,
   10, 12–16, 18, 20, 22, 29–30, 42,
   57–58, 61, 81, 86–87, 92, 106,
   125, 129–131, 150, 158–159,
   174–175, 186–188, 217,
   219–222, 229, 234, 245, 283,
   320, 348–350, 355–357, 363,
   365, 413
National Survey of Drug Use and Health
   395
Navigational error 241–**242**
NCVS 8, 10, 12–16, 18, 20, 22, 29–30,
   42, 57–58, 61, 81, 86–87, 92,
   106, 125, 129–131, 150,
   158–159, 174–175, 186–188,
   217, 219–222, 229, 234, 245,
   283, 320, 348–350, 355–357,
   363, 365, 413
Neyman allocation 120–**122**
Noncontact 166–167,**170**, 184, 186, 188,
   192–196, 204, 208
Nondirectively probing 303, 308, 312,
   315, 324
Nonignorable nonresponse **191**, 210
Nonobservation errir **40**, 54–60, 72, 152,
   162, 175, 418–419
Nonprobability sampling 409
Nonrespondents **45**, 46, 59, 173, 175,
   183, 189–191, 197, 203,
   207–208, 210, 350, 355–356,
   392, 413, 418
Nonresponse 46, 59, 166–167, 183–210
   access impediments **170**, 172
   bias **59**, 165, 167–168, 183, 189–192,
      196, 200–210, 384

Nonresponse *(continued)*
  contactability **170**
  error 48, 58, **59**–60, 153, 183–211, 375, 407, 413
  ignorable 191, 210
  inability to participate **170**
  incentives 177, 179
  item **169**, 187–188
  leverage-salience theory **176**
  maintaining interaction **190**
  missing data 45, 156, 188, 226, 247, 295
  noncontact 166–167, **170**, 184, 186, 188, 192–196, 204, 208
  nonignorable **180**, 195
  opportunity cost **198**
  oversurveying **198**
  rate 58–59, 153, 163, 181, 183–184
  refusal **170**, 173, 179
  response propensity **180**
  social isolation **176**
  tailoring **190**, 229
  topic interest **176**
  two-phase sampling **195**
  unit 45–46, 153, **169**–170, 173, 178, 180–181, 187–189, 324
  weight **326**
NSDUH 14–17, 22, 29–30, 79, 162, 249, 273

Observational
  error *(see* errors of observation)
  methods 320 unit **41**
OCR **151**
Office of Management and Budget 341
Open questions 162, 209, 223, **237**–238, 247, 329–330, 335, 412, 414
Opportunity cost 198
Optical character recognition **151**, 344
Opt-in internet panel **83**
Outlier 44, 53
  detection **44**
Overcoverage **54**, 74, 76
Oversampling 134
Oversurveying **198**

Perturbation methods **397**
Plagiarism **372**, 399
Population
  survey 111, 161, 418
  target 29, 30, **44**–48, 54–56, 60–61, 63, **69**–81, 85, 88, 91–92, 98, 129, 154, 159, 161, 164, 174, 184, 204, 208, 260–263, 270, 280, 286, 329, 348, 352, 363, 365–366, 371, 375, 396, 407, 408, 410, 412, 415
  unique **396**
Positivity bias **239**
Poststratification weight **352**
Postsurvey adjustment 42, 46–**47**, 48, 59–60, 149, 202, **208**
  first-stage ratio adjustment **348**–350, 352,
  hot-deck imputation **358**
  imputation **47**, 210, 329, 331, 345–347, **356**, 357–359, 366, 397–399
  mean value imputation 356, **357**
  nonresponse weighting **352**
Poststratification weight 348, **352**
  selection weight 78, 137, 184–185, 350, 359, 418
PPS **126**–128
Precision 91–92, **98**, 106, 109, 112, 116, 120–121, 124–125, 129, 165, 233, 299, 304, 330, 349, 366, 407, 409
Pretest 22, 47, 205, 226, 260, **265**–273, 286, 304, 365, 372, 416
Primacy effects **157**, 178, **239**, 240
Privacy 358
Probability
  proportionate to size **122**
  sample **6**, **94**, 98
  sampling **57**
  selection 94–95
Probe **286**
Probing 281–282, 286, 295
  directive 289
  nondirective 281, 286–287
Processing error 48, **53**, 316
Project Metropolitan 380
Proportionate allocation 114, 120–121, 349
Protection of human subjects 376–377
  beneficence **378**, 379, 399
  certificate of confidentiality **382**–383
  Common Rule 378–379, 381 389–390, 399
  confidentiality 155, 169, 197, 207, 376, 379–**381**, 382–400
  confidentiality assurances 390
  disclosure limitation 392, 395–396, 398
  equity 383
  informed consent 377, **378**–380, 384–390, 399–400
  Institutional Review Board **376**, 384
  justice **378**, 399
  Milgram 377–378, 386
  multiple imputation **359**

INDEX

privacy 76, 90, 153, 155–156, 169, 170, 177, 207, 210, 384, 390–393
Project Metropolitan 380
regression imputation 357–359, 398
statistical disclosure 382–383, 392
Tuskegee syphilis study 198, 377
Protocol analysis 264
Proxy 4, 76, 193, 206, 211, **246**, 248

Quality 2–3, 5, 7, 30, 33–34, 41, 46, 49, 52, 59–64, 104, 150, 153–154, 168, 170, 172, 174, 178, 183–184, 192, 197, 201, 203, 210–211, 224, 259, 264, 267, 273–275, 280, 287, 295, 300, 330–331, 334, 365, 380, 385, 399, 405–408, 412–417
Question
  aided recall 243, **245**
  analogue method 249, **250**
  bipolar approach 248–**249**
  check-all-that-apply 249–**250**, 253
  closed 169, 209, 223, 237, 243, 246, 247, 316, 330, 334, 335
  double-barreled 248, **249**, 253
  excessive complexity 227, **228**, 253
  field coding 311, 334, **335**, 338
  grammatical ambiguity **227**
  instructions 241–242, 251–253, 302
  loading **247**
  open 162, 209, 223, **237–238**, 247, 329–330, 335, 412, 414
  randomized response technique **225**
  ranking 249–**250**
  sensitive 14, 154, 156, 159, 200, **240**–241, 243, 246–247, 386
Questionnaire 2, 5, 7, 12, 44, 46–47, 97, 125, 135, 149–157, 161, 167–168, 170, 172–175, 186–188, 192–193, 195, 198, 200, 204, 206, 207, 211, **217**–220, 224, 228–230, 240–242, 246, 248, 250–253, 259, 260–263, 265–269, 273, 276, 286, 295, 300, 317, 321, 329–331, 333, 335, 344–345, 347, 355, 365, 372, 376, 381, 385, 388, 392, 410, 419
  development 262
  behavior coding 226, 260, 265, **266**–272, 304, 317, 365
  cognitive interviewing 260, 263, **264**–265, 269–273, 365, 416
  cognitive testing 260, 264–270, 273, 286, 415–416
  expert review **260**, 269–270, 273, 286, 415
  focus group 229, 260–**261**, 262–263, 265, 269–270, 273, 286, 415
  interviewer debriefing **265**
  pretest 22, 47, 260, **265**–267, 269–270, 273
  protocol analysis **264**
  think-aloud interviews 260, 264–270, 273, 286, 415–416
Quota sampling 28, 236, 409–410

Random
  digit dialing **17**, **76**, 132, 151
  selection **97**, **98**, 121–124, 163
Randomized
  experiment 32–33, 94, 169, 203, 205–206, 236, 253 **267**, 416
  response technique **241**, 248
Range edit **345**
Ranking 177–178, 249–**250**, 333
Rapport 304, 319
Rare population 84, **87**
Rate-based estimation **235**
Ratio edit **345**
Reading questions 294, 303, 305, 312, 316, 324
Realization **57**, **58**, 64, 100–114, 124, 130, 137, 189, 274, 296, 348, 359, 409
Recall-and-count **235**–236
Recency effects **144**, 159, 165, **223**
Reconstruction 229, **231**
Recontact methods 320–321
Record check study **280**
Recording 14, 43, 151, 217, 260, 262, 264, 266–267, 303, 316–317, 319–320, 373, 379, 384, 416
  answers 311–312, 316, 324
  methods **371**
Reference period **217**, 233, 235, 244–245, 365
Refusal 59, 166–167, 184–188, 192, 196–204, 207–208, 21, 319, 373, 383, 385, 391, 399, 407
Regression imputation 357–359, 398
Regulations for the Protection of Human Subjects 376–377
Rehearsal **230**, 265, 273,
Reinterview 94, **282**–283, 285–286, 365, 412
Reliability **53**, 239, 259, 264, 274, **281**–286, 343, 372
Repeated cross-section design 15, **16**, 21, 25
Replicability **312**, 366

Reporting 4, 8–14, 16, 52, 62. 71. 75. 90, 94, 156, 169, 173, 188–189. 197, 206, 209, 213, 218–220. **223**–225, 230–231, 235–236. 239, 241, 243, 246, 252, 262, 267–268, 279–283, 292–295, 300–305, 319, 323, 338, 372–373
Research misconduct 372, 373, 400
Residence rule **74**
 de facto **74**, 81
 de jure **74**, 75, 81
Respect for persons **378**–379, 399
Respondents 2, 9, 14, 30–33, 40–41, 44–**45**, 46, 51, 53, 58, 60, 62, 78, 97, 106, 136
Response bias **43**, **52**–53, 259, **279**, 280, 286, 315, 416
 acquiescence **172**, **224**, 249–250
 bounding 233–**234**, 246
 estimation 62, 95, **206**, 218–219, 275
 extremeness **172**
 false inferences 227, **228**
 faulty presupposition 227–**228**
 impression-based estimation **235**
 multiple mappings **77**
 primacy effects **157**, 178, **239**, 240
 rate-based estimation **235**
 recall-and-count **235**–236
 recency effects **157**, 177, 165, **239**, 240
 record check study **280**
 reconstruction 229, **231**
 social desirability 157–158, 162–162, **168**–170, 172, 177, 209, **224**
 telescoping 233, **234**–235, 243–246
  comprehension 94, 156, 208, 218–219, **220**, 224, 227, 229, 249, 261, 265, 271, 273, 292, 334, 386, 414
  directive probing 289
  effects **155**
  encoding **203**, 209
Response error 53, 209, 233, 283, 302
 interviewer variance **295**–299, 302–318, 323, 335
 judgment 43, 156, 206, 218, **222**–225, 234, 237–238, 252–253, 261, 276, 292–293, 376, 390, 407
 satisficing **224**, 251
Response propensity **189**, 191–192, 354
Response rate 7, 14, 46, 58–60, 97, 136, 150–151, 161–162, 165–169, 175–177, 183–191, 198, 200–211, 248, 294–295, 302, 315, 321, 348, 352, 354, 372, 375, 380, 383, 385–387, 389, 392, 413
Response variance **53**, 259, 281, 285–286
Response variation
 gross difference rate **283**, 285
 index of inconsistency **283**
 realization **57**,58, 64, 100–114, 124, 130, 137, 189, 274, 296, 348, 359, 409
 reinterview 94, **282**–283, 285–286, 365, 412
 reliability **53**, 239, 259, 264, 274, **281**–286, 343, 372
 response variance **53**, 259, 281, 282, 285–286
 simple response variance 281–**282**, 283–286, 365
 trial 50–51
Retrieval 218, **221**, 222, 224, 247, 252, 259, 363
 cue **221**, 230, 245
 failure 229, **230**
 order effects **171**–172, 240
Retrospective bias **233**
roh **110**–112
 design effect **109**–113, 116, 119–120, 318, 343
 interviewer design effect 298
Rotating panel design **12**, 18
Rotation group **131**
RDD **76**, 80, 83, 91–93, 134, 160,164, 175, 185

Sample
 area probability **6**, 11–12, 15, 21, **73**, 76, 110–111, 129–131, 164, 348–354, 408
 design 11, 15, 18, 21, 25, 28, 47, 91–92, 97–139, 149, 162–163, 347–363, 375–376, 407–409, 411, 418
 disproportionate allocation 25, 120–122
 effective sample size **112**, 138, 143
 element sample **58,** 64, 109, 115, 120, 122, 124, 139
 element variance 99, **100**, 102–105, 108–109, 116–118, 121, 137, 141, 144–145, 357
 epsem **103**, 114, 127, 138, 140, 144, 348–349
 frame 11, 15, 18, 21, 25, 28, 30, 37, 42, **45**–48, 54–60, 63–64, 66, 69–94, 113, 121, 132, 134, 137, 149, **160**–162, 164, 177, 184,

INDEX 459

190, 210, 291, 345, 348, 355, 363, 366, 375, 407–408, 412, 418
mean 49, 52, 54–55, 57–62, 87, 95, 100–138, 145, 183, 185, 189, 190, 208, 279, 299, 367, 397, 417
probability 6, 9, 11–12, 15, 21, 25, 28, 31, 35, **58**, 64, 73, 76–77, 89–90, 92, 94–95, 98–138, 158, 160, 163, 164, 169, 185, 205, 208, 248, 291, 296, 317, 321, 323, 348, 350, 352, 354, 368, 387, 391, 396, 399, 406–410
proportionate to size **126**, 127, 132
proportionate allocation 114, 120–122, 349
random digit dialing **17**, **76**, 132, 151
random selection **97**, 98, 121–124, 126–127, 138, 143–144, 163, 401
realization **57**,58, 64, 100–114, 124, 130, 137, 189, 274, 296, 348, 359, 409
segments 96, **130**–131, 138, 169, 173, 205, 231, 262, 293, 297, 317
simple random 25, **103**, 105–128, 137–139, 141–144, 296–297, 299, 360, 410, 411
stage of selection 128
strata 56, 111, **113** -122, 129–130, 138, 140–142, 144, 185, 354, 359
stratification 31, 58, 64, 81, 83, 102–138, 144, 348, 352–353, 359, 361–362, 366–367, 369, 408, 411
stratum 56, 90, **113**–122, 129–130, 138–142, 144–146, 185, 348, 354, 359, 361
strata 56, 111, 113–122, 129–130, 138, 140–142, 144, 185, 354, 359 361
systematic selection 6, **122**–124, 127
sample unique 396–397, 400
Sampling
 bias **56**–58, 64, **98**, 137, 409
 error **10**, 35, 48, **56**, 58, 63–65, 95, 97, 179, 208, 213, 375, 407
 fraction **102**–103, 113–114, 116–118, 121–122, 125, 138, 141, 185, 349,
 frame 11, 15, 18, 21, 25, 28, 30, 37, 42, **45**–48, 54–60, 63–64, 66, 69–94, 113, 121, 132, 134, 137, 149, **160**–162, 164, 177, 184, 190, 210, 291, 345, 348, 355, 363, 366, 375, 407–408, 412, 418
 margin of error 410–411
 multiple frame 72, 76, 79–80, 87, 91–93, 95, 131, 434

 multiplicity 76, **90**–91, 95, 445
 multistage 11–12, 15, 21, 111, 125, 128–129, 348, 359–360, 362
 Neyman allocation 120–121, **122**, 138, 146
 nonprobability 90, 136, 409–410
 two-stage design **125**–127, 138
Sampling variance **56**–58, 64, 98–138, 296, 299, 331, 348, 359–361, 366, 369, 407, 409, 411
 roh 12, 110–112, 138, 140–141, 143–144, 381
 without replacement **103**, 138
SAQ 152–**153**, 156–157, 172, 178
Satisficing **224**, 251, 254,
Scientific integrity 398
 fabrication **372**–374, 399–400, 445
 falsification 319–322, **372**–373, 399–400, 443
 interviewer falsification 319–322, 372, 399–400, 443
 research misconduct **372**–373, 400
 plagiarism **372**–373, 399–400
Segments 96, **130**–131, 138, 169, 173, 205, 231, 262, 293, 297, 317
Selection weight 78, 137, 184–185, **350**, 324, 359, 418,
Self-administered questionnaire 152–153, 156, 170, 200, 206, 218, 241–242, 252, 269, 294, 419
Sensitive
 information 168, 170, 241
 question 14, 154, 156, 159, 200, **240**–24, 243, 246–247, 386
Show cards 156, **172**, 177
Simple
 random sample 25, **103**, 105–128, 137–139, 141–144, 296–297, 299, 360, 410, 411
 response variance 281, **282**–286, 365,
Size of interviewer assignments 300
SOC 17–18, 24, 30, 42–46, 53, 59, 72, 82, 97, 132, 174, 188, 193, 217, 219, 222, 234, 248, 337, 339, 376, 406
Social
 desirability 157–158, 162–163, **168**–170, 172, 177, 209, **224**
 desirability and interviewers 292–294
 desirability bias 168, 170
 isolation **198**,
 presence **156**, 292
Split-ballot experiment 260, **267**–268, 273, 286
Stage of selection 128

Standard error of the mean **101**–102, 104, 108–109, 119
Standardized interviewing 305, **312**
Statistic 1, **2**, 27, **40**, 49, 54–57, 61, 87–88, 98, 102, 106–107, 109, 112, 122, 193, 196, 200–201, 208, 280–281, 298, 332, 342–343, 360–361, 407, 409, 413, 418
   analytic **2**, 62, 191, 210, 411
   descriptive **2**, 191, 210, 354, 362
Statistical
   disclosure **382**–383, 392
   error **3**, 51, 62, 324, 366
Strata 56, 90, **113**–122, 129–130, 138–142, 144–146, 185, 348, 354, 359, 361
Stratification 31, **58**, 81, 83, 102, **113**, 119–120, 122, 128, 131, 134, 137, 359, 361, 366, 408, 411
   disproportionate allocation 25, 120–122
   proportionate allocation 114, 120–121, 349
Stratum 56, 90, **113**–122, 129–130, 138–142, 144–146, 185, 348, 354, 359, 361
Structural modeling 277
Subjective constructs 280
Suppression **398**
Survey **2**
   costs 109, 173, 178, 291, 317, 412
   methodology 2, **3**, 30–32, 34, 39–41, 49, 63, 75, 87, 177, 253, 274, 331, 342, 371, 405–406, 414, 417, 419
   population **70**
   research 3–7
Survey of Consumers 17–18, 24, 30, 42–46, 53, 59, 72, 82, 97, 132, 174, 188, 193, 217, 219, 222, 234, 248, 337, 339, 376, 406
Systematic interviewer effects 292, 294
Systematic selection 6, **122**–123, 127
   fractional interval **124**

T-ACASI **151**–153
Tailoring **203**, 206
Target population 29, 30, **44**–48, 54–56, 60–61, 63, **69**–81, 85, 88, 91–92, 98, 129, 154, 159, 161, 164, 174, 184, 204, 208, 260–263, 270, 280, 286, 329, 348, 352, 363, 365–366, 371, 375, 396, 407, 408, 410, 412, 415
Taylor series **360**–361
TDE 28, **152**
Technology use 157–158
Telephone 45, 128, 138–139, 142, 147, 152–153, 158, 163, 175, 189, 283
   audio computer-assisted self-interviewing 15, **16, 151**, 152, 153, 156, 158, 169, 172, 174, 176, 224, 242
   coverage 81–83, 164–165
Telescoping 233, **234**–235, 243–246
Text-CASI 152–**153**
Text computer-assisted self-interviewing, 152–**153**
Think-aloud interviews 260, 264–270, 273, 286, 415–416
Topic interest 198
Total survey error 30, 33–34, **49**
   bias **52**–54, 60–62, 78, 118, 274, 366, 387
   coverage 4, **55**–56, 69–94, 375, 407
   errors of nonobservation **40**, 54–60, 72, 152, 162, 175, 418–419
   errors of observation **40**, 50–54, 162
   measurement error **40**, 48, **52**, 62, 63, 121–122, 135, 157, 162, 172, 175–176, 208–210, 225, 230, 253, 259, 280, 295, 372, 375, 407, 412
   nonresponse error **59**, 165, 167–168, 183, 189–192, 196, 200–210, 384
   postsurvey adjustment 42, 46–**47**, 48, 59, 149, 202, **208**
   precision **98**, 165, 375
   response error 53, 209, 233, 283, 302
   sampling **10**, 48, **56**, 58, 97–137, 375
   validity 48, **50**–51, 63, 259–273, **274**–286, 416–417
Touchtone data entry 28, **152**
Trace files 269
Trial 50
True value 3, **50**–52
Tuskegee syphilis study 198, 377
Two-phase sampling 202, 208
Two-stage design **125**–127, 138

Undercoverage **54**–55, **72**–96, 164, 165, 352
Unfamiliar term 227, **228**
Unit nonresponse **45**–46, 153, **169**–170, 173, 178, 180–181, 187–189, 324
Usual residence **74**
Usability standards **259**

Vague
   concepts 227–**228**
   quantifier 227–**228**

INDEX

Validation 246, 267–268, 417
Validity 48, **50**–51, 63, 259–273, **274**–286, 416–417
  concurrent 277
  construct **50**, 51, 63, 276, 278, 416
  multiple indicators 275–276, 282–285, 412, 416
Variable costs **173**–174
Variance 53–60, 98–137, 224, 281–286, 295–316, 329–331, 335, 342–343, 359–366, 389–391, 409, 411
  coder 335, **342**–343
  interviewer **295**–304, 312, 318, 323, 325
Variance estimation
  balanced repeated replica-tion 359–**360**, 361
  finite population correction **103**–104
  intracluster homogeneity **109**–113
  jackknife replication **360**–363
  standard error of the mean **101**–119
  Taylor series **360**–362
Video-CASI 152, **153**

Web 83–85, 152, 157–166, 365–366, 384, 410
Weight **47**, 59–61, 78–93, 114–119, 184–185, 329–331, 347–354, 359–366, 411, 417–418
  first-stage ratio adjustment **348**–353
  nonresponse **351**–352
  poststratification **352**–353
Within-household selection 134–137, 153–154, 163–164

WILEY SERIES IN SURVEY METHODOLOGY
Established in Part by WALTER A. SHEWHART AND SAMUEL S. WILKS

Editors: *Robert M. Groves, Graham Kalton, J. N. K. Rao, Norbert Schwarz, Christopher Skinner*

The *Wiley Series in Survey Methodology* covers topics of current research and practical interests in survey methodology and sampling. While the emphasis is on application, theoretical discussion is encouraged when it supports a broader understanding of the subject matter.

The authors are leading academics and researchers in survey methodology and sampling. The readership includes professionals in, and students of, the fields of applied statistics, biostatistics, public policy, and government and corporate enterprises.

ALWIN · Margins of Error: A Study of Reliability in Survey Measurement
BETHLEHEM · Applied Survey Methods: A Statistical Perspective
*BIEMER, GROVES, LYBERG, MATHIOWETZ, and SUDMAN · Measurement Errors in Surveys
BIEMER and LYBERG · Introduction to Survey Quality
BRADBURN, SUDMAN, and WANSINK ·Asking Questions: The Definitive Guide to Questionnaire Design—For Market Research, Political Polls, and Social Health Questionnaires, *Revised Edition*
BRAVERMAN and SLATER · Advances in Survey Research: New Directions for Evaluation, No. 70
CHAMBERS and SKINNER (editors · Analysis of Survey Data
COCHRAN · Sampling Techniques, *Third Edition*
CONRAD and SCHOBER · Envisioning the Survey Interview of the Future
COUPER, BAKER, BETHLEHEM, CLARK, MARTIN, NICHOLLS, and O'REILLY (editors) · Computer Assisted Survey Information Collection
COX, BINDER, CHINNAPPA, CHRISTIANSON, COLLEDGE, and KOTT (editors) · Business Survey Methods
*DEMING · Sample Design in Business Research
DILLMAN · Mail and Internet Surveys: The Tailored Design Method
GROVES and COUPER · Nonresponse in Household Interview Surveys
GROVES · Survey Errors and Survey Costs
GROVES, DILLMAN, ELTINGE, and LITTLE · Survey Nonresponse
GROVES, BIEMER, LYBERG, MASSEY, NICHOLLS, and WAKSBERG · Telephone Survey Methodology
GROVES, FOWLER, COUPER, LEPKOWSKI, SINGER, and TOURANGEAU · Survey Methodology, *Second Edition*
*HANSEN, HURWITZ, and MADOW · Sample Survey Methods and Theory, Volume I: Methods and Applications
*HANSEN, HURWITZ, and MADOW · Sample Survey Methods and Theory, Volume II: Theory
HARKNESS, VAN DE VIJVER, and MOHLER · Cross-Cultural Survey Methods
KALTON and HEERINGA · Leslie Kish Selected Papers
KISH · Statistical Design for Research
*KISH · Survey Sampling
KORN and GRAUBARD · Analysis of Health Surveys
LEPKOWSKI, TUCKER, BRICK, DE LEEUW, JAPEC, LAVRAKAS, LINK, and SANGSTER (editors) · Advances in Telephone Survey Methodology
LESSLER and KALSBEEK · Nonsampling Error in Surveys

*Now available in a lower priced paperback edition in the Wiley Classics Library.

LEVY and LEMESHOW · Sampling of Populations: Methods and Applications, *Fourth Edition*
LYBERG, BIEMER, COLLINS, de LEEUW, DIPPO, SCHWARZ, TREWIN (editors) · Survey Measurement and Process Quality
MAYNARD, HOUTKOOP-STEENSTRA, SCHAEFFER, VAN DER ZOUWEN · Standardization and Tacit Knowledge: Interaction and Practice in the Survey Interview
PORTER (editor) · Overcoming Survey Research Problems: New Directions for Institutional Research, No. 121
PRESSER, ROTHGEB, COUPER, LESSLER, MARTIN, MARTIN, and SINGER (editors) · Methods for Testing and Evaluating Survey Questionnaires
RAO · Small Area Estimation
REA and PARKER · Designing and Conducting Survey Research: A Comprehensive Guide, *Third Edition*
SARIS and GALLHOFER · Design, Evaluation, and Analysis of Questionnaires for Survey Research
SÄRNDAL and LUNDSTRÖM · Estimation in Surveys with Nonresponse
SCHWARZ and SUDMAN (editors) · Answering Questions: Methodology for Determining Cognitive and Communicative Processes in Survey Research
SIRKEN, HERRMANN, SCHECHTER, SCHWARZ, TANUR, and TOURANGEAU (editors) · Cognition and Survey Research
SUDMAN, BRADBURN, and SCHWARZ · Thinking about Answers: The Application of Cognitive Processes to Survey Methodology
UMBACH (editor) · Survey Research Emerging Issues: New Directions for Institutional Research No. 127
VALLIANT, DORFMAN, and ROYALL · Finite Population Sampling and Inference: A Prediction Approach

CPSIA information can be obtained
at www.ICGtesting.com
Printed in the USA
BVOW11s0017250117
474341BV00004B/9/P